HEAVY QUARKS
AT FIXED TARGET

HEAVY QUARKS AT FIXED TARGET

Batavia, IL   October 1998

EDITORS
Harry W. K. Cheung
Joel N. Butler
*Fermi National Accelerator Laboratory,
Batavia, IL*

AIP CONFERENCE
PROCEEDINGS 459

American Institute of Physics   Woodbury, New York

Editors:

Harry W. K. Cheung
Fermi National Accelerator Laboratory
P.O. Box 500
Batavia, IL 60510-0500
U.S.A

Email: cheung@fnal.gov

Joel N. Butler
Fermi National Accelerator Laboratory
P.O. Box 500
Batavia, IL 60510-0500
U.S.A

Email: butler@fnal.gov

Authorization to photocopy items for internal or personal use, beyond the free copying permitted under the 1978 U.S. Copyright Law (see statement below), is granted by the American Institute of Physics for users registered with the Copyright Clearance Center (CCC) Transactional Reporting Service, provided that the base fee of $15.00 per copy is paid directly to CCC, 222 Rosewood Drive, Danvers, MA 01923. For those organizations that have been granted a photocopy license by CCC, a separate system of payment has been arranged. The fee code for users of the Transactional Reporting Service is: 1-56396-864-9/ 99 /$15.00.

© 1999 American Institute of Physics

Individual readers of this volume and nonprofit libraries, acting for them, are permitted to make fair use of the material in it, such as copying an article for use in teaching or research. Permission is granted to quote from this volume in scientific work with the customary acknowledgment of the source. To reprint a figure, table, or other excerpt requires the consent of one of the original authors and notification to AIP. Republication or systematic or multiple reproduction of any material in this volume is permitted only under license from AIP. Address inquiries to Office of Rights and Permissions, 500 Sunnyside Boulevard, Woodbury, NY 11797-2999; phone: 516-576-2268; fax: 516-576-2499; e-mail: rights@aip.org.

L.C. Catalog Card No. 98-83158
ISBN 1-56396-864-9
ISSN 0094-243X
DOE CONF- 981066

Printed in the United States of America

CONTENTS

Preface ix
List of Committees xi

SPECIAL CONTRIBUTIONS

CP Violation: The Kaon Problem 3
    L. Wolfenstein
The Arrival of Charm 9
    J. L. Rosner
High Promises of Beauty 28
    A. I. Sanda

PART I
CP VIOLATION AND MIXING

Status of the HERA-B Experiment 45
    B. Schwingenheuer
**CP Violation, Mixing and Rare Decays at the Tevatron
Now and in RUN II** 55
    K. T. Pitts
$B^0 - \bar{B}^0$ Oscillations and Measurements of $|V_{ub}|/|V_{cb}|$ at LEP 66
    A. Stocchi
The PEP-II B-Factory and the BABAR Detector 77
    G. Bonneaud
B Physics in the Next Millennium 87
    M. Artuso
One Reason Why CP Violation is Way Radically Cool 97
    P. Arnold
**CP Violation in Strange Baryon Decays: A Report from
Fermilab Experiment 871** 107
    C. James, R. A. Burnstein, A. Chakravorty, A. Chan, Y. C. Chen,
    W. S. Choong, K. Clark, E. C. Dukes, C. Durandet, J. Felix, R. Fuzesy,
    G. Gidal, H. R. Gustafson, C. Ho, T. Holmstrom, M. Huang, M. Jenkins,
    D. M. Kaplan, L. M. Lederman, N. Leros, M. J. Longo, F. Lopez,
    W. Luebke, K. B. Luk, G. Moreno, K. Nelson, V. Papavassiliou,
    J. P. Perroud, D. Rajaram, H. A. Rubin, M. Sosa, P. K. Teng, B. Turko,
    J. Volk, C. G. White, S. L. White, and P. Zyla
The Status of the KLOE Experiment 116
    G. Bencivenni
**Towards Results on Direct CP Violation in Kaon Decays from the
CERN-SPS Experiment NA48** 125
    A. Ceccucci

Results on *CP* Violation from KTeV 135
    M. Arenton

PART II
HEAVY IONS

Heavy Flavour Production in Heavy Ion Collisions: Present Status
and Future Prospects.. 147
    C. Lourenço

PART III
PRODUCTION DYNAMICS AND STRUCTURE FUNCTIONS

E791 Results on Hadroproduction of Charm from 500 GeV π^--N
Interactions 159
    K. Stenson
Charm Physics Results from SELEX 168
    A. Y. Kushnirenko
Charm and Beauty Production in Experiment WA92 179
    D. Barberis, C. Gemme, and L. Malferrari
Charm Detection at HERMES.. 189
    E.-C. Aschenauer
Heavy Quark Production in Neutrino Deep-Inelastic Scattering 198
    T. Adams, A. Alton, C. G. Arroyo, S. Avvakumov, L. de Barbaro,
    P. de Barbaro, A. O. Bazarko, R. H. Bernstein, A. Bodek, T. Bolton, J. Brau,
    D. Buchholz, H. Budd, L. Bugel, J. M. Conrad, R. B. Drucker, J. A. Formaggio,
    R. Frey, J. Goldman, M. Goncharov, D. A. Harris, R. A. Johnson, J. H. Kim,
    B. J. King, T. Kinnel, S. Koutsoliotas, M. J. Lamm, W. Marsh, D. Mason,
    K. S. McFarland, C. McNulty, S. R. Mishra, D. Naples, P. Nienaber,
    A. Romosan, W. K. Sakumoto, H. M. Schellman, F. J. Sculli,
    W. G. Seligman, M. H. Shaevitz, W. H. Smith, P. Spentzouris,
    E. G. Stern, B. M. Tamminga, M. Vakili, A. Vaitaitis, V. Wu,
    U. K. Yang, J. Yu, and G. P. Zeller
Heavy Flavour Physics at HERA.. 207
    B. Naroska
New Results from E866 on Charmonium Production .... (NA)*
    T. Chang
Study of Neutrino Production of Charm Particles in CHORUS .... (NA)*
    M. Nakamura
B Production and Onium Production at the Tevatron...................... 218
    A. Zieminski
Heavy Quark Fragmentation 228
    E. H. Norrbin
Heavy Quark Production 238
    F. I. Olness

(NA)* indicates that the paper was not available at time of press

PART IV
HADRONIC DECAYS

Hadronic Charm Decays and Lifetimes from E791 251
    N. Copty
The FOCUS Spectrometer and Hadronic Decays in FOCUS 261
    J. M. Link
Dalitz Analysis, Rare Decays, and Charmed Baryons at FOCUS and E687 270
    E. W. Vaandering
Hadronic Decays of Beauty and Charm from CLEO 280
    J. L. Rodriguez
Review of Charm and Beauty Lifetimes 291
    H. W. K. Cheung

PART V
SEMILEPTONIC AND LEPTONIC DECAYS

Hyperon Physics Results from SELEX 303
    I. Eschrich
An Overview of Hyperon Physics in KTeV 314
    D. A. Jensen
Charm Semileptonic Decays from E791 324
    D. Mihalcea
Charm Semileptonic Decays from FOCUS/E687: A Survey of Previous E687 Results, and Expectations from the FOCUS Data 334
    W. E. Johns
Semileptonic B Decays from CLEO 344
    K. M. Ecklund
Lattice QCD Calculations of Leptonic and Semileptonic Decays 355
    A. S. Kronfeld

PART VI
SPECTROSCOPY

Charmonium Spectroscopy from Fermilab E835 369
    T. K. Pedlar
Charmed Baryon Spectroscopy from CLEO at CESR 379
    M. S. Alam
The COMPASS Experiment at CERN: Prospects for Charm and Structure Function Measurements 390
    A. Bravar
B and D Spectroscopy at LEP 399
    F. Muheim

Heavy Quark Spectroscopy .... (NA)*
    E. Eichten

PART VII
RARE AND FORBIDDEN DECAYS

Rare Kaon Decays: Brookhaven E871 411
    D. A. Ambrose
Evidence for $K^+ \to \pi^+ \nu \bar{\nu}$ 421
    S. Kettell
Recent Results from BNL E865 431
    A. Sher, R. Appel, G. S. Atoyan, B. Bassaleck, D. N. Brown,
    D. R. Bergman, N. Cheung, S. Dhawan, H. Do, J. Egger, S. Eilerts,
    C. Felder, H. Fischer, M. Gach, W. D. Herold, V. V. Isakov, H. Kaspar,
    D. Kraus, D. M. Lazarus, L. Leipuner, J. Lowe, J. Lozano, H. Ma,
    W. Majid, W. Menzel, S. Pislak, A. A. Poblaguev, A. L. Proskurjakov,
    P. Rehak, P. Robmann, R. Stotzer, J. A. Thompson, P. Truöl, H. Weyer,
    and M. E. Zeller
Review of Rare Decays and *CP* Violation in Charm 441
    S. W. L. Kwan
Rare *B* Decays at CLEO 451
    P. B. Gaidarev
Theoretical Issues in Rare *K*, *D* and *B* Decays 461
    G. Burdman
Rare Decays of the K_L and π^0: New Results from E799-II and E832 472
    E. D. Zimmerman

CONFERENCE SUMMARY

Perspectives on Heavy Quark 98 485
    C. Quigg

Collaborations 505
List of Participants 515
Workshop Program 519
Conference Photographs 525
Author Index 531

(NA)* indicates that the paper was not available at time of press

PREFACE

This volume contains the written contributions to the fourth in a series of conferences, "*Heavy Quarks at Fixed Target*", initiated in 1993 by the Laboratori Nazionali di Frascati (LNF) in Italy. The first three editions were held at LNF (1993); the University of Virginia, Charlottesville, USA (1994); and St. Goar, Germany (1996). This fourth edition was held at Fermi National Accelerator Laboratory in Batavia, Illinois, USA (1998). These conferences were devoted to results on states containing strange, charm and beauty quarks from past and ongoing fixed target experiments as well as discussions of future plans for fixed target facilities. We wanted to provide a comprehensive overview of the areas where Fixed Target experiments made significant contributions so that experimenters could see what work was left to be done. Since experiments at colliders contribute strongly in many of these areas, we decided to welcome the participation of collider experimenters. Hence this edition of the conference had many contributions from the collider community.

In order to try to cover the huge amount of physics included in studies of heavy quarks many of the contributions were review types, including both theoretical and experiment overviews in seven different areas: CP violation and mixing; heavy ions; production dynamics and structure functions; hadronic decays; semileptonic and leptonic decays; spectroscopy; and rare and forbidden decays. This volume is divided into these seven areas of interest.

Another trend today is the incredible growth of communications and ease with which information can be disseminated electronically. For this edition of the conference we made a special effort to try to publish these proceedings in the shortest possible time. We wish to thank all the authors who put up with our bludgeoning tactics and delivered their contributions in a timely manner. Regrettably a few contributions were not received for these proceedings.

It took the efforts of many people to make HQ98 possible. We would like to thank all our speakers for their excellent presentations. We especially want to thank Lincoln Wolfenstein, Jonathan Rosner, and Anthony Sanda for opening the Workshop with special presentations reviewing how our knowledge of strange, charm, and bottom quarks developed. We also want to thank Chris Quigg for undertaking the daunting task of conference summarizer. We could not have developed the program without the efforts of the International Advisory Committee. The local organizing committee dealt with all kinds of issues ranging from the physics program to housing and transportation. The Conference Secretariat, veterans of many such events, kept everything running smoothly while also keeping the conference chairperson (relatively) calm.

We want to thank our helpers from Glenbard East High School in Lombard, Illinois. Three members of the Glenbard East High School Performing Strings, Samantha May, Julie Barnes, and Kelly Cerf, under the leadership of orchestra director Joanne May, provided musical entertainment at the conference banquet at Chicago's Shedd Aquarium. We also want to thank our student 'scanners', Beth Greene and Stephanie Butler, for helping us get the transparencies of talks on the web almost immediately.

Finally, we would like to thank Fermi National Accelerator Laboratory and its director, John Peoples, Jr., for hosting the conference. The Fermi National Accelerator Laboratory is operated by the Universities Research Association, Inc., under contract DE-AC02-76CHO3000 with the U.S. Department of Energy. Many people in every part of the lab helped us out. We would especially like to thank the heads of the Particle Physics Division and the Computing Division for the help we received from their personnel.

Harry W. K. Cheung
Joel N. Butler
Fermilab

LIST OF COMMITTEES

International Advisory Committee

Bellini, Gianpaolo	University of Milan/INFN
Bianco, Stefano	Laboratori Nazionali di Frascati/INFN
Bigi, Ikaros	University of Notre Dame
Butler, Joel N.	Fermi National Accelerator Laboratory, Chairman
Cester, Rosanna	University of Torino/INFN
Cox, Bradley	University of Virginia
Dornan, Peter	Imperial College, London
Fabbri, Franco L.	Laboratori Nazionali di Frascati/INFN
Grancagnolo, Francesco	University of Lecce/INFN
Kleinknecht, Konrad	Mainz University
Paul, Stephan	Technische Universitaet Muenchen
Pretzl, Klaus	Bern University
Rossi, Leonardo	University of Genova/INFN

Local Organizing Committee

Appel, Jeffrey A.	Fermi National Accelerator Laboratory
Artuso, Marina	Syracuse University
Butler, Joel N.	Fermi National Accelerator Laboratory, Chairman
Cheung, Harry W. K.	Fermi National Accelerator Laboratory
Herrera, Mari	Fermi National Accelerator Laboratory
Kutschke, Rob	Fermi National Accelerator Laboratory
Mascione, Patricia	Fermi National Accelerator Laboratory
Ramberg, Erik	Fermi National Accelerator Laboratory
Sazama, Cynthia	Fermi National Accelerator Laboratory

HQ98 Session Convenors

Appel, Jeffrey A.	Fermilab	Artuso, Marina	Syracuse
Bellini, Gianpaolo	Milan/INFN	Bianco, Stefano	Frascati/INFN
Brown, Charles, N.	Fermilab	Butler, Joel N.	Fermilab
Cheung, Harry W. K.	Fermilab	Cox, Bradley	Univ. of Virginia
Fabbri, Franco, L.	Fermilab	Garbincius, Peter G.	Fermilab
Kaplan, Daniel	IIT	Kutschke, Robert	Fermilab
Ramberg, Erik	Fermilab		

Secretariat

Sazama, C. M.   Poole, P.   Mascione, P.   Herrera, M.   Brogiato, L.

Scientific Secretaries

Gottschalk, E.   Kutschke, R.   Ramberg, E.

SPECIAL CONTRIBUTIONS

CP Violation: The Kaon Problem

L. Wolfenstein

Department of Physics, Carnegie Mellon University
Pennsylvania, Pittsburgh, 15213, USA

Abstract. Three possible solutions of the CP violation in kaon decay are discussed. All CP violation could be explained by (1) the standard model or (2) the superweak theory, but an important possibility is (3) the standard model plus new physics.

In 1957 after the discovery of parity violation, a standard model of weak interactions emerged called the $V - A$ theory. The effective Hamiltonian involved maximum parity P and particle-antiparticle conjugation C violation but CP invariance was a basic feature.

When CP violation was discovered in K decays in 1964 two possibilities emerged (1) :

1. Make some small modification of the $V - A$ theory to allow for the small CP violation observed.

2. Introduce a new superweak interaction that allowed for $\Delta S = 2$ at tree level and violated CP invariance. This would contain an effective term of the form

$$G_{sdsd}\bar{s}\theta d\bar{s}\theta d \tag{1}$$

and G_{sdsd} could be as small as 10^{-11}.

The basic idea of the superweak interaction is its effect on the $K - \bar{K}$ mixing matrix. If we use the representation in which the CP eigenstates (K_1, K_2) are diagonal the mass matrix is

$$M - i\frac{\Gamma}{2} = \begin{pmatrix} M_1 & im' \\ -im' & M_2 \end{pmatrix} - i\begin{pmatrix} \Gamma_1 & 0 \\ 0 & \Gamma_2 \end{pmatrix} \tag{2}$$

The superweak CP violation is in the term m' where the phase factor i is fixed by CPT invariance. The decay eigenstates are then

$$K_S = K_1 + \tilde{\epsilon}K_2$$
$$K_L = K_2 + \tilde{\epsilon}K_1 \tag{3}$$

$$\tilde{\epsilon} = \frac{im'}{\Delta M + i\Delta\Gamma/2} \tag{4}$$

where $\Delta M = M_L - M_S$ and $\Delta\Gamma = \Gamma_S - \Gamma_L \approx \Gamma_S$. Since ΔM is a second-order weak effect the empirical value of $\tilde{\epsilon}$ of 2×10^{-3} requires that m' be only 10^{-2} to 10^{-3} times this second-order weak effect.

Of course, Eqs. (2) – (4) hold also in the standard model (or historically in any proposed model with $\Delta S = 1$ violation) but then there is also CP violation in the decay amplitude as well as the mixing. The superweak theory makes two predictions:

1. Since the only CP violation was in the mixing all CP violation in K decays would be given by $\tilde{\epsilon}$. In particular

$$\eta_{+-} = \eta_{oo} = \tilde{\epsilon}$$

so that ϵ' must vanish.

2. Because CP violation was so small there would be no place other than K^0 decays to find it.

These predictions have proven all too true for the past third of a century.

In 1973 with the discovery of weak neutral currents a new standard model of weak interactions appeared to be established. This was the spontaneously broken gauge theory of Weinberg and Salam with four quarks as required by the GIM mechanism. A fundamental feature of this theory was CP invariance. In retrospect, this is very interesting since CP violation was the most striking failure of the old $V - A$ theory and the new theory did nothing to solve it. However, CP violation was limited to one parameter in K^0 decay and this could always be blamed on some superweak interaction.

A number of suggestions as to how to modify the new standard model gradually appeared. In 1976, Weinberg proposed that there could be three Higgs bosons allowing CP violation via charged Higgs exchange. However, already in 1973 there appeared in one paragraph of a paper in Progress of Theoretical Physics the proposal that if there were six quarks instead of four (at that time even the fourth had not yet been discovered) then CP violation was allowed. The authors, of course, were Kobayashi and Maskawa.

The weak interaction may be written

$$g \, \bar{u}_j \, V_{ji} \, \gamma_\lambda \, (1 - \gamma_5) \, d_i \, W^\lambda + H.c.$$

Here $u_j = (u, c, t)$ are the up-type quarks and $d_j = (d, s, b)$ are the down type. V is the unitary CKM (Cabibbo-Kobayashi-Maskawa) matrix, the 3X3 generalization of the Cabibbo matrix.

The empirical data on the magnitudes suggests an expansion in powers of $\lambda = \sin\theta = 0.22$; for most purposes it is adequate to stop at λ^3:

$$V \begin{pmatrix} 1 - \frac{\lambda^2}{2} & \lambda & A\lambda^3(\rho - i\eta) \\ -\lambda & 1 - \frac{\lambda^2}{2} & A\lambda^2 \\ A\lambda^3(1 - \rho - i\eta) & -A\lambda^2 & 1 \end{pmatrix}$$

The analysis of experimental data from decay rates is summarized by:

$$A = 0.83 \pm 0.07,$$
$$(\rho^2 + \eta^2)^{1/2} = 0.4 \pm 0.15.$$

We may now consider three alternatives:

1. All CP violation is explained by the one parameter η in the CKM matrix.

2. η is approximately zero and all CP violation is explained by physics beyond the standard model; in particular, a superweak model.

3. η does not equal zero but there is other CP violation from new physics.

The fact that the one CP-violating observable, the parameter usually noted as ϵ, can be explained in terms of the one parameter η in the CKM matrix is a significant piece of evidence in favor of the standard model. If one tries to use data independent of CP violation, the rate of $b \to u$ decays and Δm in the B_d and B_s systems, to determine η one obtains a value that agrees with that needed to explain ϵ. On the other hand the hadronic uncertainties in the calculation of $b \to u$ decays and $B - \bar{B}$ mixing are such that the solution $\eta = 0$ could also be allowed (2).

To distinguish between the standard CKM model and superweak, one needs to look for a CP-violating effect in the $\Delta S = 1$ amplitude. This has led to a large effort to measure ϵ' where

$$\eta_{+-} \approx \epsilon + \epsilon'$$
$$\eta_{oo} \approx \epsilon - 2\epsilon'$$

Because CPT invariance tells us the phase of ϵ and ϵ' are approximately equal, to a good approximation

$$\mid \eta_{oo}/\eta_{+-} \mid^2 = 1 - 6 \mid \epsilon/\epsilon' \mid$$

Unfortunately the CKM model prediction is very uncertain because of the evaluation of hadronic matrix elements. This is particularly true because there is a cancellation between the gluonic and electroweak penguins. The predictions (3) for $|\epsilon'/\epsilon|$ vary from 10^{-4} to 2×10^{-3}. If the value is not too small it should be detected in the data now in hand from the $KTeV$ experiment or CERN NA48. A value greater than 10^{-3} would be evidence that $|\eta|$ is probably greater than 0.2 which would rule out any possibility of detecting new physics in $K - \bar{K}$ mixing. On the other hand an upper limit of 2 to 3×10^{-4} would not provide any fundamental information.

Since CP violation is now embodied in the standard model the question arises whether there is any motivation for setting η equal to zero and thus require something like a superweak alternative. A possible motivation is given by the following arguments (4):

1. CP violation in the CKM model is introduced by inserting a whole set of arbitrary phases in the Yukawa couplings of the quarks to the Higgs bosons. One would like something simpler and perhaps more fundamental as the origin of CP violation.

2. Since we believe there is some physics at some higher scale, whether the GUT scale or another, one may imagine introducing CP violation either spontaneously in the expectation values of scalar boson fields.

$$\langle S_i \rangle = v_i e^{i\theta_i}$$

or softly via a term such as

$$\mu^2 S_1 S_2 + h.\ c.$$

with μ^2 complex.

3. Such models have the great advantage that θ_{QCD} (or more accurately $\bar{\theta}$) can be set equal to zero in the Lagrangian and then will be calculable as a higher-order effect.

4. Since the original Lagrangian is CP-invariant (except possibly for soft terms) the progenitor of the Yukawa interaction has no arbitrary complex phases. Thus, the value of η in the effective Lagrangian at the electroweak scale (the "standard model) is calculable. If it is a loop effect it may be very small (5).

It should be noted that even within the scenarios above it may turn out that η is not small but may agree with that needed in the CKM model. This is the case for the Nelson-Barr models originated by Anne Nelson (6).

A question arises as to the definition of a superweak model since there exists many examples in the literature (4, 5, 7). In general a superweak model is defined by the effective Lagrangian at the electroweak scale or below:

$$L_{eff} = L_o + L_{SW}$$

where L_o is the standard model with η much less than that required to explain ϵ and L_{SW} contains various four-quark interactions with coefficients G_{ijkl} including the term (1) needed to explain ϵ.

A theory is superweak if all the G_{ijkl} are much smaller than the usual weak interactions in L_o. An extreme superweak theory would be one in which all G_{ijkl} were as small as G_{sdsd} and involved a change of flavor ΔF of 2 units. However, I would consider theories with all $G_{ijkl} \leq 10^{-6}$ as essentially superweak and allow for $\Delta F = 1$ as well as 2. This raises the interesting point of whether a value of $\epsilon'/\epsilon \sim 5 \times 10^{-4}$ (which means $\epsilon' \sim 10^{-6}$) could possibly be consistent with some theories that are essentially superweak. In fact one might say that if L_{eff} is defined below the b-mass scale then the standard model becomes practically superweak: L_o is then the CP-conserving four-quark model and L_{SW} represents the effective interactions due to b and t quarks.

The most definitive way to distinguish the standard model from superweak is in B physics. The first observation of CP violation should come from the asymmetry in the decay $B^0(\bar{B}^0) \to \Psi K_s$. Within the standard model this measures $\sin 2\beta$ (where β the phase of $1 - \rho + i\eta$), which is constrained to lie between 0.3 and 0.9. This result could also be explained by a superweak interaction with the coupling G_{bdbd} of order 10^{-7} to 10^{-8} times G_F. The superweak theory could then be ruled out if the asymmetry had a different value in some other decay. The difference would have to be blamed on a CP-violating phase difference between the two amplitudes, analogous to the measurement of ϵ'. An example would be the asymmetry in $B^0(\bar{B}^0) \to \pi^+\pi^-$. In the tree approximation this asymmetry is given by $\sin 2(\beta+\gamma)$. Independent of the tree approximation it is very likely these two will be different, although the present standard model constraints give little in the way of prediction for the $\pi\pi$ asymmetry.

Another possibility is the decay $K_L \to \pi^0 \nu \bar{\nu}$. In the standard model the branching ratio (summed over the three types of neutrinos) is

$$\text{B. R.} \approx 4 \times 10^{-10} \, A^4 \, \eta^2$$

As the standard model value of η becomes better constrained from B decays it may become possible to predict this within a factor of 3 or so. Agreement would provide further evidence for the standard model and could further constrain the magnitude of η, although not its sign. Conversely if the branching ratio is well below 10^{-11} that would serve as evidence for the need of new physics (such as superweak) to explain the value of ϵ.

Even if η is significantly different form zero there remains the important possibility that there are other sources of CP violation from physics beyond the standard model. There are a number of reasons why CP-violating effects are a good place to look for new physics:

1. In the standard model flavor - conserving CP violation is very small. This is a strong motivation for searches for the electric dipole moments of the neutron and the electron (8).

2. CP violation is studied via flavor mixing and so is sensitive to superweak interactions.

One example is $D^0 - \bar{D}^0$ mixing. In the standard model one expects $(\Delta M_D/\Gamma_D)$ to be much less than .01 and to have very little CP violation. The search for $D^0 - \bar{D}^0$ mixing thus offers an opportunity to search for superweak interactions with G_{cucu} of order 10^{-8}. It actually may be easier to observe this mixing if it has a significant amount of CP violation (9).

Assuming the large CP violation expected in the B decay to ψK_s is found, the long-term goal of B physics is to provide a quantitative test of the CKM model with the hope of detecting new physics effects as discussed in many papers (10). One needs as many constraints on the parameters (ρ, η) as possible. In this program K physics continues to play a role. The decay $K^+ \to \pi^+ \nu \bar{\nu}$ with a branching ratio between 10^{-10} and 10^{-11} is primarily sensitive to $(1-\rho)$ while, as mentioned before, $K_L \to \pi^+ \nu \bar{\nu}$ is proportional to η^2. Thus rare K decays and CP violation in B physics complement each other.

ACKNOWLEDGEMENTS

I thank Darwin Chang for discussions of superweak theories. This work was supported in part by the Department of Energy under Contract DE-FG02-91-ER-40682.

REFERENCES

1. Early papers may be found in the review volume CP Violation (L. Wolfenstein, editor) North Holland (1989)
2. R. Barbieri et al, Phys. Lett. B 425, 119 (1998)
3. G. Buchalla, A. J. Buras, and M. E. Lautenbacher, Rev. Mod. Phys. 68, 1125 (1996); S. Bertolini et al, Nuc. Phys. B 514, 93 (1998)
4. S. M. Barr, Phys. Rev. D 34, 1567 (1998)
5. H. Georgi and S. L. Glashow, hep-ph 9807399
6. A. Nelson, Phys. Lett. 136 B, 332 (1984)
7. D. Bowser-Chao, D. Chang, and W. Y. Keung, hep-ph/9803273
8. I. B. Khriplovich and S. K. Lamoreaux, CP Violation Without Strangeness, Springer-Verlag (1997)
9. L. Wolfenstein, Phys. Rev. Lett. 75, 2460 (1995)
10. L. Wolfenstein, Phys. Rev. D 57, 6857 (1998) and other references therein

The Arrival of Charm

Jonathan L. Rosner

Enrico Fermi Institute and Department of Physics
University of Chicago
5640 S. Ellis Avenue, Chicago IL 60637

Abstract. Some of the theoretical motivations and experimental developments leading to the discovery of charm are recalled.

I INTRODUCTION

The discovery of charm was an exciting chapter in elementary particle physics. The theoretical motivations were strong, the predictions were crisp, and the experimental searches ranged from inadequate to serendipitous to inspired. Perhaps we can learn something relevant to present-day searches from those experiences. I would like to describe the evolution of the case for charm, some subsequent developments, and some questions which remain nearly a quarter of a century later.

The argument for charm was most compellingly made in the context of unification of the weak and electromagnetic interactions, briefly described in Sec. 2. Parallel arguments based on currents (Sec. 3) and quark-lepton analogies (Sec. 4) also played a role, while gauge theory results (Sec. 5) strengthened the case. In the early 1970's, when electroweak theories began to be taken seriously, theorists began to exhort their experimentalist colleagues in earnest to seek charm (Sec. 6). In the fall of 1974, the discovery of the J/ψ provided a candidate for the lowest-lying spin-triplet charm-anticharm bound state, and several other circumstances hinted strongly at the existence of charm (Sec. 7). Nonetheless, not everyone was persuaded by this interpretation, and it remained for open charm to be discovered before lingering doubts were fully resolved (Sec. 8).

Some progress in the post-discovery era is briefly noted in Sec. 9, while some current questions are posed in Sec. 10. A brief epilogue in Sec. 11 asks whether the search for charm offers us any lessons for the future. Part of the author's interest in (recent) history stems from a review, undertaken with Val Fitch, of elementary particle physics in the second half of the Twentieth Century [1], which is to be issued in a second edition in a year or two.

CP459, *Heavy Quarks at Fixed Target*
edited by Harry W. K. Cheung and Joel N. Butler
© 1999 The American Institute of Physics 1-56396-864-9/99/$15.00

II ELECTROWEAK UNIFICATION

The Fermi theory of beta decay [2] involved a pointlike interaction (for example, in the decay $n \to pe^-\bar{\nu}_e$ of the neutron). This feature was eventually recognized as a serious barrier to its use in higher orders of perturbation theory. By contrast, quantum electrodynamics (QED), involving photon exchange, was successfully used for a number of higher-order calculations, particularly following its renormalization by Feynman, Schwinger, Tomonaga, and Dyson [3].

Attempts to describe the weak interactions in terms of particle exchange date back to Yukawa [4]. A theory of weak interactions involving exchange of charged spin-1 bosons was written down by Oskar Klein in 1938 [5], to some extent anticipating that of Yang and Mills [6] describing self-interacting gauge particles.

Once the $V - A$ theory of the weak interactions had been established in 1957 [7], descriptions involving exchange of charged vector bosons were proposed [8]. These tried to unify charged vector bosons (eventually called $W^\pm$) with the photon (γ) within a single SU(2) gauge symmetry. However, the (massless) photon couples to a vector current, while the (massive) W's couple to a $V - A$ current. The SU(2) symmetry was inadequate to discuss this difference. Its extension by Glashow in 1961 [9] to an SU(2) × U(1) permitted the simultaneous description of electromagnetic and charge-changing weak interactions at the price of introducing a new *neutral* massive gauge boson (now called the Z^0) which coupled to a specific mixture of V and A currents for each quark and lepton.

The Glashow theory left unanswered the mechanism by which the $W^\pm$ and Z were to acquire their masses. This was provided by Weinberg [10] and Salam [11] through the Higgs mechanism [12], whereby the SU(2) × U(1) was broken spontaneously to the U(1) of electromagnetism. Proofs of the renormalizability of this theory, due in the early 1970's to G. 't Hooft, M. Veltman, B. W. Lee, and J. Zinn-Justin [13], led to intense interest in its predictions, including the existence of charge-preserving weak interactions due to exchange of the hypothetical Z^0 boson. By 1973, a review by E. Abers and B. W. Lee [14] already was available as a guide to searches for neutral weak currents and other phenomena predicted by the new theory.

III CURRENTS

Let $Q_l^{(+)}$ be the spatial integral of the time-component of the charge-changing leptonic weak current, so that $Q_l^{(+)}|e^-\rangle_L = |\nu_e\rangle_L$, where the subscript L denotes a left-handed particle. It is a member of an SU(2) algebra, since it is just an isospin-raising operator. Defining $Q_l^{(-)} = [Q_l^{(+)}]^\dagger$, we can form $2Q_{3l} \equiv [Q_l^{(+)}, Q_l^{(-)}]$ and then find that the algebra closes: $[Q_{3l}, Q_l^{(\pm)}] = \pm Q_l^{(\pm)}$. In order to describe the decays of strange and non-strange hadrons in a unified way with a suitably normalized weak hadronic current, M. Gell-Mann and M. Lévy [15] proposed in 1960 that the corresponding hadronic charge behaved as $Q_h^{(+)}|n\cos\theta + \Lambda\sin\theta\rangle = |p\rangle$, with

$\sin\theta \simeq 0.2$. Such a current also is a member of an SU(2) algebra. This allowed one to simultaneously describe the apparent suppression of strange particle decay rates with respect to strangeness-preserving weak interactions, and to account for small violations of weak universality in beta-decay, which had become noticeable as a result of radiative corrections [16].

In 1963 N. Cabibbo adopted the idea of the Gell-Mann – Lévy current by writing the weak current as

$$J^{\mu(+)} = \cos\theta J^{\mu(+)}_{\Delta S=0} + \sin\theta J^{\mu(+)}_{\Delta S=1} \qquad (1)$$

and using the newly developed flavor-SU(3) symmetry [17] to evaluate its matrix elements between meson and baryon states. In the language of the u, d, s quarks this corresponded to writing the hadronic charge-changing weak currents as

$$Q_h^{(+)} = \begin{bmatrix} 0 & \cos\theta & \sin\theta \\ 0 & 0 & 0 \\ 0 & 0 & 0 \end{bmatrix} \quad, \quad Q_h^{(-)} = [Q_h^{(+)}]^\dagger \quad,$$

$$Q_{3h} \equiv \frac{1}{2}[Q_h^{(+)}, Q_h^{(-)}] = \begin{bmatrix} 1 & 0 & 0 \\ 0 & -\cos^2\theta & -\sin\theta\cos\theta \\ 0 & -\sin\theta\cos\theta & -\cos^2\theta \end{bmatrix} \qquad (2)$$

Again, the algebra closes: $[Q_{3h}, Q_h^{(\pm)}] = \pm Q_h^{(\pm)}$, so the Cabibbo current is suitably normalized. A good fit to weak semileptonic decays of baryons and mesons was found in this manner, with $\sin\theta \simeq 0.22$.

As a student, I sometimes asked about the interpretation of Q_{3h}, which has strangeness-changing pieces! A frequent answer, reminiscent of the Wizard of Oz, was: "Pay no attention to that [man behind the screen]!" The neutral current was supposed just to close the algebra, not to have physical significance.

IV QUARK-LEPTON ANALOGIES: QUARTET MODELS

Very shortly after the advent of the Cabibbo theory, a number of proposals [18–20] sought to draw a parallel between the weak currents of quarks and leptons in order to remove the strangeness-changing neutral currents just mentioned. Since the electron and muon each were seen to have their own distinct neutrino [21], why shouldn't quarks be paired in the same way? This involved introducing a quark with charge $Q = 2/3$, carrying its own quantum number, conserved under strong and electromagnetic but not weak interactions. As a counterpoise to the "strangeness" carried by the s quark, the new quantum number was dubbed "charm" by Bjorken and Glashow. The analogy then has the form

$$\begin{bmatrix} \nu_e \\ e^- \end{bmatrix} \begin{bmatrix} \nu_\mu \\ \mu^- \end{bmatrix} \Leftrightarrow \begin{bmatrix} u \\ d \end{bmatrix} \begin{bmatrix} c \\ s \end{bmatrix} \qquad (3)$$

The matrix elements of the hadronic $Q^{(+)}$ (we omit the subscript h) were then

$$\langle u|Q^{(+)}|d\rangle = \langle c|Q^{(+)}|s\rangle = \cos\theta \quad, \quad \langle u|Q^{(+)}|s\rangle = -\langle c|Q^{(+)}|d\rangle = \sin\theta \quad, \qquad (4)$$

while those of Q_3 were

$$\langle u|Q_3|u\rangle = \langle c|Q_3|c\rangle = -\langle d|Q_3|d\rangle = -\langle s|Q_3|s\rangle = \frac{1}{2} \quad, \qquad (5)$$

with all off-diagonal (*flavor-changing*) elements equal to zero. Here, as before, $\sin\theta \simeq 0.22$. Bjorken and Glashow were the first to call the isospin doublet of non-strange charmed mesons "D" (for "doublet"), with $D^0 = c\bar{u}$ and $D^+ = c\bar{d}$.

V GAUGE THEORY RESULTS

The promotion of electroweak unification to a genuine gauge theory permitted quantitative predictions of the properties of the fourth quark.

A The Glashow-Iliopoulos-Maiani ("GIM") paper

Taking his gauge theory of electroweak interactions seriously, Glashow in 1970 together with J. Iliopoulos and L. Maiani observed that the quartet model of weak hadronic currents banished flavor-changing neutral currents to leading order of momentum in higher orders of perturbation theory [22]. Thus, for example, higher-order contributions to K^0–$\bar{K}^0$ mixing, expected to diverge in the $V-A$ theory or in a gauge theory without the charmed quark, would now be cut off by m_c, where m_c is the mass of the charmed quark. In this manner an upper limit on the charmed quark mass of about 2 GeV was deduced. In view of the predominant coupling (4) of the charmed quark to the strange quark, charmed particles should decay mainly to strange particles, with a lifetime estimated to be about $\tau_\text{charm} \simeq 10^{-13}$ s.

The GIM paper contained a number of other specific predictions about the properties of charmed particles. Among these were:

- A branching ratio of the charmed meson $D^0 = c\bar{u}$ to $K^-\pi^+$ of a few percent;
- Strong production of charm-anticharm pairs;
- Direct leptons in charm decays;
- Charm production in neutrino reactions;
- Neutral flavor-preserving currents;
- The observability of a Z^0 in the direct channel of e^+e^- annihilations.

These were all to be borne out over the next few years. The discovery of the Z^0 took longer, and was first made in a hadron rather than a lepton collider [23].

B Anomalies

Once the electroweak theory was on firm theoretical grounds, it was noticed by several authors in 1972 [24–26] that contributions to various triangle diagrams involving fermion loops had to cancel. For the electroweak theory it was sufficient to consider the sum over all fermion species of $I_{3L}Q^2$, where I_{3L} is the weak isospin of the left-handed states and Q is their electric charge. For the first family of quarks and leptons the cancellation is arranged as follows:

Fermion:	ν_e	e^-	u	d	Sum
Contribution:	$\frac{1}{2}(0)^2$	$-\frac{1}{2}(-1)^2$	$\frac{1}{2}(3)\left(\frac{2}{3}\right)^2$	$-\frac{1}{2}(3)\left(-\frac{1}{3}\right)^2$	0
Equal to:	0	$-\frac{1}{2}$	$\frac{2}{3}$	$-\frac{1}{6}$	0

The corresponding cancellation for the second family reads

$$\nu_\mu + \mu + c + s = 0 \quad , \tag{6}$$

so that the charmed quark was *required* for such a cancellation, given the existence of the muon and the strange quark.

C Rare kaon decays

In a landmark 1973 paper, M. K. Gaillard and B. W. Lee [27] took the charmed quark seriously in calculating a host of processes involving kaons to higher order in the new electroweak theory. These included K^0–$\bar{K}^0$ mixing and numerous rare decays such as $K_L \to (\mu^+\mu^-,\ \gamma\gamma,\ \pi^0 e^+ e^-,\ \pi^0 \nu \bar\nu,\ \ldots)$ and $K^+ \to (\pi^+ e^+ e^-,\ \pi^+ \nu \bar\nu,\ \ldots)$. The analyses of K^0–$\bar{K}^0$ mixing and $K_L \to \mu^+\mu^-$ indicated that $m_c^2 - m_u^2$ obeyed a strong upper bound, while the failure of $K_L \to \gamma\gamma$ to be appreciably suppressed indicated that $m_u^2 \ll m_c^2$. Together these results supported the GIM estimate of $m_c \le 2$ GeV and considerably strengthened an earlier bound by Lee, J. Primack, and S. Treiman [28].

VI EXHORTATIONS

K. Niu and collaborators already had candidates for charm in emulsion as early as 1971 [29]. These results, taken seriously by theorists in Japan [30,31], will be mentioned again presently. Meanwhile, in the West, theorists besides GIM began to urge their experimental colleagues to find charm. C. Carlson and P. Freund [32] discussed, among other things, the properties of a narrow charm-anticharm bound state. George Snow [33] listed a number of features of charm production and decays. Through an interest in hadron spectroscopy, I became involved late in 1973 in these efforts in collaboration with Gaillard and Lee. We started to look at charm production and detection in hadron, neutrino, and electron-positron reactions. It quickly became clear that a new quark, even one as light as 2 GeV, could have been overlooked.

Glashow spoke on charm at the 1974 Conference on Experimental Meson Spectroscopy, held at Northeastern University [34]. In addition to the properties mentioned in the earlier GIM paper, he told his experimental colleagues to expect:

- Charm lifetimes ranging between 10^{-13} and 10^{-12} s;
- Comparable branching ratios for semileptonic and hadronic decays;
- An abundance of strange particles in the final state;
- Dileptons in neutrino reactions (with the second lepton due to charm decay).

He ended with the following charge to his colleagues:

WHAT TO EXPECT AT EMS-76

There are just three possibilities:
1. Charm is not found, and I eat my hat.
2. Charm is found by hadron spectroscopers, and we celebrate.
3. Charm is found by outlanders, and you eat your hats.

In the summer of 1974, Sam Treiman, then an editor of Reviews of Modern Physics, pressed Ben Lee, Mary Gaillard, and me to write up our results with the comment: "It's getting urgent." Our review of the properties of charmed particles was eventually published in the April 1975 issue [35]. Better late than never. By then we were able to add an appendix dealing with the new discoveries. The body of our article ("GLR") was written before them. Our conclusions, most of which I mentioned at a Gordon Conference late in the summer of 1974, were as follows:

We have suggested some phenomena that might be indicative of charmed particles. These include:

(a) ''direct'' lepton production,
(b) large numbers of strange particles,
(c) narrow peaks in mass spectra of hadrons,
(d) apparent strangeness violations,
(e) short tracks, indicative of particles with lifetime of order 10^{-13} sec.,
(f) di-lepton production in neutrino reactions,
(g) narrow peaks in e^+e^- or $\mu^+\mu^-$ mass spectra,
(h) transient threshold phenomena in deep inelastic leptoproduction,
(i) approach of the $(e^+e^- \to$ hadrons$)/(e^+e^- \to \mu^+\mu^-)$ ratio [''R''] to $3\frac{1}{3}$, perhaps from above, and
(h) any other phenomena that may indicate a mass scale of 2 - 10 GeV.

A couple of these bear explanation. "Apparent strangeness violations" can occur in the transitions $c \leftrightarrow d$; otherwise strangeness would directly track charm (aside from a sign; the convention is that the strangeness of an s quark is -1, while the charm of a charmed quark is $+1$). "Narrow peaks in e^+e^- or $\mu^+\mu^-$ mass spectra" were not just dreamt up out of the blue; we were aware of an effect in muon pairs at a mass around 3.5 GeV [36] which could have been the lowest spin-triplet $c\bar{c}$ bound state. John Yoh remembers hearing this interpretation from Mary K. Gaillard in the Fermilab cafeteria in August of 1974. Our estimate of the width of this state was about 2 MeV, based on extrapolating the Okubo-Iizuka-Zweig (OZI) rule [37] which suppressed "hairpin" quark diagrams. An early prediction by T. Appelquist and H. D. Politzer [38] of the properties of $c\bar{c}$ bound states used QCD to anticipate a narrower spin-triplet than GLR.

I invited Glashow to the University of Minnesota in October of 1974 to speak on charm and much else (including grand unified theories, which he was then developing with Howard Georgi [39]). An unpersuaded curmudgeonly astronomer turned to a younger colleague in the audience, whispering: "When do they bring in the men in white coats?" The timing could not have been better. Charm was to be discovered within a month.

VII HIDDEN (AND NOT-SO-HIDDEN) CHARM

As was suspected even before the days of QCD and asymptotic freedom, the ratio $R \equiv \sigma(e^+e^- \to \text{hadrons})/\sigma(e^+e^- \to \mu^+\mu^-)$ probes the sum $\sum Q^2$ of the squared charges of quarks pair-produced at a given c.m. energy. Thus, above the resonances ρ, ω, and ϕ which are features of low-energy e^+e^- annihilations into hadrons, one expected to see $R = 3[(2/3)^2 + (-1/3)^2 + (-1/3)^2] = 2$, corresponding to the three light quarks u, d, and s. With wide errors, the ADONE Collider at Frascati found this to be the case. (See [40] for earlier references.)

In 1972 the Cambridge Electron Accelerator (CEA) was converted to an electron-positron collider. At energies above 3 GeV the cross section for $e^+e^- \to$ hadrons, instead of falling with the expected $1/E_{c.m.}^2$ behavior characteristic of pointlike quarks, was found to remain approximately constant [41]. At $E_{c.m.} = 4$ GeV, R was 4.9 ± 1.1, while it rose to 6.2 ± 1.6 at $E_{c.m.} = 5$ GeV [42]. These results were confirmed, with higher statistics, at the SPEAR machine [42]. At the 1974 International Conference on High Energy Physics, Burt Richter voiced concern about the validity of the naive quark interpretation of R.

The London Conference was distinguished by various precursors of charm in addition to the rise in R just mentioned. Deep inelastic scattering of muon neutrinos was occasionally seen (in about 1% of events) to lead to a pair of oppositely-charged muons. One muon carried the lepton number of the incident neutrino; the second could be the prompt decay product of charm. This interpretation was mentioned by Ben Lee at the end of D. Cundy's rapporteur's talk [43]. Leptons produced at large transverse momenta [44] were due in part to prompt decays of charmed particles. John Iliopoulos [45] not only laid out a number of the predictions for properties of charmed particles, but bet anyone a case of wine that they would be discovered by the next (1976) International Conference on High Energy Physics. Though he recalls several takers, they never paid off.

On November 11, 1974, the simultaneous discovery of the lowest-lying $^3S_1$ charm-anticharm bound state, with a mass of 3.1 GeV/c^2, was announced by Samuel C. C. Ting and Burt Richter. Ting's group, inspired in part by the suggestion of a peak in an earlier experiment [36] and in part by an innate confidence that lepton-pair spectra would reveal new physics, collided protons produced at the Brookhaven Alternating-Gradient Synchrotron (AGS) with a beryllium target to produce electron-positron pairs whose effective mass spectrum was then studied with a high-resolution spectrometer [46]. The new particle they discovered was called "J" (the character for "Ting" in Chinese). Richter's group, working at SPEAR, wished to re-check anomalies in the cross section for electron-positron annihilations that had shown up in earlier running around a center-of-mass energy of 3 GeV. By carefully controlling the beam energy, they were able to map out the peak of a narrow resonance at 3.1 GeV [47], which they called "ψ", a continuation of the vector-meson series $\rho, \omega, \phi, \ldots$. The dual name J/ψ has been preserved. I was made aware of these discoveries by a call from Ben Lee on November 11. They certainly looked like charm to me, as well as to a number of other people [38,48].

However, a large portion of the community offered alternative interpretations [49]. Some potential objections to charm (see the next Section) were worth putting to experimental tests (e.g., by finding singly-charmed particles [50]). However, I doubt the situation was ever as grave as implied by the comment made to me in March of 1975 by Dick Blankenbecler at SLAC:

```
Don't give up the ship.  It has just begun to sink.
```

VIII   OPEN CHARM

In 1971, well before the discovery of the J/ψ, there were intimations of particles carrying a single charmed quark through the short tracks they left in emulsions, as studied by K. Niu and collaborators at Nagoya [29]. The best candidate appears now to be an example of the rare decay $D^+ \to \pi^+\pi^0$. Tony Sanda reminded us in this meeting [51] that by the 1975 International Conference on Cosmic Ray Physics this group had accumulated [52] a significant sample of such "short-lived particles."

A candidate for the charmed baryon now called Λ_c (as well as for the decay $\Sigma_c \to \Lambda_c \pi$) was reported in neutrino interactions in 1975 [53]. The properties of the Λ_c and Σ_c were very close to those anticipated by an analysis of charmed-particle spectroscopy [54] which appeared shortly after the discussion of the J/ψ.

Despite these indications, as well as the discovery of a candidate for the first radial excitation ("ψ'") of the J/ψ [55] just 10 days after the observation of the ψ in e^+e^- collisions, the charm interpretation of the J/ψ and ψ' required several key tests to be passed.

A Where was the $D \to \bar{K}\pi$ decay?

The decays of charmed nonstrange mesons, with predicted masses of nearly 2 GeV/c^2, could involve a wide variety of final states, so that any individual two-body (e.g., $D^0 \to K^-\pi^+$) or three-body (e.g., $D^+ \to K^-\pi^+\pi^+$) mode should have a branching ratio of a few percent [22].

GLR attempted to estimate this effect using a current algebra model to estimate multiple-pion production [35]. Unfortunately we used a value of the pion decay constant f_π high by $\sqrt{2}$ [56], and neglected other modes besides $\bar{K} + n\pi$ [57]. Our results implied $\mathcal{B}(D^0 \to \bar{K}\pi)$ of nearly 50% for a 2 GeV/c^2 charmed particle, clearly an overestimate both in hindsight and intuitively (see, e.g., [22]). Our result was quoted in the report [58] of an initial SPEAR search which failed to find charmed particles, and may have led to overconfidence in some other proposed experiments [59] which failed to find charm. Subsequent calculations (also taking into account non-zero pion mass), based both on the current algebra matrix element and on a statistical model [60], found smaller $D \to \bar{K}\pi$ branching ratios than GLR [56].

B Why did R rise beyond its predicted value of $3\frac{1}{3}$?

The rise in R observed at 4 GeV and higher was *too large* to account for charm, which predicted $\Delta R = 3Q_c^2 = 4/3$. The resolution of this problem was that pairs of τ leptons [61], whose threshold is $E_{\text{c.m.}} = 2m_\tau c^2 \simeq 3.56$ GeV, were also contributing to R. These τ leptons also diluted the rise in kaon multiplicity expected above charm threshold. This coincidence had all the aspects of a mystery thriller [62]; the near-degeneracy of charm and τ production thresholds is one of those effects (like the comparable masses of the muon and pion) that seems just to have been put in to make the problem harder.

The value of R is still a bit large in comparison with theoretical expectations in the range covered by SPEAR [63].

C Where were the predicted electric dipole transitions from the ψ' to P-wave levels?

The lowest P-wave charmonium levels (now called χ_c) were predicted to lie between the 1S and 2S levels [64]. Thus, one expected to be able to see the electric dipole transitions $\psi' \to \gamma\chi_c$, leading to monochromatic photons. Initial inclusive searches using a NaI(Tl) detector at SPEAR did not turn up these transitions [65], leading to some concern.

The problem turned out to be more experimentally demanding than originally suspected. By looking for the cascade transitions $\psi' \to \gamma\chi_c \to \gamma\gamma J/\psi$, the DASP group, working at the DORIS storage ring at DESY, presented the first results [66] for the $\chi_{c1} = {}^3P_1$ level (with some possible admixture of $\chi_{c2} = {}^3P_2$). By looking for events of the form $\psi' \to \gamma\chi_c \to \gamma + (\pi\pi, K\bar{K},\ldots)$ and reconstructing the mass of the final hadronic state, the Mark I group at SPEAR [67] detected states corresponding to both χ_{c2} and $\chi_{c0} = {}^3P_0$.

D Discovery of the D mesons

By 1975, estimates based on the mass of the J/ψ, on QCD [54], and on potential models incorporating coupled-channel effects [68] predicted D masses in the range of 1.8 to 1.9 GeV/c^2, so that the rise in R could, at least in part, be accounted for by $D\bar{D}$ threshold. Glashow urged Gerson Goldhaber to re-examine the negative search results [69]. Together with F. M. Pierre and other collaborators, Goldhaber incorporated time-of-flight information to improve kaon identification, and found peaks in $D^0 \to K^-\pi^+$ and $K^-\pi^+\pi^-\pi^+$ [70], corresponding to a mass which we now know to be 1.863 GeV/c^2. Low-multiplicity decays of the D^+ were also seen shortly thereafter [71].

The first discoveries of D mesons were announced in June of 1976. This would have been too late for the 1976 Meson Conference, which was traditionally held in April, so Glashow could have lost his bet made at the 1974 Conference [34]. (See, however, [53].) But meson spectroscopy was entering a slower period, and the next conference was not held until 1977. Since charm had clearly been discovered by outlanders, the participants were obliged to eat their (candy) hats, graciously distributed by the conference organizers.

E The τ as interloper

What about the τ lepton, whose appearance complicated the interpetation of the SPEAR results? It destroyed the anomaly cancellation, mentioned earlier! As a result, a new pair of quarks with charges 2/3 and $-1/3$, named top and bottom by Harari [62], had to be postulated. Just such a quark pair had been invented earlier (in 1973) by Kobayashi and Maskawa [31] in order to explain the observed CP violation in kaon decays. The discovery of these quarks is another story, of which Fermilab has a right to be proud but which we shall not mention further here.

F Total rate vs. purity in charm detection

A question which arose in the search for charmed particles is being played out again as present and future searches are planned. Is it better to work in a relatively clean environment with limited rate, or in an environment where rate is not a problem but backgrounds are high? For charm in the mid-1970's, the choice lay between the reaction $e^+e^- \to \gamma^* \to c\bar{c}$, contributing $\Delta R = 4/3$ above charm threshold, and fixed-target proton-proton collisions at 400 GeV/c^2, with $\sigma_{c\bar{c}} = \mathcal{O}(10^{-3})\sigma_{\text{tot}}$ but overall greater charm production rates than in e^+e^- collisions. (The CERN Intersecting Storage Rings (ISR) were also running at that time, providing proton-proton c.m. energies of up to 63 GeV but with limited rates compared to fixed-target experiments.)

After much time and effort, the balance eventually tipped in favor of fixed target hadron (or photon) collisions. (In photon collisions the photon can couple directly to a charm-anticharm pair via the electric charge, leading to diffractive production.) Two advances that greatly enhanced the ability to isolate charm were the use of the soft pion in $D^* \to D\pi$ decays [72] and the impressive growth in vertex detection technology [73].

Soft pion tagging. The lowest-lying $^1S_0$ and $^3S_1$ bound states of a charmed quark and a nonstrange antiquark are called D and D^*, respectively. Their masses are such that D^{*0} can decay to $D^0\gamma$ and just barely to $D^0\pi^0$, while D^{*+} can decay to $D^{*0}\gamma$ and just barely to $D^+\pi^0$ or $D^0\pi^+$. In the last case, the charged pion has a very low momentum with respect to the D^0, and can be used to "tag" it. One takes a hypothetical set of D^0 decay products and combines them with the "tagging" pion. If the decay products really came from a D^0, the difference in effective masses of the products with and without the extra pion should be $M(D^{*+}) - M(D^0) \simeq 145$ MeV/c^2. This method not only can help to see the D^0, but can tell whether it was produced as a D^0 or a $\bar{D}^0$, since the only low-mass combinations are $\pi^+ D^0$ or $\pi^- \bar{D}^0$. This distinction is important if one wishes to study D^0-$\bar{D}^0$ mixing or suppressed decay modes of the D^0 (where the flavor of the decay products does not necessarily indicate the flavor of the decaying state).

Vertex detection. The earliest technique for detecting the short tracks made by charmed particles, nuclear emulsions, was successfully used in Fermilab E-531 for the detection of charmed particles produced in neutrino interactions, has been used by Fermilab E-653 for the study of decays of charmed and B mesons, and is still in use for detecting decays of τ leptons produced in neutrino-oscillation experiments [74]. It has profited greatly from automatic scanning methods introduced by Niu's group at Nagoya. Still, it can be subject to systematic errors, such as a bias against long neutral decay paths.

When it was realized that charmed particles could have lifetimes less than 10^{-12} s, numerous attempts were made to improve the resolution of existing devices such as bubble chambers and streamer chambers. Some of these are described in [73].

In the late 1970's, electronic spectrometers such as the OMEGA spectrometer at CERN began to be equipped with new, high-resolution silicon vertex detectors. These devices had the advantages of radiation hardness, excellent spatial resolution, and electronic readout, making them *the* technique of choice for resolving the tracks of short-lived particles in the busy environments of hadro- and electroproduction. Experiments which have profited from this technique over the years include CERN WA-82, WA-89, WA-92 and Fermilab E-687, E-691, E-769, E-791, and E-831 (FOCUS).

TABLE 1. Lowest orbitally-excited charmed mesons.

j	$J = j - \frac{1}{2}$ state	$J = j + \frac{1}{2}$ state	$l(D^{(*)}\pi)$	Width
1/2	$? \to D\pi$	$? \to D^*\pi$ [a]	0	Broad
3/2	$D(2420)[\to D^*\pi]$	$D(2460)[\to D^{(*)}\pi]$	2	Narrow

[a]Candidate exists (see below).

IX EXAMPLES OF FURTHER PROGRESS

A Emulsion results

Emulsion studies of neutrino- and hadroproduction of charmed particles have displayed the variation of lifetimes among charmed particles, measured the decay constant f_{D_s} of the charmed-strange meson $c\bar{s} \equiv D_s$, and set limits on neutrino oscillations. The scanning techniques pioneered by the Nagoya group are beginning to be disseminated so that many institutions can analyze future results.

B Excited charmed mesons

A meson containing a single heavy quark and a light antiquark is like a hydrogen atom of the strong interactions. The heavy quark corresponds to the nucleus, while the antiquark (and its accompanying glue) correspond to the electron and electromagnetic field.

The lowest orbitally excited states of charmed mesons follow an interesting pattern rather different from that in charm-anticharm bound states. In $c\bar{c}$ levels, the charge-conjugation parity $C = (-1)^{L+S}$ prevents the mixing of spin-singlet and spin-triplet levels with the same L. Thus, the properties of levels are best calculated by first coupling the c and $\bar{c}$ spins to $S = 0$ or 1 and then coupling S with the orbital angular momentum L to total angular momentum J. One thus labels the states by $^{2S+1}[L]_J$, where $[L] \equiv S, P, D, F, \ldots$ for $L = 0, 1, 2, 3, \ldots$. In heavy-light states, however, nothing prevents mixing of $^1P_1$ and $^3P_1$ levels, and there is a favored pattern in the limit that the heavy quark's mass approaches infinity [54,75,76]. One first couples the light antiquark's spin $s = 1/2$ to the orbital angular momentum $L = 1$ to obtain the total angular momentum $j = 1/2, 3/2$ carried by the light quark. One then couples j to the heavy quark's spin $S_Q = 1/2$ to obtain two pairs of levels, as shown in Table 1.

The $j = 1/2$ states are expected to decay to $D^{(*)}\pi$ via S-waves and thus to be broad and hard to find, while the $j = 3/2$ states should decay via D-waves and thus should be narrower and more easily distinguished from background. The first orbitally excited charmed mesons were reported by the ARGUS Collaboration [77] in 1985. Since then, considerable progress has been made on these states by the ARGUS, CLEO, LEP, and fixed-target Fermilab collaborations, with the properties of the $j = 3/2$ states well mapped out. There is now a candidate for a broad $(j = 1/2)$ state, with spin-parity $J^P = 1^+$, mass $M = 2.461^{+0.041}_{-0.034} \pm 0.010 \pm 0.032$ GeV, and width $\Gamma = 290^{+101}_{-79} \pm 26 \pm 36$ MeV [78].

C Charmonium with antiprotons

The ability to control the energy of an antiproton beam, first in the CERN ISR [79] and then in the Fermilab Antiproton Accumulator Ring [80], permitted the study of charmonium states through direct-channel production on a gas-jet tartget. A series of experiments studied the production and decay of states like the η_c (the $^1S_0$ charmonium ground state), the J/ψ, and the χ_c levels, and led to the discovery of the h_c, the $^1P_1$ level. Precise measurements of masses and decay widths were made, and an earlier claim [81] for the 2^1S_0 level, the η'_c, has been disproved. The search for this state, as well as for possible narrow $c\bar{c}$ levels above $D\bar{D}$ threshold, continues at Fermilab as well as elsewhere (see, e.g., [82]).

D Photo- and hadroproduction with vertex detection

An impressive series of fixed-target experiments has refined the technique of vertex detection using silicon strips or pixels [83], obtaining unparalleled numbers of charmed particles. Among the significant results are detailed studies of lifetime differences among charmed particles, ranging from greater than 10^{-12} s for the D^+ to less than 10^{-13} s for the $\Omega_c = css$.

E Electron-positron collisions

The ARGUS and CLEO Collaborations continued to contribute significant results on charmed particles produced in e^+e^- collisions, with results still flowing from CLEO on such topics as the leptonic decay of the D_s [84] and the spectroscopy of charmed baryons [85].

X EXAMPLES OF CURRENT QUESTIONS

A Lifetime hierarchies

The charmed-particle lifetimes mentioned in the previous Section, with

$$\tau(\Omega_c) < \tau(\Lambda_c) < \tau(D^0) \simeq \tau(D_s) < \tau(D^+) \tag{7}$$

varying by more than a factor of 10, continue to be a mild source of concern to theorists. The above hierarchy is better-understood [86,87] than that in strange particle decays, where lifetimes vary by more than a factor of $600 \simeq \tau(K_L)/\tau(K_S)$.

However, the same methods which appear to have described the charm lifetime hierarchy do not explain why $\tau(\Lambda_b)/\tau(B^{+,0}) < 0.8$, whereas a ratio more like 0.9 to 0.95 is expected. It appears that non-perturbative effects, probably the main feature of the lifetime differences for kaons and still important for charmed particles, continue to have some residual effects even for the decays of the heavy b quark.

B Decay constants

The latest average for the D_s decay constant [88] is $f_{D_s} = 255 \pm 21 \pm 28$ MeV, based on observation of the decays $D_s \to \mu\nu$, $\tau\nu$. We still need the values of the other heavy meson decay constants: f_D, f_B, and f_{B_s}. Lattice [89] and QCD sum rule [90] predictions for these quantities exist. The value of f_{D_s} is consistent with predictions, though a bit on the high side. The value of f_D is in principle accessible with present CLEO data samples [91]. One would like to be able to distinguish between the quark-model prediction [92] $f_{D_s}/f_D \simeq 1.25$ and the lattice/sum rule predictions of this ratio, which range between 1.1 and 1.2. One may be able to isolate $D^+ \to \mu^+\nu_\mu$ via the kinematics of the decay $D^{*+} \to \pi^0 D^+$ [93].

C Excited D mesons

Using heavy-quark symmetry, we can relate the properties of a meson containing a heavy quark Q and a light antiquark $\bar{q}$ to those where Q is replaced by another heavy quark Q'. Thus, further study of excited $D = c\bar{q}$ mesons would give us information about the corresponding $\bar{B} = b\bar{q}$ states. The properties of P-wave $b\bar{q}$ ("B^{**}") mesons would be very useful for "tagging" neutral B's [94], since a $\bar{B}^0$ resonates with a π^- to form a B^{**-} while a B^0 resonates with a π^+ to form a B^{**+}.

D Charm-anticharm mixing and CP violation

Both mixing and CP-violating effects are expected to be far smaller for charmed particles than for B's [95]. Since these effects are easier to study in the charm system (at least in hadronic production, where charm production is much easier than b production), they are thus ideal for displaying beyond-standard-model physics, since the standard-model effects are so much smaller.

XI LESSONS?

Should we be learning from history, or will we always be fighting the last war? The search for charm has possible lessons, perhaps to be taken with a grain of salt, for theory, experiment, and their synthesis in the form of future searches.

A Theory

The optimism of theorists was justified in the search for charm. The charmed quark indeed was light, $m_c \simeq 1.5$ GeV/c^2. Perturbative QCD was at least a qualitative guide to the properties of charmonium and charmed particles. The discovery of the first quark with mass substantially exceeding that of the QCD scale was a tremendous boost to the idea (already strongly suggested by deep inelastic scattering) that fundamental quarks needed to be taken seriously.

B Experiment

Many searches for charmed particles were harder than people thought. Sometimes they were aided by sheer instrumental "overkill," as in the case of the superb mass resolution attained in the experiment which discovered the J particle. Sometimes the choice of a fortunate channel also helped, as in the production of the ψ by e^+e^- collisions with carefully controlled beam energies, or in the choice of the e^+e^- decay mode in which to observe the J. Advances in instrumentation proved crucial, whether in the use of particle identification to pull out the initial D^0 signal from background or the study of charmed particles in high-background environments using vertex detection.

C Future searches

I do not see as clear a path in future searches as there was toward charm. In the case of supersymmetry, for example, the landscape looks very different. There is a wide variety of predictions, and one is looking for the whole supersymmetric system at once. Alternate schemes for solving the problems addressed by supersymmetry (e.g., dynamical electroweak symmetry breaking) are not yet even formulated in a self-consistent manner. Perhaps that makes the searches for physics beyond the standard model, which will be addressed in future experiments here at Fermilab and elsewhere, even more exciting.

ACKNOWLEDGMENTS

I wish to thank Val Fitch for a pleasant collaboration on Ref. [1], Ikaros Bigi and Joel Butler for the chance to take this trip down memory lane, and Peter Cooper, Mary K. Gaillard, John Iliopoulos, Scott Menary, Chris Quigg, Tony Sanda, Lincoln Wolfenstein, and John Yoh for helpful comments. This work was supported in part the United States Department of Energy under Contract No. DE FG02 90ER40560.

REFERENCES

1. V. L. Fitch and J. L. Rosner, "Elementary Particle Physics in the Second Half of the Twentieth Century," Ch. 9 in *Twentieth Century Physics*, edited by L. M. Brown, A. Pais, and B. Pippard (AIP/IOP, New York and Bristol, 1995), pp. 635–794.
2. E. Fermi, Nuovo Cim. **11**, 1–19 (1934); Zeit. Phys. **88**, 161–71 (1934).
3. S. Schweber, *QED and the Men Who Made It* (Princeton University Press, 1994).
4. H. Yukawa, Proc. Phys.-Math. Soc. Japan **17**, 48–57 (1935).
5. O. Klein, in *Les Nouvelles Théories de la Physique* (Inst. Int. de Coöperation Intellectuelle, Paris, 1939), pp. 81–98.
6. C. N. Yang and R. L. Mills, Phys. Rev. **96**, 191–5 (1954).
7. R. P. Feynman and M. Gell-Mann, Phys. Rev. **109**, 193–8 (1958); E. C. G. Sudarshan and R. E. Marshak, in *Proc. Int. Conf. on Mesons and Newly Discovered Particles (Padua–Venice, 22–28 September, 1957)* edited by N. Zanichelli (Società Italiana di Fisica, Bologna, 1958), reprinted in *The Development of Weak Interaction Theory*, edited by P. K. Kabir (Gordon and Breach, New York, 1963), pp. 118–28; Phys. Rev. **109**, 1860–2 (1958).
8. J. Schwinger, Ann. Phys. (N.Y.) **2**, 407–34 (1957); S. A. Bludman, Nuovo Cim. **9**, 433–45 (1958); G. Feinberg, Phys. Rev. **110**, 1482–3 (1958); S. L. Glashow, Nucl. Phys. **10**, 107–17 (1959); M. Gell-Mann, Rev. Mod. Phys. **31**, 834–8 (1959); T. D. Lee and C. N. Yang, Phys. Rev. **119**, 1410–19 (1960).
9. S. L. Glashow, Nucl. Phys. **22**, 579–88 (1961).
10. S. Weinberg, Phys. Rev. Lett. **19**, 1264–6 (1967).
11. A. Salam, in *Proc. of the Eighth Nobel Symp.*, edited by N. Svartholm (Almqvist and Wiksell, Stockholm / Wiley, New York), pp 367–77.
12. P. W. Anderson, Phys. Rev. **130**, 439–42 (1963); P. W. Higgs, Phys. Lett. **12**, 132–3 (1964); Phys. Rev. Lett. **13**, 508–9 (1964); Phys. Rev. **145**, 1156–63 (1966); F. Englert and R. Brout, Phys. Rev. Lett. **13**, 321–3 (1964); G. S. Guralnik, C. R. Hagen, and T. W. B. Kibble, Phys. Rev. Lett. **13**, 585–7 (1965); T. W. B. Kibble, Phys. Rev. **155**, 1554–61 (1967).
13. G. 't Hooft, Nucl. Phys. **B33**, 173–99 (1971); Nucl. Phys. **B35**, 167–88 (1971); G. 't Hooft and M. J. G. Veltman, Nucl. Phys. **B44**, 189–213 (1972); B. W. Lee, Phys. Rev. D **5**, 823–35 (1972); B. W. Lee and J. Zinn-Justin, Phys. Rev. D **5**, 3121–37, 3137–55, 3155–60 (1972).
14. E. Abers and B. W. Lee, Phys. Rep. **9**, 1–141 (1973).
15. M. Gell-Mann and M. Lévy, Nuovo Cim. **16**, 705–25 (1960).
16. T. Kinoshita and A. Sirlin, Phys. Rev. **113**, 1652–60 (1959).
17. M. Gell-Mann, Caltech Report CTSL-20, reprinted in M. Gell-Mann and Y. Ne'eman, *The Eightfold Way* (Benjamin, New York, 1964), pp. 11–57; Y. Ne'eman, Nucl. Phys. **26**, 222–9 (1961).
18. B. J. Bjorken and S. L. Glashow, Phys. Lett. **11**, 255–7 (1964).
19. Y. Hara, Phys. Rev. **134**, B701–4 (1964).
20. Z. Maki and Y. Ohnuki, Prog. Theor. Phys. **32**, 144–58 (1964).
21. G. Danby *et al.*, Phys. Rev. Lett. **9**, 36–44 (1962).
22. S. L. Glashow, J. Iliopoulos, and L. Maiani, Phys. Rev. D **2**, 1285–92 (1970).

23. UA1 Collaboration, G. Arnison et al., Phys. Lett. **126B**, 398–410 (1983); UA2 Collaboration, P. Bagnaia et al., Phys. Lett. **129B**, 130–40 (1983).
24. C. Bouchiat, J. Iliopoulos, and Ph. Meyer, Phys. Lett. **38B**, 519–23 (1972).
25. H. Georgi and S.L. Glashow, Phys. Rev. D **6**, 429–31 (1972).
26. D. J. Gross and R. Jackiw, Phys. Rev. D **6**, 477–93 (1972).
27. M. K. Gaillard and B. W. Lee, Phys. Rev. D **10**, 897–916 (1974).
28. B. W. Lee, J. R. Primack, and S. B. Treiman, Phys. Rev. D **7**, 510–6 (1973).
29. K. Niu, E. Mikumo, and Y. Maeda, Prog. Theor. Phys. **46**, 1644–6 (1971).
30. T. Hayashi et al., Prog. Theor. Phys. **47**, 280–7, 1998-2014 (1972); **49**, 350–2, 353–4 (1973); **52**, 636–46 (1974).
31. M. Kobayashi and T. Maskawa, Prog. Theor. Phys. **49**, 652–7 (1973).
32. C. E. Carlson and P. G. O. Freund, Phys. Lett. **39B**, 349–52 (1972).
33. G. Snow, Nucl. Phys. **B55**, 445–54 (1973).
34. S. L. Glashow, in *Experimental Meson Spectroscopy—1974*, edited by D. A. Garelick (AIP, New York, 1974), pp. 387–92.
35. M. K. Gaillard, B. W. Lee, and J. L. Rosner, Rev. Mod. Phys. **47**, 277–310 (1975).
36. J. H. Christenson et al., Phys. Rev. Lett. **25**, 1523–6 (1970); Phys. Rev. D **8**, 2016–34 (1973).
37. S. Okubo, Phys. Rev. Lett. **5**, 165–8 (1963); G. Zweig, CERN Reports 8182/CERN-TH-401 and 8419/CERN-TH-412, 1974, second paper reprinted in *Devlopments in the Quark Theory of Hadrons*, vol. 1, edited by D. B. Lichtenberg and S. P. Rosen (Hadronic Press, Nonantum, MA, 1980), pp. 22–101; J. Iizuka, Prog. Theor. Phys. Suppl. **37–38**, 21–34 (1966), and references therein.
38. T. Appelquist and H. D. Politzer, Phys. Rev. Lett. **34**, 43–5 (1975).
39. H. Georgi and S. L. Glashow, Phys. Rev. Lett. **32**, 438–41 (1974); H. Georgi in *Proc. 1974 Williamsburg DPF Meeting*, edited by C. E. Carlson (AIP, New York, 1975), pp. 575–82.
40. $\gamma\gamma$2 Collaboration, C. Bacci et al., Phys. Lett. **86B**, 234–8 (1979); MEA Collaboration, B. Esposito et al., Lettere al Nuovo Cim. **19**, 21–31 (1977).
41. A. Litke et al., Phys. Rev. Lett. **30**, 1189–92, 1349(E) (1973); G. Tarnopolsky et al., Phys. Rev. Lett. **32**, 432–5 (1974).
42. B. Richter, in *Proceedings of the XVII International Conference on High Energy Physics*, London, July 1974, edited by J. R. Smith (Rutherford Laboratory, Chilton, England, 1974), pp. IV-37–55.
43. D. C. Cundy, 1974 London Conference [42], pp. IV-131–48.
44. Parallel session chaired by L. M. Lederman, 1974 London Conference [42], pp. V-1–56, and references therein.
45. J. Iliopoulos, 1974 London Conference [42], pp. III-89–116.
46. J. J. Aubert et al., Phys. Rev. Lett. **33**, 1404–6 (1974).
47. J.-E. Augustin et al., Phys. Rev. Lett. **33**, 1406–8 (1974).
48. A. De Rújula and S. L. Glashow, Phys. Rev. Lett. **34**, 46–9 (1975); S. Borchardt, V. S. Mathur, and S. Okubo, Phys. Rev. Lett. **34**, 38–40 (1975); C. G. Callan, R. L. Kingsley, S. B. Treiman, F. Wilczek, and A. Zee, Phys. Rev. Lett. **34**, 52–6 (1975); T. Appelquist, A. De Rújula, H. D. Politzer, and S. L. Glashow, Phys. Rev. Lett. **34**, 365–9 (1975); E. Eichten, K. Gottfried, T. Kinoshita, K. D. Lane, and T.-M. Yan,

Phys. Rev. Lett. **34**, 369–72 (1975).

49. A. S. Goldhaber and M. Goldhaber, Phys. Rev. Lett. **34**, 36–7 (1975); J. Schwinger, Phys. Rev. Lett. **34**, 37–8 (1975); R. M. Barnett, Phys. Rev. Lett. **34**, 41–3 (1975); H. T. Nieh, T. T. Wu, and C. N. Yang, Phys. Rev. Lett. **34**, 49–52 (1975); J. J. Sakurai, Phys. Rev. Lett. **34**, 56–8 (1975).

50. J. L. Rosner, in *Proceedings of the Annual Meeting of the Division of Particles and Fields*, American Physical Society, Seattle, Washington, August 27–29, 1975, edited by H. J. Lubatti and P. M. Mockett, University of Washington, 1975, pp. 140–58.

51. A. I. Sanda, these Proceedings.

52. K. Hoshino *et al.*, Prog. Theor. Phys. **53**, 1859–62 (1975); in *Proceedings of the Fourteenth International Cosmic Ray Conference*, München, Aug. 15–29, 1975, vol. 7, pp. 2442–7 and 2448–53.

53. E. G. Cazzoli *et al.*, Phys. Rev. Lett. **34**, 1125–8 (1975).

54. A. De Rújula, H. Georgi, and S. L. Glashow, Phys. Rev. D **12**, 147–62 (1976).

55. G. S. Abrams *et al.*, Phys. Rev. Lett. **33**, 1453–5 (1974).

56. B. W. Lee, C. Quigg, and J. L. Rosner, Phys. Rev. D **15**, 157–65 (1977); C. Quigg and J. L. Rosner, Phys. Rev. D **17**, 239–47 (1978).

57. M. B. Einhorn and C. Quigg, Phys. Rev. D **12**, 2015–30 (1975); Phys. Rev. Lett. **35**, 1114–6 (1975).

58. A. M. Boyarski *et al.*, Phys. Rev. Lett. **35**, 196–9 (1975).

59. See, e.g., the SPIRES Experiment subfile, for the years before 1980. This data base may be consulted for information about specific experiments listed elsewhere in the present review.

60. E. Fermi, Prog. Theor. Phys. **5**, 570–83 (1950).

61. M. L. Perl *et al.*, Phys. Rev. Lett. **35**, 1489–92 (1975); Phys. Lett. **63B**, 466–70 (1976); Phys. Lett. **70B**, 487–90 (1977).

62. H. Harari, in *Proc. 1975 Int. Symp. on Lepton and Photon Interactions (Stanford University, August 21–27, 1975)*, edited by W. T. Kirk (Stanford Linear Accelerator Center, Stanford, CA, 1976), pp. 317–53; H. Harari, in Proceedings of the 20th Annual SLAC Summer Institute on Particle Physics: *The Third Family and the Physics of Flavor*, edited by L. Vassilian, Stanford Linear Accelerator Center report SLAC-412, pp. 647–54.

63. Mark I Collaboration, J. L. Siegrist *et al.*, Phys. Rev. D **26**, 969-90 (1982); J. Kirkby, in *Proceedings of the 1979 International Symposium on Lepton and Photon Interactions at High Energies*, Fermilab, August 23–29, 1979, ed. by T. B. W. Kirk and H. D. I. Abarbanel (Fermi National Accelerator Laboratory, Batavia, IL, 1979, pp. 107–22, and references therein.

64. See, e.g., E. Eichten *et al.* in [48].

65. J. W. Simpson *et al.*, Phys. Rev. Lett. **35**, 699–701 (1975).

66. DASP Collaboration, W. Braunschweig *et al.*, Phys. Lett. **57B**, 407–12 (1975).

67. G. J. Feldman *et al.*, Phys. Rev. Lett. **35**, 821–4 (1975); W. M. Tanenbaum *et al.*, Phys. Rev. Lett. **35**, 1323–6 (1975).

68. E. Eichten *et al.* [48]; Phys. Rev. Lett. **36**, 500–4 (1976).

69. M. Riordan, *The Hunting of the Quark* (Simon and Schuster, New York, 1987), p. 318.

70. G. Goldhaber *et al.*, Phys. Rev. Lett. **37**, 255–9 (1976).

71. I. Peruzzi et al., Phys. Rev. Lett. **37**, 569–71 (1976).
72. S. Nussinov, Phys. Rev. Lett. **35**, 1672–5 (1975).
73. G. Bellini, I. Bigi, and P. Dornan, Phys. Rep. **289**, 1–155 (1997).
74. CHORUS Collaboration, E. Eskut et al., Phys. Lett. B **434**, 205–13 (1998).
75. J. L. Rosner, Comments on Nucl. Part. Phys. **16**, 109–30 (1985).
76. M. Lu, M. B. Wise, and N. Isgur, Phys. Rev. D **45**, 1553–6 (1992).
77. ARGUS Collaboration, H. Albrecht et al., Phys. Rev. Lett. **56**, 549–52 (1986).
78. CLEO Collaboration, presented by J. L. Rodriguez, these Proceedings.
79. CERN ISR R-704 Collaboration, C. Baglin et al., Phys. Lett. B **195**, 85–90 (1987); Nucl. Phys. **B286**, 592–634 (1987), and references therein.
80. Fermilab E-760 and E-835 Collaborations, T. Armstrong et al., Phys. Rev. D **55**, 1153–8 (1997), **56**, 2509–31 (1997); G. Zioulas, in Int. J. Mod. Phys. A **12**, 4049–58 (1997); in *Hadron 97* (Proceedings of the 7th International Conference on Hadron Spectroscopy, Upton, NY, 25–30 August 1997, AIP Conference Proceedings v. 432), edited by S. U. Chung and H. J. Willutzki, (AIP, New York, 1998), pp. 682–90, and references therein.
81. Crystal Ball Collaboration, C. Edwards et al., Phys. Rev. Lett. **48**, 70–3 (1982).
82. DELPHI Collaboration, P. Abreu et al., CERN report CERN-EP/98-151, September, 1998, submitted to Phys. Lett. B.
83. See, e.g., the SPIRES **Experiment** subfile, for experiments such as CERN WA-82, WA-89, WA-92 and Fermilab E-687, E-691, E-769, E-791, and E-831 (FOCUS).
84. CLEO Collaboration, M. Chadha et al., Phys. Rev. D **58**, 032002-1–10 (1998).
85. J. Yelton, in XXIX International Conference on High Energy Physics, Vancouver, BC, Canada, July 23–29, 1998, Proceedings.
86. See, e.g., M. Shifman, in *QCD and Beyond* (Proceedings of the Theoretical Advanced Study Institute in Elementary Particle Physics, Boulder, CO, 1995), edited by D. E. Soper (World Scientific, Singapore, 1996), pp. 409–514.
87. M. Neubert and C. Sachrajda, Nucl. Phys. **B483**, 339–70 (1997).
88. S. Stone, Syracuse University report SU-HEP-98-11, hep-ph/9809580, presented at 4th International Workshop on Particle Physics Phenomenology, Kaohsiung, Taiwan, China, 18–21 June 1998.
89. T. Draper, in *Lattice 98*, Proceedings, Boulder, CO, July 13–17, 1998, and references therein; D. Becirevic et al., Univ. of Rome (I) report ROME1 98/1227, hep-ph/9811003 (unpublished).
90. S. Narison, talk given at the QCD 98 Euroconference–Montpellier (2–8 July 1998), Univ. of Montpellier report PM-98/36, hep-ph/9811208 (1998).
91. L. K. Gibbons, private communication.
92. J. L. Rosner, Phys. Rev. D **42**, 3732–40 (1990); J. F. Amundsen et al., Phys. Rev. D **47**, 3059–62 (1993).
93. J. L. Rosner, Phys. Rev. D **47**, 3057–58 (1993).
94. M. Gronau, A. Nippe, and J. L. Rosner, Phys. Rev. D **47**, 1988–93 (1993).
95. See, e.g., E. Golowich, Nucl. Phys. Proc. Suppl. **59**, 305–15 (1997).

High Promises of Beauty

A. I. Sanda

Physics Dept., Nagoya University, Nagoya 464-01, Japan.
e-mail address: sanda@eken.phys.nagoya-u.ac.jp

Abstract. My recollections of events which took place, from the period of the theoretical search for a large **CP** asymmetry in B decay to the building of the B factory, are discussed. A special emphasis is placed on the importance of interaction between theorists and experimentalists.

I INTRODUCTION

I was asked by the organizers to talk about how interaction between theorists and experimentalists played the role in: (a) predicting large **CP** violation in the B decays; (b) going from the prediction to the completion of the B factory. It is their belief that such discussion is essential for future experimental and theoretical activities. While I am not used to giving historical talks, I thought it might be useful to do so - after all, it has been 18 years since our suggestion that there is gigantic **CP** violation in B decays, and finally, B factories are getting ready to check our predictions.

II NECESSITY OF BEAUTY

The necessity for beauty originated during 60's and 70's at E-ken (which stands for elementary particle physics laboratory) - the laboratory to which I belong at Nagoya University. At that time Sakata and his coworkers were working on the concept that all elementary particles are made up of fundamental particles [1]. They not only took the fundamental nature of these particles seriously, but also they took lepton quark symmetry seriously [2]. They argued that quarks were bound states of (B^+, B^0) and four leptons:

$$\chi^0 = (B^0 \nu_1), \quad P = (B^+ \nu_2), \quad N = (B^+ e^-), \quad \Lambda = (B^+ \mu^-), \tag{1}$$

where $\nu_{1,2}$ are two mass eigenstates [3] *:

---

*) They reasoned that what is known today as Cabibbo angle arises from the leptonic mixing. This was before Cabibbo came up with the Cabibbo angle. For this reason, we call the leptonic mixing angle Maki-Nakano-Sakata(MNS) angle.

$$|\nu_1\rangle = \cos\theta_{MNS} |\nu_e\rangle + \sin\theta_{MNS} |\nu_\mu\rangle$$
$$|\nu_2\rangle = \cos\theta_{MNS} |\nu_\mu\rangle - \sin\theta_{MNS} |\nu_e\rangle. \quad (2)$$

So, there are four fundamental objects which carry a baryon number.

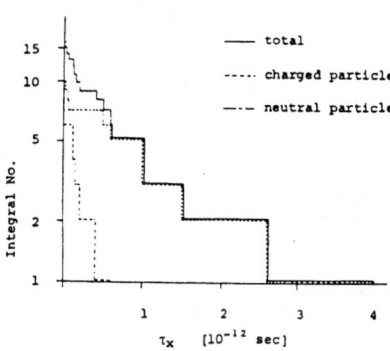

FIGURE 1. Life times of neutral and charged D are determined to be $\tau_\pm = (1 \sim 2) \times 10^{-12}$ sec and $\tau_0 = (3 \sim 4) \times 10^{-13}$ sec.

With this theoretical activity, there was also an important discovery from the experimental side. Niu and his collaborators discovered what is today known as charged and neutral D mesons [4]. The discovery of the D meson was also announced in a journal a year later [5]. All this is preceded by the discovery of a beautiful single event [6], which was believed at Nagoya to be the missing charm particle.

In this atmosphere, when Kobayashi and Maskawa needed six quarks, it was not so hard to make the jump. After all if you know that there are at least 4 quarks, there could easily be 6 quarks. Nagoya was ideally suited for their discovery.

It is ironic that around this time US researchers, for example, were sure that nuclear democracy and bootstrap ideas were correct and quarks are mere mathematical objects. This can be illustrated best from the following quotation by Gell-Mann [7]:

"In other words, we construct a mathematical theory of the strongly interacting particles, which may or may not have anything to do with reality, find suitable algebraic relations that hold in a model, postulate their validity, and then throw away the model. We may compare this process to a method sometimes employed in French cuisine: a piece of pheasant meat is cooked between two slices of veal, which are then discarded.* ···

Their non-appearance could certainly be consistent with the bootstrap idea, and also possibly with a theory containing a fundamental triplet, which is hidden, *i.e.*, has effectively infinite mass."

* I am indebted to Professor V. L. Telegdi for a discussion of this point.

Here again we see collaboration between theorists and experimentalist at work!

III GOLD MINE

In 1987, $B - \overline{B}$ mixing was discovered by the ARGUS collaboration in same sign dilepton events. But, theorists predicted it long before its discovery. The problem is that the mixing went like [9] [†]

$$x_B \equiv \frac{\Delta M_B}{\Gamma} \sim \frac{m_t^2}{700\,\text{GeV}^2}. \tag{3}$$

Because same sign dilepton rates went like x_B^2, probability of observing the effect of mixing was proportional to m_t^4. At some point, there was an experimental result that the top quark mass is bounded by 50GeV, and theorists were lead astray - we could not stick out heads our and announce that experimentalists should see the effect of mixing. We knew, however, that the mixing would be there at some level. It was also known that $\frac{\text{Im}\,M_B}{\Delta M} = \mathcal{O}(1)$. So, $\epsilon_B \gg \epsilon_K$. The problem was that both ϵ_B and ϵ_K are phase convention dependent and they are not physical observables.

So, the question was whether we could find some observable that experimentalists can measure. B mesons are produced in harsh environment - in colliders. Unlike measurement of the neutron electric dipole moment where the neutron lives practically for ever compared to the lifetime of B mesons. To overcome this difficulty, we need effects at the 10% level.

IV SEARCH WAS ON

Around 1978, Pais gave a seminar at Rockefeller University. The seminar was entitled "**CP** violation on charmed-particle decays" [10]. My recollection of how Pais started out his seminar is as follows:

"There is good news and bad news. The good news is that **CP** violation in a heavy meson system is quite similar to that of the K meson system. The bad news is that there is little distinction like K_L and K_S mass eigenstates. For heavy meson system, life times are both short."

[†] This result is based on out of date numbers for the bag parameter and f_B. But this is not important for our purpose here.

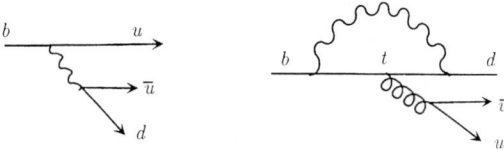

FIGURE 2. Two interfering diagrams considered by Bander, Silverman and Soni

This was the paper which stimulated my search for large **CP** violation in B decays. Soon afterwards, came the paper by Bander, Silverman, and Soni [11]. They have discussed **CP** violation in b quark decay generated by penguin amplitudes shown in Fig.2. They computed the asymmetry in quark decay rates:

$$a = \frac{\Gamma(b \to fq\bar{q}) - \Gamma(\bar{b} \to \bar{f}\bar{q}q)}{\Gamma(b \to fq\bar{q}) + \Gamma(\bar{b} \to \bar{f}\bar{q}q)}. \tag{4}$$

Their result is shown in Fig.3.

It is seen that except for $\frac{s_2}{s_3} \sim 0$, the effect is small. How can we get a big effect? This was the issue!

If ϵ_B is large, at least in some phase convention, we should look for effects which involve mixing - in spite of the fact that there is no evidence for mixing. Let us consider two diagrams, which would exist if there is mixing [12]: The problem was to have these two diagrams interfere. As these two diagrams have different final states $(s\bar{d})$ in (a) and $(\bar{s}d)$ in (b), they can not interfere. It took us one week to realize that, in fact we don't detect quarks and we detect K_S and K_L states. So, these two diagrams can be made to interfere by detecting K_S states.

First suggestion made in Ref. [12] is the asymmetry in inclusive reaction:

$$a = \frac{\Gamma(B \to c\bar{c}K_S) - \Gamma(\overline{B} \to c\bar{c}K_S)}{\Gamma(B \to c\bar{c}K_S) + \Gamma(\overline{B} \to c\bar{c}K_S)}. \tag{5}$$

Note that $B \to c\bar{c}K_S$ is one of the major decay modes of the B meson. In Fig.5, we show the numerical value of the asymmetry a for various KM parameters and the top quark mass.

Note that asymmetry could be $\mathcal{O}(1)$ at the certain region of the parameter space. Today, we know that nature chose exactly the point where the asymmetry is maximal. Otherwise I would not be talking about the historical account!

The problem was that since various decay channels have different **CP** quantum numbers and if there are equal branching ratios for **CP**=+1 states and for **CP**=-1 states, the asymmetry for the inclusive reaction Eq.(5) washes out to zero. For example, we have

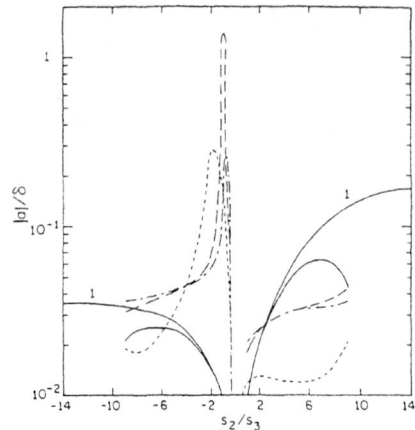

FIGURE 3. Value of the asymmetry a defined in Eq.(4). The result is valid for small δ, the **CP** violating phase in the KM matrix.

$$\mathbf{CP}|\psi K_S \, n\pi^0\rangle = -(-1)^n |\psi K_S \, n\pi^0\rangle. \tag{6}$$

where n is the number of π^0's. This was not the only problem. Unlike in the K system we don't have B or $\overline{B}$ beams. So, we can't detect the asymmetry given in Eq.(5). We first have to produce $B - \overline{B}$'s in pairs then tag one of them. For example:

$$\begin{aligned} e^+e^- \to \Upsilon(4S) \to\ & B\,\overline{B} \to c\bar{c} + X \\ & \hookrightarrow \mu^\pm + anything. \end{aligned} \tag{7}$$

So, we worked out a formula for an asymmetry

$$\frac{\Gamma([B\overline{B}]_L \to l^-c\bar{c}+X) - \Gamma([B\overline{B}]_L \to l^+c\bar{c}+X)}{\Gamma([B\overline{B}]_L \to l^-c\bar{c}+X) + \Gamma([B\overline{B}]_L \to l^+c\bar{c}+X)} = a\sin[\Delta M_B(t_1 + (-1)^L t_2)], \tag{8}$$

where L is the orbital angular momentum of the $[B\overline{B}]$ pair; t_1 and t_2 are time at which the leptonic decay and ψK_S decays are detected, respectively. To our agony, we found that the asymmetry for $L = 1$ vanished if we don't observe the decay times t_1 and t_2 - i.e. if we integrate over time. The time measurement is impossible at CSER since B's, traveled no more than 20μ cm before it decayed. So, we had to rely on the decay

$$e^+e^- \to \Upsilon(4S) \to B\overline{B}^* \to B\overline{B}\gamma. \tag{9}$$

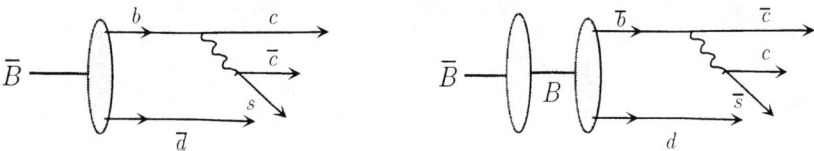

FIGURE 4. Two diagrams which may interfere if there is substantial $B - \overline{B}$ mixing.

The initial state couples to an electromagnetic current which transforms like a **C** = −1 state, the presence of a photon in the final state will guarantee that the $B\overline{B}$ pair is a **C** = +1 state or L even state. So, we hoped that B^* was light enough to be produced in $\Upsilon(4S)$ decays. I was in close contact with CLEO and CUSP collaborators who were on the lookout for almost mono-energetic 50MeV photons.

V GOLD-PLATED DECAY MODE

The collaboration between Ikaros Bigi and I started right after my seminar at CERN. We realized that statistically it is better to look for large asymmetry at the expense of smaller branching ratio than to consider a mode with large branching ratio but small asymmetry. That is, it's better to consider a mode with pure **CP** quantum number, *i.e.***CP** eigenstate. Considering an inclusive reaction to enhance the branching ratio does not help at all. So, we came up with the golden decay mode [13]:

$$B \to \psi K_S \qquad (10)$$

The asymmetry is given by

$$\frac{\Gamma(B^0(t) \to \psi K_S) - \Gamma(\overline{B}^0(t) \to \psi K_S)}{\Gamma(B^0(t) \to \psi K_S) + \Gamma(\overline{B}^0(t) \to \psi K_S)} = \text{Im}\left(\frac{q}{p}\frac{\overline{A}(B \to \psi K_S)}{A(B \to \psi K_S)}\right)\sin\Delta M_B t, \qquad (11)$$

where

$$\frac{q}{p} = \sqrt{\frac{M_{12}^* - \frac{i}{2}\Gamma_{12}^*}{M_{12} - \frac{i}{2}\Gamma_{12}}}. \qquad (12)$$

It is well known that for the B system, $|\frac{i}{2}\Gamma_{12}| \ll |M_{12}|$. So, $\frac{q}{p} = e^{-i\phi_M}$, where ϕ_M is the phase of M_{12}. Lets say that the ψK_S mode has the two contributions shown in Fig.6.

Under **CP** transformation, weak phases reverse their signs, while strong interaction phases do not. So, if ξ_i and δ_i are weak and strong phases of amplitude i, respectively, we have:

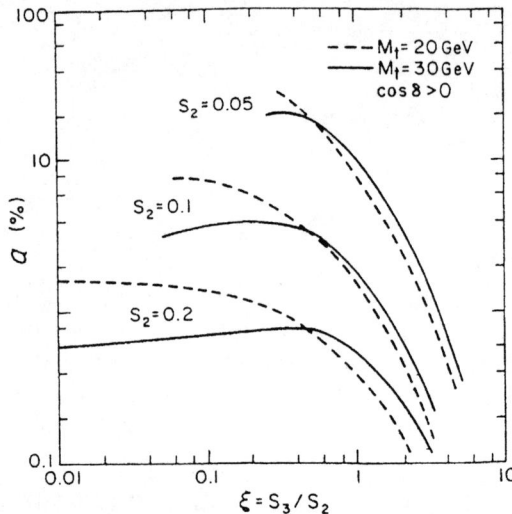

FIGURE 5. An estimate of the asymmetry as a function of $\xi = \sqrt{\rho^2 + \eta^2}$ for each $s_2 \sim V_{cb}$. A set of favoured values today is $\xi \sim .5$, $s_2 \sim .05$, and, of course, $m_t \simeq 170 GeV$.

$$A(B \to \psi K_S) = e^{i\xi_1} e^{i\delta_1} |\mathcal{A}_1| + e^{i\xi_2} e^{i\delta_2} |\mathcal{A}_2|,$$
$$A(\overline{B} \to \psi K_S) = e^{-i\xi_1} e^{i\delta_1} |\mathcal{A}_1| + e^{-i\xi_2} e^{i\delta_2} |\mathcal{A}_2|. \tag{13}$$

We see that if there is only one weak amplitude, or if $\xi_1 = \xi_2$, we have

$$\frac{\overline{A}(B \to \psi K_S)}{A(B \to \psi K_S)} = e^{-2i\xi_1} \tag{14}$$

The reasons the ψK_S mode is called a "Gold Plated" mode are two fold:

- The penguin amplitude has exactly the same weak phase so that the asymmetry is given by

$$\operatorname{Im}\left(\frac{q}{p} \frac{\overline{A}(B \to \psi K_S)}{A(B \to \psi K_S)}\right) = -\operatorname{Im}\left(\frac{\mathbf{V}_{tb}^* \mathbf{V}_{td}}{\mathbf{V}_{tb} \mathbf{V}_{td}^*} \cdot \frac{\mathbf{V}_{cb} \mathbf{V}_{cs}^*}{\mathbf{V}_{cb}^* \mathbf{V}_{cs}}\right) = \sin(2\phi_1). \tag{15}$$

- ψK_S mode has a very clear signature. $\psi \to \mu^+ \mu^-$ decay can be identified in almost any environment. Identifying $K_S \to \pi^+ \pi^-$ decay after an event is triggered by the presence of $\psi \to \mu^+ \mu^-$ is relatively easy. So, hadronic colliders like CDF, D0 and LHC can also study this decay.

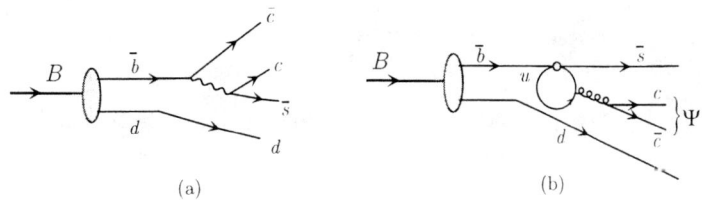

FIGURE 6. $B \to \psi K_S$ decay gets contribution from tree and penguin graphs.

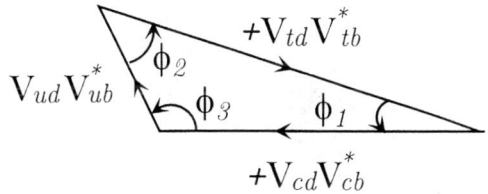

FIGURE 7. Unitarity of the KM matrix leads to a triangular relationship between elements of the KM matrix. The **CP** asymmetry can be related to angles of the triangle.

VI TECHNICAL DIFFICULTIES-NUMEROUS

There are plenty of reasons why we have not measured the ψK_S asymmetry after 18 years.

1. We need to "make" the B beam by tagging. If we tag B, say at time t, we know that the other one, at that instant, is a $\overline{B}$ as the pair is in a P-wave state. Tagging costs us in events as we have to overcome leptonic branching ratio and efficiency.

2. Let us look at the branching ratio for detecting $B \to \psi K_S$ decay while tagging the other B or $\overline{B}$ with a leptonic decay.

$$Br(B \to \psi K_S) \sim 10^{-4},$$
$$Br(B \to l\nu X) \sim 10^{-1},$$
$$Br(\psi \to l^+ l^-) \sim 10^{-1}.$$

(16)

This means that with 100% efficiency, we need 10^8 $B - \overline{B}$ pairs to have 100 tagged ψK_S events. Knowing that the cross section for $\Upsilon(4S)$ production is about a nano-barn, we must have a collider with a luminosity of $10^{34} cm^{-2} sec^{-1}$, if we want the result after about a year of running.

3. The long awaited decay mode given in Eq.(9) never happened. The threshold for $\overline{B}B^*$ production was shown to be beyond $\sqrt{s} = M(\Upsilon(4S))$ [14,15].

4. The fact that $B\overline{B}$ pair is in a P-wave state means that we need to measure the decay times and determine $t_1 - t_2$. This is impossible to accomplish with present technology in a symmetric machine.

VII DESIGNING EXPERIMENTS

To my knowledge, Bjorken was the first, in 1985 [16], to discuss how we might actually do this experiment. At the end of his discussion, he wrote:

> Should one think about following such a path? I don't know. A decision to do so requires a better understanding $\cdots$. All of this should be known better in a few years.
>
> But the real decision to follow such a path must come from those who would do the work. The task is a very long and arduous one and, even for those who would have doubts, the homework should be done. That alone leaves a lot to do for everyone.

He was right! But, this was about to change dramatically in a couple of years.

Discoveries which lead to construction of B factories:

- On the top of the list, certainly, is the discovery of $B\overline{B}$ mixing by the ARGUS collaboration [17]. As mentioned above, it was not much of a surprise for theorists, but this got experimentalists excited. In particular, at the workshop at Berkeley on Experiments, Detectors, and Experimental Areas, there were a whole group of experimentalists who took Bjorken's advice and started to do their homework. Serious attempts to design detectors for this type of physics have been made [18].

- In my mind the most important discovery is the longevity of B mesons [20,21]. An elementary particle must live long enough to show something fundamental about nature. In fact the lifetime kept on increasing every time a new experimental result came out. This was good for the asymmetry. The long lifetime means small $\mathbf{V}_{cb}$, i.e. small s_2. As you can see in Fig.5, its prediction kept on increasing.

- Advances in vertex detectors. As you will see below, measurements of particle track with a resolution of about 20μ cm is required. This was not possible when we started thinking about this experiment. Around that time, Mike Witherell and his collaborators incorporated a vertex detector in a photo-production experiment [19]. They were able to obtain beautiful results on charm particle decays. So, we got the vertex detection capabilities.

- One day, Pier Oddone, Ikaros Bigi, and I were discussing the fact that it is impossible to determine the decay time in an e^+e^- collider as $\Upsilon(4S)$ is at rest in the laboratory frame and $B's$ travel only about 20μ cm. Then Pier said, "Why not build an asymmetric collider. This will boost $\Upsilon(4S)$!" I went on to visit KEK and people at KEK assured me that if we collide electrons and positrons with different energies, the beam will blow up! But then came a bootstrap effect where KEK and SLAC machine physicists competed in improving the maximum luminosity in an asymmetric collider - on paper at least.

- In 1988, David Hitlin, Tatsuya Nakada and I were in Snowmass, Colorado for a workshop on "Summer Study on High Energy Physics in the 1990's". We asked a question [25]: "How asymmetric should an asymmetric collider be?"

  If it is too asymmetric, we would loose all the events in the beam pipe. It turned out that existing TRISTAN ring at KEK and PEP ring at SLAC would do the job!

Since 1980, I went all over the world to convince physicists that B physics was interesting. They all agreed. But, proposals were not forthcoming. Both KEK and SLAC were busy with other projects. KEK was the first, however, to announce its serious intention to build the B factory. Then there was a collapse of SSC project, followed by the B factory proposal at SLAC.

VIII ITS NOT SO EASY - SURPRISING PENGUINS

There are other **CP** eigenstates besides ψK_S. The asymmetry in $B \to \pi^+\pi^-$ allows us to determine ϕ_2, if the penguin amplitudes give negligible contribution compared to the tree graph.

If we just compute the penguin graph, without any QCD corrections, we find a suppression factor like

$$\frac{\alpha_s}{12\pi^3} \log \frac{m_t}{m_c} \sim \alpha_s \times 10^{-2}, \tag{17}$$

which is at most $\mathcal{O}(\lambda^2)$. Here $\lambda \sim \sin\theta_c \sim .23$. So, we felt that penguins will not play a crucial role. Then came the discovery of $b \to s\gamma$ decay [22]. Then CLEO Collaboration showed that [23]:

$$Br(B \to \pi\pi) < Br(B \to K\pi). \tag{18}$$

This is a very curious result.

These decays are generated by Feynman graphs shown in Fig.8. The amplitude for tree $(T(K\pi))$ and penguin$(P(K\pi))$ contributions for $K\pi$ decay mode are:

$$T(K\pi) = \frac{G_F}{\sqrt{2}} \mathbf{V}_{ub}^* \mathbf{V}_{us}[C_1(\mu)Q_{s1}^u(K\pi) + C_2(\mu)Q_{s2}^u(K\pi)],$$

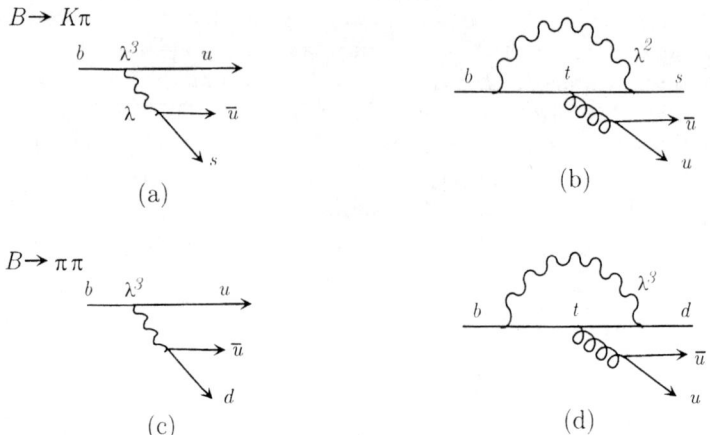

FIGURE 8. The tree and penguin graphs which contribute to $B \to \pi\pi$ and $B \to K\pi$ decays.

$$P(K\pi)_c = \frac{G_F}{\sqrt{2}} V_{cb}^* V_{cs} [C_1(\mu) Q_{s1}^c(K\pi) + C_2(\mu) Q_{s2}^c(K\pi)],$$

$$P(K\pi)_t = \frac{G_F}{\sqrt{2}} (-V_{tb}^* V_{ts}) \sum_{i=3}^{10} C_i(\mu) Q_{si}(K\pi). \tag{19}$$

For $B \to K\pi$, $P(K\pi)$ is $\mathcal{O}(\lambda^2)$ and $T(K\pi)$ is $\mathcal{O}(\lambda^4)$. For $B \to \pi\pi$, these diagrams give $T(\pi\pi) = \lambda^3 T$, $P(\pi\pi)_t = \lambda^3 P_t$, and $P(\pi\pi)_c = \lambda^3 P_c$.

Since $\frac{T(K\pi)}{T(\pi\pi)} \sim \lambda$, if the tree graph matrix elements dominate, we expect $\frac{Br(B \to K\pi)}{Br(B \to \pi\pi)} \sim \mathcal{O}(\lambda^2)$. Experimentally this is no so. This indicates that the $P(K\pi)$ amplitude is at least as large as the $T(\pi\pi)$. If $P(K\pi) \simeq T(K\pi)$, this suggests

$$\frac{[C_1(\mu) Q_{s1}^c(K\pi) + C_2(\mu) Q_{s2}^c(K\pi)]}{[C_1(\mu) Q_{s1}^u(\pi\pi) + C_2(\mu) Q_{s2}^u(\pi\pi)]} = \mathcal{O}(\lambda) \tag{20}$$

i.e., the penguin contribution is considerably larger than what a naive estimate of the loop graph suggested by Eq.(17).

Since we now have an evidence that loop graphs compete with tree graphs, we have to be prepared for a substantially more complex situation.

IX PENGUIN POLLUTION

I don't like the word "penguin pollution". Penguins are harmless and cute. In physics, it presents richness to the field. There are many interesting decays of B mesons, which we will not observe if penguin diagrams are absent. There will be many interesting **CP** asymmetries which will be generated by penguins. But, for **CP** asymmetry with $B-\overline{B}$ mixing, in particular for $B \to \pi\pi$ mode, it is an obstacle toward getting at one of the angles of the unitarity triangle.

Consider two operators differing in their KM parameters driving $B \to f$:

$$A(B \to f) = e^{i\xi_1} e^{i\delta_1} |\mathcal{A}_1| + e^{i\xi_2} e^{i\delta_2} |\mathcal{A}_2| \tag{21}$$

where δ_i and ξ_i are the strong interaction and weak phases, respectively; the moduli of the KM parameters have been incorporated into $|\mathcal{A}_i|$. We then find

$$\mathrm{Im}\frac{q}{p}\overline{\rho}(f) \sim \sin 2(\Phi_m - \xi_1) + \Delta$$

$$\Delta = -2 \left|\frac{\mathcal{A}_2}{\mathcal{A}_1}\right| \sin \Delta\xi \cos(2\Phi_m - 2\xi_1 + \Delta\delta) \tag{22}$$

where $2\Phi_m = \arg\left(\frac{q}{p}\right)$; $\Delta\xi = \xi_2 - \xi_1$, $\Delta\delta = \delta_2 - \delta_1$. In deriving Eq.(22), we have made an approximation, $|\mathcal{A}_2/\mathcal{A}_1| \ll 1$. The presence of a second weak operator poses a challenge in our ability to extract KM parameters from the data. How this difficulty be best overcome depends on the specifics of the channel under study.

For $B \to \pi^+\pi^-$ decay mode, Δ can easily be as big as the $\sin 2(\Phi_m - \xi_1) = \sin(2\phi_2)$ term. So, we must rely on the isospin analysis [24]. But the problem is that this method requires measurement of $Br(B \to \pi^0\pi^0)$ which is often swamped by backgrounds.

X SUMMARY

I have sketched events leading to the construction of B-factories - today (Dec. 1, 1998) is the scheduled day for the turn on of KEKB. I hope I have convinced you that doing real physics, physics that is related closely to experiments, is lots of fun.

There are lots of subsequent experimental findings for which I feel grateful:

- The life time of B meson turned out to be large.

- The KM matrix elements took on values which maximized the effect.

- The reaction $\Upsilon(4S) \not\to B\overline{B}^*$ does not happen. Initially, this fact disappointed us because it required us to go to the asymmetric collider. In retrospect, however, this was a blessing in disguise. It means that we have pure $B\overline{B}$ beam. If we had an admixture of $B\overline{B}$ and $B\overline{B}^*$ states, there will be additional hadronic uncertainty.

- $B\overline{B}$ mixing turned out to be maximal due to the large top quark mass.

- The $B \to \psi K_S$ decay happens.

There is lot to do. B-factories will measure ϕ_1. They will eventually get to a 10% level test of KM ansatz. To get at ϕ_2, because of large penguin amplitude, we must measure all charge modes $B \to \pi^0\pi^0$, $B^+ \to \pi^+\pi^0$, and $B \to \pi^+\pi^-$. In paticu;ar, $B \to \pi^0\pi^0$ decay is difficult because of backgrounds. If I were a young experimentalist, I would try hard to design a single purpose experiment to measure this decay mode. There are lots of very interesting physics ahead of us in b quark decays. I don't want to make a specific list of interesting physics - surely much of it will be out of date when new experimental data is available. In stead, if we take the analogy from the K meson system, we can confidently back up what I said above. The K meson was discovered in 1947. More than 50 years later, we are still designing experiments to study its properties. And these experiments probe fundamental nature of the world we live in. So, I am sure we will be talking about B physics in year 2050. Its not too late to join the quest for new knowledge through the B system.

Acknowledgements

This work has been supported in part by Grant-in-Aid for Special Project Research (Physics of **CP** violation). I wish to express my gratitude to the organizers, in particular, Ikaros Bigi, Joel Butler, and Harry Cheung for organizing an interesting workshop, their hospitality, and encouraging me to write up my lecture.

REFERENCES

1. S. Sakata, *Prog. Theor. Phys.* **16** 686 (1956) .
2. Z. Maki, *Prog. Theor. Phys.* **31** 331 (1964;) *ibid*.**31** 333 (1964) .
3. Z. Maki, M. Nakagawa, and S. Sakata, *Prog. Theor. Phys.* **30** 727 (1963) .
4. Hoshino *et al.*, 14th Cosmic Ray Conf. (Munich) 7,2442 (1975).
5. G. Goldhaber, *et al.*, *Phys. Rev. Lett.* **37** 255 (1976) .
6. K. Niu *et al.*, *Prog. Theor. Phys.* **46** 1644 (1971) .
7. M. Gell-Mman and Y. Neeman, *The Eightfold Way*, W. A. Benjamin, Inc. New York, 1964. p. 198 and p. 199.
8. H. Albrech *et al.*, *Phys. Lett.* **B192** 245 (1987) .
9. See, for example, J. Ellis, *et al.*, *Nucl. Phys.* **B133** 285 (1977) .
10. A. Pais and S. B. Treiman, *Phys. Rev.* **D12** 2744 (1975) .
11. M. Bander, D. Silverman, and A. Soni, *Phys. Rev. Lett.* **43** 242 (1979) .
12. A. B. Carter and A. I. Sanda, *Phys. Rev.* **D23** 1567 (1981) .
13. I. I. Bigi and A. I. Sanda, *Nucl. Phys.* **B193** 85 (1981) .
14. D. Andrews, *et al.Phys. Rev. Lett.* **45** 291 (1981) .
15. L. Spencer *et al.Phys. Rev. Lett.* **47** 771 (1981) .
16. J. D. Bjorken, Concluding lecture, Prceedings of Moriond Workshop on Flavour Mixing and **CP** violation Edition Frontieres Edited by J. T. T. Van Singapore (1985).
17. H. Albrech *et al.*, *Phys. Lett.* **B192** 245 (1987) .

18. K. J. Foley, *et al.*, Prceedings of Moriond Workhop on Experiments, Detectors, and Exprimental Areas, Berkeley, Edited by R. Donaldon and M. G. D. Gilchriese, World Scientific, Singapore (1988).
19. J. R. Raab *et al.*, *Phys. Rev.* **DD37** 2391 (1988) and additional references therein.
20. E. Fernandez *et al.*, *Phys. Rev. Lett.* **51** 1022 (1983) .
21. N. Lockyer *et al.*, *Phys. Rev. Lett.* **51** 1316 (1983) .
22. CLEO Collaboration M. S.Alam *et al.*, *Phys. Rev. Lett.* **74** 2885 (1995) .
23. CLEO Collaboration R. Godang *et al.*, *Phys. Rev. Lett.* **80** 3456 (1998) .
24. M. Gronau and D. London, *Phys. Rev. Lett.* **27** 3381 (1990) .
25. D. Hitlin, T. Nakada, and A. I. Sanda Proceedings of the 1988 Summer Study on High Energy Physics in the 1990's, Snowmass, Colorado, (1988).

PART I
CP VIOLATION AND MIXING

Status of the HERA-B Experiment

Bernhard Schwingenheuer
on behalf of the HERA-B Collaboration

MPI für Kernphysik, P.O.Box 103980, D-69029 Heidelberg

Abstract. The HERA-B experiment is a forward magnetic spectrometer with good particle identification for hadrons and leptons designed to study violation of CP symmetry in the neutral B meson system. Pairs of $b\bar{b}$ quarks are produced in collisions of 920 GeV/c protons of the Hera storage ring at DESY, Hamburg, with internal wire targets. Large parts of the detector are already assembled and radiation hard tracking chambers are expected to be installed in summer and fall of 1999. The current status and recent developments in the detector design are reviewed.

INTRODUCTION

Violation of CP symmetry has since its discovery in 1964 [1] been studied extensively in the neutral kaon system. Its origin is however still unknown even though a definitive answer whether CP is violated only in the mixing of K^0 and $\bar{K}^0$ or in addition in the decay amplitudes might be just ahead of us [2,3].

Extraction of the quark mixing matrix elements describing CP violation and hence a test of the Standard Model is easier in the neutral B meson system. Especially the asymmetry between $\Gamma(B^0 \to J/\Psi K_S)$ and $\Gamma(\bar{B}^0 \to J/\Psi K_S)$ is expected to be large and theoretically easy to interpret which results in a measurement of $\sin 2\beta$ with β being an angle of the unitarity triangle [4].

HERA-B [5] is a fixed target experiment with a silicon vertex detector, a charged particle spectrometer and particle identification for hadrons and leptons. Figure 1 shows an isometric view of the detector. Since the proton energy of the HERA storage ring is 920 GeV/c the $b\bar{b}$ cross section is expected to be between 8 and 27 nbarn. Compared to 13 mbarn of total inelastic cross section only about one in 10^6 collisions yields a $b\bar{b}$ pair. Hence the interaction rate has to be large (on average 4 interactions every 96 nsec bunch crossing) in order to observe a few thousand $B \to J/\Psi K_S$ decays per year. This results in severe requirements on radiation hardness and granularity for all detector components and requires the design of a very selective and efficient trigger.

To determine the flavor of the B in the $J/\Psi K_S$ decay mode the flavor of the second b quark has to be measured. Since the decay chain of the b quark contains

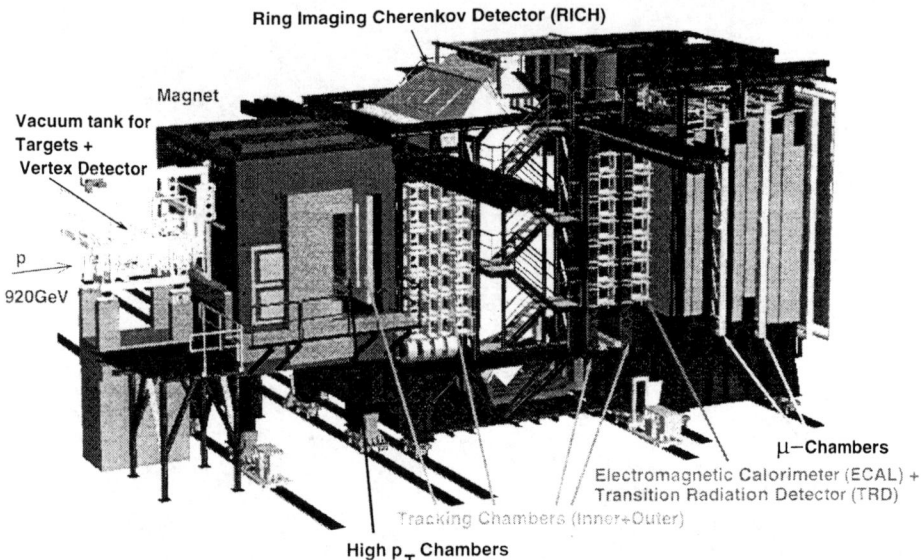

FIGURE 1. Isometric view of the HERA-B Detector.

in about 66% of all cases a K^- while the chain for $\bar{b}$ results in a K^+ identifying this K meson gives the flavor tag. The required particle identification is achieved with the RICH over a momentum range from 5 to 50 GeV/c. Other flavor tagging methods use the charge of the lepton from semileptonic decays of the second b quark or the vertex charge tag.

The main focus of the experiment is the measurement of the CP asymmetry in the B^0 and $\bar{B}^0$ decays to $J/\Psi K_S$. Special pixel and pad chambers inside and downstream of the magnet select tracks with large transverse momentum at the pre-trigger level. This allows us to trigger on many B decays and amongst others to measure the CP asymmetry in $B \to \pi^+\pi^-$.

This status report focuses on some of the more critical components of the detector.

THE TARGET

Inelastic interactions of protons occur in 50 μm wide and 500 μm thick target bands which move into the beam. For high rates the distance from the wires to the beam center is about 2 mm while the beam width is about 0.4 mm. The beam core remains therefore unaffected and passes through HERA-B in a 0.5 mm thick Aluminum beam pipe.

The target is nowadays operated at 40 MHz interaction rate routinely without causing significant background to the other experiments at HERA. Coasting beams, protons circulating in the ring at a slightly larger radius but with continuous spatial

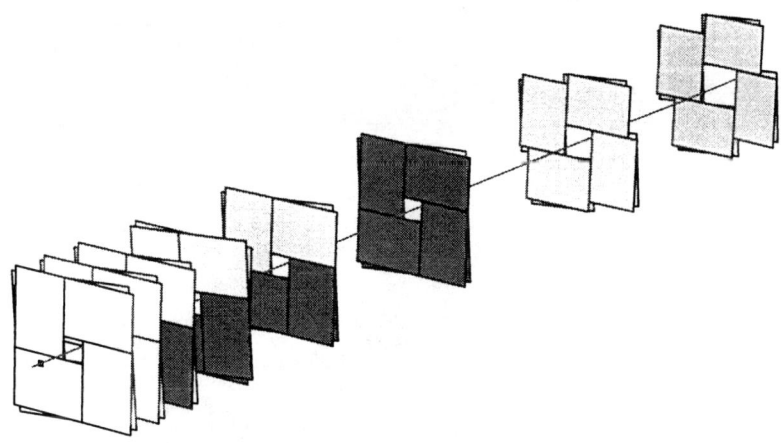

FIGURE 2. Arrangement of the 64 double sided silicon detectors in 8 superlayers. The line in the middle indicates the proton beam which enters from the left. The target wires are drawn at the beginning of the line. The currently installed detectors are dark shaded. The distance from the first to the last superlayer is 2 m.

distribution, cause off-bunch interactions in the target. Since the bunch structure is needed for proper operation of HERA-B investigations for the origin of this problem have started.

THE SILICON VERTEX DETECTOR

For background suppression it is mandatory to reconstruct the detached decay vertex of B mesons. Silicon strip detectors are the natural choice of technology for tracking to match position resolution with the B life time. We use double sided strip readout with a pitch of about 50 μm, an active area of 50×70 mm^2, polysilicon resistors for biasing and will cool the silicon close to 0° C.

During data taking the silicon detectors are positioned 10 mm from the beam while a much larger aperture is required for proton injection. Hence the silicon modules are mounted in roman pots which allow radial and lateral movements. The latter is used to distribute the radiation dosage over larger areas of the detectors. The expected flux for the innermost parts is 3×10^{14}/cm^2 MIPS per year. Even though this dosage is distributed across the detector the foreseen degradation in performance is severe enough to require yearly exchanges of all silicon modules.

Two strip detectors separated by 4 mm are mounted with stereo angles of ± 2.5 degrees in one Roman pot and comprise one quadrant of a superlayer, see figure 2.

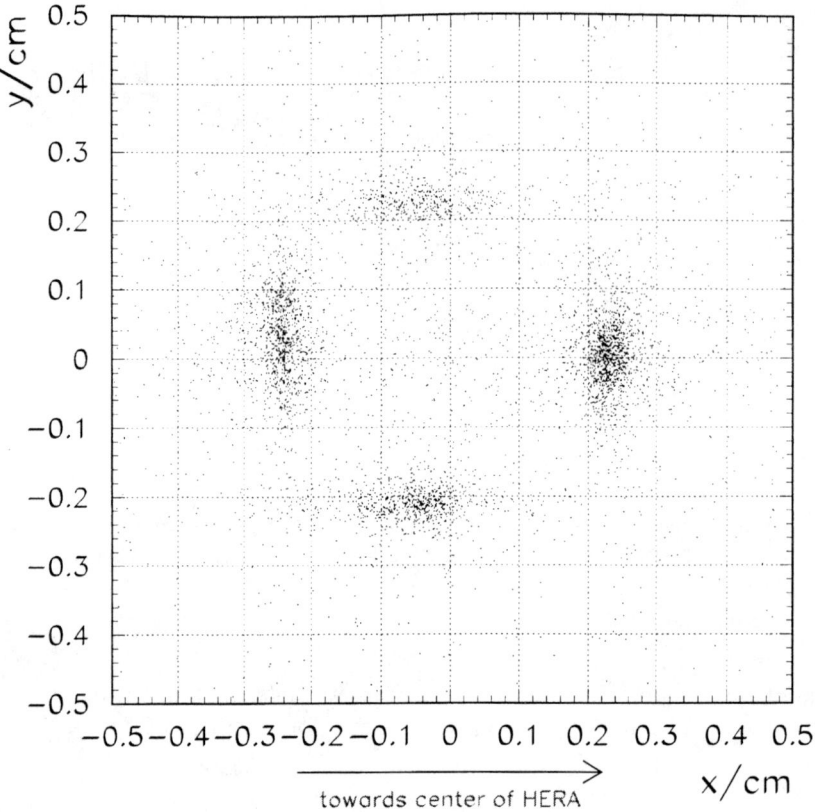

FIGURE 3. Extrapolated (x,y) position of track candidates to the z location of the target. Tracks are found with the Silicon Vertex Detector. The resolution of the "target spot" is dominated by multiple scattering since no track momentum cut was applied.

Consequently we have 3-dimensional hit reconstruction in every layer.

The Vertex Detector consists of 8 superlayers positioned such that every track passes through at least 3 layers, resulting in 12 coordinate measurements. The dark shaded modules in figure 2 are currently installed and the entire Vertex Detector is scheduled to be completed in May '99.

The silicon strips are read out with the HELIX128 chip [6,7] which was developed by the MPI in Heidelberg in collaboration with the University of Heidelberg. The ASIC development started in 1995 and the current version 2.2 fulfills all specifications. The chip has low noise preamplifiers, shapers and a pipeline for 128 input channels. The length of the pipeline allows a maximum trigger latency of 128 cycles. Upon a trigger the stored charges of all channels are multiplexed to one differential output and read out deadtimeless. Up to 8 triggers can be stored in the pipeline.

All currents and voltages needed to operate the HELIX128 are generated internally and are programmable. Thus some performance degradations due to radiation damage can be compensated. The chip gain is linear for input signals in the range of ±10 MIPs and the noise dependence on the input capacitive load is given by $345\,e^- + 38\,e^-/\text{pF}$.

Currently irradiation tests with up to 4 kGy (4× the yearly dosage at HERA-B) show that the HELIX128 is still functional afterwards. More detailed tests (especially of the analog performance) are under way.

For track candidates reconstructed with the Silicon Vertex Detector figure 3 shows their (x,y) position at the z location of the target. It is clearly visible that most of the candidates originate from the four target wires.

THE OUTER TRACKER

The main charged particle tracker has 8 superlayers (each consisting of ±5° and 0° tracking chambers) inside the magnet for particle tracing through the magnetic field to the Silicon Detector and 4 superlayers downstream in the field-free region for track finding. In addition there are two superlayers positioned between the RICH and the ECAL for triggering. Track densities close to the proton beam pipe are very high. Consequently the tracker is divided into an Inner (for radius < 20 cm and fine granularity) and an Outer Tracker (for radius > 20 cm and coarser granularity).

For ease of mass production and to avoid regions with small drift field honey comb drift chambers are the technology of choice for the Outer Tracker. The cathode is made out of conductive polycarbonate foil (Pokalon-C). In total there are 120000 channels with 5 and 10 mm drift cells which are read out with the ASD-8 chip [8].

Briefly after prototype chambers were installed and operated in HERA-B persistent currents were observed (figure 4). This problem did not show up in X-ray irradiation tests before and was found to originate from an insulating layer on the cathode (Malter effect [9]). Positive ions build up on this layer and the resulting large field across the insulation frees electrons from the cathode which ignite new avalanches. Mass production was halted and an intensive R&D program was launched in summer 1997 to find a cure to this problem.

Currently this problem (and a few more identified afterwards) seems to be solved by coating Pokalon-C with a thin layer of gold or graphite and changing the drift chamber gas from CF_4CH_4 (50:50) to $ArCF_4CO_2$ (65:30:5). In addition many other improvements in construction procedures and in the list of qualified materials have been implemented.

Further radiation tests are ongoing and the mass production is about to resume. Because of the delay the Outer Tracker will be completed last.

THE INNER TRACKER

Microstrip Gas Chambers (MSGC) with a readout pitch of 300 μm guarantee an occupancy below 5% for the Inner Tracker. Since the expected MSGC signal is sim-

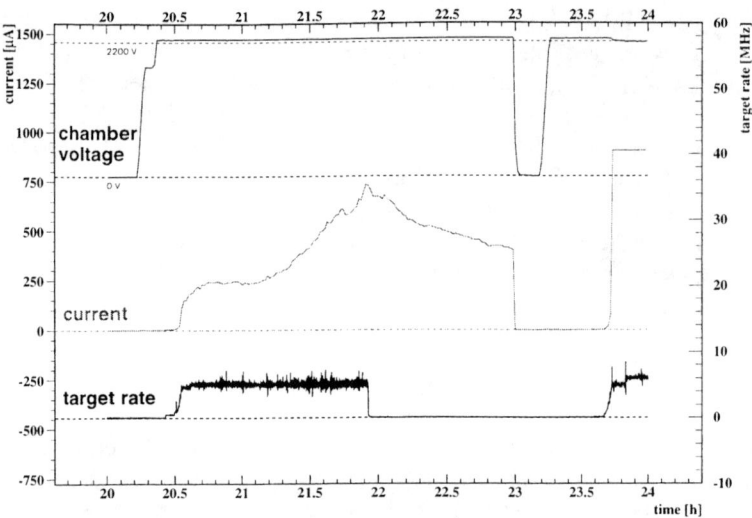

FIGURE 4. Persistent current effect: even without particles passing through the chamber a high current flows (22h < time < 23h).

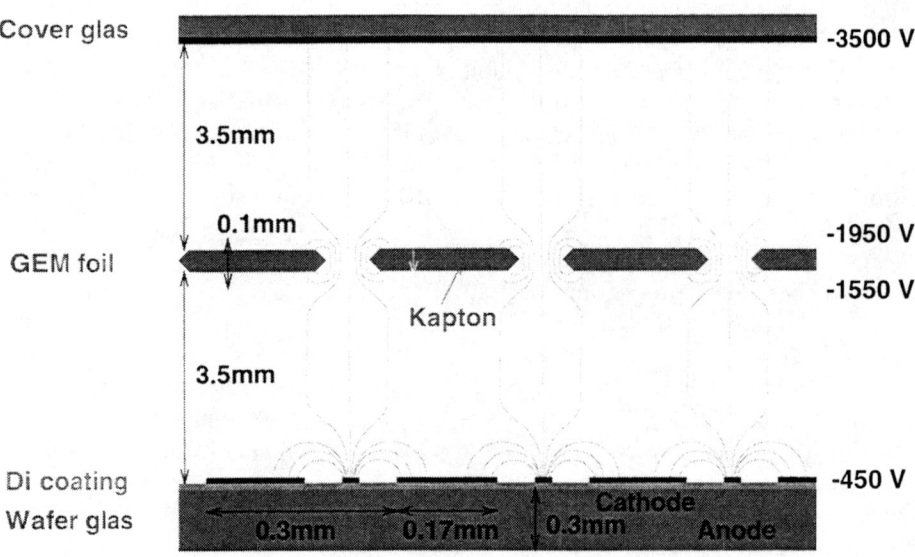

FIGURE 5. Cross section of a GEM-MSGC with two zones of moderate gas gains ($\approx$ 30 at GEM and $\approx$ 180 at MSGC).

ilar to that of the silicon detectors the same readout chip and frontend electronics will be used.

Similar to the Outer Tracker substantial changes in the detector design were needed relative to the TDR [5] to ensure radiation hardness in a hadron beam environment. First the diamond coating (see figure 5) ensures a stable homogeneous field between the anode and the cathode and thus prevents the "sudden death" of the chamber, i.e. a rapid gain drop after a dosage of a few mCb/cm.

Second at the nominal gas gain of 5000 heavy ionizing particles cause sparks (streamer discharges) which damage the anode. After a few hours of operation at HERA-B most anode strips were disconnected. The only workable solution after intensive R&D was to add a "gas electron multiplier" (GEM) foil [10] in the middle of the drift volume (see figure 5). Figure 6 shows the spark rate as a function of the MSGC gas gain for different gas mixtures. Since the gas gain of the GEM is about 30 the MSGC gain can be reduced to 180 and thus the spark rate drops by 4 orders of magnitude to a tolerable level.

A full system test was performed in October 1998 at the PSI in Villigen, Switzerland, with at a 350 MeV/c pion beam. All chambers were operated under stable conditions and hits were found in the offline analysis.

The mass production for the Inner Tracker is well under way. Since it is mechanically coupled to the Outer Tracker the installation schedule for both Trackers is linked.

OTHER SUBDETECTORS

The Ring Imaging Cherenkov Detector (RICH) is completely assembled. It uses C_4F_{10} as gas and mirrors and lenses to collect the Cherenkov photons on 27000 photo multiplier channels. With an average of 32 photons for a track with $\beta = 1$ we expect to distinguish kaons from pions in the momentum range of 5-50 GeV/c. Figure 7 shows one of the first events recorded with the RICH in August'98.

The Electromagnetic Calorimeter (ECAL) is divided into an inner, a middle and an outer section. All sections employ the same technology of a sampling scintillator/absorber sandwich structure read out by plastic wavelength shifter fibers ("shashlik"). In the inner section the absorber is tungsten and the lateral size is $(2.23 \text{ cm})^2$. The middle and outer sections use lead absorbers with lateral dimensions of $(5.575 \text{ cm})^2$ and $(11.15 \text{ cm})^2$, respectively.

The entire ECAL is assembled. Currently the available frontend electronics (about 50%) is installed and the instrumented modules are calibrated. By May'99 all electronics will be available.

The Transition Radiation Detector (TRD) covers the inner part of the acceptance to improve electron identification in the region with the largest track density. It uses foam as radiator and for mechanical support and 5 mm straw tubes as photon detectors. Test chambers were operated in HERA-B during the last years without problem and by May'99 the TRD will be operational.

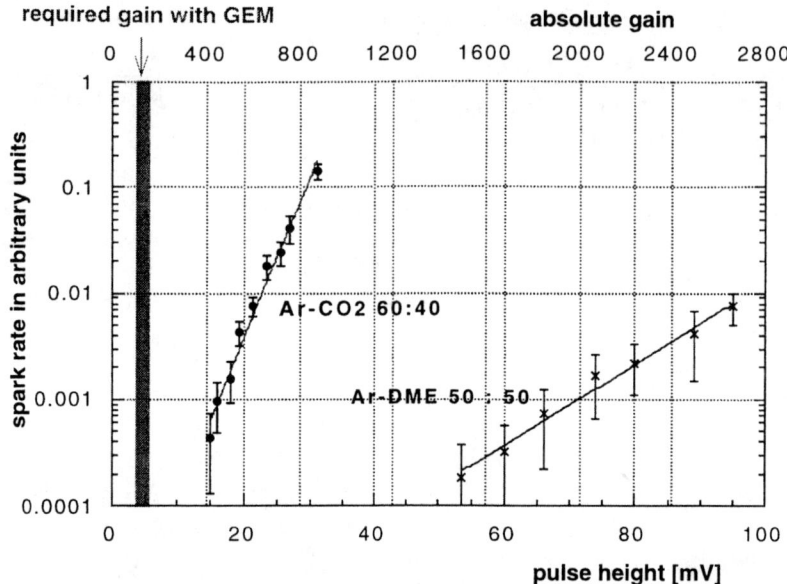

FIGURE 6. Spark rate of a MSGC without GEM as a function of the absolute gain for different gas mixtures.

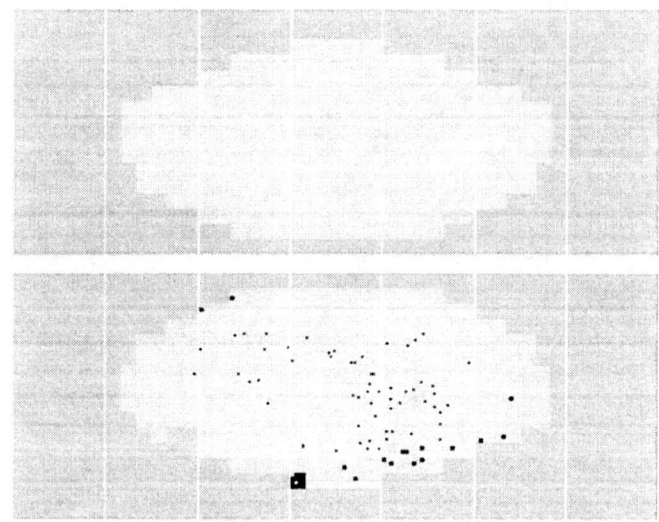

FIGURE 7. One of the first events recorded with the RICH. Only the lower half of the detector was read out. The gas was air and consequently one expects only 7 photons for a track with $\beta = 1$. Thin points represent hits in 16-channel PMTs (inner part) while thick points to hits in 4-channel PMTs correspond.

The Muon Detector consists of four superlayers of which three are used in the pre-trigger. The pre-trigger layers consist of pad chambers (at the outside) and pixel chambers (close to the beam pipe) in order to have good 2-dimensional hit resolution. Tube chambers are used otherwise. Currently about two superlayers are installed and the entire system will be operational by May'98.

The High-p_T Detector needs good 2-dimensional hit reconstruction. Therefore it employs pad and pixel chambers. Again, prototypes have been tested at HERA-B and in other beam tests and the completion of the system is expected for May'98.

THE FIRST LEVEL TRIGGER

The ratio of $B \to J/\Psi K_S \to l^+l^-\pi^+\pi^-$ to all interactions is only $\approx 10^{-11}$. Thus the success of HERA-B depends on an effective and deadtimeless trigger.

The basic scheme is to fully reconstruct the J/Ψ in its dilepton channel with the First Level Trigger and to find the detached B vertex at Level 2. Level 3 builds an event and finds detached vertices with the Silicon Vertex Detector for other triggers. At Level 4 the entire event is reconstructed and written to tape. As a consequence of this scheme the entire detector has to be calibrated online.

Levels 2, 3 and 4 use PC farms which are largely available. This is is also true for the most important software modules. The latter have been tested on Monte Carlo events and show sufficient efficiencies. Yet some more optimization is needed.

Level 1 is a custom made hardware trigger which has to carry the largest burden of all trigger levels [11]. The gigantic task of processing 150 GB/sec, performing track finding and mass reconstruction with 3% resolution in about 10 μsec and yielding a background rejection of 200 while keeping 30%-55% of the signal will be accomplished by 70 Track Finding Units (TFU), 8 Track Parameter Units (TPU) and one Trigger Decision Unit (TDU).

Four superlayers of the Outer and Inner Tracker are used by the trigger. For every layer the hit information is fed into a number of TFUs for every bunch crossing. In case the pre-trigger finds a candidate for a lepton it determines the approximate candidate's (x, y) position (called region-of-interest ROI) at the z location of the superlayer closest to the ECAL. Then the pre-trigger sends a message with these information to the TFU which stores the hits for the given ROI. In case a hit is found in the specified region the TFU calculates the candidate's ROI at the location of the next superlayer and sends a message to the appropriate TFU of that layer. This algorithm is called Kalman filtering.

After the candidate is traced through all superlayers its momentum is calculated on a TPU assuming that the track originates from the target. In case two opposite signed electrons or muons are found the TDU calculates the invariant mass.

The entire system is message driven: every board has a FIFO for incoming messages and acts on them. As a consequence any number of new messages can be generated, stored in an output FIFO and send to the next level.

At the moment two TPUs and 10 TFUs are available. However for full commissioning of the First Level Trigger the Tracker is needed.

CONCLUSION

HERA-B has mastered a number of substantial problems especially concerning the radiation hardness of the Trackers. At the moment no serious obstacles are known and the most crucial issue is the timely mass production of the Inner and Outer Tracker until fall of 1999.

Currently the full readout chain with the final DAQ architecture is being implemented. It is planned to measure the direct J/Ψ production cross section this year. This would be an important achievement in terms of system integration for the experiment and commissioning of many subdetectors.

In 1999 there will be a long shutdown in May which will be used to install half of the tracking system and to complete all other subdetectors. This setup can be used to commission the entire detector, especially the First Level Trigger. Smaller shutdowns during the summer and fall will allow us to complete both Trackers.

With the experience gained by then the first HERA-B physics run will start in fall of 1999 and will last until a longer shutdown scheduled for the year 2000 begins. By then the recorded data sample should be large enough to observe CP violation in the decay channel $B \to J/\Psi K_S$.

REFERENCES

1. J.H. Christenson, J.W. Cronin, V.L. Fitch and R. Turley, Phys. Rev. Lett. **13**, 138 (1964).
2. M. Arenton, these proceedings.
3. A. Ceccucci, these proceedings.
4. A. Stocchi, these proceedings.
5. E. Hartouni et al., Technical Design Report of HERA-B, DESY-PRC 95/01.
6. W. Fallot-Burghardt, Ph.D. Thesis, University of Heidelberg (1998), available from http://wwwasic.ihep. uni-heidelberg.de/asic/notes_engl.html.
7. W. Fallot-Burghardt et al., Helix128-2.x User Manual, HD-ASIC-33-0697, available from http://wwwasic.ihep.uni-heidelberg.de/herab.
8. F.M. Newcomer et al., IEEE Trans. Nucl. Sci. **40**, 630 (1993).
9. L. Malter, Phys. Rev. **50**, 48 (1936).
10. F. Sauli, CERN-EP/98-51; B. Schmidt, physics/9804035.
11. T. Fuljahn et al., Proc. Xth IEEE RealTime'97 conference, Beaune, France, September 1997; to be published in IEEE Trans. Nucl. Sci **RT-97**.

CP Violation, Mixing and Rare Decays at the Tevatron Now and in Run II

Kevin T. Pitts

Fermi National Accelerator Laboratory
Batavia, IL 60510 USA
email: kpitts@fnal.gov

Abstract. We review the status of current B mixing and CP violation measurements from the Fermilab Tevatron. With the existing data sample, the CDF collaboration has made competitive measurements of B_d mixing; set limits on B_s^0 mixing; and has begun a direct search for CP violation in the B system through the decay $B^0/\overline{B^0} \to J/\psi K_s^0$. The prospects for future b measurements at the Tevatron are discussed.

I INTRODUCTION

Over the course of the last 10 years, the B-physics program at the Tevatron has been very rich. Beginning with the observation of exclusive B decays in the 1980s, the program really blossomed with the introduction of silicon microvertex detectors in the 1990s.

The goal of this paper is to outline some of the measurements of the B quark sector performed at the Tevatron. The CDF and DØ experiments have learned a great deal about making these measurements in the challenging environment of a hadron collider. With forthcoming accelerator and detector upgrades, the future prospects for new and more precise measurements of CP violation, B_s^0 mixing and rare B decays is very bright.

II OVERVIEW: B PHYSICS AT THE TEVATRON

The Fermilab Tevatron offers some unique opportunities in b-physics which are not available elsewhere. The proton-antiproton collisions at $\sqrt{s} = 1.8\,\text{TeV}$ create $b\bar{b}$ pairs with a cross section of approximately $100\,\mu\text{b}$. The b quarks can hadronize into all species, including b-baryons, and B_s^0 and B_c mesons.

With an inelastic $p\bar{p}$ cross section which is approximately 1000 times larger than the $b\bar{b}$ cross section, it has so far been necessary to trigger on leptons from the B-hadron decays. The most common trigger paths are: $b \to \ell\nu c$ and $b \to \psi X$, $\psi \to \mu^+\mu^-$. The signal-to-noise can be improved significantly by constructing a detector that has excellent mass resolution and can exploit the long b lifetime

($\tau_B = 1.564 \pm 0.014$ ps, $c\tau_B = 469\,\mu$m [1].) Outer tracking detectors along with a solenoid magnetic field offer good transverse momentum (p_T) and therefore good mass resolution. Silicon microvertex detectors can resolve tracks originating from long lived particles with high efficiency. Even though the center of mass energy of the $p\bar{p}$ system is very large, the B hadron spectrum is relatively soft. For the decay $B \to \psi K_S^0$ with a 2 GeV/c transverse momentum requirement on both trigger muons, the mean p_T of the B is about 10 GeV/c.

III RUN I MEASUREMENTS

In the period 1992-1996, the CDF and DØ collaborations measured approximately 110 pb^{-1} of $p\bar{p}$ collisions at $\sqrt{s} = 1.8$ TeV. In the following sections, we present a few of the many measurements which have come from these data samples.

A B_d^0 Mixing

CDF has performed a number of measurements of time dependent B_d^0 mixing. For these analyses, the trigger path takes advantage of the $b \to \ell\nu c$ decays where an energetic lepton can be identified. In addition, the charge of this lepton provides the tag of the flavor of the B at the time of decay. Corrections must be applied for fake leptons and leptons from sequential $b \to c \to \ell\nu s$ decays. A precision measurement of the decay length in the transverse plane is made using information from the silicon microvertex detector. This is converted to a B lifetime with an approximate correction for the velocity ($\beta\gamma$) of the B.

To measure mixing, the flavor of the B meson (that is, whether it contains a b-quark or a $\bar{b}$-quark) must be identified ("tagged") at the time of production and at the time of decay. Since tagging algorithms are far from perfect, the true asymmetry is "diluted" by mistagging $B^0/\bar{B^0}$: $A_{obs} = DA_{true}$, where A_{obs} is the observed asymmetry and D is the "tagging dilution", defined as the asymmetry between the number of correct tags and incorrect tags: $D = (N_R - N_W)/(N_R + N_W)$ where $N_R(N_W)$ = number of correct (incorrect) tags. The dilution is related to the mistag rate in the following way: $D = 1 - 2w$, where w is the total fraction of incorrect tags.

Figure 2 shows the CDF measurement of B_d^0 mixing using the same-side tagging method. In this analysis, lepton triggered events are reconstructed in the $\ell D(D^*)$ mode. The flavor of the B at the time of decay is established from the charm (D) decay, while the flavor at production is inferred from the charge correlation of tracks near the B hadron with the flavor of the b-quark. This "same-side" correlation may arise through the fragmentation process or through excited (B^{**}) states [2].

Also shown in Figure 2 is the mixing result using jet charge and soft lepton flavor tagging algorithms. Here, the flavor of the B at the time of production is inferred from the second B hadron in the event. This is known as "opposite-side" tagging. The problem with opposite side tagging is that quite often ($\sim 50\%$ of the time) the second B hadron is boosted forward in the lab frame, outside the acceptance of the

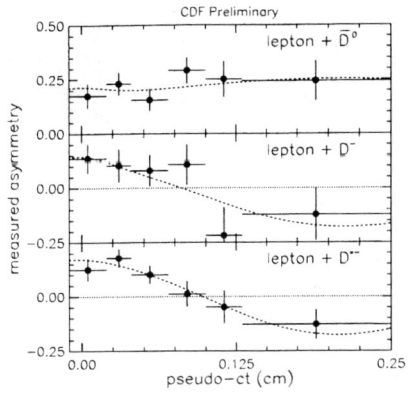
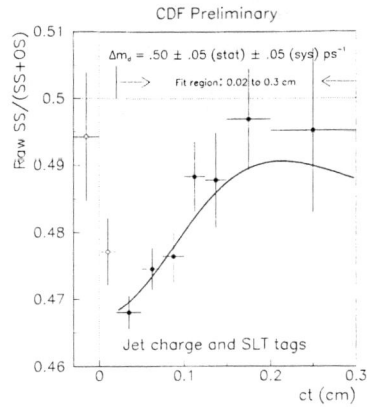

FIGURE 1. Time dependent $B^0/\overline{B^0}$ mixing as measured from: **Left:** $B^0/\overline{B^0} \to \ell^\pm \nu D^\mp (D^{*\mp})$. The top plot shows the B-π charge correlation in the $B^\pm$ sample, where no mixing takes place. The middle and bottom plots show the correlation for $B^0/\overline{B^0}$ events, where the oscillatory behavior can be seen. **Right:** an inclusive lepton sample using both jet charge and soft lepton tagging. The B events are identified with a secondary vertex. Explicit charge/neutral B separation is not performed.

detector. In addition, if the second B hadronizes as a B_d^0 or a B_s^0, then mixing can further confuse the tag. In the jet charge/soft lepton analysis, there is no explicit D reconstruction. The B sample is identified by requiring a secondary vertex to be reconstructed in conjunction with the trigger lepton. Corrections are required for direct $c\bar{c}$ production and sequential $b \to c \to \ell \nu s$ decays.

B $\sin 2\beta$

CP violation manifests itself as an asymmetry in the decay rate of particle versus antiparticle. In the case of $B^0/\overline{B}^0 \to \psi K_S^0$:

$$A_{CP} = \frac{N(\overline{B^0} \to \psi K_S^0) - N(B^0 \to \psi K_S^0)}{N(\overline{B^0} \to \psi K_S^0) + N(B^0 \to \psi K_S^0)}$$

which can be either a time dependent or time-integrated quantity. In the Standard Model, the CP asymmetry in this mode is proportional to $\sin 2\beta$: $A_{CP}(t) = \sin 2\beta \times \sin(\Delta mt)$, where the second term is the time-dependent evolution of $B^0/\overline{B^0}$ mixing. The magnitude of the CP violation shows up as an amplitude of the mixing term.

Integrating over time, the statistical error on $\sin 2\beta$ can be written as: $\delta \sin(2\beta) \approx \frac{1+x_d^2}{x_d} \frac{1}{\sqrt{\epsilon \mathcal{D}^2 S}} \sqrt{\frac{S+B}{S}}$, where ϵ is fraction of events which can be tagged (the "tagging efficiency") and $\mathcal{D}$ is the dilution. The quantity x_d is the $B^0/\overline{B^0}$ mixing parameter,

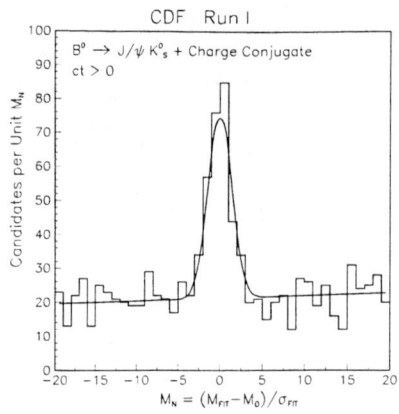
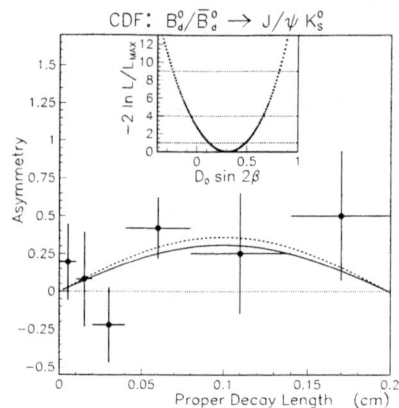

FIGURE 2. Left: Normalized mass distribution $((M_{fit} - M_B)/\sigma_{fit})$ for the fully reconstructed $B \to J/\psi K_s^0$ decay mode. There are 200 events in the mass peak. **Right:** The time dependent asymmetry as measured from same-side tagging. The solid curve represents the result of the full unbinned likelihood fit, the dashed curve is a simple fit to the points. The amplitude of the sine curve is $\mathcal{D} \sin 2\beta$.

$x_d = \Delta m_d/\Gamma$ with $\Gamma = \hbar/\tau$ the average lifetime of the heavy and light states. The signal (S) and background (B) comprise the sample of N total events, $N = S + B$. The dilution is the crucial factor in this equation as it comes in as $\mathcal{D}^2$. This is true because a mistagged event not only is absent from the correct tagging bin, it is also present in the incorrect tagging bin.

CDF has made a direct measurement of the quantity $\sin 2\beta$ using 200 $B \to J/\psi K_s^0$ decays where both muons are reconstructed in the silicon microvertex detector. The flavor of the B meson at the time of production is measured using same-side tagging. The advantage of using events with well-measured lifetime is that the time-dependent analysis utilizes more information on an event-by-event basis than does the time integrated analysis. The asymmetry versus lifetime is a sine wave with frequency Δm and amplitude $\sin 2\beta$. Most of the background is from prompt J/ψ production. Since the B vertex in the $J/\psi K_s^0$ decay is defined by the muons from the J/ψ, backgrounds from prompt J/ψ production look like short-lived B decays. At low lifetime, the asymmetry is small. At longer lifetime, the signal-to-noise is much greater and the asymmetry is larger. Overall, the time dependent analysis offers about a 30% reduction in $\delta(\sin 2\beta)$ relative to the time-integrated analysis for a given sample of B decays.

The analysis involves an unbinned likelihood fit in which the events are weighted depending upon their mass, lifetime and tag. Corrections are made for the intrinsic charge asymmetry of the detector. The dilution is required as an external input and extracted from the ℓD mixing analysis with the help of Monte Carlo. The same side tagging dilution for $B \to J/\psi K_s^0$ decays is measured to be $\mathcal{D} = 0.166 \pm 0.018(\text{stat.}) \pm 0.013(\text{syst.})$ Combining this with the measured asymmetry results in: $\sin 2\beta = 1.8 \pm 1.1$ (stat.) ± 0.3(syst.) [3]. The result is statistics limited. The dominant systematic uncertainty arises from the error in the dilution.

IV TEVATRON RUN II

For the upcoming Tevatron "Run II", the accelerator complex will be upgraded significantly with the construction of two new components: the Main Injector and the Recycler Ring. The instantaneous luminosity is expected to reach $\mathcal{L} = 1 \times 10^{32}\,\text{cm}^{-2}\text{s}^{-1}$ with a 396 ns bunch spacing and eventually improve to $\mathcal{L} = 2 \times 10^{32}\,\text{cm}^{-2}\text{s}^{-1}$ with a 132 ns bunch spacing. The integrated luminosity for Run II is anticipated to be $2\,\text{fb}^{-1}$ in two years of running.

V DØ AND CDF UPGRADES

Both experiments are undergoing significant upgrades in order to take advantage of the major increase in luminosity foreseen for Run II. The scope of these upgrades as they apply to the b-physics program will be discussed here.

For b-physics at a hadron collider, the most important aspects of the detector are the microvertex detector, tracking chamber and high-rate trigger and data acquisition system. Additionally, it is important to be able to accurately and quickly identify leptons (e and μ) for triggering. Both experiments are replacing completely their front-end electronics and trigger systems in order to handle the high event rates of Run II. The triggers will be pipelined and multi-staged, so that the lower trigger levels can process incoming events while higher level trigger decisions are made on previous events.

DØ is installing a 2T superconducting solenoid magnet which will surround a 4 layer silicon microvertex detector and a scintillating fiber tracker. The microvertex detector will include disks for forward tracking. The fiber tracker will cover out to 1.7 units of pseudorapidity (η) [4]. CDF is building a new gas-wire drift chamber to replace the existing central tracking chamber. The drift chamber will be able to perform particle identification using specific ionization (dE/dx.) A new silicon microvertex detector is being constructed. Additional strips (located cylindrically between drift chamber and the inner silicon strips) will be installed for forward tracking. The readout chip for the silicon system is a custom chip which will allow for simultaneous digitization of data from a previous event while acquiring data into the pipeline on the current beam crossing. This "deadtimeless" mode makes way for a trigger based upon the two-dimensional distance of closest approach (impact parameter) of tracks to the interaction point. The impact parameter information in real-time opens the possibility of triggering on B hadrons decaying to all-hadronic final states through displaced tracks [5].

In addition to the baseline detector upgrades for Run II, both experiments are now proposing additional upgrades which would significantly enhance the B physics programs. DØ is proposing to add an impact parameter trigger based upon hit information from the silicon microvertex detector. CDF is proposing two additional detector elements: a time-of-flight system and an additional layer of silicon very near the beamline.

VI   RUN II B PHYSICS

A $\sin 2\beta$

The expectations for our reach in $\sin 2\beta$ in Run II can be a direct extrapolation from existing measurements of tagging dilutions and event yields. Improvements over the Run I yield are expected from a) improved muon coverage and b) an improved signal-to-noise from the additional microvertex detector coverage. Based upon the improvements listed here and the Run I data sample shown in in Figure 4, the Run II yield estimate is 10,000 events in $2\,\text{fb}^{-1}$. The sample could be more than doubled by lowering the muon trigger p_T thresholds from $2.0\,\text{GeV}/c$ to $1.5\,\text{GeV}/c$ and triggering on $\psi \to e^+ e^-$ [5].

Given this large sample of events, the statistical reach in $\sin 2\beta$ will depend largely on the "effective tagging efficiency", $\epsilon \mathcal{D}^2$, as outlined earlier. The existing CDF measurement of $\sin 2\beta$ uses only one tagging algorithm, same-side tagging. Work is ongoing to incorporate information from lepton tagging and jet charge tagging into the Run I analysis. The tagging efficiencies for several methods have been measured in the context of $B^0/\overline{B^0}$ mixing. Table 1 shows measured and expected tagging efficiencies, along with the relevant detector upgrades which will improve the efficiencies.

TABLE 1. "Effective tagging efficiencies" for different flavor tagging methods as measured by CDF. The last two columns show the the expected improvements due to the detector upgrade. Many of the increases are due to an improved acceptance.

tagger	$\epsilon \mathcal{D}^2$ [%] measured Run I	$\epsilon \mathcal{D}^2$ [%] expected Run II	Relevant upgrade
Same Side π	1.8 ± 0.4	2.0	new tracking
Soft μ	0.72 ± 0.12	1.0	extend coverage
Soft e	0.35 ± 0.04	0.7	new tracking
Jet Charge	0.78 ± 0.15	3.0	new tracking
Opp.Side Kaon	–	2.4	Time-Of-Flight
All combined	3.7	6.7 (9.1)	

For Run II, the estimated error on $\sin 2\beta$ for CDF is approximately $\delta \sin 2\beta \simeq 0.08$. This error will improve significantly if triggering on low p_T muons and $\psi \to e^+ e^-$ is achieved with reasonable efficiency. A time-of-flight system would significantly improve the effective tagging efficiency by introducing a "kaon" tag.

It has been shown that tagging charged kaons from the b decays is a very powerful tagging method, due to the cascade $b \to c \to s$ decay [7]. The dE/dx particle identification of CDF does not offer sufficient π-K separation at low momenta ($p_T < 1.5\,\text{GeV}/c$) to tag efficiently with kaons. The addition of a time-of-flight system would significantly enhance this ability and effectively increase the statistics by 35%.

The Run II measurement of $\sin 2\beta$ will remain statistics limited. The dominant systematic uncertainty will again arise from the uncertainty in the tagging dilution. The dilutions for this measurement are calibrated from the exclusive decays $B^\pm \to J/\psi K^\pm$ and $B^0 \to J/\psi K^{*0}$. Given that the uncertainty on the dilution arises from the statistics of the calibration samples, the relative size of the statistical versus systematic errors in $\sin 2\beta$ will remain roughly constant. Estimates from DØ suggest an error on $\sin 2\beta$ in the range of $0.12 - 0.15$ in $2\,\text{fb}^{-1}$.

B A_{CP} in $B_d^0 \to \pi^+\pi^-$

CP violation in the decay $B^0/\overline{B^0} \to \pi^+\pi^-$ is related to the angle α in the unitarity triangle. This all hadronic decay mode is very challenging at a hadron collider. The small branching ratio ($< 8.4 \times 10^{-6}$ [8]) means that a significant sample can not be reconstructed opposite a $b \to \ell$ trigger.

CDF is implementing a secondary vertex trigger at Level 2 to separate hadronic B decays from inelastic (prompt) background. The information from the track trigger processor is combined with hits from the microvertex detector to measure the impact parameter of the tracks. The impact parameter resolution of this device is approximately $35\,\mu$m and the impact parameter information will be supplied to the trigger decision processor in less than $15\,\mu$s. This device requires beam position stability both during the store and from store-to-store. Real time beam position information will be fed back to the accelerator so that the position of the interaction region can be maintained over the course of the store.

After triggering on the $B \to$ two-track final state, the $B_d^0 \to \pi^+\pi^-$ final state must be isolated from the physics backgrounds from $B_d^0 \to K\pi$, $B_s^0 \to K\pi$ and $B_s^0 \to KK$. The mass resolution of the tracking system ($20\,\text{MeV}/c^2$ for $B_d^0 \to \pi^+\pi^-$) is not sufficient to isolate the signal from these backgrounds. In particular, the $B_s^0 \to KK$ final state reflects directly under the $B_d^0 \to \pi^+\pi^-$ mass peak when the two kaons are assumed to be pions.

To further isolate the $B_d^0 \to \pi^+\pi^-$ signal from these physics background, π-K separation is required. CDF will use dE/dx information from the central tracking chamber to separate the signal from the backgrounds on a statistical (not event-by-event) basis. The system will yield $> 1\sigma$ π-K separation for $p_T > 2\,\text{GeV}/c$. The proposed time-of-flight system, although very useful for flavor tagging, will not be capable of π-K separation at these higher momenta.

The yield estimate is 9k events in $2\,\text{fb}^{-1}$. Including all possible tagging modes (i.e. using the $\epsilon\mathcal{D}^2 = 6.7\%$ from Table 1), the estimated error on the CP asymmetry

TABLE 2. Comparison of experimental uncertainties on $\sin 2\beta$ and $\sin 2\alpha$ from the upcoming generation of experiments. The anticipated results are listed for one year of running at design luminosity. Please note that there are a number of caveats and assumptions which go into each of these projections. The numbers are tabulated here only to give the reader a feel for not only how well each of the experiments expect to ultimately perform, but to also show the complementarity of the different experiments.

		BELLE [9]	BaBar [10]	Hera-B [11]	DØ	CDF
$\int \mathcal{L} dt$ (fb^{-1})		100	30	100	1	1
$N(B_d^0 \to \psi K_S^0)$		2000	1100	1500	4k	5k
$\delta(\sin 2\beta)$	ψK_S^0	0.080	0.098	0.13	0.20	0.10
	all modes	0.062	0.059	0.12	0.20	0.10
$N(B_d^0 \to \pi^+\pi^-)$		650	350	800	–	4.6k
$BR(B_d^0 \to \pi^+\pi^-)^*$ ($\times 10^{-5}$)		1.3	1.2	1.5	–	0.5
$\delta(\sin 2\alpha)^\dagger$	$\pi^+\pi^-$	0.147	0.20	0.16	–	0.12
	all modes	0.089	0.085	0.16	–	0.12

* Assumed branching ratio.
† Assuming that contamination from penguin decays can be unfolded.

in $B_d^0 \to \pi^+\pi^-$ is approximately 0.10. If the contribution from the penguin decays were small, then the CP asymmetry measured in this mode would be $\sin 2\alpha$. The exact precision on $\sin 2\alpha$ will depend upon how well the contamination from the penguin mode can be unfolded.

C Comparing with Projections from Other Experiments

Several experiments will be collecting data in the period 1999-2002. Two e^+e^- B-factories at SLAC(BaBar) and KEK(BELLE), as well as a internal target production experiment at HERA(HERA-B) all intend to make many measurements in the B sector, including measurements of $\sin 2\beta$ and $\sin 2\alpha$. Table 2 shows the expected reach of each of these experiments in approximately one year of running at the design luminosity of each machine.

It is interesting to note that although the expected reach is similar for all of the experiments, the Tevatron measurements are in marked contrast to the measurements at BELLE, BaBar and HERA-B. Those experiments will have significantly smaller data samples but significantly better tagging efficiencies. This, along with different modes of production will allow all of these measurements to complement one another. This also points out how crucial flavor tagging is at the Tevatron.

Please also note that all of these experiments plan a rich program beyond the measurements of these two angles in the unitarity triangle. Also, CLEO-III will be running during this period, and will produce a large number of measurements of their own, including information which will help unfold the penguin contributions to $B_d^0 \to \pi^+\pi^-$ [12].

D Time Dependent B_s^0 Mixing

An observation of $B_s^0/\overline{B_s^0}$ mixing would lead to a measurement of the ratio of CKM elements V_{td} and V_{ts}. Semileptonic decays which include $B_s^0 \to D_s \ell \nu$ and $B_s^0 \to \phi \ell \nu X$ have the advantage of large statistics, but the drawback of the missing neutrino leading to poor $\beta\gamma$ (boost) resolution. CDF has established a limit on B_s^0 mixing $\Delta m_s > 5.8\,\mathrm{ps}^{-1}$ at 95% CL) using the 1068 $B_s^0 \to \phi \ell \nu X$ decays with a B_s^0 purity of 61%. For Run II, the proper decay time resolution becomes the limiting factor in semileptonic decays.

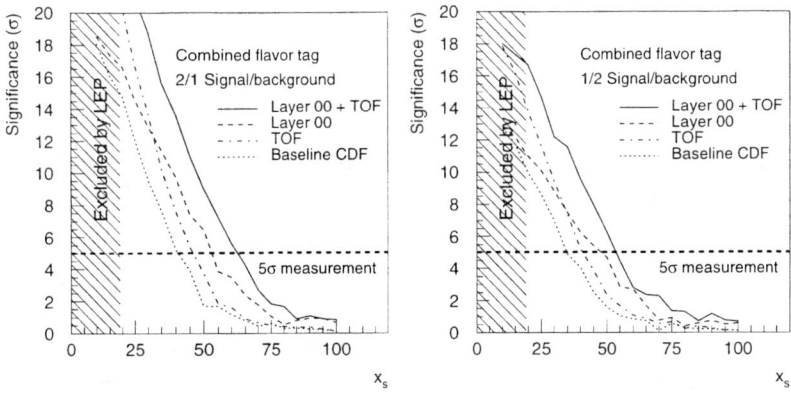

FIGURE 3. CDF Run II sensitivity to B_s^0 mixing using fully reconstructed $B_s^0 \to D_s \pi(3\pi)$. The plots show the significance of the measurement in $2\,\mathrm{fb}^{-1}$ of data for two different assumptions of S:N. The sensitivity quoted in the text comes from the point where the curves cross the 5σ line. For the baseline detector, the reach is in the range $x_s \sim 30\text{-}40$. With the additional upgrades discussed in the text, this reach is extended to $x_s \sim 55\text{-}65$.

Fully reconstructed decays offer the advantage of much better boost resolution over the semileptonic modes. The drawback is that decay modes without leptons in the final state have not yet been triggered upon (and isolated) at a hadron collider. This means that the $B_s^0 \to D_s \pi(3\pi)$ must be either reconstructed opposite a trigger $B \to \ell$ or else an impact parameter trigger must be used to attempt to separate the all hadronic final state. In either case, the statistics of the fully reconstructed mode will be significantly reduced from the semileptonic modes. However, studies performed by both DØ and CDF have shown that the ultimate reach in x_s will be better for the fully reconstructed mode than it will be for the semileptonic mode.

Figure 5 shows the anticipated reach in time dependent B_s^0 mixing using the fully reconstructed modes. The yield estimate is $20k$ events in $2\,\mathrm{fb}^{-1}$. Less certain is the signal-to-noise, so the result is plotted with two different $S:N$ assumptions. Using the tagging dilutions outlined in Table 1, the ultimate reach with the baseline CDF Run II detector is $x_s \simeq 40$, but note the significant improvement in reach when the proposed upgrades (an additional layer of silicon near the beamline and a time-of-flight detector) are included.[1] If these upgrades are approved and included at the beginning of Run II, the ultimate reach in x_s could be as high as 65 [6].

E B_c, Radiative and Rare Decays

The B_c is an interesting $q\bar{q}'$ system because of the two unequal heavy quark masses. CDF has reported and observation of the B_c in the decay $B_c \to J/\psi \ell \nu$ [16]. The lifetime measurement from this analysis, $\tau(B_c) = 0.46^{+0.18}_{-0.16}(\text{stat.}) \pm 0.05(\text{syst.})$ ps indicates a more "charm-like" lifetime for this system than expected. Detailed study of this system will have to await the additional statistics of Run II. Measurement of the B_c production cross section, mass and lifetime are of particular interest.

Although B_c decay modes involving the ψ are substantial, the dominant decay mode is expected to be the $c \to s$ transition, giving $B_c \to B_s^0 X$. Given enough statistics, the B_c could become a very powerful flavor tagging tool for B_s^0 mixing. As an example, if a sample of $B_c^\pm \to B_s^0 \pi^\pm$ could be isolated, the flavor of the B_s^0 would be unambiguously tagged at the time of production by the charge of the associated pion from the B_c decay.

Radiative Decays. In the absence of long distance effects, $|V_{td}/V_{ts}|$ can be probed via $B \to \rho\gamma$ and $B \to K^*\gamma$. CDF installed a dedicated trigger for radiative B decays for the latter part of Run I. With 23 pb^{-1} of data taken with this trigger a limit of $BR(B^0 \to K^{*0}\gamma) < 2.2 \times 10^{-4}$ was established. If the rates are near Standard Model expectations, approximately 2500 $K^*\gamma$ events should be seen. Additionally, a signal should be seen in the decay mode $B_s^0 \to \phi\gamma$.

Radiative decays detected via photon conversion will also be pursued. The conversion rate is small (few percent) and depends upon the amount material in the inner detector regions. The advantage in using conversions arises from a) lower trigger thresholds for electrons versus photons and b) much improved π^0 rejection.

Rare Decays. As in the kaon system, rare decays are sensitive to physics beyond the Standard Model. Both CDF and DØ will be well suited to search for (and likely observe some) of the following rare decays: $B^+ \to \mu^+\mu^- K^+$, $B^0 \to \mu^+\mu^- K^{*0}$, $B^0 \to \mu^+\mu^-$, $B_s^0 \to \mu^+\mu^-$ and $B_s^0 \to e^+\mu^-$. Good muon coverage and mass resolution enhance the reach of searches in the dimuon modes. Based upon Standard Model rates, it is likely that the decay modes $\mu^+\mu^- K^+$ and $\mu^+\mu^- K^{*0}$ will be observed. The Standard Model expectations for $B^0 \to \mu^+\mu^-$ and $B_s^0 \to \mu^+\mu^-$ are well below the expected Run II reach.

---

[1] The opposite-side tagging algorithms will have the same dilution for B_s^0 as they do for B_d^0. This is not the case for same-side tagging because the B_s^0 fragmentation chain yields many more kaons the does the fragmentation chain for the B_d^0. Without a time-of-flight system, the dilution for same-side tagging is worse ($\mathcal{D} \simeq 5\%$ due to same-side K^* and wrong-sign π production) than it is for B_d^0. Including the time-of-flight system to reject non-K tracks improves the same-side tagging dilution for B_s^0 to $\mathcal{D} = 40\%$.

TABLE 3. Summary of rare B decay searches.

B Decay Mode	Standard Model	CDF Run I	CDF II
$\mu^+\mu^-K^+$	$(2-5)\times 10^{-7}$	5.4×10^{-6}	2×10^{-7}
$\mu^+\mu^-K^{*0}$	$(1-2)\times 10^{-6}$	4.1×10^{-6}	2×10^{-7}
$B_d^0 \to \mu^+\mu^-$	$(0.6-2.4)\times 10^{-10}$	6.8×10^{-7}	3×10^{-8}
$B_s^0 \to \mu^+\mu^-$	$(2.5-4.5)\times 10^{-9}$	2.0×10^{-6}	1×10^{-7}
$B_s^0 \to e^+\mu^-$	forbidden	6.1×10^{-6}	3×10^{-7}

VII CONCLUSION

The CDF experiment has made competitive measurements in B_d^0 mixing and rare B decay searches and also made one of the first attempts at a direct measurement of $\sin 2\beta$. This important work has shown that world-class B physics can be done at a hadron collider and lays an important foundation for Run II. The outlook for B physics in the near future is very bright. Several experiments will produce measurements which will significantly constrain the CKM matrix and further our understanding of CKM physics and CP violation. Both CDF and DØ will play an active roll in these CP violation measurements.

The Tevatron program will also yield important measurements in the B_s^0 system with either a direct or indirect measure of B_s^0 mixing, as well as study of the B_c system. These mesons will not be accessible at the $\Upsilon(4s)$ B-factories. In addition, if the past is any guide, additional methods and measurements will be developed as experience is gained with the upgraded detectors and larger data samples.

REFERENCES

1. The Particle Data Group, C. Caso et al., European Physical Journal **C3**, 1 (1998).
2. CDF Collaboration, F. Abe et al., Phys. Rev. Lett. **80**, 2057 (1998).
3. CDF Collaboration, F. Abe et al., FERMILAB-PUB-98-189-E (1998).
4. The DØ Collaboration, FERMILAB-Pub-96/357-E (1996).
5. The CDF-II Collaboration, FERMILAB-Pub-96/390-E (1996).
6. The CDF-II Collaboration, Fermilab-Proposal-909 (1998).
7. ARGUS Collaboration, H. Albrecht et al., Z. Phys. C 62, 371-381 (1994).
8. CLEO Collaboration, CLEO CONF 98-20 (1998).
9. BELLE Collaboration, KEK Report 95-1 (1995).
10. BaBar Collaboration, SLAC-Report-95-457 (1995).
11. HERA-B Collaboration, DESY-PRC 95/01 (1995).
12. CLEO-III Collaboration, CLNS 94/1277 (1994).
13. ALEPH Collaboration, D. Buskulic et al.,Phys. Lett. **B384**, 205 (1996).
14. J. Hagelin, Nucl. Phys. **B193**, 123 (1981).
15. A. Dighe, I. Dunietz, H. Lipkin and J. Rosner, Phys. Lett. **B369**, 144 (1996).
16. CDF Collaboration, F. Abe et al., Phys. Rev. Lett. **81**, 2432 (1998).

$B^0 - \overline{B}^0$ oscillations and measurements of $|V_{ub}|/|V_{cb}|$ at LEP

Achille Stocchi

Laboratoire de l'accélérateur linéaire
B.P. 34 - 91898 Orsay Cedex

Abstract. In this paper a review of the LEP analyses on $B^0 - \overline{B}^0$ oscillations and on the measurement of $|V_{ub}|/|V_{cb}|$ is presented. These measurements are of fundamental importance in constraining the ρ and η parameters of the CKM matrix. A review of the current status of the V_{CKM} matrix determination is also given.

INTRODUCTION

The data registration at the Z^0 pole has stopped at the end of 1995. The four LEP experiments (ALEPH, DELPHI, L3 and OPAL) have collected about 4M hadronic Z^0 decays per experiment.

In the past three years the quality of the data analysis has continously improved, thanks to a better understanding of the behaviour of all components of the detector. At the same time, new ideas, and then, new analyses have been tried. A more performant statistical treatment of the information has been also developed. As a result, the precision on the Δm_d parameter has been improved and above all, the sensitivity for the Δm_s parameter has been tremendously increased. The new and precise LEP analyses on $|V_{ub}|/|V_{cb}|$ are also a consequence of these improvements. Many analyses described in this paper have been presented at the last '98 Summer Conferences and are still preliminary. This paper is organized as follows. The first sections are dedicated to the oscillations and $|V_{ub}|/|V_{cb}|$ analyses. In the last section the present status of the V_{CKM} matrix is given with a special emphasis placed on the impact of the measurements presented in this paper.

THE OSCILLATION ANALYSES

The probability that a B^0 meson oscillates into a $\overline{B}^0$ or stays as a B^0 is given by:

$$P_{B_q^0 \to B_q^0(\overline{B}_q^0)} = \frac{1}{2}e^{-t/\tau_q}(1 \pm cos\Delta m_q t) \qquad (1)$$

where the effect of CP violation has been neglected. τ_q is the lifetime of the B_q^0 meson, $\Delta m_q = m_{B_1^0} - m_{B_2^0}$ is the mass difference between the two mass eigenstates[1] and gives the period of the time oscillations (the effect of a lifetime difference between the two states has been also neglected).

The Standard Model predicts:

$$\Delta m_d \propto A^2 \lambda^6 [(1-\rho)^2 + \eta^2] f_{B_d}^2 B_d \; ; \; \Delta m_s \propto A^2 \lambda^4 f_{B_s}^2 B_s \qquad (2)$$

The difference in the λ dependence of these expressions ($\lambda \sim 0.22$) implies that $\Delta m_s \sim 20 \, \Delta m_d$. It is then clear that a very good proper time resolution is needed to measure the Δm_s parameter. A time dependent study of $B^0 - \overline{B}^0$ oscillations requires:
- the measurement of the decay proper time,
- to know if a B^0 or a $\overline{B}^0$ decays at $t = t_o$ (decay tag)
- to know if a b or a $\overline{b}$ quark has been produced at $t = 0$ (production tag).

The precision on the Δm measurement is given by the following relation:

$$\text{error} = \left(\sqrt{N} f_{B_{d(s)}^0} (2\varepsilon_1 - 1)(2\varepsilon_2 - 1) e^{-\left(\frac{\Delta m_{d(s)} \sigma_t}{2}\right)^2} \right)^{-1} \qquad (3)$$

where N is the total number of events in the sample; $f_{B_{d(s)}^0}$ is the fraction of events in which a $B_{d(s)}^0$ meson has been produced; $\varepsilon_2, \varepsilon_1$ are the tagging purities at the decay and production times respectively, defined as $\varepsilon = \frac{N_{\text{right}}}{N_{\text{right}} + N_{\text{wrong}}}$, where $N_{\text{right}}(N_{\text{wrong}})$ are the numbers of correctly (incorrectly) tagged events and σ_t is the proper time resolution given, approximately, as $\sigma_t = \sqrt{\left(\frac{m^2}{p^2}\right)\sigma_L^2 + \left(\frac{\sigma_p}{p}\right)^2 t^2}$, where σ_L and σ_p are the decay length and the momentum resolutions respectively.

Δm_d measurements

A lot of analyses have been performed since 1994. A typical time distribution is shown in Figure 1. The time dependence behaviour with frequency $\Delta m_d \sim 0.470$ ps^{-1}, for the $B_d^0 - \overline{B}_d^0$ oscillation is clearly visible. This will be a textbook plot ! The present summary of the results on Δm_d, as given by [1], is shown in Figure 1. Combining LEP/CDF and SLD measurements it follows that:

$$\Delta m_d = (0.477 \pm 0.017) \text{ps}^{-1} \qquad (4)$$

Δm_d is known with a precision of 3.4% relative error.

Analyses on Δm_s

Four types of analyses have been performed.

---

[1] Δm_q is usually given in ps^{-1}. 1 ps^{-1} corresponds to 6.58 10^{-4}eV.

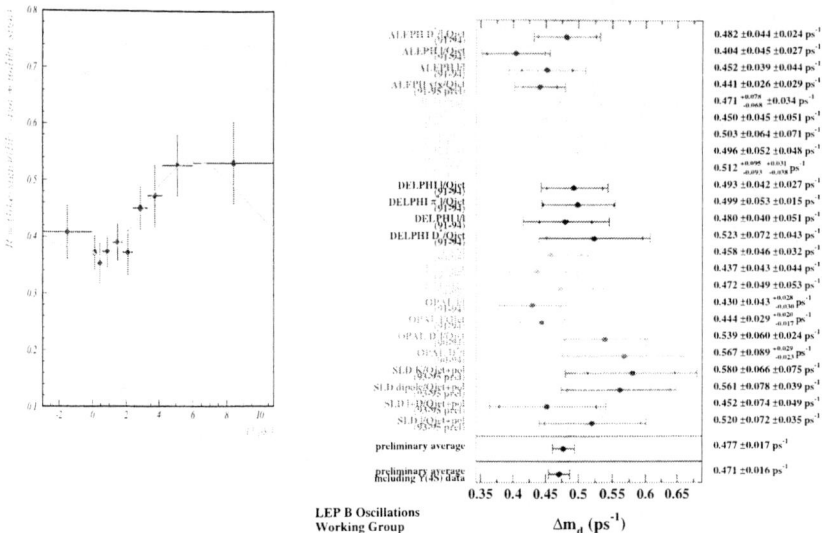

FIGURE 1. The plot on the left shows the time dependence behaviour of the $B_d^0-\bar{B}_d^0$ oscillation. The points with error bars are the data. The curve shows the result of the fit using $\Delta m_d = 0.47$ ps^{-1}. On the right, the summary of the Δm_d results from the LEP, SLD and CDF Collaborations are given. Details on how the different results have been combined are given in [1].

For all of them, the latest analyses make use of the combined tag method for tagging a b or a $\bar{b}$ at production time. At LEP, the produced b and $\bar{b}$ quarks fragment independently and the events can be divided in two separate hemispheres. If the measurement of the proper time is performed in one of those (same hemisphere), the other (opposite hemisphere) can be used to determine if a b or a $\bar{b}$ quark was produced in that hemisphere. Several variables are considered in the opposite hemisphere:

- $Q = \frac{\sum_{i=1}^{n} q_i(\vec{p}_i \cdot \vec{e}_S)^{0.6}}{\sum_{i=1}^{n}(\vec{p}_i \cdot \vec{e}_S)^{0.6}}$ the hemisphere charge, defined as the charge of all (n) charged tracks (q_i) present in the hemisphere, weighted by their momentum (p_i) projected along the thrust axis ($\vec{e}_S$) with a chosen value for the exponent (0.6),
- the hemisphere charge, considering only identified kaons,
- the charge of primary and secondary vertices,
- the presence of high p_t leptons.

The use of these variables allow to have a tagging purity of the order of 70%.

Tracks in the same hemisphere can be used also. This procedure is peculiarly clean if all the tracks from the B_s^0 have been reconstructed (as for $D_s^{\pm}\ell^{\mp}$ and exclusive B_s analyses). In this case, tracks from the B_s^0 decay can be removed and the

TABLE 1. The characteristics of the different analyses are given in terms of statistics (N), B_s^0 purity (f_{B_s}), tagging purity at the production and decay time ($\varepsilon_1, \varepsilon_2$) and time resolution in the first pico-second

Analysis	N(events)	f_{B_s}	ε_1	ε_2	$\sigma_t(t < 1\text{ps})$
Inclusive lepton	~ 50000	$\sim 10\%$	$\sim 70\%$	$\sim 90\%$	~ 0.25 ps
$D_s^\pm h^\mp$	~ 3000	$\sim 15\%$	$\sim 72\%$	$\sim 90\%$	~ 0.22 ps
$D_s^\pm \ell^\mp$	~ 400	$\sim 60\%$	$\sim 78\%$	$\sim 90\%$	~ 0.18 ps
Exclusive B_s^0	~ 25	$\sim 70\%$	$\sim 78\%$	$\sim 100\%$	~ 0.08 ps

others, coming from the primary vertex can be used. The addition of informations from the same hemisphere allows to reach a tagging purity of 74%. Finally the use of all these informations on an event by event basis gives a purity of 78%.

The tagging of a B or a $\overline{B}$ meson at decay time depends on the specific analysis and will be given in the following. Before describing the different analyses, the method used to measure or put a limit on Δm_s is briefly discussed.

The amplitude method

The method used to measure or to put a limit on Δm_s consists in modifying equation 1 in the following way: $1 \pm cos\Delta m_s t \to 1 \pm A cos\Delta m_s t$. A and σ_A are measured at fixed values of Δm_s instead of Δm_s itself. In case of a clear oscillation signal, at given Δm_s, the value of the amplitude is compatible with A = 1 for this Δm_s and with A = 0 elsewhere. With this method it is also easy to set a limit. The values of Δm_s excluded at 95% C.L. are those satisfying the condition $A(\Delta m_s) + 1.645\, \sigma_A(\Delta m_s) < 1$.

With this method it is easy to combine different experiments and to treat systematic uncertainties in an usual way since, at each value of Δm_s, a value for A with a gaussian error σ_A, is measured. Furthermore the sensitivity of the experiment can be defined as the value of Δm_s corresponding to $1.645\, \sigma_A(\Delta m_s) = 1$ (for $A(\Delta m_s) = 0$, namely supposing that the "true" value of Δm_s is well above the measurable value of Δm_s). The sensitivity is the limit which would be reached in 50% of the experiments.

The inclusive lepton/combined tag analysis

This analysis uses high p_t leptons which are mainly coming from direct b semileptonic decays ($b \to \ell$). The sign of the lepton tags the B_s^0 at decay time. The initial sample consists in 80% leptons from B decays (and among those 90% $b \to \ell$ (direct) and 10% $b \to c \to \overline{\ell}$ (cascade)) and of 20% leptons from charm decays or misidentification. The events $b \to c \to \overline{\ell}$ give the wrong tag for the $\overline{B}_s^0$ meson at decay time.

To reconstruct a B decay proper time, algorithms have been developed which aim at identifying charged (neutral) tracks which are more likely to come from the B_s^0 decays. As result, in more than 50% of the cases, the error on the decay length is $\sigma_L \sim 250\mu m$ and the relative error on the B energy is better than 10%, resulting in an error on the proper time of the order of 0.25 ps in the first pico-second.

A second crucial point for this analysis consists in trying to increase the B_s^0 purity of the sample (the natural B_s^0 purity of b events is around 10%) and to reduce the contribution from cascade decays. To enrich the sample in direct b semileptonic decays and, among those, in events coming from B_s^0 decays, several variables have been used as the momentum and transverse momentum of the lepton, the impact parameters of all tracks in the opposite hemisphere relative to the main event vertex, the kaons in primary and secondary vertices in the same hemisphere, and the charge of the secondary vertex.

The result of this procedure is to increase the B_s^0 purity by 30% and to reach more than 90% purity for the tagging at the decay time.

$D_s^\pm \ell^\mp$/combined tag analysis

The use of events in which a reconstructed D_s is accompanied by a high p_t lepton with an electric charge opposite in sign allows to select a sample having 60% B_s purity. The proper time resolution benefits also from the fact that the only missing particle is the neutrino: $\overline{B}_s^0 \to D_s^+ e^- \overline{\nu}_e$. In the first pico-second the time resolution is about 0.18 ps in more than 80% of the events.

The limiting factor is the available statistics because accessible D_s branching fractions are quite small (between $\sim 1\%$ and $\sim 5\%$). Several decay modes have to be selected. Figure 2 shows an example in which six hadronic and two semileptonic decay modes have been reconstructed.

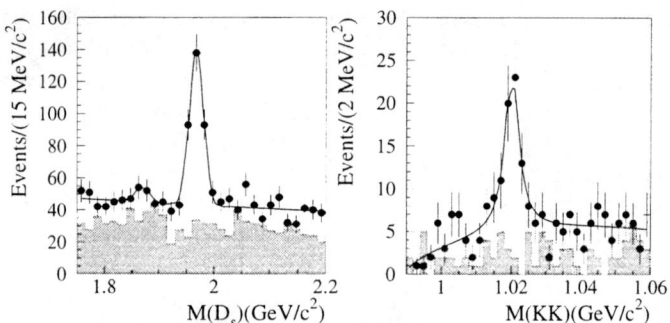

FIGURE 2. DELPHI $D_s^\pm \ell^\mp$ candidates. The figure on the left shows the D_s mass spectrum reconstructed from the following decay modes: $D_s^+ \to \phi\pi^+, \phi\pi^+\pi^0, \phi\pi^+\pi^-\pi^+, \overline{K}^{*0}K^+, \overline{K}^{*0}K^+$ and $K_S^0 K^+$. The figure on the right shows the ϕ mass spectrum from the decays $D_s^+ \to \phi e^+ \nu_e$ and $\phi \mu^+ \overline{\nu}_\mu$. The sum of the two samples gives 230 ± 18 B_s^0 candidates.

Exclusive B_s^0/combined tag analysis

At the 1998 Moriond Conference, the DELPHI Collaboration has proposed the use of exclusively reconstructed B_s^0 decays for Δm_s analyses. These events have an excellent proper time resolution $\sigma_t \sim 0.08$ ps and provide a gain in sensitivity at high values of Δm_s (equation 3). Figure 3 shows the B_s^0 mass spectrum using the decay modes: $B_s^0 \to D_s \pi$ (or a_1) and $B_s^0 \to D^0 K \pi$ (or a_1). The D_s has been reconstructed in six hadronic decay modes, as in the $D_s^{\pm} \ell^{\mp}$ analysis, and the D^0 is observed using $K\pi$ and $K\pi\pi\pi$ decay modes. 17 ± 8 events have been reconstructed in the B_s^0 mass region. The combinatorial background is estimated to be 35%.

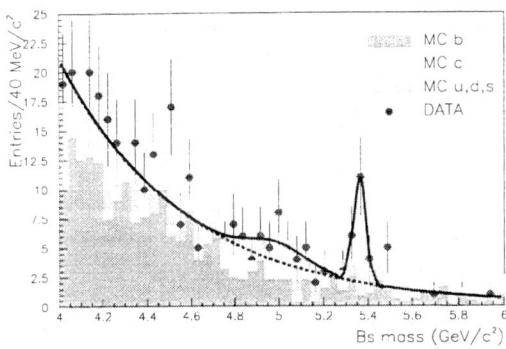

FIGURE 3. The B_s^0 mass spectrum obtained by the DELPHI Collaboration. The points with the error bars are the data with the fit superimposed. The contributions from non-B_s^0 decays, as given by the Monte Carlo simulation, are also shown.

Summary of Δm_s analyses

The combined result of LEP/SLD/CDF [1] analyses is shown in Figure 4 and is:

$$\Delta m_s > 12.4 \text{ ps}^{-1} \text{ at } 95\% \text{ C.L.}$$

The sensitivity is at 13.8 ps^{-1}. LEP alone has a limit at 11.5 ps^{-1} at 95% C.L., with a sensitivity at 12.9 ps^{-1}. $\Delta m_s = 0$ is excluded between 14.5 ps^{-1} and 16.5 ps^{-1} with a 2σ significance at 15 ps^{-1}. The present summary of the results is given in Figure 4.

$|V_{UB}|/|V_{CB}|$ MEASUREMENT

The presence of leptons above the kinematical limit for those produced in the decay $B \to D \ell \bar{\nu}_{\ell}$ ($b \to c$ transition proportional to the $|V_{cb}|$ CKM matrix element)

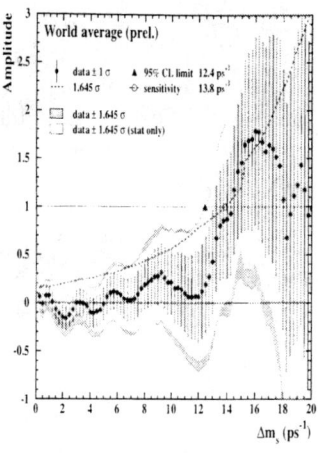 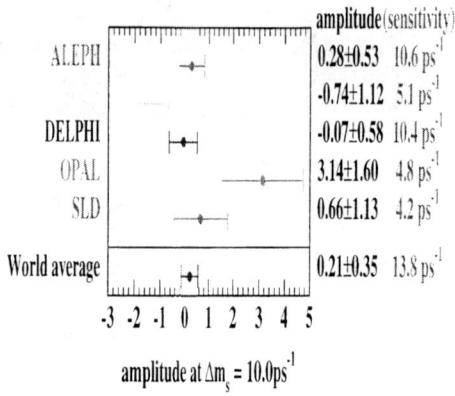

FIGURE 4. The plot on the left shows the combined Δm_s results from LEP/SLD/CDF analyses shown in an amplitude versus Δm_s plot. The point with error bars are the data; the lines show the 95% C.L. curves (in dark the systematics have been included). The dotted curve shows the sensitivity. The plot on the right shows the summary of the Δm_s results per experiment. The error are given at $\Delta m_s = 10$ ps^{-1} (the sensitivity is also given). The way in which the combined value is obtained is described in [1].

is attributed to the transition $b \to u\ell\bar{\nu}_\ell$ (proportional to the $|V_{ub}|$ CKM matrix element).

The CLEO and ARGUS Collaborations have been pioneers in this measurement. Nevertheless, as only a small fraction of the energy spectrum of these leptons is measurable, the systematic uncertainties in the modelling of the $b \to u$ transition to evaluate the ratio $|V_{ub}|/|V_{cb}|$ are quite large (of the order of 20%-25% relative error). Recently LEP experiments have shown their capabilities of measuring $|V_{ub}|$ with a statistical precision similar to the one from CLEO and with reduced systematic uncertainties. They use several kinematical variables, in events with an identified high transverse momentum lepton, which have a distinctive power to discriminate between $b \to c$ and $b \to u$ transitions. The first measurement has been performed by the ALEPH Collaboration by means of a neural network discriminating method.

The DELPHI measurement is simpler. With respect to the ALEPH analysis the information from the presence of a secondary vertex from the D decay is used. In $b \to u$ transitions, all tracks are coming from the B decay vertex. The presence of kaons at the D meson vertex is also used. The method is based on the fact that the hadronic system recoiling against the lepton in $b \to u\ell\nu$ decays is expected to have an invariant mass lower than the charm mass [2]. The sample is finally divided into a b → u enriched and a b → u depleted components and the energy of the lepton in the B rest frame is calculated. The result is shown in Figure 5 together with the summary of the results on $|V_{ub}|$.

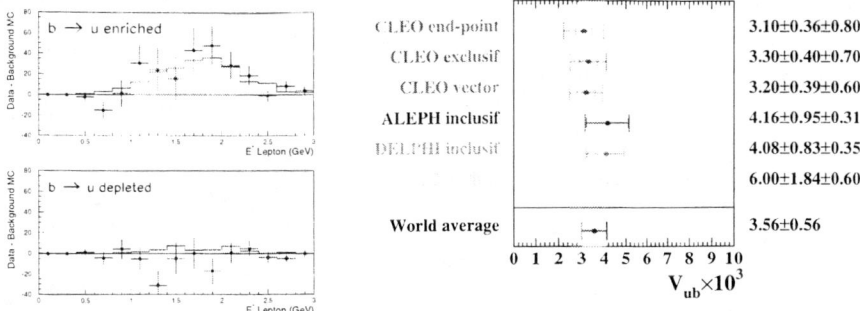

FIGURE 5. The plots on the left show the energy of the lepton in the B rest frame after the background subtraction for the $b \to u$ enriched and $b \to u$ depleted samples. On the right the summary of $|V_{ub}|$ results is given.

STATUS OF THE V_{VCM} MATRIX

TABLE 2. The four constraints which allow, at present, to define the accessible region for the ρ and η parameters are listed in the first column. In the second column the dependence of these constraints relative to the different parameters is given. The last column gives the explicit dependence in terms of $\overline{\rho}$ and $\overline{\eta}$.

Measurement	$V_{CKM} \times$ other	Constraint
$b \to u / b \to c$	$(|V_{ub}|/|V_{cb}|)^2$	$\overline{\rho}^2 + \overline{\eta}^2$
Δm_d	$|V_{td}|^2 f_{B_d}^2 B_{B_d} f(m_t)$	$(1-\overline{\rho})^2 + \overline{\eta}^2$
$\Delta m_d / \Delta m_s$	$\left\|\frac{V_{td}}{V_{ts}}\right\|^2 \frac{f_{B_d}^2 B_{B_d}}{f_{B_d}^2 B_{B_s}}$	$(1-\overline{\rho})^2 + \overline{\eta}^2$
ε_K	$f(A, \overline{\eta}, \overline{\rho}, B_K)$	$\sim \overline{\eta}(1-\overline{\rho})$

The V_{CKM} matrix can be parametrized in terms of four parameters: λ, A, ρ and η (the Wolfenstein parametrization [3]). The Standard Model predicts relations between the different processes which depend on these parameters. The unitarity of the V_{CKM} matrix can be visualized as a triangle in the $\rho-\eta$ plane. Several quantities which depend on ρ and η have to be measured and, if Standard Model is correct, they must define compatible values for the two parameters inside measurement errors and theoretical uncertainties. The measurement of $b \to u / b \to c$ transitions gives a constraint of the form $\bar{\rho}^2+\bar{\eta}^2$. [2] The measurement of Δm_d gives a constraint of the form $(1 - \bar{\rho})^2 + \bar{\eta}^2$. A measurement of the ratio $\Delta m_d/\Delta m_s$ gives the same type of constraint in the $\bar{\rho} - \bar{\eta}$ plane, as a measurement of Δm_d, but this ratio is expected to have smaller theoretical uncertainties since the ratio $f_{B_s}^2 B_{B_s}/f_{B_d}^2 B_{B_d}$ is better known than the absolute value $f_{B_d}^2 B_{B_d}$.

All details of the analysis presented here can be found in [4]. Using the available and most recent measurements and up to date theoretical calculations [4] the allowed region in the $\bar{\rho} - \bar{\eta}$ plane can be determined. It is shown in Figure 6 and corresponds to:

$$\bar{\rho} = 0.189 \pm 0.074 \ ; \ \bar{\eta} = 0.354 \pm 0.045$$

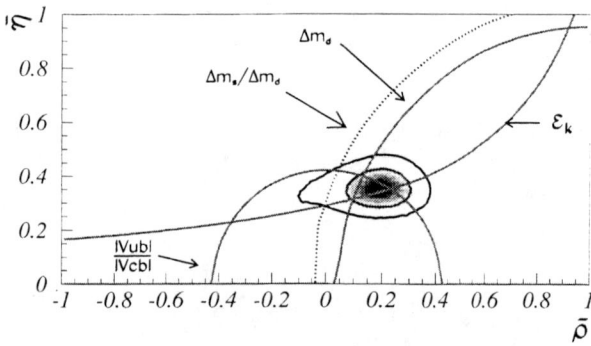

FIGURE 6. The $\bar{\rho} - \bar{\eta}$ allowed region. The contours at 68% and 95% C.L. are shown. The continous lines correspond to the constraints obtained from the measurements of $\frac{|V_{ub}|}{|V_{cb}|}$, Δm_d, and ε_K. The dotted curve corresponds to the 95% C.L. limit obtained from the experimental limit on Δm_s.

It is of interest to determine the central values and the uncertainties on the quantities $\sin 2\alpha$, $\sin 2\beta$ and γ which will be directly measured at future B-factories or LHC experiments. The result is shown in Figure 7 and is:

$$\sin 2\beta = 0.73 \pm 0.08 \ ; \ \sin 2\alpha = -0.15 \pm 0.30 \ ; \ \gamma = (62 \pm 10)^0 \quad (5)$$

The value of $\sin 2\beta$ is rather precisely determined with an accuracy already at the level expected after the first years of running at B factories. Finally it is possible to remove from the calculation the information of one of the constraint and to obtain its probability density function. The result for Δm_s and $|V_{ub}|/|V_{cb}|$ is shown in Figure 8 and summarized in Table 3.

From these results the important impact of these two measurements in the determination of the allowed region for ρ and η is clear. Furthermore the expected probability distribution for Δm_s shows that present analyses are exploring the one sigma region.

---

[2] $\bar{\rho}(\bar{\eta}) = \rho(\eta)(1 - \lambda^2/2)$

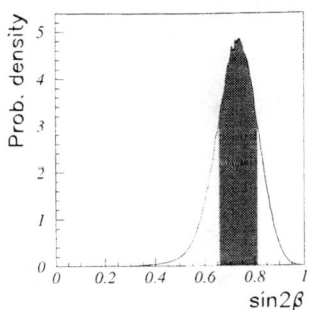 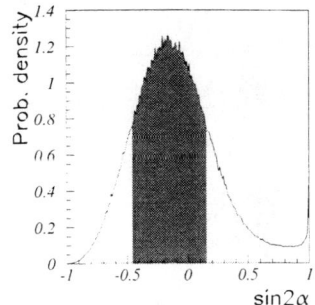

FIGURE 7. The sin 2β and sin 2α probability density distributions. The dark-shaded and the clear shaded intervals correspond to 68% and 95% C.L. regions respectively.

TABLE 3. The Δm_s and $|V_{ub}|/|V_{cb}|$ measured values are compared with those obtained using the fitting procedure after having removed them from the fit.

Quantity	Measured value	Fitted value
Δm_s	> 12.4 ps^{-1} at 95% C.L.	[9.5 - 17] ps^{-1} 68% C.L.
$|V_{ub}|/|V_{cb}|$	0.093 ± 0.014	$0.085^{+0.037}_{-0.023}$

CONCLUSIONS

Important improvements have been obtained in the last two years in the analyses of $B^0 - \overline{B}^0$ oscillations. Combining LEP results with those from SLD and CDF, Δm_d frequency is presently known with a 3.4% relative error ($\Delta m_d = 0.477 \pm 0.017$ ps^{-1}). The sensitivity on Δm_s is at 13.8 ps^{-1} and, the actual LEP/SLD/CDF combined limit, of 12.4 ps^{-1} at 95% of C.L., is exploring the region where Δm_s is expected to be according to the analysis [4]. The measurement of Δm_s is still a challenge for LEP collaborations, $|V_{ub}|$ has been measured at LEP with about the same experimental precision as the one obtained by CLEO and with a reduced dependence on theoretical models.

The phenomenologial analysis presented in this paper gives:

$$\overline{\rho} = 0.189 \pm 0.074 \; ; \; \overline{\eta} = 0.354 \pm 0.045$$

and, in an indirect way:

$$sin2\beta = 0.73 \pm 0.08 \; ; \; sin2\alpha = -0.15 \pm 0.30 \; ; \; \gamma = (62 \pm 10)^0$$

The situation will still be improved, at least until the next summer '99, before the starting of B-factories.

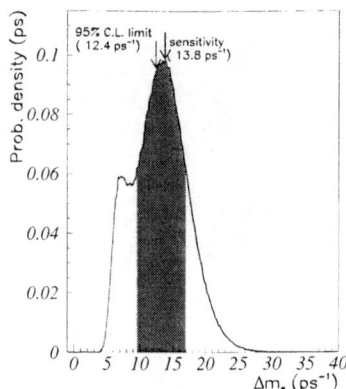 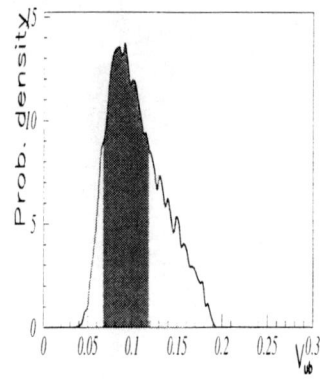

FIGURE 8. The left and the right plots show the probability density distributions for Δm_s and $|V_{ub}|/|V_{cb}|$ respectively. The dark-shaded and the clear shaded intervals correspond to 68% and 95% C.L. regions respectively.

ACKNOWLEDGEMENT

I would like to thank the organisers of HQ98 for the warm and nice atmosphere during the conference and for the unforgettable banquet at the Shedd Aquarium. Many thanks to Fabrizio Parodi and Patrick Roudeau for their help in the preparation and redaction of this contribution. Finally a grand merci to Jocelyne Brosselard, kind and efficient as usual in the preparation of this manuscript.

REFERENCES

1. The LEP B Oscillation Working Group "Combined Results on B^0 Oscillations: update for the summer 1998 Conferences" **LEPBOSC 98/2**
2. Barger, V., Kim, C.S., and Phillips, R.J.N. *Phys. Lett.* **B251 (1990) 629**
   Falk, A.F., Ligeti, Z., and Wise, M.B. **CALT-68-2110, hep/9705235**
   Bigi, I., Dikeman, R.D. and Uraltsev, N. **TPI-MINN-97/21-T, hep-ph/9706250**
3. Wolfenstein, L. **Phys. Rev. Lett. 51 (1983) 1945**
4. Paganini P., Parodi, F., Roudeau, P., and Stocchi, A., **LAL (97-79), hep-ph/9711261** *submitted to Physica Scripta*
   Parodi, F., Roudeau, P. and Stocchi, A. **LAL 98-49, hep-ph/9802289**
   Parodi, F., Roudeau, P. and Stocchi, A. paper 586 contributed to the ICHEP98 Conference (Vancouver 23th-29th July 1998)

THE PEP-II B-FACTORY AND THE BABAR DETECTOR

Gérard Bonneaud
Ecole Polytechnique, LPNHE
F-91128 Palaiseau, France

(on behalf of the BABAR Collaboration)

October 10-12, 1998

Abstract. I summarize the physics goals of the BABAR project. I give a brief status report on the PEP-II B-Factory progress. I then describe the requirements and chosen design for each of the BABAR detection systems. In conclusion, I give projections for the physics performance expected from the BABAR experiment.

THE BABAR PHYSICS PROGRAM

The primary physics goal of the BABAR experiment - the PEP-II B-Factory and the BABAR Detector - is the detailed study (1) of the « natural » explanation for CP violation as provided, within the Standard Model, by the complex phase in the CKM matrix.

The experimental signature is the existence of large, predictable asymmetries in the decays of the B^0 meson to CP eigenstates. The particular channels in which it is hoped that BABAR will be able to measure such asymmetries include the following :

- for $\sin 2\beta$, $B^0 \to J/\Psi\ K^0_S$, $B^0 \to J/\Psi\ K^0_L$, $B^0 \to J/\Psi\ K^{*0}$,

  $B^0 \to D^+ D^-$, $B^0 \to D^{*+} D^{*-}$, etc ;

- for $\sin 2\alpha$, $B^0 \to \pi^+ \pi^-$, $B^0 \to \pi^+ \pi^- \pi^0$, $B^0 \to a_1 \pi$, etc,

where α and β are two angles of the unitarity triangle (Figure 1).

However, if the CP asymmetries in question are expected to be quite large, the branching ratios to reconstructible final states are very small (~10^{-5} for $J/\Psi\ K^0_S$ and for $\pi^+ \pi^-$). The consequence is that in excess of 10^7 $B^0 \overline{B}^0$ pairs will have to be produced in order to measure the asymmetries with errors at the 10% level or better.

To achieve these measurements, in addition to a need for a high luminosity, three ingredients will play a vital role. First, the full reconstruction of exclusive final states is needed (this leads to strong requirements on the charged particle momentum resolution, on the photon detection efficiency and energy and position resolution, and on the capabilities of particle identification). Second, the flavor (beauty or anti-beauty) of the decaying particle needs to be tagged (with electrons and muons, and with charged kaons). Finally, the proper time of the B^0 decay with respect to its production needs to be measured (vertex reconstruction, also to unambiguously ascertain the time order of the B^0, $\overline{B}^0$ decay).

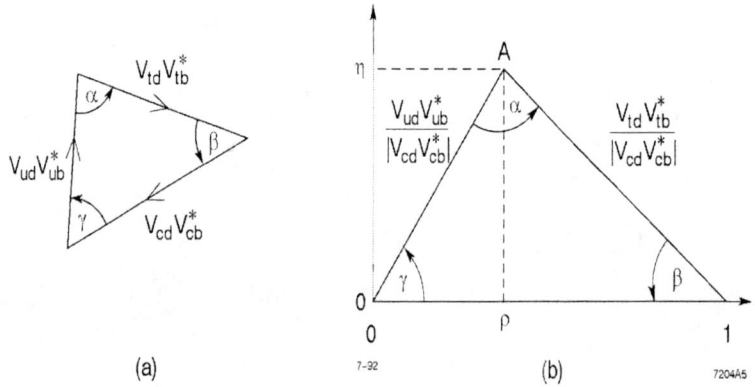

FIGURE 1. The Unitarity Triangle (a), rescaled (b) by choosing a phase convention such that $(V_{cd}V^*_{cb})$ is real and dividing the lengths of all sides by $\left| V_{cd}V^*_{cb} \right|$.

The PEP-II B-Factory (2) and the BABAR Detector (3) are designed and optimized to achieve the needs and goals specified above.

In addition to CP asymmetries, BABAR will be able to make also sensitive measurements of CKM elements in B decays, to search for rare B decays and to study the physics of charm and of the tau lepton, and two-photon physics.

THE PEP-II B-FACTORY

The main challenge of an e^+e^- B-factory like PEP-II is to reach the unprecedented high luminosity of 3×10^{33} cm^{-2} s^{-1}, an order of magnitude or higher than any existing colliders. The luminosity for optimized e^+e^- colliders has the remarkably simple form

$$L(cm^{-2}s^{-1}) \propto \frac{\Delta v . E(GeV) . I(Amp)}{\beta^*(cm)}.$$

Since the energy, E, is effectively fixed by the need to produce the Y(4S) and the tune shift limit, Δv, is more or less also given, the parameters that can be pushed to achieve the maximum luminosity are the total current, I, and the beta function or focussing strength, β^*, at the interaction point.

The PEP-II B-Factory is being constructed by a collaboration of SLAC, LBL and LLNL (2) on the SLAC site. The machine is asymmetric with energies of 3.1 GeV for the positron beam (Low Energy Ring – LER) and of 9.0 GeV for the electron beam (High Energy Ring – HER). It reuses the tunnel and the magnets of the old PEP machine for the HER. The LER is new and is put in place over the HER as shown in Figure 2. Beams are stored in the two separated rings allowing PEP-II to have the very large number of 1658 bunches (about 4 ns spacing between bunches). The collisions are head-on, requiring strong dipoles to be placed very close to the interaction point. This scheme avoids the potential instabilities due to a crossing angle but, in turn, might lead to potentially high backgrounds coming from synchrotron radiation and beam debris swept into the detector. It requests also a not-too-small spacing (1.2 m) between bunches to avoid parasitic collisions. Both the HER and the LER use room temperature RF, with copper cavities and waveguides to remove the higher order modes that would otherwise coupled the bunches.

FIGURE 2. PEP-II B-Factory with the LER on top of the HER.

With a center of mass energy of 10.58 GeV, corresponding to a moving center of mass of $\beta\gamma = 0.56$, PEP-II provides the needed feature of asymmetric collisions to measure the time dependent CP violating asymmetries. It corresponds to an average separation of $\beta\gamma c\tau = 250$ μm between the two B vertices, which is crucial

TABLE 1. Main parameters of the PEP-II collider and best commissioning results (up to October 1st, 1998)

Machine		HER		LER	
Parameter	Units	Design	July 31 '98	Design	Aug. 1 '98
Energy	GeV	9.0	9.0	3.1	3.1
Single Bunch Current	mA	0.6	12.0	1.3	5.0
Number of Bunches		1658	1658	1658	1658
Total Beam Current	A	0.995	0.75	2.14	0.053
Beam Lifetime	hours	4	12 @ 50mA 2.5 @ 725 mA	4	3min@10mA

for studying the cleanest and most promising CP violating modes. Parameters of the PEP-II B-Factory together with some achieved results are shown in Table 1. The PEP-II B-Factory has currently entered its commissioning phase. Starting in February 1999, BABAR will move onto the beam line and physics collisions will begin early May 1999.

THE BABAR DETECTOR

In order to achieve its physics program (1), the BABAR detector needs:

- The maximum possible acceptance in the center-of-mass system. Although the forward boost of the decay products in the laboratory frame is rather a small one, optimizing the detector acceptance leads to an asymmetric detector.

- To accommodate machine components close to the interaction region (high luminosity requirement).

- Excellent vertex resolution (in particular its z-component): the decay time difference ($t_{CP} - t_{tag}$) will be measured via the difference in the z-component of decay positions of the B mesons (since they travel almost parallel to the z-axis). Also the best possible vertex resolution is needed in order to discriminate between beauty, charm and light quark vertices.

- Good momentum resolution for kinematic reconstruction and tracking over the range ~ 60 MeV/c < p_t < ~ 4 GeV/c.

- To discriminate between e, μ, π, K and p over a wide kinematic range. The tagging of the flavor of B-meson decays can be done with high efficiency and purity only if electrons, muons and kaons can be well identified. In addition, π–K discrimination up to about 4 GeV/c is essential in order to discriminate between decay channels like $B^0 \to \pi^+\pi^-$ and $B^0 \to K^{\pm}\pi^{\mp}$, $B^0 \to \rho^+\pi^-$ and $B^0 \to K\rho$ and $B^0 \to K^*\pi$.

- To detect γ's and π^0's over the wide energy range ~ 20 MeV < E < ~ 5 GeV.

- To have neutral hadron identification functionality ($B^0 \to J/\Psi\ K^0_L$).

- To be able to record data at a 100 Hz trigger rate (up to about 2 kHz at the first level trigger).

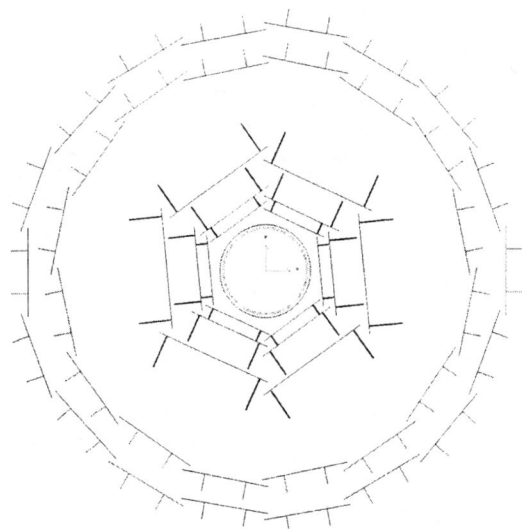

FIGURE 3. Layout of the BABAR silicon vertex tracker. Cross-sectional view in the plane orthogonal to the beam axis.

To provide all the above features, the major subsystems (3) of the BABAR detector are :

- A silicon vertex detector (the SVT) whom the main task is to reconstruct the decay vertices of the two primary B mesons in order to determine the time between the two decays. This determination will allow the measurement of the time dependent asymmetries. The SVT, with five concentric cylindrical layers of double-sided silicon detectors with 90^0 stereo (see Figure 3), provides also the complete tracking information for charged particles with $p_t \leq 100$ MeV/c (which cannot reach the drift chamber). The readout pitch in the three inner layers is 50 μm in φ, 100 μm in z with one floating strip. The readout electronics use a custom rad-hard IC and time-over-threshold analog readout for pulse height measurements.

• A drift chamber (the DCH) provides up to 40 measurements of space coordinate per track, ensuring high reconstruction efficiency for tracks with $p^t > 100$ MeV/c.

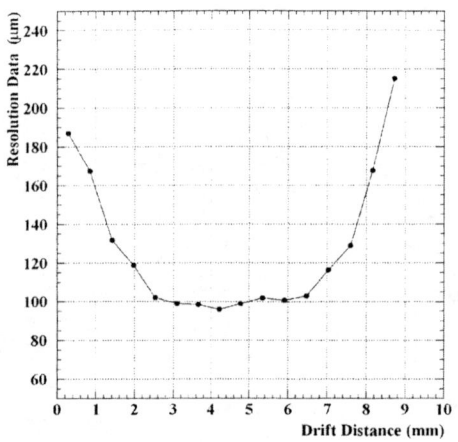

FIGURE 4. Spatial resolution obtained in the BABAR full-length prototype drift chamber with sense wire voltage at 1960 V.

The 7104 hexagonal drift cells (typical dimensions of 1.2x1.8 cm^2) are arranged in superlayers of 4 layers each (axial and stereo superlayers alternate). The chosen gas mixture, helium-isobutane (80% :20%), together with the 1.5 T magnetic field of the superconducting solenoid, provides good spatial (see Figure 4) and dE/dx resolutions (6.8% predicted resolution). Finally, the DCH provides the prompt charged trigger signals to the BABAR Level 1 Trigger system at a sampling frequency of 3.75 MHz.

• A particle identification system (the DIRC – acronym for Detection of Internally Reflected Cherenkov light) will provide excellent kaon identification up to about 4.0 GeV/c in order, in particular, to distinguish between the two-body decay modes $B^0 \rightarrow \pi^+\pi^-$ and $B^0 \rightarrow K^+\pi^-$.

FIGURE 5. View of the DIRC Standoff Box – the SOB, filled with water, and of the 10,752 PMTs that cover the detection area (also the front-end electronics crates mounted on the SOB can be seen, one crate per SOB's sector to read 896 PMTs).

The DIRC concept relies on the detection of Cherenkov photons trapped in the radiator (144 long, straight bars of synthetic quartz with rectangular section, arranged in a 12-sided polygonal barrel). The Cherenkov light, through successive total internal reflections, is brought outside the BABAR tracking and magnetic volumes and is detected by a close-packed array of linear focused 2.82 cm diameter photomultiplier tubes (see Figure 5). The DIRC technique has the advantage to occupy only a small (8 cm) radial space while keeping only 14% of an X_0 at normal incidence. The photoelectron yield is high in the DIRC, with from 20 to 50 photoelectrons at various angles. The π/K separation is always >3σ within the kinematic limits for particles from B decays.

- An electromagnetic calorimeter (the EMC) based on quasi-projective CsI(Tl) crystals. It consists of a cylindrical barrel (5760 crystals) and a forward conic endcap (820 crystals) as shown in Figure 6. The EMC detects π^0's with very high efficiency and low background for CP decays such as $B^0 \to J/\Psi K^0_S$, $K^0_S \to \pi^0\pi^0$, etc. It identifies electrons down to 500 MeV/c and supplements the information for muon and K^0_L identification. Finally the EMC provides information for the neutral trigger. For photons at a polar angle of 90^0, the energy resolution is

$$\sigma_E/E = 1\% / (E(GeV))^{1/4} \oplus 1.2\%$$

and the angular resolution is

$$\sigma_{\theta,\phi} = 3 \text{ mr} / (E(GeV))^{1/2} \oplus 2 \text{ mr}.$$

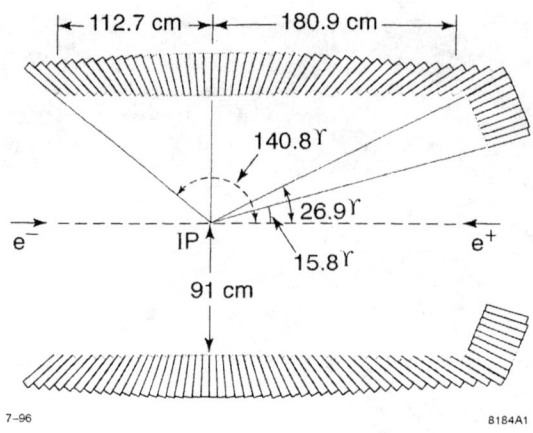

FIGURE 6. Side view of the BABAR electromagnetic calorimeter layout.

- A muon and neutral hadron identification system (The IFR – acronym for Instrumented Flux Return) uses the large iron structure needed as a magnet yoke. It consists of a central part (Figure 7) and two end caps which complete the solid angle coverage down to 300 mr in the forward direction and 400 mr in the backward direction.

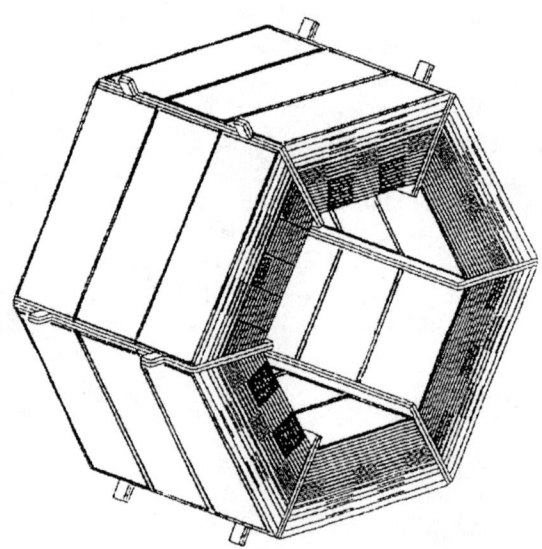

FIGURE 7. Layout of the BABAR IFR barrel.

A feature of the IFR system is the graded segmentation of the iron, which varies from 2 to 10 cm, increasing with the radial distance from the interaction region. The BABAR IFR uses plastic resistive plate counters filled with a gas mixture based on comparable quantities of argon and freon, plus a small amount (a few %) of isobutane. The primary goals of the IFR are to reduce the lower momentum limit for detecting muons to about 0.6 GeV/c and to detect $B^0 \rightarrow J/\Psi K^0_L$.

CONCLUSIONS

The PEP-II B-Factory has achieved its first collisions in July 1998. From October 1998 until mid-February 1999, the commissioning of the full machine, including the final interaction region, will continue with special emphasis on the understanding of machine related backgrounds and of luminosity optimisation. The BABAR Detector (see Figure 8) has completed its installation phase. Until end of January 1999, it will be commissioned with cosmic ray data taking. Starting early February 1999, BABAR will be moved on the PEP-II beamline and the vertex detector will be installed. The first beam data are expected to be taken early May 1999.

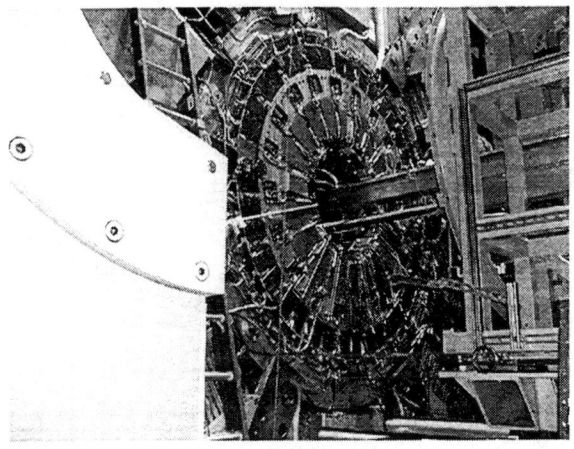

FIGURE 8. View of the forward part of the BABAR detector.

As an illustration of the BABAR physics sensitivity, the resolution on the β angle (see Figure 1) is given in Table 2 for some CP states. It is expected (1) that the

TABLE 2. Error in the measurement of $\sin 2\beta$ (assuming $\sin 2\beta = 0.7$) for an integrated luminosity of 30 fb^{-1}

Mode	Br. (10^{-4})	Rec. eff.	N_B/N_S	Error
$J/\Psi\ (l^+l^-)\ K^0_S(\pi^+\pi^-)$	4.25	0.60	0.06	0.12
$J/\Psi\ (l^+l^-)\ K^0_S(\pi^0\pi^0)$	4.25	0.21	0.06	0.30
$J/\Psi\ (l^+l^-)\ K^0_L$	4.25	0.41	0.59	0.15
$J/\Psi\ (l^+l^-)\ K^{*0}(\pi^+\pi^-\pi^0)$	13.2	0.09	0.18	0.50
D^+D^- (6D modes)	4.5	0.24	2.80	0.48
$D^{*+}D^{*-}$ (4D modes)	9.7	0.05	0.23	0.44

combined sensitivity for $\sin 2\beta$, for the same integrated luminosity and for all major CP modes, will likely be on the order of 0.08.

ACKNOWLEDGMENTS

I would like to thank David Hitlin and Jonathan Dorfan for useful input. I acknowledge the members of the BABAR Collaboration who contributed to the BABAR Physics Book (1), from which most of the material for this talk was drawn. I would like to thank also Christophe Renard and Diana Rogers for their precious help in editing this document. This work was supported by CNRS/IN2P3, France.

REFERENCES

1. *The BABAR Physics Book*, the BABAR Collaboration, SLAC-R-504 (1998).
2. *An Asymmetric B Factory Based on PEP*, The Conceptual Design Report for PEP-II, LBL Pub 5303, SLAC-372 (1991).
3. *BABAR Technical Design Report*, the BABAR Collaboration, SLAC-R-95-457 (1995).

B Physics in the Next Millennium

Marina Artuso[1]

Syracuse University
Syracuse NY 13244

Abstract. As we approach the turn of the century, the Standard Model is still consistent with all our experimental observations and the path to a more complete picture of the fundamental constituents and their interactions has yet to be clearly identified. Beauty flavored hadrons have provided crucial experimental information on several fundamental parameters of the Standard Model and may lead to one of the most challenging test of its validity and provide some clues on the path towards a more complete theory. Several experiments will try to explore this rich phenomenology in the next few years. Their physics goals and discovery potential will be compared.

INTRODUCTION

The investigation of B meson decays has provided a wealth of information on one of the least understood aspects of the Standard Model: the quark mixing that underlies the complex pattern of charge-changing transition in the quark sector. This pattern is summarized by a 3 x 3 unitary matrix, the Cabibbo-Kobayashi-Maskawa (CKM) matrix:

$$V_{CKM} = \begin{pmatrix} V_{ud} & V_{us} & V_{ub} \\ V_{cd} & V_{cs} & V_{cb} \\ V_{td} & V_{ts} & V_{tb} \end{pmatrix}. \quad (1)$$

A commonly used approximate parameterization was originally proposed by Wolfenstein [1]. It reflects the hierarchy between the magnitude of matrix elements belonging to different diagonals. The 3 diagonal elements and the 2 elements just above the diagonal are real and positive. It is defined in first order as:

$$V_{CKM} = \begin{pmatrix} 1 - \lambda^2/2 & \lambda & A\lambda^3(\rho - i\eta) \\ -\lambda & 1 - \lambda^2/2 & A\lambda^2 \\ A\lambda^3(1 - \rho - i\eta) & -A\lambda^2 & 1 \end{pmatrix}. \quad (2)$$

---

[1] Work supported by the National Science Foundation.

There are several reasons why the experimental determination of the CKM parameters is interesting. On one hand, it is important to test that it is indeed a unitary matrix, as dictated by the Standard Model. On the other hand, the complex phase that is inherent in the 3-generation CKM matrix can be an explanation for the phenomenon of CP violation, so far observed only in the neutral K meson system. This violation is expected to be responsible for baryon dominance in our world and thus the understanding of its mechanisms has profound implications for our understanding of the origin and evolution of the universe.

The b quark provides a unique opportunity to study several CKM parameters. The study of semileptonic decays allows the measurement of $|V_{cb}|$ and $|V_{ub}|$. $B^o - \bar{B}^o$ mixing provides information on V_{td} and V_{ts}. These different measurements provide independent constraints on the 'unitarity triangle' shown in Fig. 1. A more accurate determination of the magnitude of V_{ub}/V_{cb} and of the mixing parameters can pin down two sides of this triangle. Note that one side has a length equal to one by construction. In addition, several experiments will try to get some information on the angles α, β and γ. The knowledge of these angles will answer some very fundamental questions:
(1) Is the CKM phase of the three generation Standard Model the only source of CP violation?
(2) Is there new physics in the quark sector?

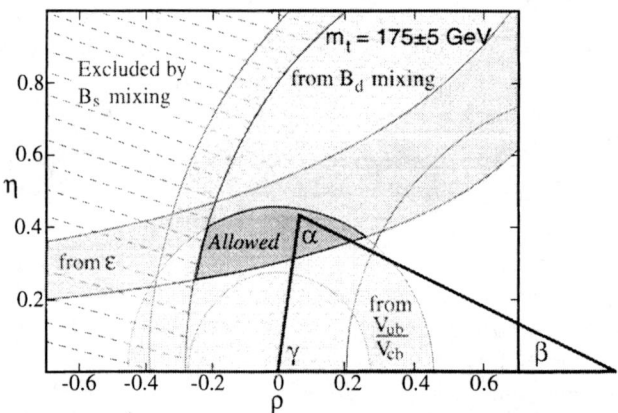

FIGURE 1. The regions in $\rho - \eta$ space (shaded) consistent with measurements of CP violation in K_L^o decay (ϵ), V_{ub}/V_{cb} in semileptonic B decay, B_d^o mixing, and the excluded region from limits on B_s^o mixing. The allowed region is defined by the overlap of the 3 permitted areas, and is where the apex of the CKM triangle sits. The bands represent $\pm 1\sigma$ errors. The large width of the B_d mixing band is dominated by the uncertainty in the parameter f_B. Here the range is taken as $240 > f_B > 160$ MeV.

TABLE 1. B experiments in the near future.

	CLEO III	**BaBar-Belle**	**HERA-B**	**CDF-D0**
$L(\text{cm}^{-2}\text{s}^{-1})$	1.7×10^{33}	3×10^{33}	(Int.Rate)40 MHz	2×10^{32}
$\sigma_{b\bar{b}}$	1.15 nb	1.15 nb	$\approx$ 10 nb	100 μb
$\sigma_{b\bar{b}}/\sigma_{had}$	0.25	0.25	$\approx 10^{-6}$	$\approx 10^{-3}$
Trigger	all B's	all B's	ψ	high p_t μ's
Time res.	very modest	modest	good	good
PID	e,μ,π,K,p	e,μ,π,K,p	e,μ,π,K,p	e,μ

TABLE 2. B experiments starting around 2005.

	e^+e^- **b-factories**	**ATLAS/CMS**	**LHCB**	**BTeV**
$L(\text{cm}^{-2}\text{s}^{-1})$	10^{34}	10^{33}(first run)	1.5×10^{32}	2×10^{32}
$\sigma_{b\bar{b}}$	1.15 nb	500μb	500μb	100 μb
$\sigma_{b\bar{b}}/\sigma_{had}$	0.25	$\approx 5\times 10^{-3}$	$\approx 5\times 10^{-3}$	$\approx 10^{-3}$
L1 Trigger	all B's	high $p_t\mu$'s	medium p_t μ,e,h'	detached vertices
Time res.	modest	good	very good	very good
PID	e,μ,π,K,p	e,μ	e,μ,π,K,p	e,μ,π,K,p

FUTURE FACILITIES FOR B PHYSICS

The next decade will see a blooming of experimental facilities planning to explore the B decay phenomenology with increasing sensitivity to different observables and various final states.

Table 1 summarizes the main properties of the experiments that will take data in the near future. Among them there are upgraded versions of previous experiments that have contributed to our present knowledge of B physics and some new facilities, HERA-B, a fixed target experiment at HERA, Hamburg, Germany, and the two experiments taking data at the new asymmetric e^+e^- B-factories, BaBar at SLAC and Belle at KEK.

A few years later, ATLAS and CMS should start taking data at the new LHC pp collider and they are also planning to address some of the B physics issues discussed below. The experiments at hadronic machines discussed so far are pursuing B physics, but they have all been optimized for their main goal, high p_t physics. They take advantage of the high cross section for b production, but have not been designed to study b decays. Two experiments have been proposed to exploit the full discovery potential offered by the high cross section and richness of final states accessible at hadron machines: LHCb, approved to take data at LHC, and BTeV, planning to take data at Fermilab. BTeV is an official R&D project at Fermilab. Table 2 summarizes the most distinctive features of this next round of experiments that are expected to take data around the year 2005.

The traditional advantage of e^+e^- machines operating at the $\Upsilon(4S)$ are their low non-B background. In addition, the final state is composed just of a $B\bar{B}$ meson pair, making it easy to apply powerful kinematical constraints to identify specific final states or to reconstruct inclusive decays, like $b\to s\gamma$ or decays with missing particles like neutrinos. On the other hand, in order to measure rare decays or tiny CP asymmetries it is necessary to collect huge data samples, posing a significant challenge to the accelerator physicists striving to design machines of ever increasing luminosity.

Experiments taking place at hadronic facilities have the advantage of copious production of b-flavored hadrons. On the other hand, their main challenge is the identification of the interesting b events from the much more frequent 'minimum bias' events. The key detector element in this endeavour has been a high resolution vertex detector, as the distinctive feature of b decays in this environment is that their lifetime is longer than the one of the light quark products. CDF has been quite successful in exploiting this feature as a tool for b physics.

Experiments at hadron machines designed for b physics have observed that the forward region is the most favorable to the study of b decays. Several characteristics of hadronic b production favor the forward direction as the region of choice for the detector acceptance. Fig. 2 shows that the b quarks are produced at the Tevatron with a relatively flat pseudo-rapidity distribution, where the pseudo-rapidity η is defined as $\eta = -ln(tan(\theta/2))$ where θ is the polar angle with respect to the beam axis. Note that the more forward the b, the higher its Lorentz boost is, as shown in Fig. 2, thus making it easier to identify the b decays by their detached decay vertices. Finally, Fig. 3 shows the correlation in the production angles of the b and $\bar{b}$ quarks. Note that in the forward directions the pair shows a strong correlation in their production angle. On the other hand, in the central region there isn't any correlation.

BTeV and LHCb have several features in common. In particular, both experiments take advantage of the open detector geometry in the forward direction to include a Ring Imaging Cherenkov detector, allowing an excellent discrimination

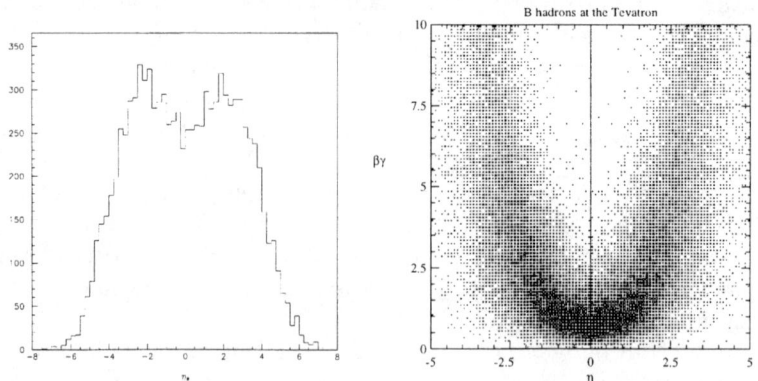

FIGURE 2. The B yield plotted versus η (left). $\beta\gamma$ of the B plotted versus η (right). Both plots are for the Tevatron.

between π's, p's and K's, crucial to some of the measurements discussed below. The most distinctive feature of the BTeV experiment is a unique vertex detector, based on high resolution pixel devices inside a dipole magnetic field, associated with fast trigger processors that will provide the detached vertex information at the first level trigger. This feature is critical to achieve high efficiency for hadronic decay modes, like $B \to \pi^+\pi^-$. In addition, it is a two arm spectrometer and takes advantage of the extended luminous region of the Tevatron ($\sigma_Z \approx 30$ cm) to utilize also multiple interactions per crossing. These features more than compensate the lower $\sigma_{b\bar{b}}$ than LHCb and the two experiments are quite competitive.

A SCENARIO FOR THE UNFOLDING OF THE CP ASYMMETRIES

In the Standard Model, CP violation in B decays may occur whenever there are at least two weak decay amplitudes with different CKM coefficients that lead to a given final state. In the charged B decay the two decay mechanisms are provided by two competing decay diagrams, for instance the spectator quark decay and the so called 'penguin' diagrams. On the other hand, in neutral B decays, because of $B^o - \bar{B}^o$ mixing, a B^o may decay to a final state f via two paths: $B^o \to B^o \to f$ or $B^o \to \bar{B}^o \to f$. The phases in the second path differ from the phases in the first because of the phase in the mixing diagram and sometimes because of the phase difference between $B^o \to f$ and $\bar{B}^o \to f$. In the case of charged B decays, we have only direct CP violation, whereas in the case of neutral B decays we can have both CP violation induced by mixing and direct CP violation. In this paper we cannot summarize all the facets of this rich phenomenology, discussed in other excellent reviews [2], [3]. As an illustration of the challenges involved in the measurement,

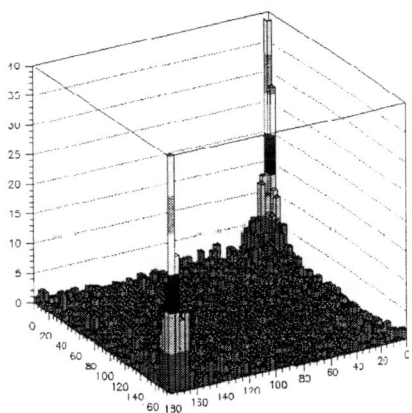

FIGURE 3. The production angle (in degrees) for a hadron containing a b quark plotted versus the production angle (in degrees) for a hadron containing a $\bar{b}$ quark.

TABLE 3. Summary of ϵD^2 available at different facilities: forward and central refer to the detector geometry at a hadron collider.

Method	Forward [4]	Central [5]	e^+e^-b-factories [6]
$K^\pm$	5%	0%-2.4%	9.6%
$\mu^\pm$	1.6 %	1.0	5.4%
$e^\pm$	1.0%	0.7%	8.4%
SST	> 2%	2%	-
Jet Charge	6.5 %	3%	-

we recall that for neutral B decays to CP eigenstates, only the mixing contribution is present and the time dependent asymmetry can be expressed as:

$$A(f_{CP}) \equiv \frac{\Gamma(B^o(t) \to f_{CP}) - \Gamma(\bar{B}^o(t) \to f_{CP})}{\Gamma(B^o(t) \to f_{CP}) + \Gamma(\bar{B}^o(t) \to f_{CP})} = \chi sin(2\phi_i - \Phi_M)\frac{x}{1+x^2} \quad (3)$$

where $\chi = \pm 1$ is the sign of the CP parity of the eigenstate, ϕ_i are the CP violating phases related to the relevant CKM parameters, Φ_M is the phase in the $B^o - \bar{B}^o$ mixing and $x \equiv \Delta M/\Gamma$ characterizing $B^o - \bar{B}^o$ mixing. The difference $2\phi_i - \Phi_M$ is related to the quark mixing parameters: different CP asymmetries will provide information on different angles of the unitarity triangle.

In order to measure the CP asymmetries there are three crucial ingredients: adequate data samples, as often it is necessary to measure tiny differences between final states that have a quite small branching fraction, the ability of tagging the flavor of the initial meson and finally an adequate suppression of backgrounds.

The development of a variety of flavor tagging techniques in the different experimental configurations has been one of the most active area of investigation towards the development of the physics analysis tools at different facilities. Some of the tagging techniques, like the charge of the μ's produced in semi-leptonic decays, are common to most experiments, others are environment-specific. For example, e^+e^- b-factories can try to take advantage of the low momentum leptons produced in the cascade $B \to D \to K^{(*)}\ell\nu$. On the other hand, hadronic machines can take advantage of same side tagging, exploiting the correlation between the flavor of the B hadron and the charge of the pion produced in close association with it. Table 3 illustrates a comparison between the tagging efficiency of different approaches.

Table 4 shows the prospects for the $\sin 2\beta$ measurement by the different experiments. Note that the projections are taken from proposals and simulation studies performed by the proponents of the various experiments and do not necessarily share the same level of realism. The data shown here and in the following discussion should be taken as indicating some trends rather than as a detailed quantitative comparison.

The determination of the angle $\sin 2\alpha$ is a much more complex issue. The 'golden

TABLE 4. Prospects for $\sin 2\beta$ with 1 year of running at the nominal luminosity

Experiment	$\delta \sin 2\beta$	Remarks
BaBar	±0.09	using ψK^0 [6] assuming $sin(2\beta)=0.7$
Belle	±0.11	using ψK^0 [7]
HERA-B	±0.13	using ψK_S [8]
CDF	±0.09	using ψK_S [5]
D0	±0.11	using ψK_S [9]
BTeV	±0.013	using ψK_S [4]
LHCb	±0.017-0.011	using ψK_S and $sin(2\beta)=0$-0.866 [10]
ATLAS	±0.018	using ψK_S [11]
CMS	±0.058	using ψK_S [12]

TABLE 5. Prospects for $\sin 2\alpha$ with 1 year of running at nominal luminosity

Experiment	$\delta \sin 2\alpha$	Remarks
BaBar	±0.29	using $\pi^+\pi^-$ [6]
Belle	±0.27	using $\pi^+\pi^-$ (assuming no penguin) [7]
CDF	±0.22	using $\pi^+\pi^-$ and assuming PID=TOF+dE/dx [5]
BTeV	±0.026	using $\pi^+\pi^-$ (no penguin) [4]
LHCb	±0.05	using $\pi^+\pi^-$ (no penguin) [10]
ATLAS	±0.1 $\sin(2\alpha) \oplus 0.011$	using $\pi^+\pi^-$ (no penguin,) [11]
CMS	±0.067	using $\pi^+\pi^-$ (no penguin, statistical error only) [12]

mode' in this case used to be the decay $B^0 \to \pi^+\pi^-$. However 'penguin pollution' [13] complicates the relationship between the measured asymmetries and $\sin 2\alpha$.

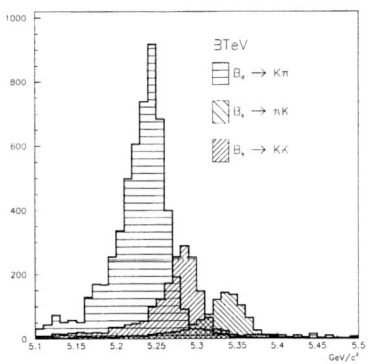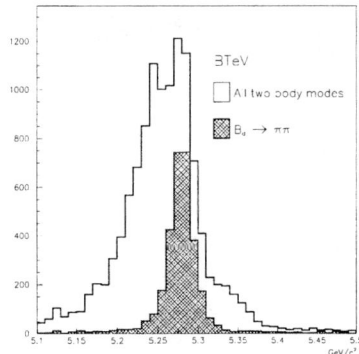

FIGURE 4. Invariant mass distribution for all the $B \to h^+h^-$ final states, where h denotes either a π or a K, and the mass is computed assuming that both the particles are π's. The plot on the left shows all the individual background contributions and the plot on the right shows the sum of all the channels, properly normalized (see text). The plot refers to the Tevatron.

In addition to the spectator diagram, that would provide a contribution to A_{CP} according to the formulation discussed above, the penguin diagram adds a term proportional to $cos(xt)$. The fraction of penguin contribution needs to be known to extract α [14]. Note that the different experiments simulate the effects of penguin pollution with very different degrees of approximation. In general, the label 'no penguin' refers to simulations that assume that the penguin pollution is negligible. Moreover, recent CLEO results [15] have shown that the branching fraction for this decay may be quite smaller than anticipated, its upper limit at 90% C.L. being 0.8×10^{-5}, thus seriously affecting the prospects of e^+e^- b-factories. Lastly, a state of the art hadron identification system is necessary to single out this final state from other two-body decay modes of the B_d and B_s^0 mesons, as illustrated by Fig. 4. The data are taken from the BTeV simulation studies but apply to all the experiments taking data at hadronic facilities, thus showing that the excellent particle identification featured by LHCb and BTeV are crucial to make a reliable measurement. The normalization between the different decay modes is obtained assuming $\mathcal{B}(B \to K^+\pi^-) = 1.5 \times 10^{-5}$ and $\mathcal{B}(B \to \pi^+\pi^-) = 0.75 \times 10^{-5}$, according to the recent CLEO results [15], and $\mathcal{B}(B_s \to K^+K^-) = \mathcal{B}(B_d \to K^+\pi^-)$ and $\mathcal{B}(B_s \to K^+\pi^-) = \mathcal{B}(B_d \to \pi^+\pi^-)$.

The determination of the angle γ introduces new challenges. In principle, decay modes of the B_S meson to CP eigenstates, like ρK_S, could be used. However the same 'penguin pollution' problems alluded to in the discussion of $\sin 2\alpha$ are present here and they are even more difficult to be disentangled than in the previous case because of the vector-pseudoscalar nature of the final state. An alternative approach can be used to extract the angle γ from the decays $B_S \to D_S^{\pm}K^{\mp}$, where a time-dependent CP violation can result from the interference between the direct decay and the mixing induced decays [16]. BTeV and LHCb have studied the possibility of extracting the angle γ with this approach. They project errors of $\pm 8°$ and $\pm 10°$, respectively.

Another method for extracting γ has been proposed by Atwood, Dunietz and Soni [17], who refined a method suggested originally by Gronau and Wyler [18]. A large CP asymmetry can result from the interference between the decays $B^- \to K^-D^o, D^o \to f$, where f is a doubly-suppressed Cabibbo decay of the D^o (for example, $f = K^+\pi^-$, and $B^- \to K^-\bar{D}^o$, $D^o \to f$,). Since $B^- \to K^-\bar{D}^o$ is color-suppressed and $B^- \to K^-D^o$ is color allowed, the overall amplitude for the two decays are expected to be approximately equal in magnitude. The weak phase between them is γ. The subtleties of extracting the CKM angle from the measurements of two different states are discussed in Ref. [4].

Finally, Gronau and Rosner [19] originally proposed a method based on the study of the decays $B \to K\pi$. The use of these decay modes is complicated by rescattering processes and $SU(3)$ breaking effects, as pointed out in several subsequent papers. However the intense theoretical effort to understand these decay modes of the neutral and charged B meson can ultimately provide a good strategy to extract the angle γ. A recent analysis by A. Buras and G. Fleischer [20] examines all the hadronic effects in great detail and gives some promising strategies to use this approach more effectively.

A complementary constraint to the unitarity triangle is provided by the measurement of $B_s \bar{B}_s$ mixing, using the ratio:

$$\left|\frac{V_{td}}{V_{ts}}\right|^2 = \xi^2 \frac{m_{B_s}}{m_{B_d}} \times \frac{\Delta m_d}{\Delta m_s} \qquad (4)$$

where $\xi = f_{B_s}\sqrt{\mathcal{B}_{B_s}}/f_B\sqrt{\mathcal{B}_B} = 1.15 \pm 0.05$ is the SU(3) breaking term, estimated from lattice and QCD sum rules. The time resolution is crucial in this measurement. The projected proper time resolution for BTeV is about 30 fs [4], and for LHCb is 43 fs [10], whereas for CDF is 60 fs (possibly down to 46 fs with an additional silicon layer) [21] and for ATLAS is 64 fs. BTeV expects to be able to measure Δm_s at least up to 51 ps^{-1} within a reasonable time scale. LHCb expects to measure Δm_s with a statistical significance of at least 5 σ if the true value of $\Delta m_s \leq 48$ ps^{-1} or exclude values of ΔM_s at 95% C.L. up to 58 ps^{-1} (corresponding to a value of $x_s \equiv \Delta M_s/\Gamma_s = 91$). For comparison, CDF claims to be able to make this measurement if $x_s \leq 20$, D0 claims to be able to make this measurement if $x_s \leq 16$ and ATLAS and CMS if $x_s \leq 38$. Note that the extraction of $|V_{td}/V_{ts}|$ with this method minimizes the theoretical uncertainty and therefore this observation will have a significant impact on our understanding of the CKM matrix.

CONCLUSIONS

This paper illustrates how different experiments will contribute to a precision determination of the angles and sides involved in the unitarity triangle and thus provide a crucial test of the validity of the CKM picture of quark mixing and CP violation. In addition, they will perform detailed studies of rare B decays, thus providing complementary tests of the Standard Model and useful constraints on more exotic models, like SUSY or a more complex Higgs sector. In the next year, experiments at asymmetric e^+e^- B factories, Babar and Belle, will start collecting data. They are likely to make the first significant measurements of $\sin 2\beta$. In the same time period, the symmetric B factory experiment CLEO will start with its III upgraded version. If it proves easier to make luminosity with a single ring symmetric energy machine, they may be the first to see direct CP violation and will continue their measurements of rare B decays that have already provided quite interesting results [15]. With the start of Tevatron Run II, CDF and D0 will try to beat the the e^+e^- machines to the first measurements of $\sin 2\beta$. HERA-B will also enter the race. Ultimately, crucial measurements on B_s mixing, $\sin 2\alpha$, γ and very rare B decays are likely to be made at forward experiments at hadron machines, LHCb or BTeV, where the B rates are large, the vertexing is accurate and the particle identification is excellent. These measurements are crucial to a complete and accurate picture of this complex phenomenology.

ACKNOWLEDGEMENTS

The author would like to thank J. Butler and H. Cheung, as they distinguished themselves among the organizers for their indefatigable dedication to the rich scientific program and the smooth running of this very enjoyable conference. In addition, many thanks are due to S. Stone, I. Bigi and A.I. Sanda for interesting discussions. Lastly I would like to thank Julia Stone for many pleasant interludes during the writing of this manuscript.

REFERENCES

1. L. Wolfenstein *Phys. Rev. Lett.* **51**, 1945 (1983).
2. I.I Bigi, V. A. Khoze, N.G. Uraltsev and A.I. Sanda, in *CP Violation*, ed. C. Jarlskog (World Scientific, Sinagpaore, 1989), p. 175
3. Y. Nir and H.R. Quinn, in *B Decays*, revised 2nd edition, ed. S. Stone (World Scientific, 1994), p.520.
4. A. Kulyavtsev *et al.*, *The BTeV Program at Fermilab*, submitted to the XXIX International Conference on High Energy Physics, Vancouver, Canada, July 1998.
5. CDF II, submission to Fermilab PAC, Fermilab Pub 96/390-E
6. The BaBar Physics Book, 1998.
7. Belle Technical Design Report, KEK Report 95-1.
8. I. Abt, *Nucl. Instr. Meth. A* **384**, 185 (1996).
9. D0 Upgrade, submission to Fermilab PAC, Fermilab Pub 96/357-E.
10. LHCb Technical Proposal.
11. P. Eerola, *Nucl. Instr. Meth. A* **384**, 93 (1996).
12. A. Kharchilava, *Nucl. Instr. Meth. A* **384**, 100 (1996).
13. M. Gronau, *Phys. Rev. Lett.* **63**, 1451 (1989).
14. I. Dunietz, in *B Decays*, revised 2nd edition, ed. S. Stone (World Scientific, 1994), p.550.
15. P. Gaidarev, *Rare B Decays and CP Violation from CLEO*, in these proceedings.
16. R. Aleksan *et al.*, *Z. Phys. C* **54**, 653 (1992).
17. D. Atwood, I. Dunietz and A. Soni, *Phys. Rev. Lett.* **78**, 3257 (1997).
18. M. Gronaw and D. Wyler, *Phys Lett. B* **265**, 172 (1991).
19. M. Gronau and J. Rosner, EFI-98-23, hep-ph/9806348 (1998).
20. A.I. Buras and R. Fleischer, CERN-TH/98-319 (1998).
21. Fermilab-Proposal-909, October 23, 1998.

One Reason Why CP Violation is Way Radically Cool

Peter Arnold

Department of Physics
University of Virginia, Charlottesville, VA 22911

Abstract. There are a lot more baryons in the universe today than anti-baryons. I review the necessary conditions for any explanation of this asymmetry in the abundance of baryons and anti-baryons, and I review how all the necessary elements can be found in the physics of the Standard Model and the yet-unknown Higgs sector.

INTRODUCTION

The organizers wanted to provide entertainment for this conference. Unfortunately, they were unable to book the magic act or the dancing bear, so you're stuck with me instead.

I want to discuss the results of a very simple experiment that can be done here at Fermilab and that has consequences for CP violation. The experiment is to look around and notice that Fermilab is not exploding. This immediately tells us that, even though there may be a lot more anti-matter at Fermilab than elsewhere, there's not in fact very much of it at all. In fact, a basic observation about our universe is that there seems to be a lot more baryons in it than anti-baryons.[1] Another way to phrase this is that the net baryon number of the universe (the number of baryons minus the number of anti-baryons) is positive. Or, more meaningfully, that the baryon number *density* is positive.

Today, the asymmetry seems very pronounced, because there are essentially no anti-baryons at all. However, when the universe was very young—a millionth of a second or so old—there were in fact many anti-quarks bouncing around. That's because, even if you imagined for a moment that there weren't any, they would be instantly created through processes such as

---

[1] This is at least true locally, at the scale of the galactic cluster, since otherwise we'd see annihilations in inter-galactic gas. It turns out that it's much less awkward to understand theoretically why matter might predominate everywhere than to understand why matter and anti-matter might instead be separated over cosmic distance scales, so I won't pursue the separation possibility any further.

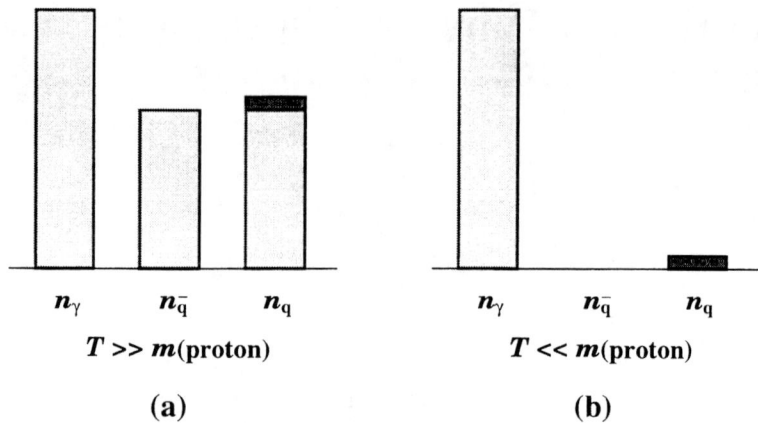

FIGURE 1. An impressionistic depiction of the relatives densities of photons, quarks, and anti-quarks (a) well before and (b) well after the universe cools to temperatures of order $m(\text{proton})$.

$$\gamma\gamma \rightleftharpoons q\bar{q} \tag{1}$$

if the temperature were large enough that the mass energy associated with baryons was negligible. In fact, such processes cause the density of anti-quarks to be the same order of magnitude as the density of quarks or the density of photons or anything else:

$$n_q \sim n_{\bar{q}} \sim n_\gamma \sim n(\text{anything else}). \tag{2}$$

Here and henceforth, I'll only be worried about orders of magnitude in my equalities and not in factors of 2.

Now imagine that early on the universe had only a very small asymmetry between the number densities of quarks and anti-quarks, as depicted schematically by the bar graphs in figure 1a. As the universe cools, eventually the energies of typical photons become smaller than the mass energy associated with quarks or baryons or whatever, and the process (1) can no longer proceed in the forward direction. Annihilation still takes place, however, so one by one the anti-quarks in the universe find a quark, annihilate with it, and are not replaced. This continues until all the anti-quarks have paired up with quarks, and there are no anti-quarks left. All that will be left is that small excess of quarks we started with, as depicted in fig. 1b, since those quarks have no one to annihilate with. Those quarks are all that's left today.

It's possible to extrapolate backward and figure out how big the original excess had to be. Let's concentrate on its relative size in the Early universe: $(n_q - n_{\bar{q}})/n_q$. I already mentioned that everything has roughly the same density in the very hot early universe, so

$$\frac{n_q - n_{\bar{q}}}{n_q} \text{ (Early universe)} \sim \frac{n_q - n_{\bar{q}}}{n_\gamma} \text{ (Early universe)}. \tag{3}$$

Now think about what happens as the quarks and anti-quarks pair up and annihilate. Since they annihilate in pairs, the difference $n_q - n_{\bar{q}}$ is not changed. The photons produced by the annihilation might increase n_γ by a factor of 2 or something, but it won't change the order of magnitude of n_γ. So the order of magnitude of the right-hand side of (3) is not changed by the annihilation. As the universe cools further, it expands and so dilutes number densities like n_γ. However, this dilution by expansion affects all the number densities equally, and the effect cancels out in ratios such as $(n_q - n_{\bar{q}})/n_\gamma$. The upshot is that the right-hand side of (3) is the same today as it was in the early universe, and so

$$\frac{n_q - n_{\bar{q}}}{n_q} \text{ (Early universe)} \sim \frac{n_q - n_{\bar{q}}}{n_\gamma} \text{ (Today)}. \sim \frac{n_B}{n_\gamma} \text{ (Today)}, \tag{4}$$

where n_B is the density of baryons. We can now get a number by measuring n_B and n_γ today.[2] n_γ can be obtained from the cosmic microwave background, which is the remnant of the original gas of photons in the Big Bang—one just looks up in the sky and counts photons. n_B can be estimated within a factor of 100 or so just by looking up, naively counting stars, and estimating how massive they are. A more accurate determination of n_B comes from primordial nucleosynthesis, as n_B turns out to affect some of the primordial light element abundances, which one can measure by looking at primordial gas clouds. In any case, the result turns out to be that

$$\frac{n_q - n_{\bar{q}}}{n_q} \text{ (Early universe)} \sim \frac{n_B}{n_\gamma} \text{ (Today)} \sim 10^{-9}. \tag{5}$$

This means that the original asymmetry between the abundance of quarks and anti-quarks in the early universe was just *one part in a billion!* It's that tiny but crucial one part in a billion excess that survived to become us, the planets, and the stars.

SAKHAROV'S CONDITIONS

Why did the early universe have 1,000,000,001 quarks for every 1,000,000,000 anti-quarks? If baryon number is exactly conserved [as it is in (1) and any other process that's ever been observed experimentally], then there is *no* explanation at all. If baryon number is exactly conserved, then the excess can never change and is whatever it was at $t = 0$; the excess is then just an initial condition—an input into our theories of nature rather than an output. So a necessary element of any attempt to instead *explain* the baryon excess must be the violation of baryon number (B). In fact this is just one of three classic requirements for any explanation, attributed to Sakharov. The requirements are that there must be

---

[2] For a nice introduction to all this sort of stuff, see Kolb and Turner's book [1].

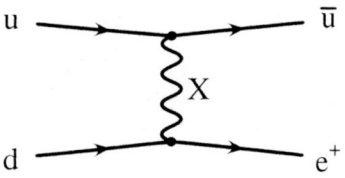

FIGURE 2. Baryon number violation mediated by a super-heavy GUT boson X.

- B violation,
- C and CP violation,
- disequilibrium (at some relevant time in the early universe).

C and CP violation are required for the following reason. Imagine that some B violating process occurred somewhere in the early universe, causing, say, a quark to be produced without an accompanying anti-quark. If C or CP were good symmetries, then there would be an exactly equal probability that somewhere else a process occurs which instead makes an anti-quark without an accompanying quark. The result would be that, even though B violation might be taking place, there would be nothing to drive it on net in one direction over the other.

Disequilibrium is required because of the CPT theorem. CPT tells us that the masses of baryons are the same as those of anti-baryons. More generally, it tells us that the energy of a configuration of baryons is the same as the energy of a corresponding configuration of anti-baryons. In *equilibrium*, the probability of configurations are just weighted by their energy, as $\exp(-E/T)$. As a result, in equilibrium, the thermal ensemble gives equal weight to configurations of anti-baryons and baryons. That is, the average baryon number in equilibrium is simply zero. If you try to start with more baryons than anti-baryons, and if there are indeed processes which can violate baryon number, then entropy likes to even out the numbers, yielding $B = 0$.

GUT BARYOGENESIS

The first well motivated theories which satisfied all of the previous conditions were grand unified theories (GUTs). In GUTs, there are very heavy gauge bosons that mediate explicit violation of baryon number, through processes like that shown in fig. 2. When the universe was so young that the temperature was of order the GUT scale,

$$k_\mathrm{B} T \gtrsim M_X \sim 10^{16} \text{ GeV}, \tag{6}$$

then these heavy X bosons were copiously produced and there was a lot of baryon number violation.

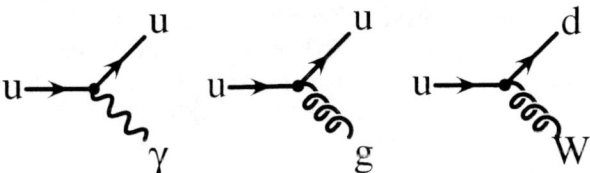

FIGURE 3. Gauge interactions of the Standard Model.

CP violation in GUT theories consists of the usual CKM stuff *plus* extra CP violating phases associated with the heavy GUT-scale physics.

The source of disequilibrium needed for a GUT explanation of the baryon excess comes from the simple fact that, back when the universe was so incredibly young and hot, it was also expanding very rapidly. The expansion of the universe itself provides a source of disequilibrium.

There are two sad aspects of a GUT explanation for the baryon asymmetry, however. The first is that there is not yet any direct experimental evidence for GUTs (*e.g.* proton decay, magnetic monopoles). The second is that, if we really wanted to post-dict the 10^{-9} baryon to photon ratio of today's universe, then we would have to first measure *all* the relevant parameters of the theory—for instance, all the CP violating phases associated with super-heavy GUT interactions. To measure that physics directly would require an accelerator 10^{13} times more energetic than the Tevatron. It could be a long wait. For this reason, I now want to turn instead to a much more exotic model than GUTs. This model is known as

THE STANDARD MODEL.

Fig. 3 depicts some of the interactions of the Standard Model. None of them violate baryon number, and baryon number violation was one of the fundamental requirements for any explanation of the baryon asymmetry of the universe. The idea of baryon number violation in the standard model is obviously absurd—a look at the Lagrangian of the theory shows that every term manifestly conserves baryon number, and there is no way to draw a Feynman diagram that violates baryon number. The fact that something is clearly absurd is, however, no impediment to theorists.

B is an anomalous symmetry

It turns out that baryon number is an *anomalous* symmetry of the standard model. An anomalous symmetry is one that is an exact symmetry of the classical Lagrangian, that's not broken spontaneously or anything ordinary and familiar like that, but which is necessarily broken by the act of quantizing and regularizing the theory.

It's actually possible to give a physical picture of why baryon number is an anomalous symmetry of the standard model, without resorting to any complicated theoretical mumbo jumbo, but I don't have time to do so in the present talk.[3] Instead, I am only going to mention some examples of anomalous symmetries that have direct relevance to high energy experiments, to try to reassure you that anomalous symmetries are not just a crazy theoretical idea. The first example is to consider QED or QCD at very high energies—energies large enough that electron or baryon masses are negligible. If the masses are negligible, then one makes a negligible error by throwing away the masses and considering instead massless QED or massless QCD. But these latter theories are classically scale invariant—the only parameter in the Lagrangian is the dimensionless coupling g. There is nothing in the theory which sets a scale. Nonetheless, we know that the actual behavior of high-energy QED or QCD is that there is logarithmic scaling violation and that effective couplings run logarithmically with energy. Scale invariance is an anomalous symmetry of massless QED or QCD. It appears to be a symmetry of the classical theory, but it is broken when you quantize the theory.

Another example from everyday high energy physics that I won't explain but will only mention in passing is the large rate for the decay of $\pi^0 \to \gamma\gamma$. It turns out that the rate would not be nearly as large as it is if were not for the fact that a particular symmetry of QED (axial T_3 of isospin) is an anomalous symmetry.

B violation is non-perturbative

There is no way to draw a Feynman diagram that violates baryon number. However, Feynman diagrams are merely a tool for understanding perturbation theory. It turns out that B is violated in the standard model by *non-perturbative* electroweak processes.

The rate

Under normal conditions, such as those in this room or in a proton decay detector buried in a mine, the rate for standard model violation is of order

$$\text{rate} \sim e^{-4\pi/\alpha_w} \sim 10^{-170} = \text{zero}, \tag{7}$$

where α_w is the weak-interaction fine structure constant. 10^{-170} is so infinitesimal that it doesn't even matter that I haven't told you what the units are. At this rate, the process would never have occurred even once in the entire history of the entire observable universe.

---

[3] Try sec. 1 of an old sumer school lecture of mine [2].

Standard model B violation is saved from the black depths of theoretical obscurity, however, because it turns out that under extreme conditions, such as those in the very, very early universe (specifically temperatures large compared to M_w/α_w), theory predicts that the rate was *not* suppressed![4] Such temperatures correspond to when the universe was very roughly a billionth of a second old.

I don't have time to give any better explanation of standard model B violation, so I'm going to leave it at this and proceed.

THE ELECTROWEAK PHASE TRANSITION

The third of Sakharov's conditions was the requirement of disequilibrium during the generation of the baryon excess. In GUTs, that disequilibrium came from the rapid expansion of the universe when it was at GUT temperatures. In the standard model scenario, however, the B violation is associated with electroweak physics, and baryogenesis takes place at temperatures of order the electroweak scale. By the standard of the problem at hand, the universe is very old and lethargic at such "cold" temperatures as 100 GeV; the expansion of the universe has slowed down too much to have any hope of providing enough disequilibrium to generate as many baryons as we observe today.

As it turns out, something else interesting is potentially happening at temperatures of order the electroweak scale: electroweak theory is going through a phase transition. Think, for a moment, on the classic example of spontaneous symmetry breaking that's always used to introduce the subject: a ferromagnet. The Hamiltonian of a classical ferromagnet is manifestly rotation invariant—interactions only depend on the *relative* angle between neighboring spins. Nonetheless, the ground state spontaneously breaks rotational invariance; the spins choose a random direction to line up in. If you heat a ferromagnet, however, the spins start to jiggle around relative to each other, and their average net alignment is reduced. If you heat it up enough, the spins jiggle so much that they stop being aligned at all, and the magnetization M disappears. There is a phase transition between the low-temperature symmetry-broken phase ($\langle M \rangle \neq 0$) and the high-temperature symmetric phase ($\langle M \rangle = 0$).

Electroweak theory is the same, except that the Higgs field ϕ plays the roll of the magnetization M. At low temperature, $\langle \phi \rangle \neq 0$ and SU(2) symmetry is broken. At high temperature, however, SU(2) symmetry is restored[5] and $\langle \phi \rangle = 0$. Depending on the parameters of the Higgs sector, the phase transition that occurs between these phases as the universe cools could be a first-order phase transition. Depending on the parameters of the Higgs sector, it could in fact be a *super-cooled* first order phase transition: the universe could get temporarily trapped in the symmetric phase even after the symmetry-broken phase has become the preferred

---

[4] Again, try sec. 1 of ref. [2] for an introduction. Or else see the original discussion in refs. [3]
[5] There are all sorts of technical qualifications inherent in these statements that are not the least bit interesting for the purpose at hand.

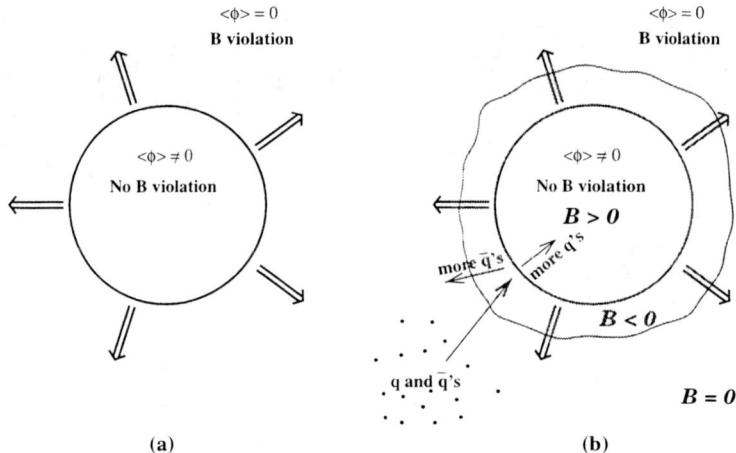

FIGURE 4. Gauge interactions of the Standard Model.

one. The characteristic of super-cooled phase transitions is that they are violent and non-equilibrium. They occur through the spontaneous nucleation of bubble of the low-temperature phase inside the high-temperature phase. Those bubbles then expand violently and eventually collide, percolating and converting the entire universe into the high-temperature phase.

ELECTROWEAK BARYOGENESIS

Now we're ready to put all the elements together.[6] Consider the super-cooled electroweak transition and a bubble of the symmetry-broken phase expanding inside the symmetric phase, as depicted in fig. 4a. I told you before that under normal conditions, such as those in the universe today, standard model B violation is insignificant, whereas in the hot early universe it was substantial. The symmetry-broken phase inside the bubble is qualitatively similar to the universe today, and it turns out that, if the first order transition is strong enough, the rate of B violation inside the bubble is therefore insignificant. The rate in the symmetric phase outside, however, turns out to be unsuppressed. (I said before the rate was unsuppressed for $T \gg M_w/\alpha_w$. But the W only gets a mass through symmetry breaking: M_w

---

[6] This scenario was invented and developed over the years by many people, sometimes working independently, sometimes building on each other's work. Attribution of credit is a delicate issue, because any listing of names is bound to annoy some subset of those involved. Here's a list of some of the people in random order: M. Shaposhnikov, A. Nelson, D. Kaplan, A. Cohen, M. Dine, P. Huet, L. Susskind, B. Singleton, L. McLerran, N. Turok, J. Zadrozny, M. Joyce, T. Prokopec, M. Voloshin, B. Liu. For reviews, see ref. [4].

is proportional to ϕ. So it's easy to see that this temperature condition could be satisfied outside the bubble but not inside, if ϕ is big enough inside.)

The bubble does not expand in a vacuum; the hot, early universe is filled with a hot plasma of quarks, gluons, W bosons, and so forth. Now imagine what happens as quarks and anti-quarks in the plasma outside the bubble wall collide with the expanding bubble wall. In each collision, there will be some transmission probability that the particle makes it through the wall, and some reflection probability that it bounces back away from it. If C and CP are violated, then there is no reason that the reflection probability for quarks has to be exactly the same as that for anti-quarks. Let's imagine that things work out so that anti-quarks are reflected with slightly higher probability than quarks. Corresponding, quarks would be transmitted with slightly higher probability than anti-quarks. After a little bit, an excess of anti-quarks over quarks will build up in a diffusive shell leading the expanding bubble wall, while there will be an excess of quarks over anti-quarks inside the bubble. This is depicted by the regions "$B > 0$" and "$B < 0$" in fig. 4b.

If baryon number were conserved in the standard model, nothing interesting would result from all this. The total B in the universe wouldn't change, and eventually everything would get smoothed out again after the transition was over. But B is violated. Remember also that *in equilibrium* B gets driven to zero. Now imagine a tiny region of space inside the diffusive shell of anti-quark excess depicted in fig. 4b. The processes taking place there are doing their best to *locally* restore equilibrium—that is, to restore $B = 0$. So B violation is biased inside the diffusive shell to convert anti-quarks into quarks.

You might at first imagine that the exact opposite is happening inside the bubble: that B violation is trying to restore $B = 0$ there by turning the excess quarks into anti-quarks. But remember that the rate of B violation is insignificant in the symmetry-broken phase, so this doesn't happen. The whole scenario of fig. 4b is therefore an engine that biases the B violation that's taking place during the transition to be converting more anti-quarks into quarks than visa versa. As a result, having originally started in equilibrium with $B = 0$ before the transition, we'll end up with a net, positive B once the transition is complete. That's all there is to standard model baryogenesis.

IMPLICATION FOR CP VIOLATION

Does the forgoing mean that the CP violating phase in the CKM matrix ends up ultimately predicting the baryon to photon ratio of today's universe, $n_B/n_\gamma \sim 10^{-9}$? Unfortunately not. It seems that new sources of CP violation are required in the Higgs sector if the Standard Model is to explain baryogenesis. The CP violation associated with the quarks just doesn't couple strongly enough to the physics of the expanding bubble wall to produce enough baryons in the universe today.

This leads to a quibble about what I meant when I called these scenarios "Standard Model" baryogenesis. What's ruled out as an explanation of baryogenesis is the Minimal Standard Model—the Standard Model with a single doublet Higgs. For me, "Standard Model" means the standard forces plus God-knows-what for the Higgs sector. As it turns out, new CP violation in the Higgs sector is a natural possibility if one introduces more than a single Higgs doublet. It appears that the CP violation one needs can be accommodated in two-doublet Higgs models. It can also be accommodated in the Minimal Supersymmetric Standard Model (which, by the way, is required to have two Higgs doublets).

Actually, it turns out that a non-minimal Higgs sector is required for another reason, if Standard Model baryogenesis is to work. It turns out that, in the Minimal Standard Model, the current experimental limit on the Higgs mass rules out the possibility of a supercooled first-order phase transition. But it's perfectly possible to have such a transition in non-minimal models.

For more on the subject of this talk, try the reviews in ref. [4].

REFERENCES

1. E. Kolb and M. Turner, *The Early Universe*: Addisson-Wesley, 1990.
2. P. Arnold, "An Introduction to Baryon Violation in Standard Electroweak Theory" in *Testing the Standard Model: Proceedings of the 1990 Theoretical Advanced Study Institute in Elementary Particle Physics*, eds. M. Cvetič and P. Langacker: Singapore, 1991.
3. N. Manton, *Phys. Rev.* **D28**, 2019 (1983); V. Kuzmin, V. Rubakov, and M. Shaposhnikov, *Phys. Lett.* **B155**, 36 (1985).
4. A. Cohen, D. Kaplan and A. Nelson, *Annu. Rev. Nucl. Part. Sci.* **43**, 27 (1993); V. Rubakov and M. Shaposhnikov, hep-ph/9603208, *Usp. Fiz. Nauk* **166**, 493 (1996) [*Phys. Usp.* **39**, 461 (1996)]; M. Trodden, hep-ph/9803479, Case Western report no. CWRU-P6-98.

CP Violation in Strange Baryon Decays: A Report from Fermilab Experiment 871

C. James,[a] R. A. Burnstein,[e] A. Chakravorty,[e] A. Chan,[b]
Y. C. Chen,[b] W. S. Choong,[c] K. Clark,[i] E. C. Dukes,[j] C. Durandet,[j]
J. Felix,[d] R. Fuzesy,[c] G. Gidal,[c] H. R. Gustafson,[g] C. Ho,[b]
T. Holmstrom,[j] M. Huang,[j] M. Jenkins,[i] D. M. Kaplan,[e]
L. M. Lederman,[e] N. Leros,[f] M. J. Longo,[g] F. Lopez,[g] W. Luebke,[e]
K. B. Luk,[c] G. Moreno,[d] K. Nelson,[j] V. Papavassiliou,[h]
J. P. Perroud,[f] D. Rajaram,[e] H. A. Rubin,[e] M. Sosa,[d] P. K. Teng,[b]
B. Turko,[c] J. Volk,[a] C. G. White,[e] S. L. White,[e] and P. Zyla[c]

[a] *Fermi National Accelerator Laboratory, Batavia, Illinois 60510*
[b] *Academia Sinica, Nankang, Taipei 11529, Taiwan, R.O.C.*
[c] *University of California and Lawrence Berkeley Laboratory, Berkeley, California 94720*
[d] *University of Guanajuato, 37000 Leon, Mexico*
[e] *Illinois Institute of Technology, Chicago, IL 60616*
[f] *University of Lausanne, CH-1015 Lausanne, Switzerland*
[g] *University of Michigan, Ann Arbor, Michigan 48109*
[h] *New Mexico State University, Las Cruces, New Mexico, 88003*
[i] *University of South Alabama, Mobile, Alabama 36688*
[j] *University of Virginia, Charlottesville, Virginia 22901*

Abstract. Fermilab experiment 871, *HyperCP*, is a search for direct CP violation in Ξ and Λ hyperon decays. A non-zero value in the asymmetry parameter $\mathcal{A}$, defined in terms of the decay parameter products $\alpha_\Xi \alpha_\Lambda$ and $\alpha_{\bar{\Xi}} \alpha_{\bar{\Lambda}}$, would be unambiguous evidence for direct CP violation. The first data-taking run finished at the end of 1997 and accumulated over one billion Ξ^- and $\bar{\Xi}^+$ decays. A sensitivity in $\mathcal{A}$ of $\approx 10^{-4}$ is expected. A review of CP violation in hyperon decays is given, the *HyperCP* detector is described, and the status of the data analysis is discussed.

INTRODUCTION

Using the hyperon system to look for evidence of T or CP symmetry violation is not a new idea. In fact it is an idea that has been around some 40 years. In the late 1950s, parity non-conservation was discovered [1] and the impact on hyperon decays was developed and shown [2]. An idea to look for CP violation in hyperon decays [3] led to a few experimental searches after CP violation was discovered in K^0 decays in the early 1960s [4]. Practical experimental efforts were hampered, however, by statistics: making quantities of hyperons was difficult, given the accelerators available in the 1960s and 1970s, and given the hyperon production cross section at those accelerator energies. The high statistics are required to get to the level of CP-odd effects predicted by most models. Within the past 15 years, there have been three experiments which searched for CP violation in Λ decay [5–7]. But none were designed specifically for studying CP effects, and the results are statistics limited. With the current generation of proton accelerators, along with recent improvements in detector hardware, it is possible to collect large numbers of hyperons in a reasonably short time. *HyperCP* is the first *dedicated* hyperon CP violation experiment.

PHENOMENOLOGY

All hyperons, spin-$\frac{1}{2}$ strange baryons, decay into a spin-$\frac{1}{2}$ baryon and a pion, reducing strangeness by 1 [1]. Both *S*- and *P*-wave final states are allowed by conservation of angular momentum, and only the *P*-wave final state is allowed by conservation of parity. But parity is not conserved in these weak decays, so the decay goes into a mixture of *S*- and *P*-wave final states. Hyperon non-leptonic decays are conventionally described by the Lee-Yang variables, defined in terms of these *S*- and *P*-wave amplitudes [8].

$$\alpha = \frac{2Re(S^*P)}{|S|^2 + |P|^2} \quad (1)$$

$$\beta = \frac{2Im(S^*P)}{|S|^2 + |P|^2} \quad (2)$$

$$\gamma = \frac{|S|^2 - |P|^2}{|S|^2 + |P|^2}, \quad (3)$$

where the decay parameters $\alpha^2 + \beta^2 + \gamma^2 = 1$. The parity violation in the decays shows up as a forward-backward asymmetry in the distribution of the daughters, quantified by the parameter α :

$$\frac{dN}{d\Omega} = \frac{1}{4\pi}(1 + \alpha \vec{P}_p \cdot \hat{p}_d) \quad (4)$$

where $\vec{P}_p$ is the parent hyperon polarization, and $\frac{dN}{d\Omega}$ and $\hat{p}_d$ are the daughter baryon distribution and the daughter baryon momentum unit vector, both in the parent's rest frame. The polarization of the daughter baryon is in general given by:

---

[1] Except the Σ^0, which decays electromagnetically into Λ-γ with $\Delta S = 0$.

$$\vec{P}_d = \frac{(\alpha + \vec{P}_p \cdot \hat{p}_d)\hat{p}_d + \beta(\vec{P}_p \times \hat{p}_d) + \gamma(\hat{p}_d \times [\vec{P}_p \times \hat{p}_d])}{1 + \alpha \vec{P}_p \cdot \hat{p}_d}, \tag{5}$$

which simplifies to

$$\vec{P}_d = \alpha \, \hat{p}_d \tag{6}$$

if the parent is unpolarized.

Under a CP operation, a daughter baryon emitted in the forward direction (as defined by the parent polarization vector) is transformed into a daughter antibaryon emitted in the backward direction. So if CP is conserved, $\alpha = -\overline{\alpha}$, and, following a similar argument, $\beta = -\overline{\beta}$. Differences in the magnitude of either α or β for a particle-antiparticle pair indicate direct CP violation.

In practice, experimental measurements of β to the necessary precision are difficult, requiring the measurement of both the daughter and parent polarization. Measurement of α can be made simple by utilizing a feature of hyperon production dynamics. In a fixed target experiment using protons, Ξs are produced through the reaction $p + \text{Nucleon} \to \Xi^\pm + X$. A Ξ emerging with $p_t \cong 0$ is unpolarized due to parity conservation in the strong interaction which produced it. The Ξ decays to a Λ and π, with, following equation (6), the Λ in a helicity state with known polarization:

$$\vec{P}_\Lambda = \alpha_\Xi \hat{p}_\Lambda. \tag{7}$$

The Λ decays in turn to a proton and π, with the proton distribution given by equation (4), which is rewritten as

$$\frac{dN}{d\cos\theta} = \frac{1}{2}(1 + \alpha_\Lambda \alpha_\Xi \cos\theta) \tag{8}$$

where θ is the angle between the proton momentum and Λ polarization vectors in the Λ rest frame. The decay parameter product $\alpha_\Xi \alpha_\Lambda$ is extracted by measuring the slope of the decay angle θ distribution. A CP observable $\mathcal{A}$ is defined as

$$\mathcal{A} = \frac{\alpha_\Xi \alpha_\Lambda - \alpha_{\overline{\Xi}} \alpha_{\overline{\Lambda}}}{\alpha_\Xi \alpha_\Lambda + \alpha_{\overline{\Xi}} \alpha_{\overline{\Lambda}}} \simeq A_\Xi + A_\Lambda. \tag{9}$$

$\mathcal{A}$ is determined by measurement of $\alpha_\Xi \alpha_\Lambda$ from Ξ^- decays and $\alpha_{\overline{\Xi}} \alpha_{\overline{\Lambda}}$ from $\overline{\Xi}^+$ decays. Using this method, one cannot distinguish whether an observed asymmetry originates in the decay of the Ξ or the Λ, since α_Ξ and α_Λ are not measured separately.

Calculating A_Ξ and A_Λ is difficult and model dependent, requiring knowledge of poorly understood hadronic matrix elements [9]. Published predictions for A_Ξ and A_Λ range in magnitude from 1×10^{-5} to 6×10^{-4} [10–15]. Models with no $\Delta S = 1$ CP-odd effects, such as the superweak model of Wolfenstein [16], predict no asymmetry.

THE *HYPERCP* EXPERIMENT

CP violation in hyperon decay is best studied using a simple, high rate spectrometer. High rate is required to achieve the required statistics, while simplicity is demanded to keep systematic effects small and controllable. The HyperCP spectrometer is shown schematically in Figure 1.

The Spectrometer

Ξs are produced by steering an 800 GeV proton beam into a 2 mm square target. Immediately following the target is a curved collimator imbedded in a 6 m long 1.6-Tesla dipole magnet. The collimator serves the dual purpose of selecting a narrow momentum range of charged particles while stopping all primary protons which did not interact in the target. The mean momentum selected is about 160 GeV/c, at which the Ξ to charged particle ratio is approximately maximal. The charge of the secondary beam is selected with the sign of the dipole field; Ξ^- and $\bar{\Xi}^+$ decays cannot be collected simultaneously. A typical primary beam intensity of 1.5×10^{11} protons per 20s spill gives a secondary beam rate of 20 MHz, dominated by protons, pions, and kaons. By aiming the primary protons at 0 degrees wrt the long axis of the collimator, the Ξ accepted through the channel will have zero polarization.

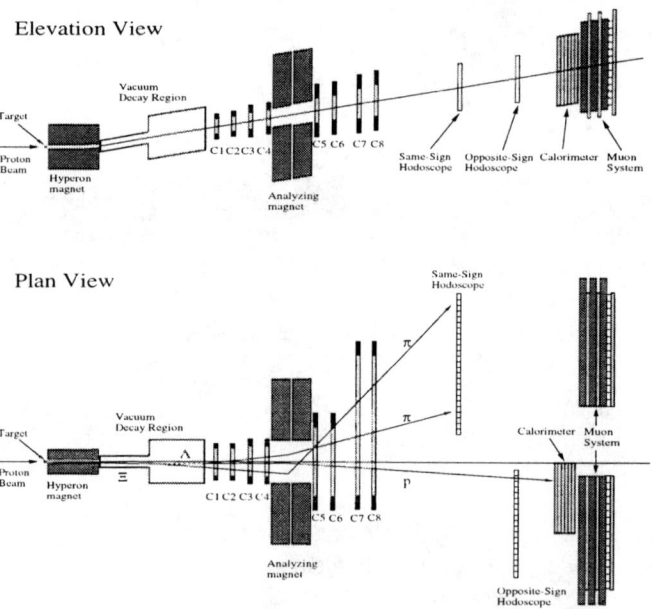

FIGURE 1. Schematic diagram for the *HyperCP* spectrometer.

Upon exiting the collimator, the secondary beam traverses a 13 meter evacuated decay region where most of the hyperons decay. After the decay region, the beam enters the spectrometer proper, made from high rate, narrow pitch proportional wire chambers. The analyzing magnet has sufficient strength to insure that the proton from the Λ decay is always bent to one side (the 'opposite sign' side) and the two pions from the Ξ and Λ decays are always bent to the other ('same sign') side, and both are well separated from the intense secondary beam. Farther downstream are two scintillation hodoscopes, a hadronic calorimeter used as a trigger element, and a muon detection station. Helium bags are positioned between all the detector components to reduce multiple scattering and secondary interactions.

The Ξ trigger simply requires a coincidence of at least 1 hit in each of the hodoscopes, along with a calorimeter energy signature; the calorimeter threshold is set not only to exclude muons, but also to reduce triggers caused by low energy hadrons produced from beam interactions in the spectrometer material. By reversing the polarities of both the collimator and analysis magnets, Ξ^- and $\bar\Xi^+$ decays are collected with the identical trigger. A typical Ξ trigger rate is 30 kHz. Additional triggers are used to select kaon and muon events, for detector monitoring and systematics studies. A typical overall trigger rate is 75 kHz.

The sheer size of the data set to be recorded mandates a highly capable data acquisition system. Simplicity and parallelism in the readout and DAQ are the hallmarks [17]. Average dead time per event is 3 μs. The average event size during standard running is 550 bytes. With the trigger rate listed above, $HyperCP$ produces data at 41 MB/sec during spill. Spill is delivered 19 seconds every minute for an average data to tape rate of 13 MB/sec. The maximum sustainable data rate to tape achieved by $HyperCP$ is 17 MB/sec (a world record for a HEP experiment). Despite the high rate of the secondary beam and trigger, most events reconstruct cleanly, as shown in Figure 2.

Status of the Analysis

The 1996-1997 fixed target program at FNAL came to an end on September 5th 1997. By that time, $HyperCP$ had written to tape 75 billion triggers of all types. 39 billion Ξ triggers have been recorded on 38 TB of storage. Preliminary analysis of $\sim$ 10 percent of the data indicates that we have accumulated 1.2 billion $\Xi^- \to \Lambda\pi^-$ decays and 0.3 billion $\bar\Xi^+ \to \bar\Lambda\pi^+$ decays. 280 million $K^\pm \to 3\pi$ decays, as well as a large number of $\Omega \to \Lambda K$ decays were also collected. The estimated statistical uncertainty for the CP violation observable $\mathcal{A}$ with this data set is $\sim 2 \times 10^{-4}$. Additional smaller data sets are available with different target materials (Cu, W, and Be) to study A dependence in hyperon yield, and with non-zero production angles to study polarized hyperons. Event reconstruction of the full data set is currently underway. A preliminary CP measurement is not yet available. However, to give an idea of the quality of the data, Figure 3 shows the reconstructed $\Lambda\pi$ invariant masses from about 20 million Ξ events of both polarities. The mass resolution is about 1.6 GeV/c^2. There is good agreement between the Ξ^- and $\bar{\Xi}^+$ distributions, and low background.

Systematics

A precision measurement can only be successful if systematic effects can be controlled and understood. In general, the potential sources of bias are: acceptance differences between Ξ^- and $\bar{\Xi}^+$ decays, non-zero polarization of the parent Ξ, and differences in the particle anti-particle interaction rate within the spectrometer (i.e. p versus $\bar{p}$ interaction cross sections), and differences in the backgrounds under the Ξ^- and $\bar{\Xi}^+$ and the Λ and $\bar{\Lambda}$ mass distributions.

Because the α parameters for the Ξ and Λ both change sign under CP, the proton decay distribution in the Λ rest frame should be identical to the anti-proton decay distribution in the $\bar{\Lambda}$ rest frame (assuming the parent Ξs are unpolarized and CP is conserved). One might think no acceptance corrections are required; however the production mechanisms and therefore the momentum spectra of Ξ^- and $\bar{\Xi}^+$ are different, so at least that correction is required. Implicit in the goal to minimize other acceptance corrections is that the magnetic fields, detector components, and reconstruction efficiencies remain stable between positive running ($\bar{\Xi}^+$ data) and negative running (Ξ^- data). To facilitate stability, the running polarity is changed frequently (about every 4 hours at full beam intensity), the magnetic fields are monitored with high precision Hall probes, and a uniform secondary beam flux is

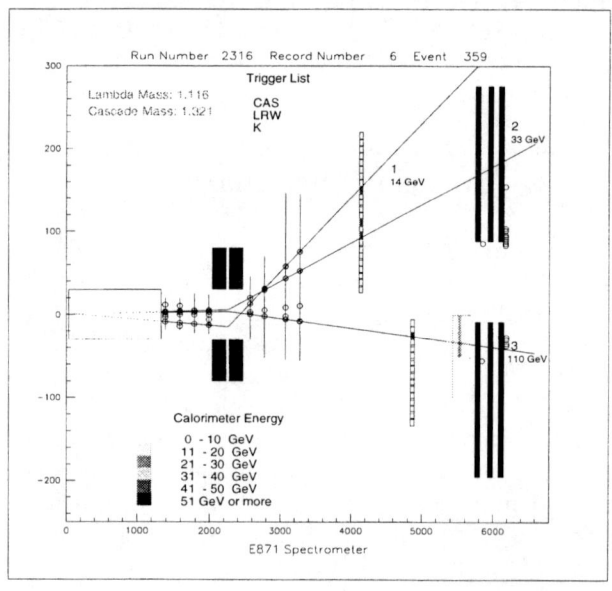

FIGURE 2. Event display for a reconstructed Ξ decay.

maintained by using different length targets for the two beam polarities. Given that real experiments are never ideal, acceptance variations will be tracked and corrected as the data analysis proceeds.

The $\Xi \to \Lambda\pi$; $\Lambda \to p\pi$ decay topology naturally reduces systematic acceptance errors because the Λ polarization vector is not fixed with respect to the lab coordinates. In the Ξ rest frame, the Λ decays throughout the 4π solid angle. Each value of θ in equation (8), therefore, probes a large geometric region within the spectrometer, such that local chamber inefficiencies and geometric acceptance variations map across a wide range of $\cos\theta$. This effect, along with high acceptance for all Ξ decays, significantly dilutes potential biases due to acceptance differences.

Due to mistargeting and the finite size of the hyperon channel, some parent Ξs may be produced with non-zero polarization. The magnitude of this polarization, although expected to be quite small, will be measured and any bias can, in principle, be removed (if required). For these events, the parent Ξ will have a small, fixed polarization in the lab frame; however, in the Λ rest frame, the effects of this polarization will be diluted as discussed above.

A real-life estimate of experimental errors using differing samples of Ξ^- decays confirms the ruggedness of this method. Pairs of data sets taken with various values of p_t and $\vec{p}_\Xi$ are compared without making any acceptance corrections. Since only Ξ^- events are used, all observed difference are due to biases. When the samples differ significantly in mean momentum, biases are observed; however, as these differences diminish (e.g. $\Delta p < 20$ GeV/c), no statistical differences in the

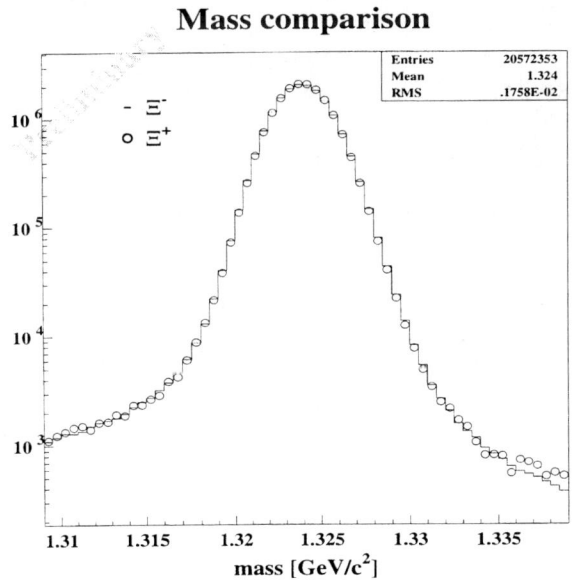

FIGURE 3. The $\Lambda\pi$ invariant mass for positive and negative Ξ trigger events.

$\cos\theta$ distributions are present down to a few parts per thousand. No statistical differences are found in samples with similar Ξ momentum, yet widely differing polarizations. For example, Figure 4 shows the $\cos\theta$ distributions in the Λ helicity frame for two data sets where the Ξ^- were produced at opposite non-zero production angles, giving them opposite and relatively large polarizations. Studies so far show that without corrections, most biases should not contribute to a false asymmetry in $\mathcal{A}$ above the 10^{-4} level.

Finally, differences in particle and anti-particle interaction cross sections are currently being studied. Corrections will be required; however, we expect to understand this bias to $\sim 1 \times 10^{-4}$. Differences in the backgrounds of the reconstructed hyperons and anti-hyperons are small, and are also being studied.

FUTURE PROSPECTS AND ACKNOWLEDGEMENTS

The *HyperCP* collaboration is preparing for a second run in 1999. Modest upgrades to the detector and DAQ should enable a four-fold increase in statistics resulting in a sensitivity in $\mathcal{A}$ of better than 1×10^{-4}.

The *HyperCP* collaboration gratefully acknowledges the efforts of the FNAL staff in providing us with a good run in 1997, and for their assistance in preparations for 1999 running. This work was supported in part by the U.S. Dept. of Energy, the National Science Foundation, and the National Science Council of Taiwan, R.O.C.

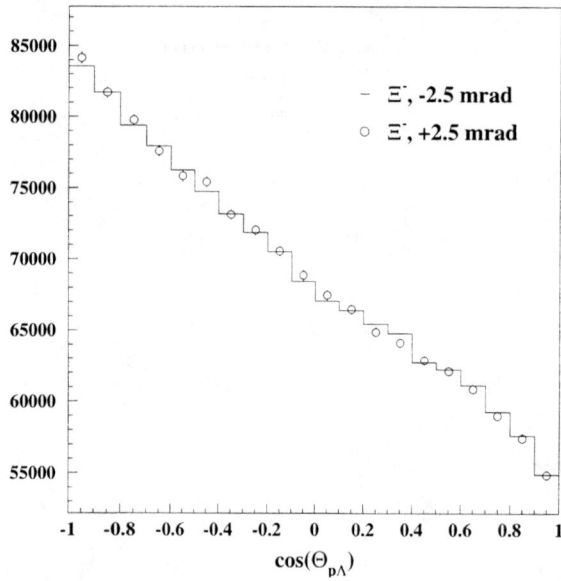

FIGURE 4. The proton $\cos\theta$ distributions in the Λ helicity frame for Ξ^- events from ± 2.5 mrad production angles.

REFERENCES

1. C.S. Wu et al., Phys. Rev. **105** 1413 (1957).
2. T.D. Lee and C.N. Yang, Phys. Rev. **104** 254 (1956), F. Crawford et al., Phys. Rev. **108** 1102 (1957), F. Eisler et al., Phys. Rev. **108** 1353 (1957), R. Adair and L. Leipuner, Phys. Rev. **109** 1358 (1958).
3. A. Pais, Phys. Rev. Lett. **3**, 242 (1959).
4. J.H. Christenson et al., Phys. Rev. Lett. **13** 138 (1964).
5. P. Chauvat et al., Phys. Lett. B **163**, 273 (1985).
6. M.H. Tixier et al., Phys. Lett. B **212**, 523 (1988).
7. P.D. Barnes et al., Phys. Lett. B **199**, 147 (1987); P.D. Barnes et al., Phys. Rev. C **54**, 1877 (1996).
8. T.D. Lee and C.N. Yang, Phys. Rev. **108** 1645 (1957).
9. X.G. He and G. Valencia, Phys. Rev. **D52**, 5257 (1995).
10. J.F. Donoghue, and S. Pakvasa, Phys. Rev. Lett. **55** 162 (1985).
11. J.F. Donoghue, X.G. He, and S. Pakvasa, Phys. Rev **D34** 833 (1986).
12. D. Chang, X.G. He, and S. Pakvasa, Phys. Rev. Lett. **74** 3927 (1995).
13. X.G. He, H. Steger, and G. Valencia, Phys. Lett. **B272** 411 (1991).
14. J.F. Donoghue, Third Conf. on the Intersections between Particle and Nuclear Physics, Rockport ME, May 1988.
15. X.G. He, Proceedings of DPF94, Albuquerque, New Mexico, August 1994.
16. L. Wolfenstein, Phys. Rev. Lett. **13**, 562 (1964).
17. D.M. Kaplan et al., Sixth Annual LeCroy Conference on Electronics for Particle Physics, Chestnut Ridge, NY, May 28-29, 1997.

The status of the KLOE experiment

Giovanni Bencivenni

INFN - Laboratori Nazionali di Frascati,
Via Enrico Fermi 40, 00044 Frascati, Italy
E-mail: Giovanni.Bencivenni@lnf.infn.it

representing the KLOE Collaboration:

M. Adinolfi, F. Ambrosino, A. Aloisio, A. Andryakov, A. Angeletti, A. Antonelli,
C. Bacci, R. Baldini-Ferroli, G. Barbiellini, G. Bencivenni, S. Bertolucci, C. Bini,
C. Bloise, V. Bocci, F. Bossi, P. Branchini, G. Cabibbo, A. Calcaterra, R. Caloi,
P. Campana, G. Capon, G. Carboni, M. Carboni, A. Cardini, C. Carusotti, G. Cataldi,
F. Ceradini, F. Cervelli, F. Cevenini, G. Chiefari, P. Ciambrone, S. Conticelli, E. De Lucia,
R. de Sangro, P. De Simone, G. De Zorzi, S. Dell'Agnello, A. Denig, A. Di Domenico,
S. Di Falco, A. Doria, F. Donno, E. Drago, V. Elia, L. Entesano, O. Erriquez, A. Farilla,
G. Felici, A. Ferrari, M. L. Ferrer, G. Finocchiaro, D. Fiore, C. Forti, G. Foti,
A. Franceschi, P. Franzini, M. L. Gao, C. Gatti, P. Gauzzi, S. Giovannella, V. Golovatyuk,
E. Gorini, F. Grancagnolo, E. Graziani, P. Guarnaccia, U. v. Hagel, H. G. Han, S. W. Han,
M. Incagli, L. Ingrosso, Y. Y. Jiang, W. Kim, W. Kluge, V. Kulikov, F. Lacava,
G. Lanfranchi, J. Lee-Franzini, T. Lomtadze, C. Luisi, C. S. Mao, A. Martini, W. Mei,
L. Merola, R. Messi, S. Miscetti, A. Moalem, S. Moccia, F. Murtas, M. Napolitano,
A. Nedosekin, P. Pagès, M. Palutan, L. Paoluzi, E. Pasqualucci, L. Passalacqua,
M. Passaseo, A. Passeri, V. Patera, E. Petrolo, G. Petrucci, D. Picca, M. Piccolo,
G. Pirozzi, L. Pontecorvo, M. Primavera, F. Ruggieri, P. Santangelo, E. Santovetti,
G. Saracino, R. D. Schamberger, C. Schwick, B. Sciascia, A. Sciubba, F. Scuri, I. Sfiligoi,
S. Sinibaldi, T. Spadaro, S. Spagnolo, E. Spiriti, C. Stanescu, L. Tortora, E. Valente,
P. Valente, G. Venanzoni, S. Veneziano, D. Vettoretti, S. Weseler, Y. G. Xie, C. D. Zhang,
J. Q. Zhang, P. P. Zhao

Abstract. The main physics objectives of the KLOE experiment at DAΦNE are reviewed. The design and status of the KLOE detector and the DAΦNE collider are discussed.

INTRODUCTION

A high luminosity ϕ-factory is a powerful laboratory for the study of *CP*, *T* and *CPT* physics in *K* decays. The Frascati ϕ-factory, DAΦNE, and the KLOE detector are almost complete, and data taking will start early 1999. DAΦNE is an e^+e^- collider designed to achieve a peak luminosity of more than 10^{32} cm^{-2}s^{-1} at the ϕ resonance ($E \simeq 1020$ MeV), corresponding to a ϕ production rate of a few kHz. Kaons, K_L-K_S, from ϕ decays are practically monochromatic, back-to-back and in a pure quantum state. A very interesting consequence (unique to a ϕ-factory) of these properties is the availability, for the first time, of a pure, tagged K_S beam.

KLOE is a general purpose detector optimised to measure the real part of ε'/ε with statistical and systematic accuracies of O(10^{-4}), with a complementary approach with respect to experiments at hadron machines. In addition KLOE will allow a whole spectrum of other measurements, such as T and CPT tests. Because of the tagged K_S beam, studies of semileptonic and CP violating K_S decays as well as several rare K_S decays will be possible.

In this paper, after a short review of the main physics objectives of the KLOE experiment, I will discuss the design and status of the KLOE detector as well as the DAΦNE machine.

THE DAΦNE ϕ-FACTORY

The measurement of the real part of ε'/ε to a statistical accuracy of 10^{-4} requires a DAΦNE luminosity of $\sim 5 \times 10^{32}$ cm^{-2}s^{-1}. The strategy to achieve this luminosity is to circulate up to 120 + 120 bunches in two separate rings interesecting in two regions only, in order to minimise beam-beam interaction. The accelerator complex [1] consists of a superconducting LINAC, which accelerates electrons and positrons to 510 MeV, followed by the accumulator ring, which stores up to 120 bunches of each type of particles.

During the machine commissioning the maximum current obtained in single bunch mode was about 110 mA for both positrons and electrons, without any active feedback. The design current in each bunch for two beams operation is 44 mA. In multi-bunch mode up to $\sim$ 0.5 A have been stored in about 25 ÷ 30 bunches for both electrons and positrons. Stable collisions have been reached in single bunch mode with a luminosity of 1.4×10^{30} cm^{-2}s^{-1}, to be compared with the design luminosity of 4×10^{30} cm^{-2}s^{-1}. In multi-bunch mode, with 13 bunches stored in each beam (with a maximum current of 200 mA per beam), a luminosity in the range of 10^{31} cm^{-2}s^{-1} has been obtained.

PHYSICS AT DAΦNE

At DAΦNE ϕ mesons are produced in a very clean environment with a cross section of $\sim$ 4.5 μb, to be compared with a $\sigma(e^+e^- \to hadrons) \sim 0.17$ μb. Bhabha events, even though copiously produced ($\sigma(Bhabha) \sim$ 30.7 μb in the range $10° \leq \theta \leq 170°$), will be easily rejected or down-scaled at the trigger level. At full DAΦNE luminosity KLOE will collect about 2000 K_L-K_S pairs per second. K_L-K_S beams will be practically monochromatic, back-to-back and with identical and known intensity. In addition, thanks to the clean separation between the K_L-K_S fiducial volumes ($\lambda_L \sim 3.5$ m, $\lambda_S \sim 6$ mm), a very efficient and high purity kaon tagging will be possible: the detection of a K_S (K_L) implies the presence of a K_L (K_S) in the opposite direction. Moreover, because ϕ mesons are produced in a pure C-odd quantum state, and C is conserved in the strong decay of the ϕ, the final state is an antisymmetric superposition of $K_L K_S$ and $K_S K_L$. As a consequence the final decay products will display quantum mechanical interference patterns [2]. By studying the time dependence of these patterns it is possible to isolate CP, T and CPT violating effects in the decays of the K_L and K_S, in a way that is unique to a ϕ-factory and complementary to the CP experiments at hadronic machines. Of course symmetry violation tests can be performed also by studying the usual time-integrated asymmetries. In particular the direct CP violation parameter, $\Re(\varepsilon'/\varepsilon)$, can be measured by means of the well known double ratio.

Besides the study of the standard CP physics, DAΦNE will be competitive in measuring all $K^0\bar{K}^0$ phenomenological parameters, to search for K_S rare decays at the 10^{-8} single event sensitivity, to observe (for the first time) CP violation in the $K_S \to \pi^0 \pi^0 \pi^0$. For the canonical value of the BR($K_S \to \pi^0 \pi^0 \pi^0$) $\sim 10^{-9}$, only ~ 30 events are expected.

Symmetry violation by quantum interferometry

Given two generic final states f_1 and f_2 into which the two kaons decay, at times t_1 and t_2 respectively, the decay intensity as a function of the time difference $\Delta t = t_1 - t_2$ is given, for $\Delta t > 0$, by:

$$I(f_1, f_2; \Delta t > 0) = \frac{|\langle f_1|K_S\rangle\langle f_2|K_L\rangle|^2}{2\Gamma} \times \{|\eta_1|^2 e^{-\Gamma_L} + |\eta_2|^2 e^{-\Gamma_S} - 2|\eta_1||\eta_2|e^{-\Gamma|\Delta t|/2}\cos(\Delta m|\Delta t| + \varphi_1 - \varphi_2)\} \quad (1)$$

with $\Gamma = \Gamma_S + \Gamma_L$, $\eta_j \equiv |\eta_j|e^{i\varphi_j} = \langle K_L|f_i\rangle/\langle K_S|f_i\rangle$. A similar expression holds for $\Delta t < 0$. A whole spectrum of precision kaon-interferometry experiments can be performed at DAΦNE by measuring the above decay intensity for appropriate choices of the final states f_1 and f_2. With $f_1 = \pi^+\pi^-$ and $f_2 = \pi^0\pi^0$ (fig. 1) one can measure the CP violating parameters: $\Re(\varepsilon'/\varepsilon)$, at $|\Delta t| \gg 0$, and $\Im \varepsilon'/\varepsilon$, at $|\Delta t| \sim 0$.

T and CPT violation tests can be performed by studying the interference patterns for $f_1 = \pi\pi$ and $f_2 = \pi^\pm l^\pm \nu$; CPT can be studied using semileptonic final states (fig. 2): $f_1 = \pi^+ l^- \bar{\nu}$ and $f_1 = \pi^- l^+ \nu$.

Time integrated asymmetries

Besides time-dependent studies, in a ϕ-factory one can measure directly some symmetry violation parameters by looking at differences in the semileptonic decay rates of the kaons.

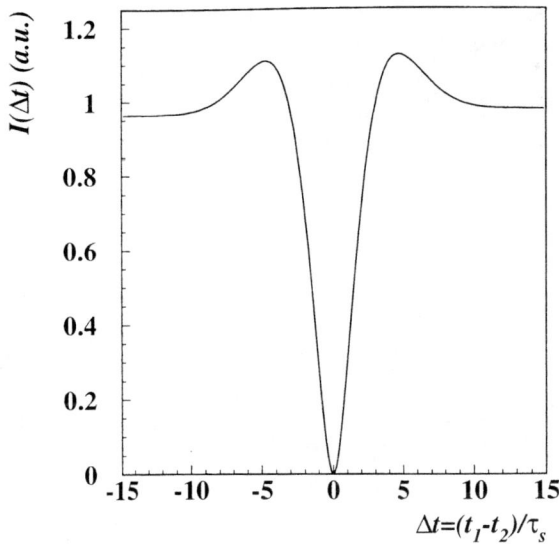

FIGURE 1. Interference pattern from $\phi \to K_S K_L \to \pi^+\pi^-\pi^0\pi^0$.

At present, only the CP-violating asymmetry in the rates of the semileptonic decays of the K_L, has been measured: $\mathcal{A}_L=(3.30\pm 0.12)\cdot 10^{-3}$. At a ϕ-factory, detecting the presence of a K_L, K_S tag, it will be possible, for the first time, to measure the semileptonic decay rates of the K_S (the expected $BR(K_{S,l3}) \sim 10^{-3}$) and the corresponding K_S asymmetry.

Assuming the validity of the $\Delta S = \Delta Q$ the semileptonic amplitudes for neutral kaons, can be written as:

$$A(K^0 \to \ell^+) = a + b \qquad A(\overline{K}^0 \to \ell^-) = a^* - b^* \qquad (2)$$

where $\Re(a) \neq 0$ conserves and $\Im(a) \neq 0$ violates both CP and CPT, $\Re(b) \neq 0$ violates CP and CPT and $\Im(b) \neq 0$ conserves CP and violates CPT. Using eqn. 2, the K_L and K_S asymmetries can be written as a function of the CP and CPT violating parameters:

$$\mathcal{A}_L = 2\Re\varepsilon - 2\Re\delta + 2\frac{\Re b}{\Re a} \quad and \quad \mathcal{A}_S = 2\Re\varepsilon + 2\Re\delta + 2\frac{\Re b}{\Re a} \qquad (3)$$

where $2\Re\delta$ and $2\frac{\Re b}{\Re a}$ account for CPT violation in the mass matrix and for "direct" CPT violation, respectively. If CPT is conserved, $\delta=0$ and $\mathcal{A}_S = \mathcal{A}_L$, otherwise:

$$\mathcal{A}_L - \mathcal{A}_S = -4\Re\delta \qquad (4)$$

At full DAΦNE luminosity, assuming a K_S tagging efficiency of $\sim 70\%$, KLOE will collect $\sim 10^7$ K_S semileptonic decays, then $\mathcal{A}_S$ will be measured with an accuracy

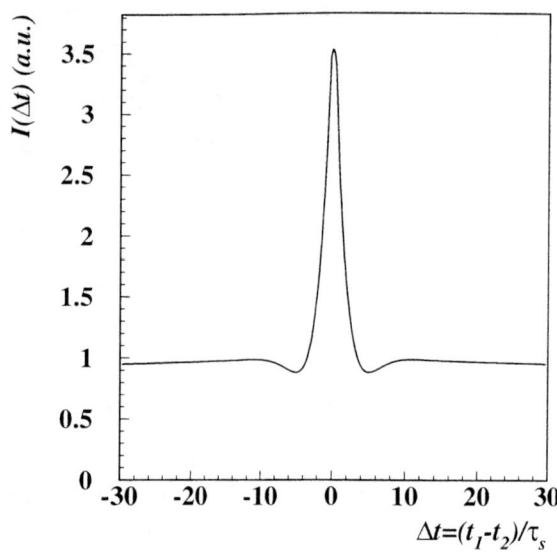

FIGURE 2. Interference pattern from $\phi \to K_S K_L \to \pi \ell \nu \pi \ell \nu$.

of $\sim 2.5 \times 10^{-4}$ (while $\mathcal{A}_L$ to $\sim 2 \times 10^{-5}$, about 5 times better than present measurement). Consequently $\Re\delta$ will be measured to an accuracy of $\sim 0.6 \times 10^{-4}$, almost ten times better than the *CPLEAR* measurement (albeit assuming the validity of the $\Delta S = \Delta Q$ rule). The $\Delta S = \Delta Q$ rule can be tested using strangeness tagged $K^0(\overline{K}^0)$ beams from charge exchange of $K^+(K^-)$ interactions in the beam pipe or the internal wall of the drift chamber ($\sim 10^8$ events expected in 1 year at full machine luminosity, ~ 10 times the *CPLEAR* statistics).

THE KLOE DETECTOR DESIGN

The Kloe detector [3,4] has been designed to study mainly the *CP* violation in the K_L ($K_L \to 2\pi$) and then to measure $\Re(\varepsilon'/\varepsilon)$ to a statistical and systematic accuracy of 10^{-4} via the double ratio (DR):

$$\mathcal{R} \equiv \frac{\mathcal{R}^{+-}}{\mathcal{R}^{00}} = \frac{\Gamma(K_L \to \pi^+\pi^-)}{\Gamma(K_S \to \pi^+\pi^-)} \cdot \frac{\Gamma(K_S \to \pi^0\pi^0)}{\Gamma(K_L \to \pi^0\pi^0)} = 1 + 6\Re\left(\frac{\varepsilon'}{\varepsilon}\right) \qquad (5)$$

The size of the detector, defining the fiducial volume (FV) for K_L decays, determines the fraction of collected events and then the statistical error on the measurement. The detector should measure the K_S and K_L path length with a good and uniform vertex resolution (for $\pi^0\pi^0$ - by detecting photons hitting the calorimeter - and $\pi^+\pi^-$ - by reconstructing the pion tracks), in order to count the events inside the defined fiducial volumes.

TABLE 1. backgrounds

Signal	Background	Signal/Bkgd	Rejection
$K_L \to \pi^0 \pi^0$	(1) $K_L \to \pi^0 \pi^0 \pi^0$	1/240	1.2×10^{-5}
	(2) $K_L \to \pi \mu \nu$	1/135	1.6×10^{-5}
$K_L \to \pi^+ \pi^-$	(3) $K_L \to \pi e \nu$	1/190	1.1×10^{-5}
	(4) $K_L \to \pi^+ \pi^- \pi^0$	1/60	5×10^{-5}

The non CP violating background must be rejected to a level of 10^{-5}. Each single systematical contribution that can affect the measurement of the DR, such as the knowledge of the combined efficiency (geometrical, detection, reconstruction), the error due to event miscounting in the proximity of the FV boundary, and the possible difference in the distance scales for charged and neutral decays (essentially for K_L), must be kept to a $10^{-4} \div 10^{-5}$ level.

For K_S the efficiencies are measured on the K_S data sample itself, and, because the FV can be considered infinite, there is no contribution to systematics from K_S event miscounting. For K_L the knowledge and the equalisation of the neutral and charged FVs can be performed exploiting the high statistics decay channels, $K_L \to \pi^- \pi^+ \pi^0$ and/or $K^\pm \to \pi^\pm \pi^0$ for which both neutral and charged decay vertices are present. The same reactions can be used to measure, and also monitor, the photon efficiency (once the charged vertex has been reconstructed) as well as the tracking efficiency (when photon from π^0 has been detected).

In Table.1 the main sources of background are given. The required rejection factors are achieved with a careful design and construction of the detector components: i) the rejection of the background reactions 1 and 4 asks for a very hermetic calorimeter, a high efficiency in the detection of low energy (E down to 20 MeV) photons, and very good energy resolution; ii) the K_{l3} background must be rejected on the basis of kinematical constraints, requiring a good and uniform tracking efficiencies and good momentum resolution at low momentum. An additional requirement is to minimise the K_S regeneration in the beam pipe or in the proximity of the internal wall of the tracking detector and to choose the FV in order to make negligible the regeneration effects on the measurement of the DR.

It must be noted that in the measurement of $\Re(\varepsilon'/\varepsilon)$ by the DR, the tagging efficiencies and the kaon beam intensities cancel completely, because they do not depend on the specific decay channel.

Status of the KLOE detector

The KLOE detector (fig. 3) is composed of a large cilindrical drift chamber surrounded by an electromagnetic calorimeter (barrel + end-caps), both immersed in a longitudinal magnetic field of ~ 0.6 T. Inside the internal cylinder of the drift chamber two small annular calorimeters, placed symmetrically with respect to the interaction point, have been installed around the permanent quadrupoles, to increase the $K_L \to \pi^0 \pi^0 \pi^0$ background reduction (a factor ~ 5 improved) [5]. The

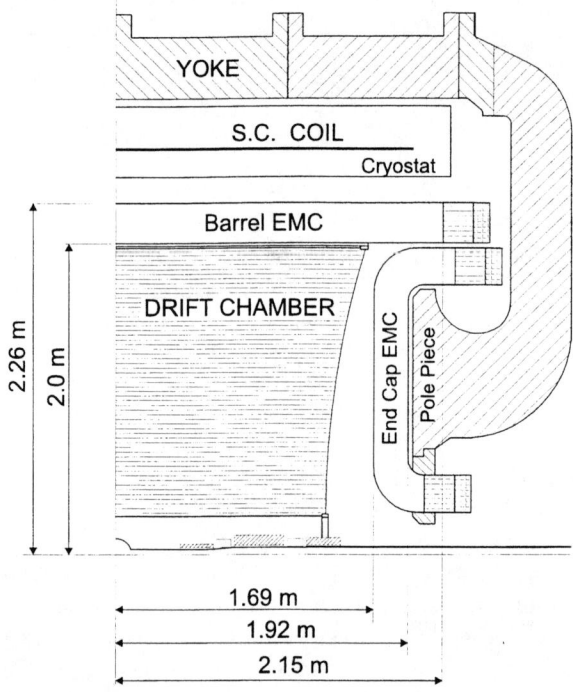

FIGURE 3. Schematic section of the KLOE detector.

beam pipe, realized in an aluminum-beryllium alloy, is 500 μm thick and has been designed with a spherical shape, 10 cm diameter, in order to work as an almost infinite fiducial volume for K_S.

The size of the drift chamber [6,7] is dictated by the need to collect as many K_L events as possible. The quasi isotropic angular distribution of the charged products requires a uniform filling of the sensitive tracking volume. In addition the tracking detector should be able to reconstruct the vertex of the charged decays of kaons at a level of ∼ 1 mm; to measure pion momenta, at 0.1% level, in the range 50 - 300 MeV/c^2, minimising multiple scattering; to be as transparent as possible to low energy photons (20 MeV) and to K_S in order to minimise K_S regeneration.

The tracking detector is a large cylindrical drift chamber (DC), ∼4 m diameter and 3 m length, with an almost square drift cell (∼ 2×2 cm² for the innermost 12 layers and ∼ 3×3 cm² for the outermost 46 layers) organized in 58 concentric alternating stereo layers. The whole mechanical structure is made of carbon fiber (X_0 ∼ 25 cm): the end-plates are spherical, 8 mm thick; the internal tube is 700 μm thick; the external 12 panels are realized as a carbon fiber/honeycomb/carbon fiber (2 mm/10mm/2 mm) sandwich. The global thickness, including the low mass front-end electronics (HV cards and pre-amplifiers), is less than 0.1 radiation lengths. The chamber is operated in a He-isobutane gas mixture (90-10); the ∼ 12600 sense wires, 25 μm diameter, are gold plated tungsten wires, while, to minimise multiple scattering, the ∼ 40000 field (+guard) wires, 80 μm diameter, are made of silver plated aluminum. The gas mixture and the wires have an overall radiation length of about 900 m.

The DC has been strung at the Laboratori Nazionali di Frascati, with a dedicated semi-automatic machine. The whole operation has been performed in a clean room and took about 11 months; the chamber was then closed and sealed to helium, and then moved to the assembly hall, where it has been inserted inside the rest of the detector. After the FEE installation and cabling the chamber has been debugged and tested, for about three months, with cosmic rays, using the calorimeter as trigger.

The main task of the electromagnetic calorimeter (EmC) is to measure the $K_L \to \pi^0 \pi^0$ and reject the $K_L \to \pi^0 \pi^0 \pi^0$. Hermeticity, high sergmentation and high efficiency in the detection of low energy photons (down to 20 MeV) are the main requirements for the calorimeter. The EmC [4] is a fine sampling lead/scintillating fiber calorimeter, with good energy resolution ($\Delta E/E = 4.7\%/\sqrt{E(\text{GeV})}$) and excellent time resolution ($\Delta t = 55\,ps/\sqrt{E(\text{GeV})}$). The lead absorber, 0.5 mm thick, is suitably shaped in order to embed the fibers, simulating a quasi-homogeneous structure. The barrel is composed by 24 modules, 4.5 m long; the end caps are divided into 32 C-shaped modules of different length. The overall coverage is $\simeq$ 99% of the solid angle. Each EmC module is read-out from both sides by fine-mesh photomultipliers: one every 4.5×4.5 cm^2, for a total number of ∼ 5000. The double side read-out allows the measurement of the z-coordinate of the photon impact point to ∼ 1 cm. The $K_L \to \pi^0 \pi^0$ decay vertex is measured with an accuracy of better than 1 cm using the time of flight of the photons and exploiting the low (∼ 0.2) K_L velocity, and the knowledge of the K_L direction from the detected $K_S \to \pi^+ \pi^-$. The EmC installed by the end of the 1997, is under continuous test with cosmic rays.

At the design luminosity the required trigger efficiency for the ϕ decays (∼ 5 kHz expected) must be > 99% . The Bhabha events, as well as the machine background and the cosmic rays must be rejected or down scaled by the trigger to a few kHz level. The trigger [8], working asynchronously (the beam crossing time is about 3 ns), is organised in two levels, using the local energy deposit on the EmC and the hit multiplicity in the DC. The level-1 trigger occurs within 150 ns after the beam-crossing and is mainly generated by the EmC (about 80% EmC + 20% DC), while the level-2 occurs after 850 ns from the level-1 and is based mainly on the DC information. The trigger rate, after the level-2, will be ∼ 10 kHz, while keeping a very high efficiency for CP violating events.

The DAQ system [9] has to handle about 25000 FEE channels, organized in 10 parallel VME crate chains, and has to handle up to 50 Mbytes/s. The overall DAQ and trigger system have been extensively tested during the last months, proving to largely fulfill the design requirements.

CONCLUSIONS

The commisionning of KLOE has been completed and the detector roll-in will begin by the end '98. Machine operation is expected to resume by early 1999 with a luminosity $>10^{31}$ cm^{-2}s^{-1}.

During its shake-down run KLOE will collect $>5\times10^8$ ϕ decays (about 1% of the final data), which will lead to the first physics results, while tuning-up the detector. Such a data will allow a measurement of the $\Re(\varepsilon'/\varepsilon)$ to 10^{-3} accuracy. At full DAΦNE luminosity CP violation will be measured to an accuracy of 10^{-4} with a new approach complementary to KTEV and NA48. The availability of a tagged K_S beam will allow, for the first time, measurements of the K_S semileptonic asymmetry, the observation of the CP violation in K_S decays and the search for K_S rare decays.

REFERENCES

1. G. Vignola in *Workshop on Physics and Detectors for DAΦNE 95*, ed. R. Baldini *et al.*, (SIS-Pubblicazioni, Frascati, 1995).
2. I. Dunietz, J. Hanser and J. Rosner, *Phys. Rev.* D **35**, 2166 (1987).
3. The KLOE Collaboration, *A General Purpose Detector for DAΦNE*, LNF-92/019 (1992).
4. The KLOE Collaboration, *The KLOE detector, Technical Proposal*, LNF-93/002 (1993).
5. The KLOE collaboration, submitted paper # 633, ICHEP98, Vancouver, Canada, 1998.
6. The KLOE Collaboration, *The KLOE Central Drift Chamber, Addendum to the Technical Proposal*, LNF-94/028 (1994).
7. G. Bencivenni for the KLOE Drift Chamber group, *The KLOE Drift Chamber*, presented at the VIII Vienna Wire Conference, Vienna, Feb. 23-27, 1998, and references therein.
8. The KLOE Collaboration, *The KLOE Trigger System, Addendum to the Technical Proposal*, LNF-96/043 (1996).
9. The KLOE Collaboration, *The KLOE Data Acquisition System, Addendum to the Technical Proposal*, LNF-95/014 (1995).

Towards results on direct CP violation in Kaon decays from the CERN-SPS experiment NA48

Augusto Ceccucci *

CERN CH-1211, Genève 23

*On behalf of the NA48 Collaboration: Cagliari, Cambridge, CERN, Dubna, Edinburgh, Ferrara, Firenze, Mainz, Orsay, Perugia, Pisa, Saclay, Siegen, Torino, Vienna and Warsaw.

Abstract. The NA48 experiment aims to measure the direct CP violation parameter Re(ε'/ε) using the method of the double ratio with a precision of 2×10^{-4}. The experiment has collected data during 1997 and during 1998 and plans to produce a result based on the 1997 statistics soon. The status of the analysis and the perspectives for the future are outlined in this paper.

INTRODUCTION

The NA48 experiment [1] aims to measure the direct CP violation in the decay of neutral kaons into two pions. Defining the usual amplitude parameters [2]:

$$\frac{<\pi^+\pi^-|K_L>}{<\pi^+\pi^-|K_S>} = \eta_{+-} \sim \varepsilon + \varepsilon', \quad \frac{<\pi^0\pi^0|K_L>}{<\pi^0\pi^0|K_S>} = \eta_{00} \sim \varepsilon - 2\varepsilon' \quad (1)$$

one gets the following double ratio expression for the direct CP violating parameter Re(ε'/ε):

$$\text{Re}(\varepsilon'/\varepsilon) \sim \frac{1}{6}\left(1 - \left|\frac{\eta_{00}}{\eta_{+-}}\right|^2\right) = \frac{1}{6}\left\{1 - \frac{\Gamma(K_L \to \pi^0\pi^0)}{\Gamma(K_S \to \pi^0\pi^0)} \Big/ \frac{\Gamma(K_L \to \pi^+\pi^-)}{\Gamma(K_S \to \pi^+\pi^-)}\right\}. \quad (2)$$

A non zero value of Re(ε'/ε) signals CP violation in a decay amplitude. Using the technique of the double ratio, NA48 plans to achieve a total error of 2×10^{-4} on Re(ε'/ε). Precision measurements available so far come from the CERN-NA31 Collaboration [3] which reported Re(ε'/ε) = $(2.0 \pm 0.7) \times 10^{-3}$, a 3σ effect, and from the FNAL-E731 Collaboration which on the other hand [4] found, with comparable precision, a result compatible with no direct CP violation: Re(ε'/ε) = $(7.4 \pm 5.9) \times 10^{-4}$. NA48 collects the four decay modes concurrently,

CP459, *Heavy Quarks at Fixed Target*
edited by Harry W. K. Cheung and Joel N. Butler
© 1999 The American Institute of Physics 1-56396-864-9/99/$15.00

in the same detector and from the same decay region. To achieve the required statistical precision we have to collect a few million $K_L \to 2\pi^0$ decays. With a $BR(K_L \to \pi^0\pi^0) = (9.36 \pm 0.20) \times 10^{-4}$ this is the statistical limitation of the experiment.

THE K_L AND K_S BEAMS

Clean, intense neutral kaon beams are necessary to accumulate the required statistics [5]. The main parameters for the two beams are reported in Table 1. An intense proton beam of 1.5×10^{12} protons per pulse (ppp) strikes a one interaction length Be target. A neutral beam is selected by collimation with a production angle of 2.4 mrad. A small fraction of the non-interacting protons are turned away from the dump by means of channelling in a bent silicon crystal [6]. These protons are steered on the K_L beam axis and aimed to a second target $\sim$ 120 m downstream. By then the K_S component of the long neutral beam has decayed. The second neutral beam is selected with a production angle of 4.2 mrad. At the centre of the K_S target the two beams are vertically separated by 72 mm. The 0.6 mrad angle between the two beams allows them to converge at the detector. The K_S is identified by the time coincidence between the detectors and the tagging counter placed on the proton beam upstream of the K_S target. Two pions decays from the short beam come mostly from K_S decays. The K_L contamination in one K_S lifetime is in fact suppressed by a factor $\Gamma(K_L \to 2\pi)/\Gamma(K_S \to 2\pi) \sim 5 \times 10^{-6}$. The choice of the targeting angle and the distance between the K_S target and the final collimator contribute to make the momentum spectra of the accepted K_S and K_L quite similar.

THE DETECTORS

Charged decays are reconstructed by a magnetic spectrometer formed by four large drift chambers [7] and by a dipole magnet which provides a p_T kick of 250 MeV/c. A plastic scintillator hodoscope is used to provide the charged first level trigger and the charged event time. A hadronic calorimeter made of Fe and plastic scintillator sandwiches is used as part of the charged first level trigger. Three large plastic scintillation planes separated from each other by about 0.8 m of iron are used to reject $K_{\mu 3}$ decays.

Neutral decays are reconstructed by means of a liquid krypton electro-magnetic calorimeter [8]. This is formed by 13212 squared towers of $\sim$ 4 cm^2 cross section. The geometry of the calorimeter points to the centre of the K_L, K_S decay region $\sim$ 110 m upstream. The calorimeter is read out by FADCs at 40 MHz sampling frequency [9]. The energy resolution of the calorimeter is better than 1 % for photons above 20 GeV. The position resolution is better than 1 mm and the time resolution for a single photon is of the order of 200 ps [10]. A scintillating fibre

TABLE 1. Characteristics of the K_L and K_S beams

Beam Parameter	K_L	K_S
Primary protons per pulse on target	1.5×10^{12}	3×10^7
Proton Momentum, p_0 (GeV/c)	450	450
Production angle of K^0 beam (mrad)	2.4	4.2
Length of K^0 beam:		
target to final collimator/AKS (m)	126.00	6.07
target to front of e.m. calorimeter (m)	241.10	121.10
Angle of convergence of beam (mrad)	0.0	-0.6
Angular acceptance of beam (mrad)	± 0.15	± 0.375
R.M.S. radius at e.m. calorimeter (mm)	∼ 26	∼ 39
K^0 flux per pulse at exit final coll.	$\sim 2 \times 10^7$	$\sim 2 \times 10^2$
decays between collimators and detector	$\sim 1.4 \times 10^6$	$\sim 2 \times 10^2$
K^0 flux per pulse in useful p_K range 70 170 GeV/c	6.4×10^6	1.5×10^2
decay per pulse		
($70 < p_K < 170$ GeV/c and $c\tau < 4\lambda_S$)	4.4×10^4	1.5×10^2
decay per pulse to $\pi^0\pi^0$		
($70 < p_K < 170$ GeV/c and $c\tau < 4\lambda_S$)	40	45
Detector acceptance for $\pi^0\pi^0$ decays	∼ 0.20	∼ 0.20
Useful $K^0 \to \pi^0\pi^0$ per pulse	∼ 8	∼ 9

hodoscope is inserted in the LKr structure at approximately the shower maximum depth. This device provides a down-scaled trigger to measure the efficiency of the main $\pi^0\pi^0$ trigger. In addition it provides a cross-check for the time reconstruction of the calorimeter to tag the $K_S \to \pi^0\pi^0$ events.

TRIGGER AND DAQ

The experiment runs in a high rate K_L decay environment. Only one in a thousand of the decayed K_L represents an interesting CP violating decay. In order to measure the trigger efficiency, down-scaled trigger with relaxed criteria are logged.

A charged level one trigger is issued when hits in opposite quadrants of the charged hodoscope are found in coincidence with a large energy deposit in the calorimeters ($E_{TOT} > 30$ GeV). The second level trigger is a hardware based coordinate builder using data from the first, second and fourth drift chamber followed by a microprocessor based system [11]. Data from the first two chambers are used to reconstruct the position of the decay vertex. Events with an opening angle greater than 15 mrad are rejected. Data from the fourth chamber, placed downstream of the dipole magnet, is used to calculate the momentum of the tracks and the invariant mass of the two pairs in the hypotheses of a $\pi^+\pi^-$ decay. Events with $c\tau$ shorter than 4.5 K_S lifetimes and with an invariant mass larger than 472 MeV/c^2 are kept.

The 40 MHz pipelined neutral trigger [12] sums the energy deposits in the electromagnetic calorimeter in X and Y projections of 2 cm width. The trigger criteria require the total energy for a decay (E_K) to be larger than 50 GeV, the centre of gravity (COG) to be less than 15 cm and $c\tau$ to be shorter than 4.5 K_S lifetimes (5.5 during the 1997 run). To further reduce the $3\pi^0$ background the number of on time energy peaks found in each of the two projections must be smaller than six (seven for the X projection during the 1997 run).

The NA48 DAQ system is described elsewhere [14]. During the 1998 run a PC farm was used as event builder [15].

STATISTICS COLLECTED SO FAR

During the 1997 run 650,000 $K_L \rightarrow \pi^0\pi^0$ events were collected over four K_S lifetimes in the momentum range between 70 and 170 GeV/c. The statistical precision on $Re(\varepsilon'/\varepsilon)$ from this data sample is expected to be $4 \div 5 \times 10^{-4}$. This makes a preliminary result based on the 1997 data competitive with respect to the already published values. A second physics run took place in 1998. During this run in excess of two million $K_L \rightarrow \pi^0\pi^0$ were collected over a decay region of 3.5 K_S lifetimes. A new run will begin in May 1999 and should yield an amount of decays comparable to the 1998 statistics. An option for $Re(\varepsilon'/\varepsilon)$ running during the year 2000 is available.

SYSTEMATIC ERRORS

Most of the potential sources of errors cancel out in the double ratio. However to reach the required high precision the careful design of the experiment is essential. In addition, great care is placed to control the tagging errors, the K_L background subtraction, the relative (charged versus neutral) momentum/distance and lateral scales, and the accidental correction. In the next subsections the most relevant aspects of the $Re(\varepsilon'/\varepsilon)$ analysis are outlined, highlighting the key features of the experiment to keep the systematic error small.

Tagging

A K_S is identified by the time coincidence between the tagging counter [13] and the detectors. For a $\pi^0\pi^0$ decay the quantity $\Delta t = t_{LKr} - t_{TAG}$ is defined. For a $\pi^+\pi^-$ event the charged hodoscope instead of the calorimeter is used: $\Delta t = t_{HOD} - t_{TAG}$. In the above t_{TAG} is the time of the nearest proton in the tagging counter with respect to the event time. A K_S is defined by the relation: $|\Delta t| \leq 2$ ns. Tagging inefficiencies are a few 10^{-4}. For charged decays they are measured directly distinguishing K_S and K_L by the vertical position of the decay vertex. For the neutral events checks using $K_S \rightarrow \pi^0 e^+e^-\gamma$ events and K_S *only* runs are made. Only an uncontrolled inefficiency due to the detectors can create an artificial $Re(\varepsilon'/\varepsilon)$ different from zero. An inefficiency in the tagging counter itself is not harmful since the tagging counter does not know if the K_S will decay in a charged or in a neutral final state. If $|\Delta t| > 2$ ns the event is defined to be a K_L decay. Accidental activity in the tagging counter causes $K_L \to K_S$ migration which dilutes $\mathrm{Re}(\varepsilon'/\varepsilon)$. The dilution ($\sim 11.5$ %) is precisely measured selecting K_S charged decays by the vertex vertical position.

Background subtractions

Semileptonic $K_{\mu 3}$ and K_{e3} are respectively ~ 131 and ~ 188 times more frequent than $K_L \to \pi^+\pi^-$ decays. The background under the signal region has to be kept to a few per mill. Selected samples of K_S, K_{e3} and $K_{\mu 3}$ are used to fit the K_L distributions. The background extrapolated underneath the $K_L \to \pi^+\pi^-$ peak is $\sim 3 \times 10^{-3}$ with a statistical error ten times smaller. Figure 2 shows the p_T^2 distribution for $K_L \to \pi^+\pi^-$ events together with the distributions for signed $K_{\mu 3}$ and K_{e3}. Also the sum of the two backgrounds is shown. The background curve with the steeper p_T^2 dependence is the the $K_{\mu 3}$ component.

$K_L \to 3\pi^0$ decay mode is ~ 221 times more frequent than $\pi^0\pi^0$ and can mimic it if two photons are missed. The kaon mass is used as constraint to calculate the neutral decay vertex. Photons are then paired in order to form the best $\pi^0\pi^0$ combination. Events with five photons are allowed only if the 5th cluster is more than 3 ns apart from the event time. The signal region is defined as $R_{ell} < 1.5$, where R_{ell} is:

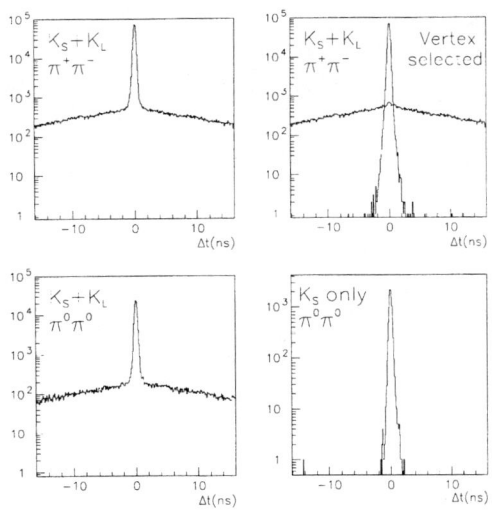

FIGURE 1. Proton tagging

$$R_{ell} \equiv \frac{1}{9}\left\{\left(\frac{\frac{m_1+m_2}{2} - m_{\pi^0}}{\sigma_{\frac{m_1+m_2}{2}}}\right)^2 + \left(\frac{\frac{m_1-m_2}{2}}{\sigma_{\frac{m_1-m_2}{2}}}\right)^2\right\}. \qquad (3)$$

The background is $\sim 3 \times 10^{-3}$ if no lifetime weighting is applied. Applying weighting the background drops to $\sim 1 \times 10^{-3}$. This is because $3\pi^0$ decays with missing photons tend to be reconstructed closer to the calorimeter.

Lifetime weighting and acceptance corrections

To make the vertex distributions for K_L and K_S similar we apply a momentum dependent statistical weight to the K_L events:

$$\exp(-z/\lambda(p)), \quad \lambda(p) = (p/m_K) \cdot c\tau_S/(1 - \tau_L/\tau_S). \qquad (4)$$

Applying this technique, the acceptance formally cancels except for the leftover 0.6 mrad angle between the two beams. This is important for the $\pi^+\pi^-$ decays since the two tracks obey strict two body kinematics and therefore the K_S and K_L illumination is not completely symmetric in the area around the beam hole. To minimise this acceptance correction, an asymmetry cut is applied on the relative difference between the two charged tracks momenta. This cut also effectively removes the $\Lambda \to \pi^- p$ and $\bar{\Lambda} \to \pi^+ \bar{p}$ backgrounds which affect the K_S beam.

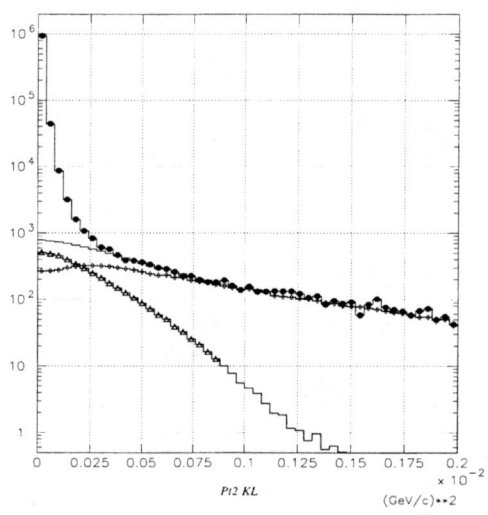

FIGURE 2. Charged background

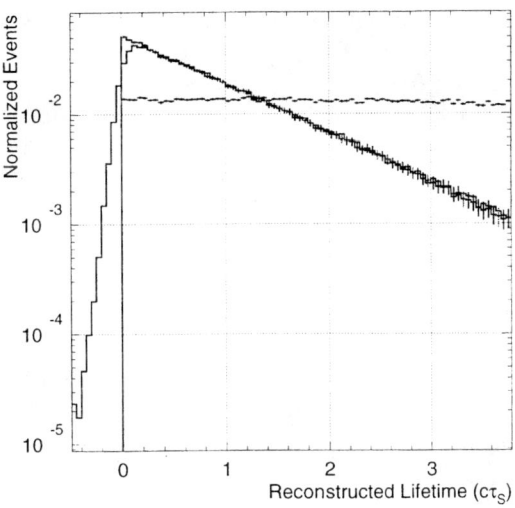

FIGURE 3. Lifetime weighting

Relative momentum scale

The momentum of the charged events is reconstructed using the opening angle of the two charged tracks, the kaon and the charged pion masses, and the ratio of the two charged track momenta (r) as measured by the magnetic spectrometer:

$$E_{\pi^+\pi^-} = \frac{\sqrt{(M_K^2 - m_\pi^2 R) R}}{\theta} , \qquad (5)$$

where θ is the opening angle formed by the two charged tracks and $R = \frac{1}{r} + r + 2$. In addition to the kaon and charged pion masses, the charged momentum scale is therefore determined by the transverse scale of the drift chambers, and by the longitudinal distance between the first two drift chambers.

For the neutral decays one measures the sum of the photon energies to get the kaon momentum. To measure the the distance (D) between the decay vertex and the calorimeter one uses the impact points of the photons, the photon energies, and the kaon mass:

$$D = \frac{1}{m_K} \sqrt{\sum_{i \neq j} E_i E_j d_{i,j}^2} , \qquad d_{i,j}^2 = (x_i - x_j)^2 + (y_i - y_j)^2 . \qquad (6)$$

To calibrate the neutral momentum scale one fits the upstream vertex distribution for $K_S \to \pi^0\pi^0$ decays requiring no hits in the K_S anti-counter (AKS). This counter is placed on the K_S beam just downstream of the final collimator. It consists of a photon converter followed by plastic scintillators and vetoes the K_S decaying before it. The vertex distribution is fitted by an exponential convoluted with a Gaussian smearing. The fitted position of the AKS is adjusted to its known position.

A powerful cross-check for the momentum scale is provided by special runs during which a π^- beam is sent to two short polyethylene targets placed at precisely known positions. The first target is placed just downstream of the K_L final collimator and the other is placed towards the end of the decay region. The prompt decays of $\pi^0 \to \gamma\gamma$, $\eta \to \gamma\gamma$ and $\eta \to 3\pi^0 \to 6\gamma$ allow the cross check of the distance/momentum scale. In this case the function to be fitted is just a Gaussian with no convoluted exponential. The position of the charged vertex reconstruction is also checked collecting events with two or more charged tracks. The downstream target is important to check the linearity of the calorimeter.

To measure the double ratio one has to define a useful momentum range and a decay region. The lower and upper bounds of the accepted momentum range are optimised to minimise the sensitivity to the relative momentum scale uncertainty. A good choice for NA48 is to accept kaon momenta from 70 to 170 GeV/c. A standard choice for the decay region is to accept kaon decays between 0 and 3.5 K_S lifetimes. A cut on the reconstructed lifetime is applied to the K_L events at both the upstream and the downstream end of the decay region. An error on the relative momentum scale shifts the weights applied to the K_L events effectively shifting the selected decay region. Since the K_L vertex distribution is reasonably flat, the gains or losses of K_L events due to the shift of the upstream lifetime cut are compensated, to large extent, by the gains and losses due to the shift of the downstream lifetime cut. Quite different is the situation for the K_S events. Since the vertex distribution is quite steep, the effect of displacing the upstream cut is not compensated by the same shift of the downstream end. To keep under control the systematics, the beginning of the decay region for K_S decays is defined requiring no hits in the AKS counter.

RELATIVE TRANSVERSE SCALE

All of the above assumes that the transverse position scales for the drift chambers and the calorimeter are the same. To cross-check the relative lateral scale electrons from K_{e3} decays and/or electrons from the e^- calibration beam are used. The shower maximum depth for electrons and photons of the same energy is displaced by roughly one radiation length. This implies a correction to the reconstructed position of photons with respect to electrons if the particles impinge on the calorimeter with a nonzero angle. The fact that the calorimeter electrode structure points to the centre of the decay region, $\sim$ 110 m upstream, greatly reduces this correction.

Accidental correction

Neutral and charged decays are collected simultaneously and the losses and gains of events due to accidental kaon decays and muons cancel to first order. Since however we employ two beams and two targets, the correlation of the relative intensity of the two beam must be demonstrated. In order to do this we collect events proportional to K_S and K_L beam intensities but uncorrelated with detectors activities. From the study of these events the accidental K_S and K_L gains and losses are evaluated.

CONCLUSIONS ON THE $\text{Re}(\varepsilon'/\varepsilon)$ ANALYSIS

A new result on $\text{Re}(\varepsilon'/\varepsilon)$ competitive with the already published data should be available soon. In the mean time we keep accumulating statistics and we are confident to achieve the proposed precision in the $\text{Re}(\varepsilon'/\varepsilon)$ determination.

K_L RARE DECAYS AND DIRECT CP VIOLATION

In the CKM description of CP violation, some very rare K_L decays have a large CP violating component. The Standard Model prediction for the direct CP violating contribution to $\text{BR}(K_L \to \pi^0 e^+ e^-)$ is a few 10^{-12} [16]. The single event sensi-

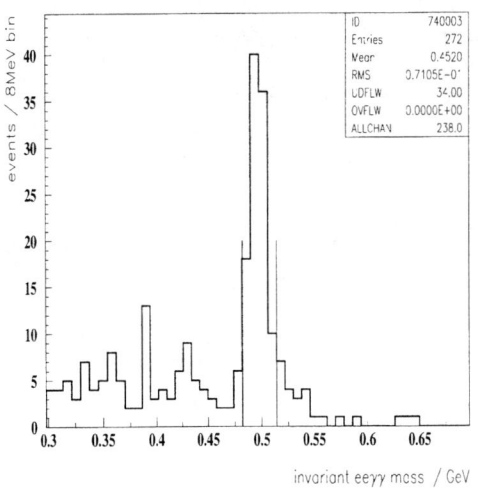

FIGURE 4. $K_L \to e^+ e^- \gamma \gamma$

tivity for such a small branching ratio is out of range for the NA48 experiment in its current configuration. However NA48 can measure with precision rare K_L decays which represent a serious background for future attempts to measure direct CP violation in the $K_L \to \pi^0 e^+ e^-$ decay. They are the $K_L \to \pi^0 \gamma\gamma$ (which contributes to the CP conserving amplitude to $K_L \to \pi^0 e^+ e^-$) and the $K_L \to e^+ e^- \gamma\gamma$ [17]. On this latter decay mode, NA48 presents a preliminary result extracted from the 1997 data sample and based on 106 events in the signal region shown in figure 4. The Preliminary result is BR($K_L \to e^+ e^- \gamma\gamma$) = $(4.6 \pm 0.7 \pm 1.4) \times 10^{-7}$ with the conditions $E_\gamma > 5$ MeV and $m_{e\gamma} > 1$ MeV.

REFERENCES

1. Barr D.G. et al., CERN/SPSC/90-22/P253.
2. Wu T. T., and Yang C. N., *Phys. Rev. Lett.* **13**, 380 (1964).
3. Barr G. D. et. al, *Phys. Lett.* **B317**, 233, (1993).
4. Gibbons L. K. et al., *Phys. Rev. Lett.* **70**, 1203 (1993).
5. Biino C., Doble N., Gatignon L., Grafström P., and H. Wahl, The Simoultaneous and Nearly-collinear K^0 Beams for Experiment NA48, presented at the 6th European Particle Accelerator Conference, Stockholm, Sweden, 22-26 June 1998.
6. Doble N., Gatignon L., and Grafström P., *Nucl. Instr. and Meth.* **B119** (1996) 181.
7. Bédérède D. et al., *Nucl. Instr. and Meth.* **A367**, 88 (1995).
8. Barr G.D. et al., *Nucl. Instr. and Meth.* **A370**, 413 (1996).
9. Hallgren B. et al., The NA48 LKr Calorimeter Digitizer Electronics Chain, presented at the Wirechamber Conference 1998, Vienna, Austria, February 23-27, 1998 / CERN-EP/98-48.
10. Martini M., Performance of the NA48 Liquid Krypton Calorimeter for the Re(ε'/ε) Measurement at the CERN SPS, presentet at the CALOR97 Conference, Tucson, AZ, USA, November 9-14, 1997.
11. Anvar S. et al., *IEEE Transactions on Nuclear Science* **45**, 1776 (1998).
12. Fischer G. et al., A 40 MHz-Pipelined Trigger for $K^0 \to 2\pi^0$ Decays for the CERN NA48 Experiment, presented at the Wire Chamber Conference 1998, Vienna, Austria, February 23-27, 1998.
13. T. Beier et al, *Nucl. Instr. and Meth.* **A360**, 390 (1995).
14. Bal F. et al., The NA48 Data Acquisition System, presented at the 10th IEEE Real Time Conference, Beaune, France, September 22-26, 1997.
15. Luitz S., The NA48 Data Acquisition PC Farm, presented at the CHEP98 Conference, Chicago, IL, USA, August 31 - September 4, 1998.
16. Donoghue J.F. and Gabbiani F., *Phys. Rev.* **D51**, 2187 (1995).
17. Greenlee H. B. , *Phys. Rev.* **D42**, 3724 (1990).

Results on CP Violation from KTeV

Michael Arenton, for the KTeV Collaboration

Physics Department and Institute of Nuclear and Particle Physics, University of Virginia Charlottesville, VA, 22901

Abstract. The status of CP violation measurements in the KTeV experiment is presented. We present the current status of the measurement of $Re(\epsilon'/\epsilon)$. Preliminary results on a T-odd angular asymmetry in the rare decay $K_L \to \pi^+\pi^-e^+e^-$ are presented.

INTRODUCTION

This paper presents the status and a preliminary result on CP violation from the KTeV experiment. First we discuss the status of the major goal of the experiment, the precise measurement of $Re(\epsilon'/\epsilon)$, the ratio of direct to indirect CP violation in K decay. Following this we present preliminary results on the observation of a new indirect CP violating effect in the rare decay $K_L \to \pi^+\pi^-e^+e^-$, consisting of the observation of an asymmetry in a T-odd kinematic quantity.

STATUS OF $Re(\epsilon'/\epsilon)$ MEASUREMENT

The principal objective of the KTeV experiment is the precise measurement of the direct CP violation parameter $Re(\epsilon'/\epsilon)$. The method is based on the measurement of the double ratio of the decay rates of K_L and K_S into $\pi^+\pi^-$ and $\pi^0\pi^0$ pairs:

$$\frac{\Gamma(K_L \to \pi^+\pi^-)/\Gamma(K_S \to \pi^+\pi^-)}{\Gamma(K_L \to \pi^0\pi^0)/\Gamma(K_S \to \pi^0\pi^0)} \approx 1 + 6\mathrm{Re}\,(\epsilon'/\epsilon)$$

The KTeV apparatus is shown in Figure 1 The experiment is done with two neutral beams produced by 800 GeV/c protons on a BeO target. At the distance of the experiment the kaon component of these beams is predominately K_L. K_S are produced by regeneration of K_L striking a regenerator placed in one of the beams. To remove apparatus asymmetries the regenerator is moved from one beam to the other every beam spill.

Data were taken in 1996 and 1997. The focus at present is on 20% of the data. This includes $\pi^0\pi^0$ data from 7 weeks of running in 1996 (20% of the total) and $\pi^+\pi^-$ data from 3 weeks of running in 1997 (17% of the total). We are not using

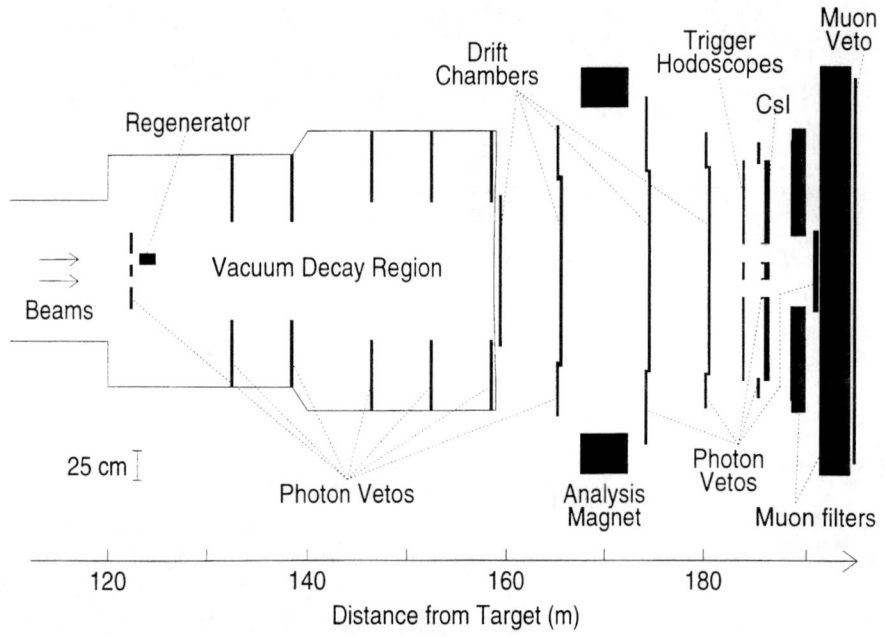

FIGURE 1. Plan view of the KTeV spectrometer (ϵ'/ϵ configuration)

the $\pi^+\pi^-$ data from 1996 because a drift chamber pathology combined with our Level 3 online software trigger rejected 20% of the events. Level 3 was modified in 1997 to avoid this loss.

The statistics used in the present analysis are:

ϵ'/ϵ mode	events	bkg/signal
Vac ($K_L \to$) $\pi^0\pi^0$	1.0×10^6	0.8%
Reg ($K_S \to$) $\pi^0\pi^0$	1.6×10^6	1.5%
Vac ($K_L \to$) $\pi^+\pi^-$	1.9×10^6	0.1%
Reg ($K_S \to$) $\pi^+\pi^-$	3.9×10^6	0.1%

This sample will lead to a statistical error on ϵ'/ϵ of about 3×10^{-4}. We also recorded large samples of other decay modes for calibrations and systematic studies, for instance in the 1996 running period we have 200 million $K \to \pi e\nu$ events for calibration of the CsI calorimeter, and 20 million $K \to \pi^0\pi^0\pi^0$ events for "neutral" systematic studies.

We are currently studying systematic effects, with the expectation that we will reduce systematics effects on ϵ'/ϵ to less than 2×10^{-4} in the present 20% sample. Some of the systematics being studied include the effects of early "accidental" activity in the detector, drift chamber and tracking algorithm inefficiencies, simulation of the trigger hardware, biases in the neutral reconstruction and backgrounds to $K_{L,S} \to \pi^0\pi^0$.

The analyis is being done in a "blind" manner, where the fitting program hides the value of ϵ'/ϵ until we believe we have accounted for all possible biases. We expect to have a result from this 20% analysis within a few months.

CP VIOLATION IN $K_L \to \pi^+\pi^-e^+e^-$

The realization that it would be possible to observe a CP violating effect in $K_L \to \pi^+\pi^-e^+e^-$ began several years ago with the first observations of the γ energy distribution in the decays $K_{L,S} \to \pi^+\pi^-\gamma$ [1] The K_S decay is consistent with a CP conserving Inner Bremsstralung mechanism. K_L on the other hand appears to have two components, an Inner Bremsstralung term, which here is CP violating, and a CP conserving Direct Emission term. Interference of these two terms results in a polarization of the γ.

It was soon realized that in the related process $K_L \to \pi^+\pi^-e^+e^-$ where the e^+e^- pair corresponds to internal conversion of the γ in $K_L \to \pi^+\pi^-\gamma$, the angular distribution of the normal to the plane defined by the e^+e^- pair with respect to the plane defined by the $\pi^+\pi^-$ pair has an asymmetric distribution reflecting the γ polarization. Several calculations were done [2] [3] [4] [5] and asymmetries on the order of 10% were predicted.

Define ϕ as the angle between the planes defined by the $\pi^+ \times \pi^-$ and $e^+ \times e^-$ momentum vectors in the K_L rest frame.

$$\sin\phi\cos\phi = (\mathbf{n}_l \times \mathbf{n}_\pi) \cdot \hat{\mathbf{z}}(\mathbf{n}_l \cdot \mathbf{n}_\pi)$$

where

$$\mathbf{n}_l = (p_{e^+} \times p_{e^-})/|p_{e^+} \times p_{e^-}|$$

$$\mathbf{n}_\pi = (p_{\pi^+} \times p_{\pi^-})/|p_{\pi^+} \times p_{\pi^-}|$$

are the unit vectors normal to the planes of the e^+e^- and $\pi^+\pi^-$ pairs, respectively, and

$$\hat{\mathbf{z}} = (p_{\pi^+} + p_{\pi^-})/|p_{\pi^+} + p_{\pi^-}|.$$

Under T reversal, $\sin\phi\cos\phi$ changes sign.

Integrating over all other variables the ϕ dependence of the decay rate is predicted to have the form:

$$\frac{d\Gamma}{d\phi} = \Gamma_1 cos^2\phi + \Gamma_2 sin^2\phi + \Gamma_3 sin\phi cos\phi$$

The term in $sin\phi cos\phi$ is CP violating. Sehgal and Wanninger predict an asymmetry (ϕ quadrants $1 + 3 - 2 - 4$) of $\approx 14\%$. The branching fraction is predicted to be about 3×10^{-7}.

Experimental Method

To detect the predicted asymmetry we need several thousand cleanly reconstructed events. The KTeV (E-799) experiment used a new high acceptance, high rate spectrometer in a new high intensity, very clean beam. The apparatus was the same as in Figure 1 except that there was no regenerator, larger, higher intensity beams were used, the analysing magnet was run at a lower p_t kick to increase acceptance, and there was a set of transition radiation detectors (not used in this analysis) for further electron identification. [6]

The trigger for 4 track events required equal or greater than 3 hits in a charged particle hodoscope, greater than 11 GeV energy deposit in the CsI calorimeter, 2 or more clusters with energy greater than 2 GeV in the calorimeter, at least 3 or 4 good hits in each drift chamber in the non-bend view, no hits in the muon detector, and no hits in the veto counters defining the outside aperture. The resulting trigger rate was about 10% of the total taken during the experiment. The total run was sensitive to about $2.5 \times 10^{11} K_L$ decays. 1.3×10^8 4 track triggers were written to tape after level 3 processing.

The offline analysis required first the reconstruction of 4 tracks forming a good vertex. Two of the tracks are required to be electrons, where E/p in the CsI calorimeter in the range $0.95 \leq$ E/p ≤ 1.05 identifies electrons. Pions are identified by E/p ≤ 0.85. The event must then also have low p_t^2 with respect to the beam. Even at this stage the signal is clearly visible in a plot of $m_{\pi^+\pi^-e^+e^-}$ vs p_t^2. The principal background (and also the flux normalization mode) is the decay $K_L \to \pi^+\pi^-\pi_D^0$ where π_D^0 denotes the Dalitz decay $\pi^0 \to e^+e^-\gamma$. This background may be suppressed by cutting on the variable:

$$P_{\pi^0}^2 = \frac{[(M_K^2 - M_{\pi^0}^2 - M_c^2)^2 - 4M_{\pi^0}^2 M_c^2 - 4M_K^2(P_T^2)_c]}{4[(P_T^2)_c + M_c^2]}$$

where M_c and $(P_T^2)_c$ are the mass and p_t^2 of the two charged pions. This variable is the momentum that the π^0 from a $K_L \to \pi^+\pi^-\pi_D^0$ decay would have. Except for resolution effects it is positive definite for Dalitz events and its distribution for $K_L \to \pi^+\pi^-e^+e^-$ is mostly negative. A cut is made requiring $P_{\pi^0}^2 \leq -0.006$. Other backgrounds include two overlapping $K_L \to \pi^\pm e^\mp \nu$ decays and also hyperon decays. These backgrounds are suppressed by data quality cuts.

A background that is indistinguishable from the signal is $K_L \to \pi^+\pi^-\gamma$ where the γ converts to an e^+e^- pair in the apparatus, mostly in the window of the vacuum decay region. These events have, however, very low e^+e^- mass and are removed with a mass cut of 0.002 GeV. Signal events are lost by this cut as well. Also at small $m_{e^+e^-}$, which corresponds to a small opening angle of the e^+e^- in the lab, the measured ϕ may have poor resolution, which would smear out an asymmetry. Monte Carlo studies show only small effects above $m_{e^+e^-} \geq 0.002$ GeV. Eventually we will unfold this resolution effect, but this has not yet been done in the results presented here.

Results

Figure 2 shows the final $K_L \to \pi^+\pi^- e^+ e^-$ signal. Fitting the background in regions away from the peak, and subtracting the background fit from under the peak we find a signal of 1811 ± 42 events with a signal/background of 40 ± 6 in the mass range $0.492 \leq m_{\pi^+\pi^- e^+ e^-} \leq 0.504$. We have previously published [7] based on 2% of the data a branching fraction of $(3.2 \pm 0.4(stat) \pm 0.6(syst)) \times 10^{-7}$. Based on 60% of the data we reported at the Vancouver conference [8] a branching fraction of $(3.32 \pm 0.14(stat) \pm 0.28(syst)) \times 10^{-7}$.

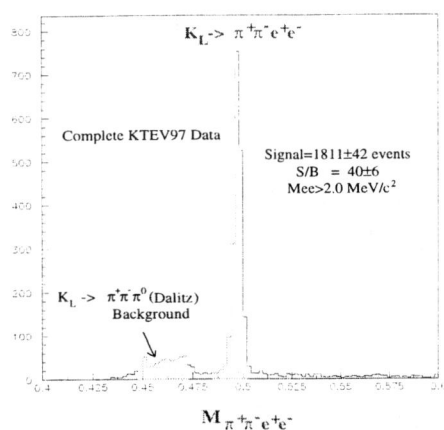

FIGURE 2. Mass spectrum of four track events with an identified $\pi^+\pi^- e^+ e^-$ topology

The acceptance corrections have an important influence on the extraction of the asymmetry. To calculate the acceptance we use a Monte Carlo generator based on the theory of Sehgal and Wanninger. [2] The matrix element is a function of 5 variables, m_{ee}, $m_{\pi\pi}$, $cos\theta_e$, $cos\theta_\pi$ and ϕ. The acceptance as functions of m_{ee} and $m_{\pi\pi}$ is shown in Figure 3. The most significant acceptance effect is a substantial drop in acceptance at high $m_{\pi\pi}$. The distributions of the data agree with the Monte Carlo with the exception that the $m_{\pi\pi}$ distributions is shifted to higher masses in the data. To account for this we have inserted a form factor into the matrix element such as was used in the earlier result on $K_L \to \pi^+\pi^-\gamma$. [1]

$$F = a_1[(M_\rho^2 - M_K^2) + 2M_K(E_{e^+} + E_{e^-})]^{-1} + a_2$$

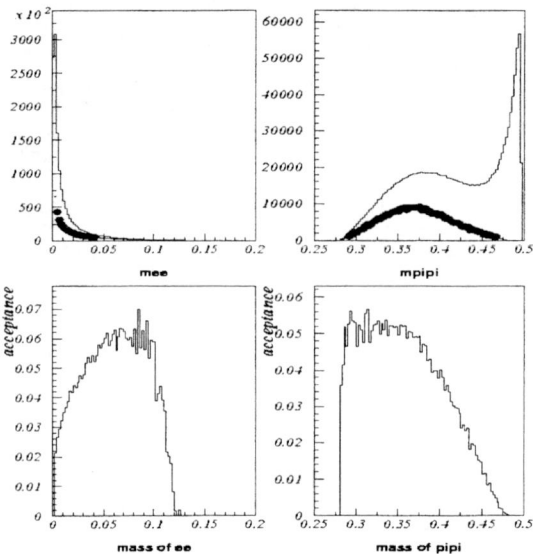

FIGURE 3. Monte Carlo acceptance as functions of m_{ee} and $m_{\pi\pi}$. The upper plots show the generated and reconstructed distributions (with the reconstructed points multiplied by 10 for clarity). The lower plots show the corresponding acceptance.

We use $a_1/a_2 = -1.8 \pm 0.2$ which is a theoretically motivated value that gives better agreement with the data than the matrix element without a form factor. In the final analysis we will fit this form factor from the data.

It is important to consider the effect of the acceptance on the measured ϕ distribution and its asymmetry. The acceptance is not flat in ϕ, with lower acceptance near $\pi/2$ and $3\pi/2$. This is shown in Figure 4 where Monte Carlo has been generated with zero asymmetry (by artificially fixing the Inner Bremsstralung and Direct Emission to be out of phase). The acceptance in this case does not introduce any asymmetry.

If however we generate with the nominal matrix element, the accepted asymmetry is different (and in fact larger) than the initial asymmetry, as is shown in Figure 5. Basically this is because the ϕ asymmetry varies as a function of the other kinematic variables and the acceptance preferentially selects regions where the asymmetry is larger than average. Thus to correct the measured asymmetry it is important to use a Monte Carlo that agrees with the data in all the kinematic variables. One can also do an analysis where the acceptance is calculated for each event in a small kinematic region. Such an analysis is in progress but will not be reported here.

A check on whether the acceptance introduces any false asymmetry is to calculate

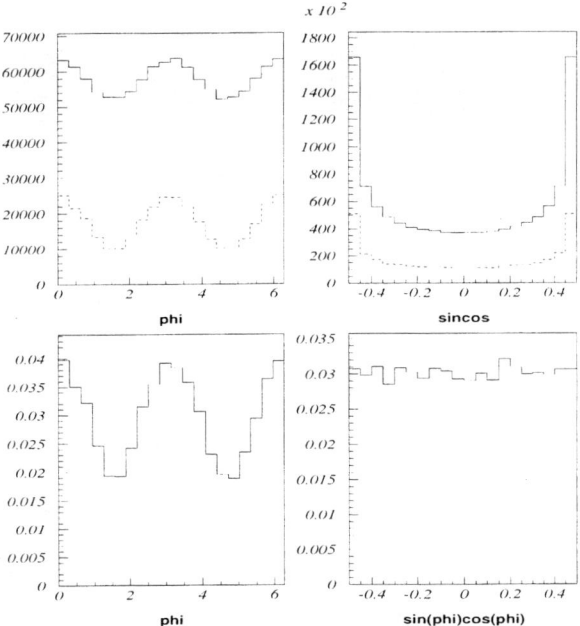

FIGURE 4. Monte Carlo acceptance calculation for a model with zero asymmetry. The upper plots show the generated and accepted ϕ and $\sin\phi\cos\phi$ distributions (the accepted distributions have been multiplied by 10 for clarity), the lower plots the corresponding acceptance.

the asymmetry for $K_L \to \pi^+\pi^-\pi_D^0$ data. Based on a sample of nearly 5 million events, the reconstructed asymmetry is $(-0.02 \pm 0.05)\%$.

We determine the asymmetry from either the ϕ or $sin\phi cos\phi$ distributions, subtracting background distributions determined from the wings of the mass peak. The ϕ and $sin\phi cos\phi$ distributions are shown in Figure 6.

We find an asymmetry before acceptance correction of $(23.3\pm2.3(stat))\%$ We currently estimate systematic errors as follows, with the expectation that most of these effects will be reduced with further analysis. Resolution effects contribute about $\pm 0.5\%$, backgrounds and variations in the analysis cuts about $\pm 2\%$, and variation in the acceptance Monte Carlo parameters $\pm 1.3\%$. Taking these in quadrature leads to a systematic error of $\pm 3\%$. Correcting for the acceptance yields a preliminary value for the asymmetry of $(13.5 \pm 2.5(stat) \pm 3.0(syst))\%$. This is in good agreement with theoretical expectations. We can also examine the uncorrected asymmetry as functions of m_{ee} and $m_{\pi\pi}$, as is shown in Figure 7.

Further analysis should reduce systematic effects. A new run scheduled for the spring of 1999 should triple the data sample. Finally it may be possible to search for direct CP violation, which would be observable as asymmetries in $cos\theta_e$ and $cos\theta_\pi$. Standard expectations predict only very small effects, however.

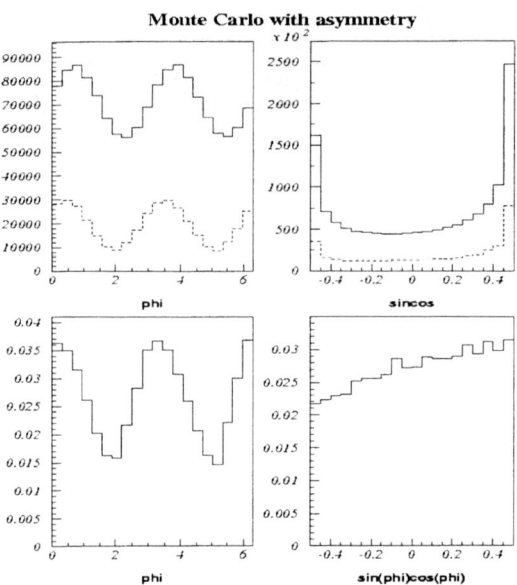

FIGURE 5. Monte Carlo acceptance calculation for the matrix element of ref [2]. The upper plots show the generated and accepted ϕ and $\sin\phi\cos\phi$ distributions (the accepted distributions have been multiplied by 10 for clarity), the lower plots the corresponding acceptance.

We thank the Fermilab staff and the technical staffs of the participating institions for their vital contributions. This work was supported in part by the U.S. Department of Energy, National Science Foundation and the Ministry of Education and Science of Japan.

REFERENCES

1. E.J.Ramberg et. al., *Phys. Rev. Lett.* **70**, 2525 (1993).
2. L.M.Sehgal and M.Wanninger, *Phys. Rev.* **D46**, 1035 (1992), **D46**,5209(E).
3. P.Heiliger and L.M.Sehgal, *Phys. Rev.* **D48**,4146 (1993).
4. J.K.Elwood, M.B.Wise and M.J.Savage, *Phys. Rev.* **D52**, 5095 (1995), **D53**, 2855(E) (1996).
5. J.K.Elwood, M.B.Wise, M.J.Savage, and J.W.Walden, *Phys. Rev.* **D53**, 4078 (1996).
6. see also the talk by E.D.Zimmerman in these proceedings.
7. J.Adams et. al., *Phys. Rev. Lett.* **80**, 4123 (1998).
8. A.R.Barker, to be published in Proceedings of XXIX International Conference on High Energy Physics, Vancouver, 1998.

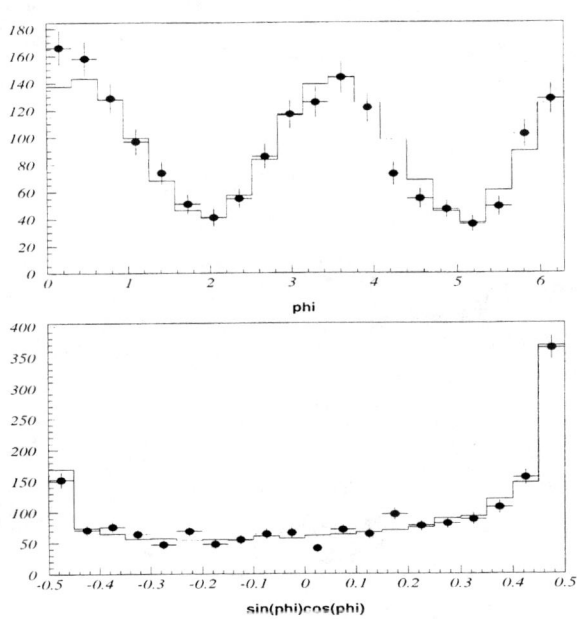

FIGURE 6. "Raw" ϕ and $sin\phi cos\phi$ data distributions (points) compared to the Monte Carlo (histogram)

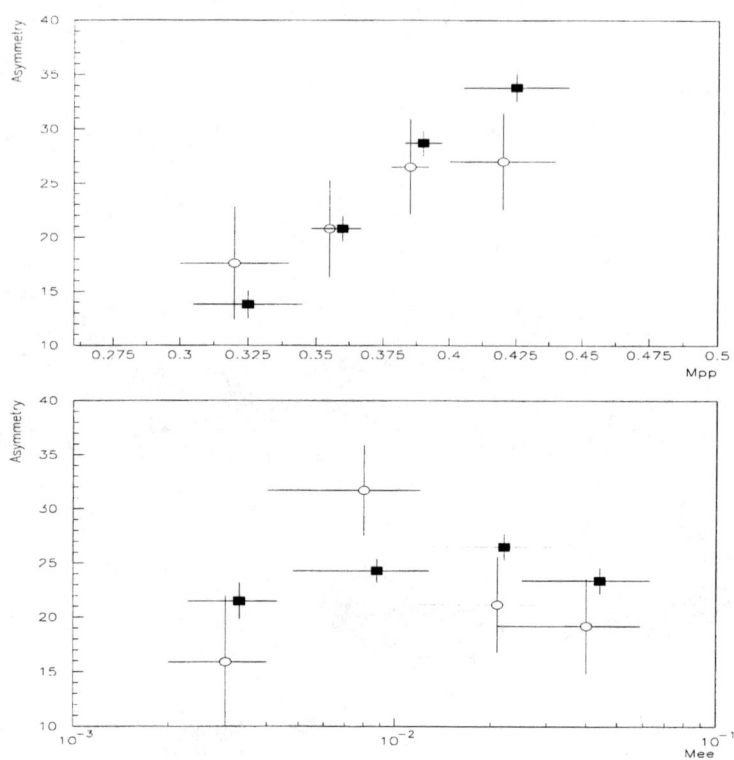

FIGURE 7. "Raw" dependence of the asymmetry on both $M_{\pi\pi}$ (top) and M_{ee} (bottom). Data (open circles) compared to the Monte Carlo (filled squares).

PART II
HEAVY IONS

Heavy Flavour Production in Heavy Ion Collisions
Present status and future prospects

C. Lourenço

CERN - EP
CH-1211 Geneva 23

Abstract. Heavy quark production in nucleus-nucleus collisions has been studied in the SPS experiments WA97 (strangeness) and NA38/NA50 (dimuons, charmonia). We present some of the results currently available from these experiments. We will also review future prospects for open charm measurements, made possible by major developments recently done in the radiation tolerance of silicon detectors.

INTRODUCTION

The main goal of the several high energy heavy ion experiments, taking data since 1986 at BNL and at CERN, is to observe a phase transition, from the ordinary hadronic matter to a new phase of partonic degrees of freedom, where quarks and gluons are no longer confined to specific hadrons. The formation of such a state of matter, at very high densities or temperatures, was predicted in the framework of lattice QCD calculations and is expected to induce remarkable changes in the normal pattern of heavy flavour production. The proof of existence of such a deconfined phase and the study of its properties are key issues in QCD, for the understanding of confinement and of chiral-symmetry. However, heavy ion collisions result in very complex systems to study, with hundreds of final-state particles. Multi-strange hyperons and heavy quarkonia production, being relatively rare processes, offer particularly clear probes, in terms of measurement and interpretation.

STRANGENESS ENHANCEMENT

Strangeness production in heavy ion collisions provides pertinent information on the nature of the matter created in these interactions. An enhanced yield of multi-strange hyperons in (central) nucleus-nucleus reactions with respect to proton-nucleus interactions is expected in case of a phase transition to a state of deconfined quarks and gluons, the "quark-gluon plasma".

CP459, *Heavy Quarks at Fixed Target*
edited by Harry W. K. Cheung and Joel N. Butler
© 1999 The American Institute of Physics 1-56396-864-9/99/$15.00

The WA97 experiment was designed to study the (multi-)strange particle yields as a function of the number of nucleons taking part in the collision. The experimental setup was mainly composed of a silicon telescope placed 60 cm downstream of the target, inside the 1.8 T magnetic field of the CERN Omega magnet. It was made of 7 planes of 75×500 μm^2 pixels and 10 planes of 50 μm pitch microstrips (a total of 500 000 channels). The details of the data analysis are presented in Ref. [1].

Figure 1 shows the strange particle yields per event for p-Pb and Pb-Pb interactions, expressed in units of the p-Pb yield, as a function of the number of participant nucleons. The yields were extrapolated to the c.m.s. rapidity window $-0.5 < y^* < 0.5$ and were obtained from a fraction of the available data sample. The error bars show statistical uncertainties only. All Pb-Pb yields show a steady increase with N_{part}, up to the most central events, and the enhancements exhibit a clear hierarchy: $\Omega > \Xi > \Lambda$.

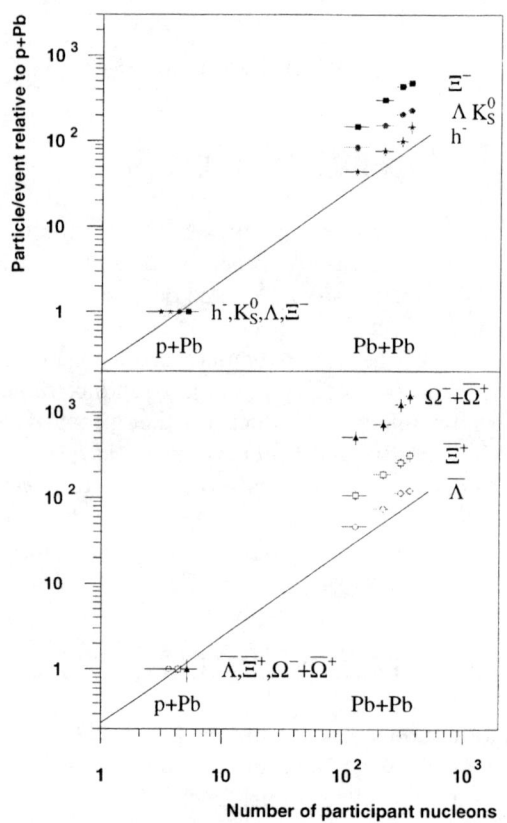

FIGURE 1. Evolution of strange particle yields in Pb-Pb collisions, relative to the p-Pb value, as a function of the number of nucleons actively participating in the reaction.

J/ψ SUPPRESSION

The formation of a deconfined medium should induce a considerable suppression of the J/ψ production yield, due to colour screening of the $c\bar{c}$ potential (ψ melting). This effect can be described in a microscopic way by noticing that only scatterings with deconfined gluons are hard enough to break the $c\bar{c}$ bound state. See Ref. [2] for a recent description of these scenarios.

Charmonia production in heavy ion collisions has been studied by the NA38/NA50 experiments, at the CERN SPS. The NA38/NA50 experimental apparatus is based on the NA10 dimuon spectrometer. It consists of a toroidal magnet, 8 MWPC's and 4 trigger hodoscopes. It is separated from the target region by a hadron absorber made of 4 m of Carbon followed by 80 cm of Iron. A segmented active target does the primary vertex identification thanks to a system of quartz blades located after each sub-target. A beam hodoscope counts the incident ions and rejects beam pile-up. An electromagnetic calorimeter measures the neutral transverse energy while two silicon strip detectors count the charged particle multiplicity and a ZDC measures the forward energy. Figure 2 shows the dimuon mass distribution produced in Pb-Pb collisions and measured by NA50.

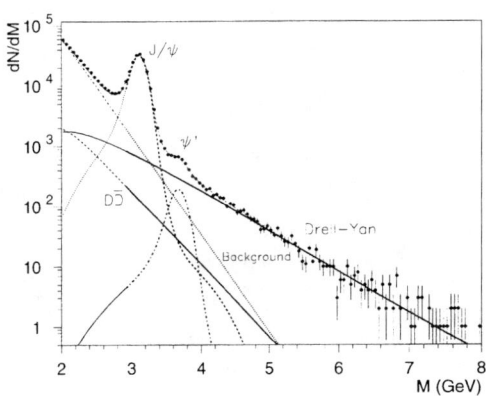

FIGURE 2. Dimuon mass distribution produced in Pb-Pb collisions, as measured in NA50.

The first data from the NA38 experiment [3] showed that there was indeed a strong suppression of the J/ψ yield in O-U and S-U interactions, as a function of the centrality of the collision (parametrised via the released transverse energy) and from p-U to nucleus-nucleus collisions (integrated over the impact parameter). These observations generated quite some interest but new p-A data revealed that the nucleus-nucleus values followed the normal pattern, as defined by the p-A systematics. See Ref. [4] for a description of the historical evolution of this field.

To clearly identify properties specific to the very dense matter produced in the most central heavy ion collisions it is fundamental to establish a reference baseline, from p-A and light ions. It is also very important to have a reference process, like Drell-Yan production, insensitive to the formation of a deconfined phase, that shows what is the "normal" behaviour of the signal we are studying.

The presently available measurements of Drell-Yan production absolute cross sections are collected in Fig. 3. This figure will be completed in the near future with points measured from high statistics p-A data sets (with Be, Al, Cu, Ag and W targets) recently collected by the NA50 experiment.

The corresponding systematics of J/ψ production is presented in Fig. 4, revealing a very significative decrease of the J/ψ production cross section per nucleon, in the Pb-Pb collision system [5]. Recently, it has been observed a step in the J/ψ suppression pattern, versus transverse energy, within the Pb-Pb system.

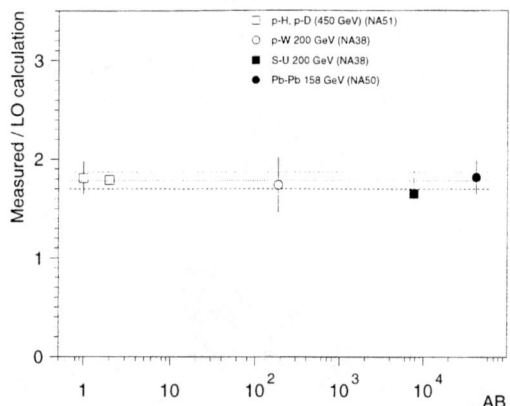

FIGURE 3. Measured Drell-Yan cross section, relative to a LO calculation, as a function of the product of the mass numbers of the projectile and target nuclei.

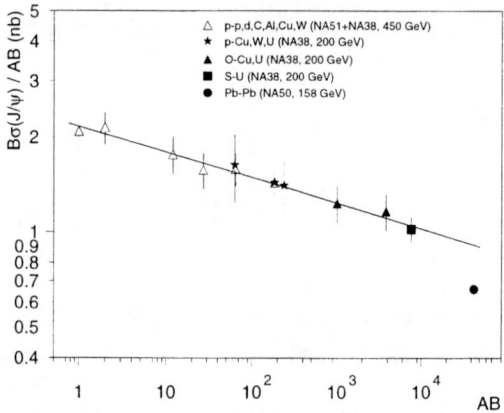

FIGURE 4. Dependence of the J/ψ absolute cross section per nucleon (times branching ratio into muons) on the product of the mass numbers of the projectile and target nuclei.

ENHANCEMENT OF CONTINUUM DIMUONS

The NA38/NA50 experiments have also studied [6] the production of dileptons in the mass window between the ϕ and the J/ψ, $1.5 < M < 2.5$ GeV, as a superposition of Drell-Yan dimuons and simultaneous semileptonic decays of D and $\bar{D}$ mesons, after subtraction of the decay background from pions and kaons.

The Drell-Yan and open charm contributions were calculated with the PYTHIA event generator [7] with the MRS A set of PDFs [8], with $m_c = 1.5$ GeV and $\langle k_T^2 \rangle = 0.8$ GeV^2. PYTHIA describes reasonably well [9] the $\sqrt{s}$ dependence of D meson cross sections, as well as their kinematical distributions and the lepton distributions from their semi-leptonic decays.

The high dimuon mass region (above 4.5 GeV) defines the Drell-Yan normalisation. The open charm cross-section that provides the best fit to the dimuon mass spectra measured by NA50, in p-A collisions, is in good agreement with the direct measurements of other experiments, as shown in Fig. 5. The dotted line in this figure represents the LO calculation of PYTHIA, scaled up by a K-factor.

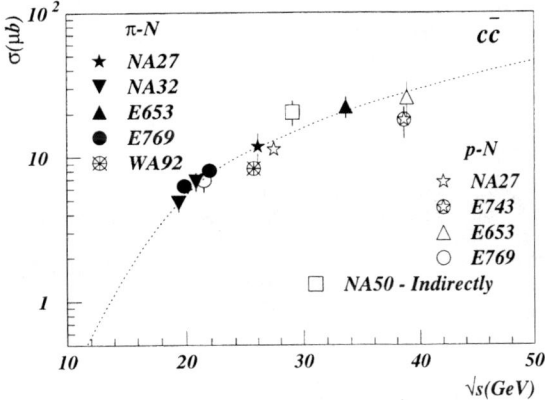

FIGURE 5. Evolution of the $c\bar{c}$ cross section on $\sqrt{s}$.

While in proton-nucleus collisions the intermediate dimuon mass region can be very well described as a superposition of Drell-Yan and open charm, with the expected absolute cross sections, in nucleus-nucleus collisions the open charm yield must be increased relative to the expected linear A-dependence, to correctly fit the measured data (see Fig. 6). This enhancement of intermediate mass dimuons is seen to increase with the size of the collision system, parametrised by the number of participant nucleons N_{part}, reaching a factor of ~ 3 in central Pb-Pb collisions, as can be seen in Fig. 7.

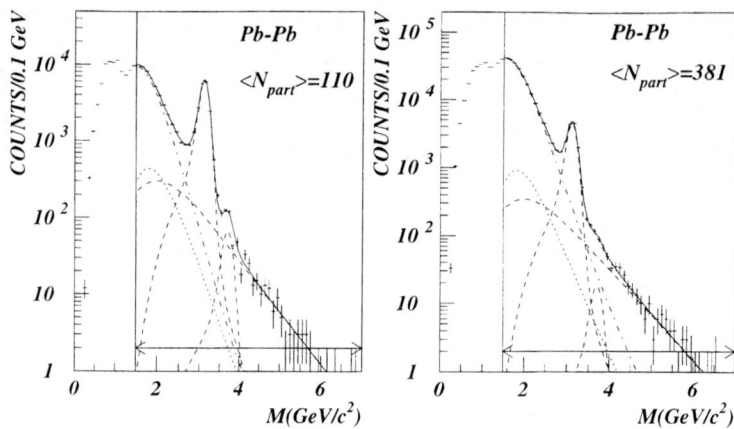

FIGURE 6. Dimuon mass distributions in peripheral (left) and central (right) Pb-Pb collisions.

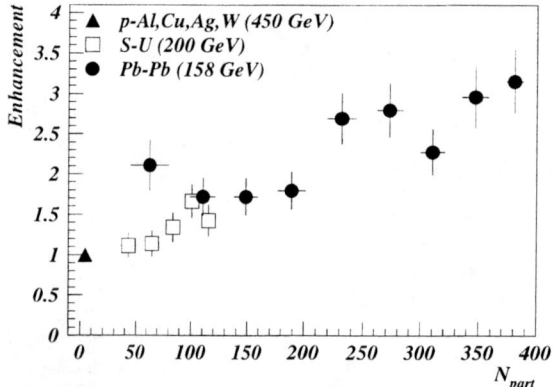

FIGURE 7. Enhancement factor of the open charm contribution, relative to the value expected from the linear A dependence, normalised to the NA50 p-A data.

OPEN CHARM IN HEAVY ION COLLISIONS

The results presented in the previous section emphasise the importance of having a direct measurement of open charm production in heavy ion collisions. Such a measurement requires a dedicated SPS experiment, that can cope with the very high particle multiplicities reached in the most central Pb-Pb collisions (around 500 charged particles per unit rapidity at midrapidity) and with the rather small D production cross section. The NA50 experiment uses CERN's highest intensity heavy ion beam (more than 10^7 Pb ions per second) and has a very clean dimuon trigger, quite appropriate to look for rare processes.

Recent simulations have indicated that it is possible to upgrade the present NA50 muon spectrometer with a silicon vertex telescope, made of pixel planes, to reach a resolution around 50 μm in the offset of the muon tracks, relative to the interaction point, in each transverse direction of the collision plane. That should be sufficient to tag events with muon pairs from $D\bar{D}$ decays.

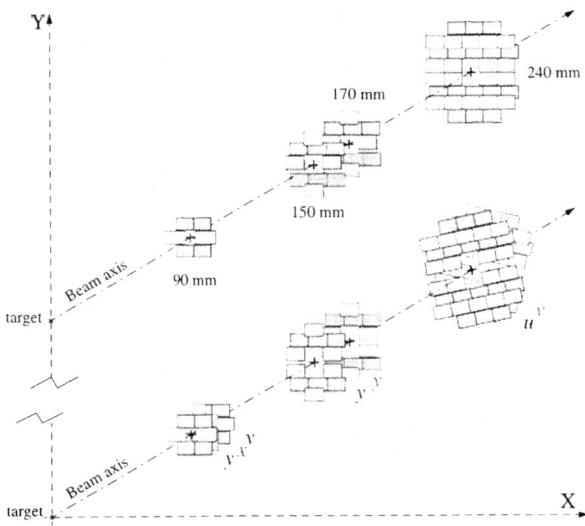

FIGURE 8. Two layouts under study for the silicon vertex telescope being added to the standard NA50 muon spectrometer.

Two layouts of such a vertex telescope, to be placed in the 25 cm long interval between the target and the hadron absorber, in a 1.8 T dipole field, are sketched in Fig. 8. The simpler version is meant to improve the dimuon mass resolution and the signal to background ratio in the low mass region, by matching the muon tracks from the muon spectrometer through the hadron absorber (matching the momentum and the angles). The expected mass resolution (see Fig. 9) should allow to separately measure ρ and ω production. Preliminary results from a test run done in April 1998, with p-Be collisions, include the observation of an ω peak with a width of $\sigma \sim 20$ MeV, in good agreement with the simulated values [10].

The lower version illustrates a more complex layout, able to extrapolate the tracks to the interaction point (plane) and measure their impact parameter with a good enough precision to separate decays of D mesons from prompt dimuons. Figure 10 shows the simulated mass spectra corresponding to two event sub-samples, distinguished by the muon offset values. By selecting events with both muons missing the interaction vertex (the common origin of the produced hadrons) by an offset between 90 and 500 μm, the dimuon mass distribution between 1.2 and 2.0 GeV/c^2 becomes dominated by the $D\bar{D}$ contribution ($\sim 90\%$ of the measured signal).

The pixel planes closer to the target (and to the beam axis) will have to stand a rather heavy and inhomogeneous radiation dose, since the dose depends on the

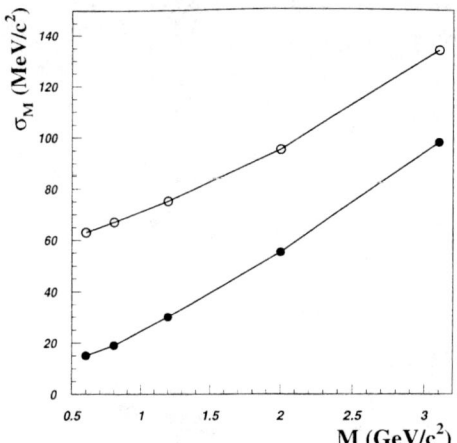

FIGURE 9. Dimuon mass resolution obtained with the standard NA50 muon spectrometer (open circles) and with the upgraded detector (close circles).

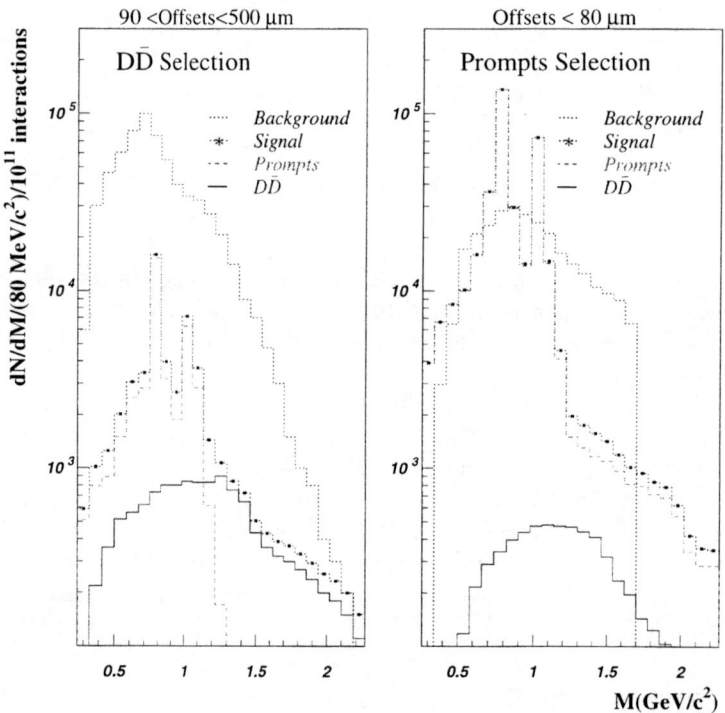

FIGURE 10. The dimuon mass spectra can be strongly enriched in $D\bar{D}$ decays (left) or in prompt muon pairs (right), tagging the events according to the impact parameter of the muon tracks.

square of the transverse distance from the beam axis, varying by more than an order of magnitude over the area of the most exposed chips. This will not be a problem for the ALICE2 chip, which will have, for each pixel, an automatic leakage current compensation and a low threshold (~ 1000 e) individually adjustable. This CMOS readout chip, developed for the Alice and LHC B experiments, uses a new (enclosed geometry) transistor layout and is produced (by IBM) in deep submicron (0.25 μm) technology. A first prototype has been found to survive up to ~ 30 Mrad [11]. The final chip will have 300×50 μm^2 pixels and a time resolution of 25 ns.

The upgraded NA50 experiment will be ready to take Pb-Pb data in the year 2000. Besides being the first experiment to profit from the radiation hardness of the deep submicron pixel readout chips, it will also use the first detector based on cryogenic operation of silicon, in the form of a micro beam hodoscope. This detector will measure the transverse coordinates of each incident Pb ion, to facilitate the determination of the interaction point. It will probably consist of 20 μm pitch microstrip planes, operated at liquid nitrogen temperatures. Thanks to the 'Lazarus effect' [12], we expect this detector to be able to stand the very high radiation dose induced by the intense Pb beam going through it. First tests in the lead beam serving the NA50 experiment will be performed in November 1998.

CONCLUSIONS

The mid-rapidity production of (multi-)strange (anti)hyperons is strongly enhanced in Pb-Pb collisions with respect to p-Pb collisions. The yields increase with centrality and the enhancements exhibit a marked $\Omega > \Xi > \Lambda$ hierarchy. These results can be interpreted as indicating the formation of a quark-gluon plasma in Pb-Pb collisions. Strange quark pairs can be easily created in a gluon rich environment, mainly via gluon-gluon interactions, while in a purely hadronic system, via consecutive rescattering processes, the observed enhancement hierarchy is not easily understandable.

Charmonium suppression beyond the normal nuclear absorption seen in p-A and light ion collisions was predicted to be a clear signature of a transition from confined to deconfined matter. A sudden change in the J/ψ production yield has, indeed, been observed in Pb-Pb collisions, by the NA50 experiment. While the peripheral Pb-Pb collisions follow the normal (nuclear absorption) behaviour, the head on collisions seem to produce a substantially different state of matter.

The intermediate mass dimuon continuum measured by the NA38/NA50 experiments in p-A, S-U and Pb-Pb collisions is well described as a superposition of muon pairs from correlated D and $\bar{D}$ semi-leptonic decays and Drell-Yan dimuons, after combinatorial background subtraction. While in p-A collisions the open charm cross-section scales linearly with A, as expected, in S-U and Pb-Pb the dimuon yield from open charm must be enhanced with respect to the expected values, in order to properly reproduce the data. The enhancement factor increases linearly with the number of nucleons participating in the collision, reaching a factor of ~ 3

in the most central Pb-Pb collisions.

These observations suffer from a rather unfavourable signal to background ratio and their clarification requires a dedicated experiment. An upgraded NA50 detector, by adding a state of the art pixel vertex telescope to the existing muon spectrometer, is presently being studied and could be ready to take data by the year 2000. This will probably be the first experiment to make use of the radiation tolerance of the new generation of pixels. Besides, work is going on to prepare a silicon detector operated at $T \sim 77$ K, so as to withstand extremely high radiation doses.

This paper has benefited from very useful discussions with many people, including F. Antinori, M. Campbell, L. Casagrande, L. Kluberg, R. Lietava, V. Palmieri, E. Quercigh, H. Satz, J. Schukraft, E. Scomparin, R. Shahoyan, W. Snoeys, C. Soave and P. Sonderegger.

REFERENCES

1. E. Andersen et al. (WA97 Coll.), to be published in Phys. Lett. B; CERN-EP/98-64
2. H. Satz, in "International School of Subnuclear Physics", A. Zichichi (Ed.), Erice, Italy, 1997; BI-TP-97-47, hep-ph/9711289.
3. C. Baglin et al. (NA38 Coll.), Phys. Lett. **B 220** (1989) 471;
   C. Baglin et al. (NA38 Coll.), Phys. Lett. **B 255** (1991) 459.
4. H. Satz, Proceedings of the RHIC/INT Workshop "Quarkonium Production in Relativistic Nuclear Collisions", INT, Seattle, USA, May 1998; hep-ph/9806319, BI-TP 98/16.
5. M.C. Abreu et al. (NA50 Coll.), Phys. Lett. **B 410** (1997) 327;
   M.C. Abreu et al. (NA50 Coll.), Phys. Lett. **B 410** (1997) 337.
6. E. Scomparin et al. (NA50 Coll.), Proceedings of the "Strangeness in Quark Matter 98" Conference, Padova, Italy, July 1998;
   C. Soave et al. (NA50 Coll.), Proceedings of the "XXIX International Conference on High Energy Physics, CHEP'98", Vancouver, Canada, July 1998.
7. T. Sjostrand, Computer Physics Commun. **82** (1994) 74.
8. A.D. Martin et al., Phys. Rev. **D51** (1995) 4756.
9. P. Braun-Munzinger et al., Eur. Phys. J. **C1** (1998) 123.
10. P. Sonderegger et al., Proceedings of the "Vertex 98" Conference, Santorini, Greece, September 1998.
11. W. Snoeys et al., Proceedings of the "Fourth Workshop on Electronics for LHC experiments", Rome, Italy, September 1998.
    M. Campbell et al., Proceedings of the "IEEE Nuclear Science Symposium '98", Toronto, Canada, November 1998.
12. V.G. Palmieri et al., Nucl. Instrum. and Meth. in Phys. Res. **A 413** (1998) 475.
    L. Casagrande et al., Proceedings of the "IEEE Nuclear Science Symposium '98", Toronto, Canada, November 1998.
    http://www.cern.ch/RD39

PART III
PRODUCTION DYNAMICS AND STRUCTURE FUNCTIONS

E791 Results on Hadroproduction of Charm from 500 GeV π^-–N Interactions

Kevin Stenson[1]

University of Wisconsin, Department of Physics, Madison, WI 53706[2]

Abstract. E791 took data with a 500 GeV π^- beam on a nuclear target during the 1991 Fermilab fixed-target run. We present a sample of charm production results from these data. We measure production asymmetries of $D^\pm$ and $D_s^\pm$ mesons, Λ_c baryons and the Λ, Ξ, and Ω hyperons. These asymmetries provide insight into the fragmentation process. We also present results from the E791 analysis of events with two fully reconstructed D mesons. By looking at correlations between charm and anticharm particles in the same event, we learn about the charm quark production mechanism as well as the hadronization process. Finally we present new results on D^0 distributions versus x_F and p_T^2 from approximately 100,000 $D^0 \to K\pi, K\pi\pi\pi$ decays.

INTRODUCTION

Hadroproduction of charm is factorizable into three parts as shown in Fig. 1. The first part is composed of the parton distribution functions (PDF's) which provide information about the flavor and momentum content of the partons. Specifically, $f_i^A(x_A, Q^2)$ is the probability of finding parton type i inside hadron A with momentum fraction x_A. These PDF's are generally derived from deep inelastic scattering data and prompt photon or lepton data. The interaction of parton i and parton j to produce a charm quark with four-momentum p_c is governed by the short distance (high Q^2) QCD interaction $\hat{\sigma}_{ij}(p_c, Q^2)$. Since this interaction takes place at a scale on the order of the charm quark mass (~ 1.5 GeV) which is above the QCD scale $\Lambda_{QCD} \sim 0.3$ GeV, this process should be calculable by perturbative QCD but perhaps not too reliably since 1.5 GeV $\not\gg$ 0.3 GeV. The third and final part of the production involves the fragmentation of the charm quarks into charm hadrons. This occurs at the QCD scale and is therefore incalculable by perturbative QCD. In addition, since the charm quarks are tied by color strings to the remnants of the beam and target hadrons, the simple Peterson fragmentation prescription is not expected to work well. A more involved fragmentation scheme such as the Lund string fragmentation model may be able to model these effects.

---

[1] for the E791 Collaboration
[2] now at Vanderbilt University, Department of Physics & Astronomy, Nashville, TN 37235

CP459, *Heavy Quarks at Fixed Target*
edited by Harry W. K. Cheung and Joel N. Butler
© 1999 The American Institute of Physics 1-56396-864-9/99/$15.00

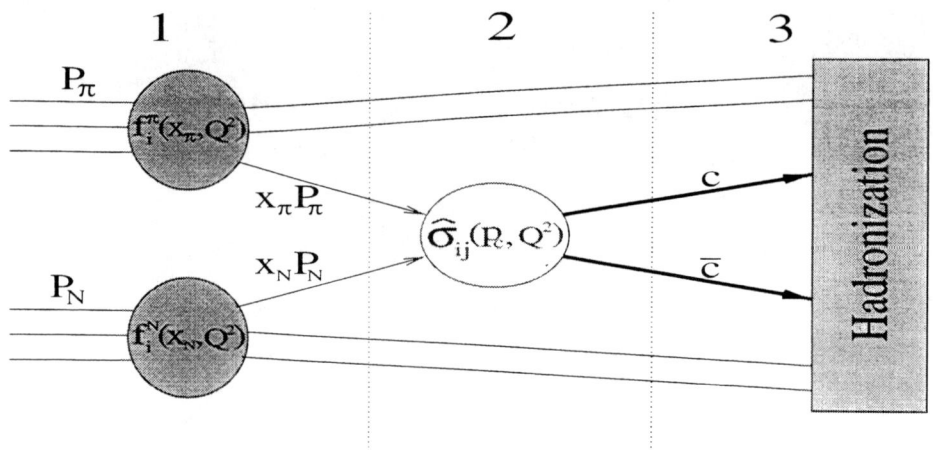

FIGURE 1. A schematic view of charm production from π^--N interactions.

By measuring various charm production properties we hope to arrive at a coherent picture of charm production and perhaps test QCD or, at least our knowledge of QCD. The goal of the E791 experiment was a high precision measurement of production and decay properties of charm particles. The E791 experiment was built upon the previous Tagged Photon Laboratory (TPL) experiments, E516, E691, and E769. The 500 GeV π^- beam was tracked with 8 planes of PWC's and 6 planes of silicon strip detectors and impinged on one platinum and four diamond targets. The downstream spectrometer, depicted in Fig 2, consisted of 17 planes of silicon strip detectors for vertexing and tracking along with 35 planes of drift chambers, 2 PWC planes and 2 analysis magnets for tracking and momentum measurement. Two threshold Čerenkov counters, an electromagnetic calorimeter, a hadronic calorimeter, and a muon wall provided particle identification. The trigger required a beam particle, a hadronic interaction, and a loose transverse energy requirement. The E791 experiment wrote 2×10^{10} events to 24,000 8 mm tapes during the 1991 fixed-target run. From this data set we have fully reconstructed more than 200,000 charm decays.

ASYMMETRIES

The asymmetries measured here are the production asymmetries of a particle versus its antiparticle as a function of some kinematic variables. Specifically, we measure $A \equiv \frac{\sigma_X - \sigma_{\overline{X}}}{\sigma_X + \sigma_{\overline{X}}}$ vs p_T^2 and $x_F \equiv \frac{p_L^*}{p_{L_{max}}^*}$. These asymmetries are quite interesting because they are expected to be nearly zero for charm quark production. Therefore, even the existence of an asymmetry gives information about the hadronization process, which is further elucidated by measuring the magnitude and kinematic dependence of the effect.

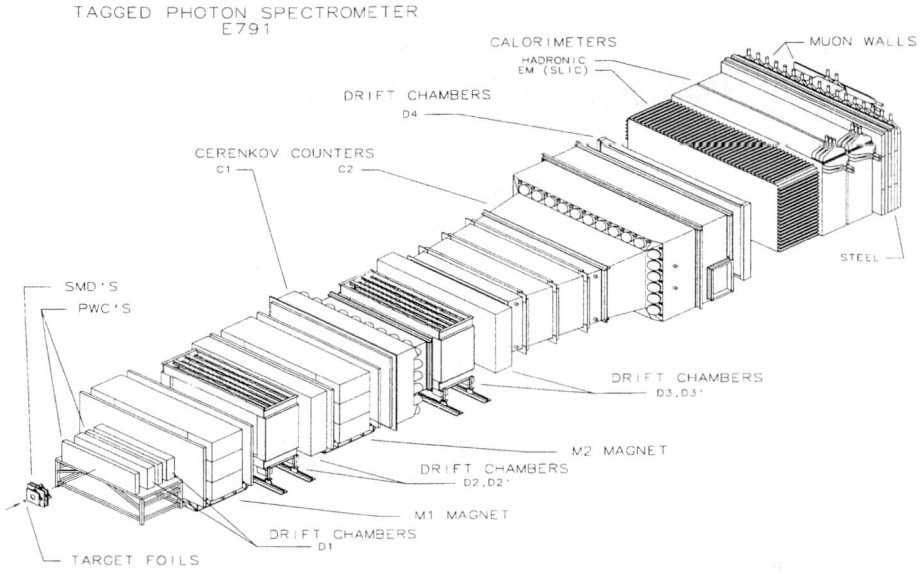

FIGURE 2. The E791 spectrometer.

A potential source of asymmetries comes from the fragmentation of the string between the charm quark and the beam/target fragments. As shown in Fig. 3, a $D^-(\bar{c}d)$ meson which has a d quark in common with the $\pi^-(\bar{u}d)$ beam can be formed from a coalescence of the π^- d quark and the $\bar{c}$ quark. Since the $D^+(c\bar{d})$ does not contain a quark in common with the beam, it cannot be produced the same way. Therefore, we might expect an excess of D^- mesons to D^+ mesons, especially at high x_F. This "leading particle effect" is clearly seen in the E791 data, as shown in Fig. 4 and reported in Ref. [1]. While the default version of Pythia predicts too much coalescence at low x_F, a modified ("tuned") version of Pythia provides

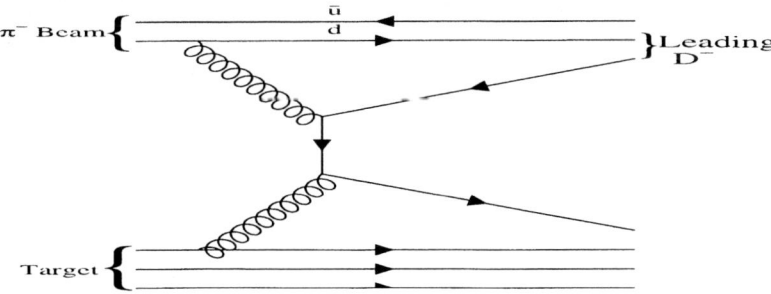

FIGURE 3. Coalescence of a valence d quark from the pion with the produced $\bar{c}$ quark leads to a high momentum $D^-(\bar{c}d)$.

good agreement with the data. The modifications to Pythia included increasing the average intrinsic k_t of the partons from 0.44 GeV/c to 1.00 GeV/c and increasing the charm quark mass from 1.35 GeV/c^2 to 1.70 GeV/c^2. This modified version of Pythia also appears to give the correct result for the p_T^2 asymmetry.

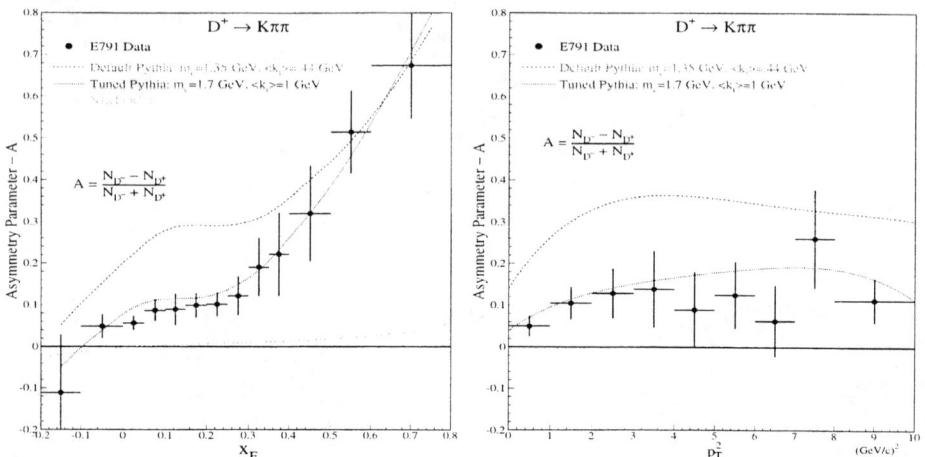

FIGURE 4. $D^\pm$ asymmetry versus x_F and p_T^2 compared to the default and "tuned" Pythia predictions as well as the next-to-leading order $c\bar{c}$ prediction.

Neither charge state of the $D_s(c\bar{s})$ meson has a quark in common with the beam or target hadrons. Therefore, we expect a flat, approximately zero, asymmetry. The data are consistent with this prediction as well as with the slight asymmetry predicted by the default Pythia program as seen in Fig. 5 and reported in Ref. [2].

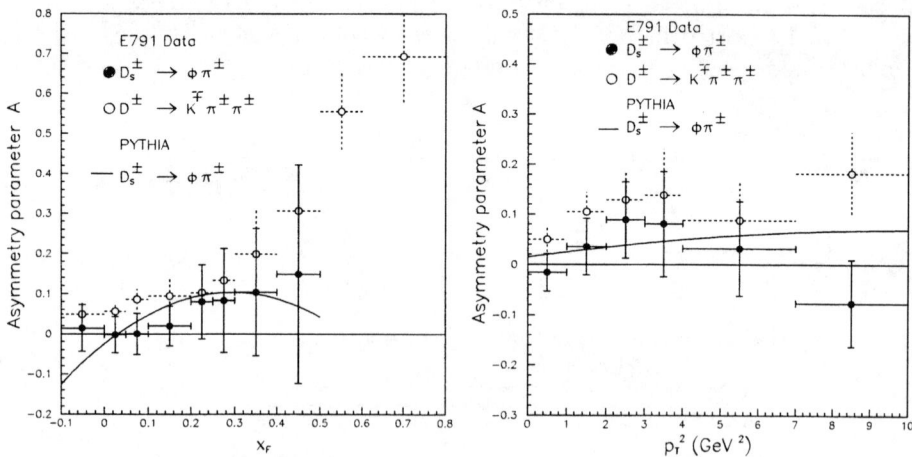

FIGURE 5. $D_s^\pm$ asymmetry versus x_F and p_T^2 compared to the default Pythia prediction.

Unlike the $D^\pm$ meson, the $\Lambda_c^\pm$ asymmetry is expected to be dominated by target, rather than beam, dragging. That is, the $\Lambda_c^+(udc)$ can come from a coalescence of a nucleon ud diquark and a c quark. This would give an increase in asymmetry as x_F decreases. The data shown in Fig. 6 are consistent with an increase at $x_F<0$, as predicted by the default Pythia program, and also consistent with a flat asymmetry. Clearly, more data are needed to determine the extent of the asymmetry.

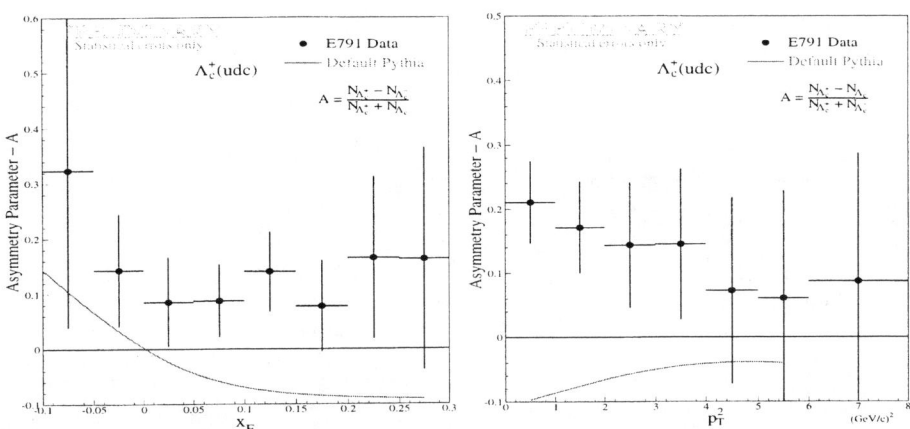

FIGURE 6. $\Lambda_c^\pm$ asymmetry versus x_F and p_T^2 compared to the default Pythia prediction.

To look for this target dragging effect with higher statistics we can measure the asymmetry for the Λ, Ξ, and Ω hyperons. As seen in Fig. 7, the magnitude of the asymmetry at negative x_F clearly increases with the number of non-strange quarks in the produced baryon, that is, the number of quarks the hyperon has in common with the target nucleons. In addition, the Ξ asymmetry increases at positive x_F relative to the Λ asymmetry; perhaps due to dragging by the $\pi^-(\bar{u}d)$ beam.

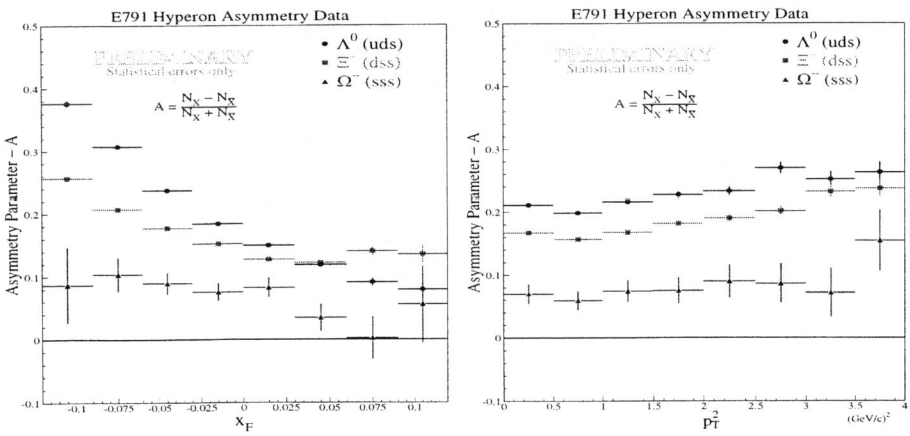

FIGURE 7. Hyperon asymmetry versus x_F and p_T^2.

CHARM-PAIR CORRELATIONS

Charm-pair correlations allow one to measure aspects of the charm quark production as well as the hadronization. The correlations are obtained from events in which both the charm and anticharm particles are reconstructed. We can construct transverse correlations such as $\Delta\phi$, $\Delta p_t^2 \equiv |p_{t,D}^2 - p_{t,\overline{D}}^2|$, and $p_{t,D\overline{D}}^2 \equiv |\vec{p}_{t,D} + \vec{p}_{t,\overline{D}}|^2$. These correlations are sensitive to the charm quark production mechanism and relatively insensitive to hadronization effects. The longitudinal correlations such as Δx_F are quite sensitive to hadronization effects and less so to the charm quark production.

The E791 analysis finds 791±44 fully reconstructed charm-pair events which, as shown in Table 1, is an order of magnitude more than any other hadroproduction experiment . The scatter plot of events as well as the function used to fit the data are shown in Fig. 8.

TABLE 1. Summary of fully-reconstructed and partially-reconstructed charm-pair samples from hadroproduction and photoproduction fixed-target experiments.

Expt	Energy	Beam	Target	Number of Pairs			
E791 [3]	500 GeV	π^-	Pt, C	791	fully		
NA27 [12]	360 GeV	π^-	H_2	12	fully	53	partially
WA92 [7]	350 GeV	π^-	Cu			475	partially
WA75 [8]	350 GeV	π^-	emulsion			177	partially
NA32 [9] [10]	230 GeV	π^-	Cu	20	fully	642	partially
E653 [11]	800 GeV	p	emulsion			35	partially
NA27 [6]	400 GeV	p	H_2	17	fully	107	partially
E687 [13]	200 GeV	γ	Be	325	fully	4534	partially
NA14 [14]	100 GeV	γ	Si	22	fully		

We compare the data to results from a next-to-leading order (NLO) calculation by Mangano, Nason, and Ridolfi (MNR) [4] and Pythia [5] using the default parameters for both. Some of the results are shown in Fig. 9; full details can be found in Ref. [3]. In leading order perturbative QCD, $\Delta\phi$ is predicted to be a delta function at 180°. While NLO terms smear the distribution, it is clear this is not enough to agree with the data. Adding intrinsic k_t to the NLO prediction helps improve agreement. While Pythia only uses leading order matrix elements in the cross section calculation, the intrinsic k_t and parton showers smear the $\Delta\phi$ distribution even more than the NLO prediction, although still not enough to account for the data. The $p_{t,D\overline{D}}^2$ distributions show similar discrepancies between theory and data. The Δp_t^2 distributions, however, are consistent between data and theory. For all three transverse variables shown, the Pythia hadronization has little effect. For the longitudinal correlations, however, this is not the case. The Δx_F distributions are approximately the same between the two $c\bar{c}$ predictions and come close to matching the data. After hadronization, however, the agreement is much

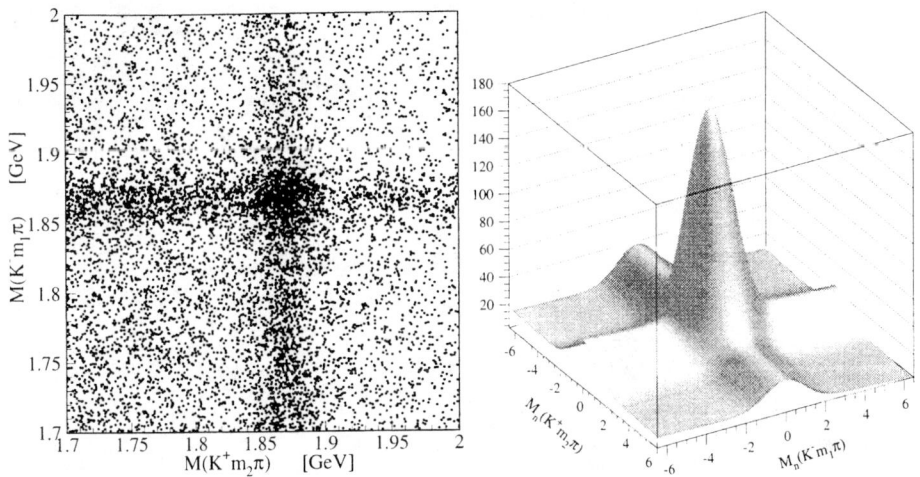

FIGURE 8. Scatter plot of the D-candidate mass versus the $\overline{D}$-candidate mass for the final unweighted charm-pair sample (left) and the function that maximizes the joint probability of the unweighted charm-pair candidates (right).

worse. The beam/target dragging in Pythia pulls the D mesons too far apart, creating the disagreement. Therefore, while we need the hadronization to explain the asymmetry data, it creates discrepancies in the longitudinal correlations of the charm and anticharm particles.

DIFFERENTIAL CROSS SECTIONS

The differential cross sections are cross sections measured as a function of kinematic variables such as x_F and p_T^2. The shapes of these distributions are sensitive to both the charm quark production and the subsequent hadronization. Figure 10 shows preliminary results for x_F and p_T^2 distributions from ~100,000 $D^0 \to K\pi, K\pi\pi\pi$ events. The high statistics from this open geometry experiment clearly shows the x_F turnover point is above zero. This is expected since the pion partons are generally harder than the nucleon partons and therefore the parton center-of-mass is boosted forward relative to the hadron center-of-mass (where x_F is measured). The high statistics also allow a good p_T^2 measurement out to 18 $(\text{GeV}/c)^2$. A comparison to predictions from the MNR [4] NLO calculation for charm quarks and a Pythia prediction for D^0 mesons is also shown in Fig. 10. The default parameters are used for both and the normalization adjusted to provide the best fit to the data. For the x_F and p_T^2 distributions, the NLO charm quark prediction is much closer to the data than the Pythia D^0 meson prediction. This implies a cancelation of non-perturbative effects such as intrinsic k_t and fragmentation. Fragmentation will soften the p_T^2 spectrum while intrinsic k_t will counter

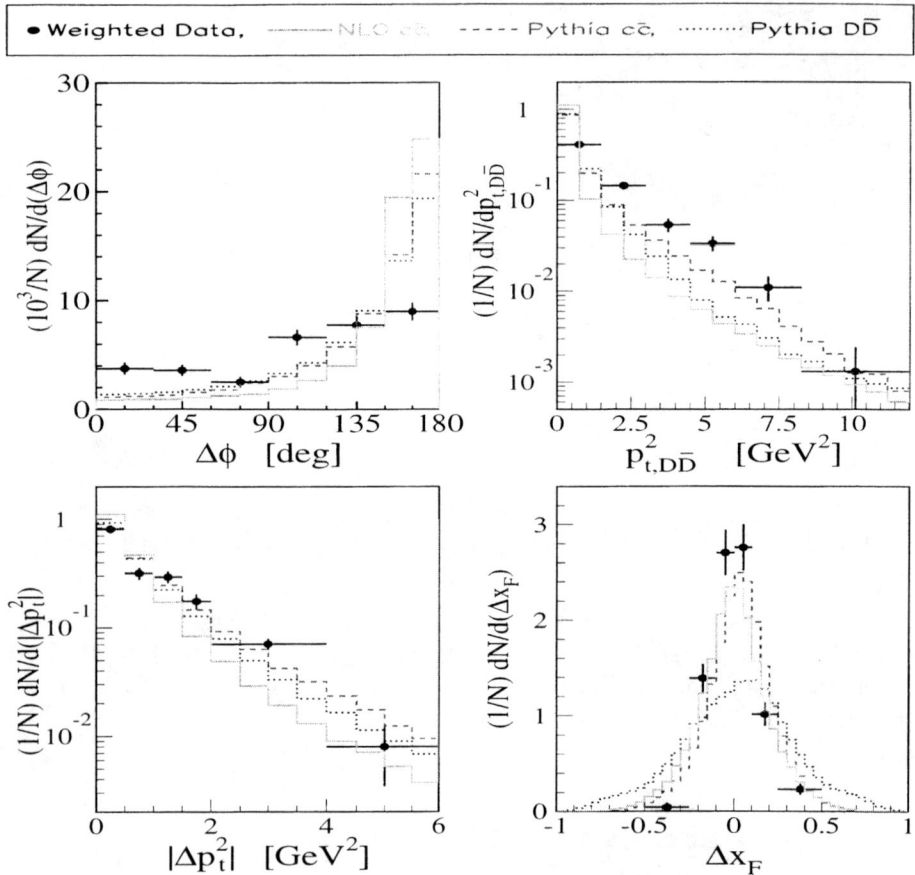

FIGURE 9. Charm-pair correlation results versus theoretical predictions from next-to-leading order $c\bar{c}$ calculations, Pythia $c\bar{c}$ results, and Pythia $D\bar{D}$ results.

that trend. Pythia hadronization appears to soften the p_T^2 spectrum too much. In addition, the excess of events in the Pythia x_F spectrum at $x_F > 0.2$ is probably due to too much beam dragging as has been seen in the $D^{\pm}$ asymmetry and charm-pairs Δx_F data.

CONCLUSIONS

E791 has provided, and will continue to provide, many results on charm hadroproduction. Large theoretical uncertainties imply that no single result, or even single experiment, can uniquely determine the physics of charm production. A synthesis of the results from E791 and other experiments may give a coherent picture.

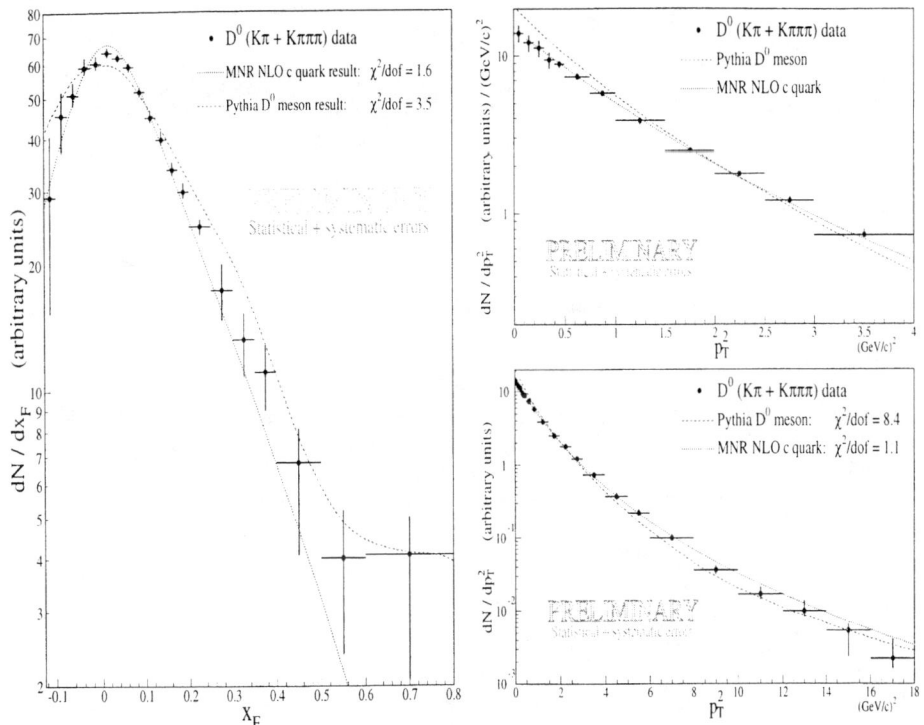

FIGURE 10. D^0 x_F and p_T^2 distributions compared to NLO c quark and Pythia D^0 predictions.

REFERENCES

1. Aitala, E. M. et al., E791 Collaboration, *Phys. Lett.* **B 371** 157 (1996).
2. Aitala, E. M. et al., E791 Collaboration, *Phys. Lett.* **B 411** 230 (1997).
3. Aitala, E. M. et al., E791 Collaboration, submitted to *Phys. Rev. D.*, hep-ex/9809029.
4. M. Mangano, P. Nason and G. Ridolfi, *Nucl. Phys.* B **405**, 507 (1993).
5. H.-U. Bengtsson and T. Sjöstrand, *Computer Physics Commun.* **46**, 43 (1987).
6. NA27 Collaboration, M. Aguilar-Benitez et al., *Z. Phys.* **C 40**, 321 (1988).
7. M. Adamovich et al., WA92 Collaboration, *Phys. Lett.* **B 348**, 256 (1995); *Nucl. Phys.* **B 495**, 3 (1997); *Phys. Lett.* **B 385**, 487 (1996).
8. WA75 Collaboration, S. Aoki et al., *Phys. Lett.* **B 209**, 113 (1988); *Progress in Theoretical Physics* **87**, 1315 (1992).
9. S. Barlag et al., NA32 Collaboration, *Phys. Lett.* **B 257**, 519 (1991); *Phys. Lett.* **B 302**, 112 (1993).
10. K. Rybicki and R. Rylko, *Phys. Lett.* **B 353**, 547 (1995).
11. K. Kodama et al., E653 Collaboration, *Phys. Lett.* **B 263**, 579 (1991).
12. NA27 Collaboration, M. Aguilar-Benitez et al., *Phys. Lett.* **B 164**, 404 (1985).
13. E687 Collaboration, P. L. Frabetti et al., *Phys. Lett.* **B 308**, 193 (1993).
14. NA14 Collaboration, M. P. Alvarez et al., *Phys. Lett.* **B 278**, 385 (1992).

Charm Physics Results from SELEX

The SELEX Collaboration:
presented by Alexander Y. Kushnirenko[†]

[†]*Physics Department, Carnegie Mellon University, Pittsburgh, PA, 15213, USA*

Abstract. The SELEX experiment (E781) [1] at Fermilab is a new fixed target multi-stage spectrometer with high acceptance for forward interactions and decays. It took data in 1996-97 with 600 GeV Σ^-, π^- and 540 GeV p beams, collecting large sample of charm decays.

Preliminary results on charm - anticharm production asymmetries, Λ_c^+ production x_F dependence in different beams, Λ_c^+ lifetime, and the first observation of the Cabibbo-suppressed decay $\Xi_c^+ \to pK^-\pi^+$ are presented.

I INTRODUCTION

Charm hadroproduction is still one of the most challenging problems in both experimental and theoretical physics [2]. The SELEX experiment has a unique opportunity to resolve many experimental aspects of this problem. It used a variety of beam particles and different target materials. Its design provided the capability to identify many different charm states and provided high sensitivity to charm particles produced at large x_F.

Another goal of SELEX is to study the charm baryon sector, especially charm-strange baryons, the least well-understood charm hadrons. Precision measurement of their lifetimes and studies of their excited states are important inputs for further development of the QCD $1/M_Q$ expansion.

II EXPERIMENT OVERVIEW

The SELEX experiment used the Fermilab Charged Hyperon beam, composed of 50% Σ^-, 50% π^- for negative polarity and 92% p, 8% π^+ for positive polarity. The beam was run at 0 mrad production angle. Beam particles were identified by Beam TRD detectors. The experiment was designed to have high acceptance and resolution in the x_F region $0.1 < x_F < 1$. The spectrometer layout is shown in Fig. 1.

Interactions happened in 5 foil targets: 3 diamond and 2 copper. The total interaction length of all targets was about 5% for protons. Trigger requirements

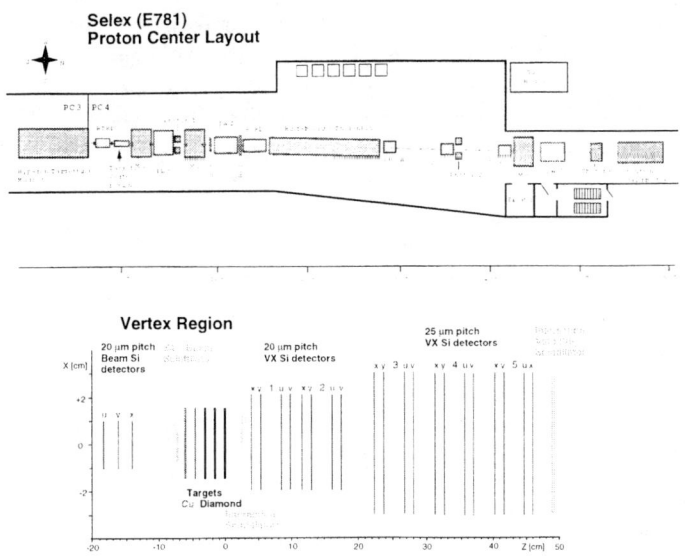

FIGURE 1. E781 layout

were very loose: > 4 charged tracks in the forward 150-mrad cone and > 2 hits in a counter hodoscope located after the second analyzing magnet. We triggered on about 30% of inelastic interactions.

Downstream of the target region, charged reaction products were measured in 20 vertex silicon planes having position resolution of about $4\mu m$ for infinite momentum tracks. Beam tracks were measured in 8 planes of $20\mu m$ pitch beam silicon. The downstream tracking system included 18 silicon planes for precise measurement of high momentum (100-600 GeV) tracks. In total the SELEX silicon system had 74000 strips with analog and digital information.

Another distinct feature of SELEX was an extensive particle ID system, which included

- 3000 phototube RICH with K/π separation up to 165 GeV [3].
- Beam TRD for beam tagging ($\Sigma^-/\pi^-, p/\pi^+$)
- Downstream electron TRD to identify secondary $e^\pm$ for semileptonic decay studies.
- 3 lead-glass photon calorimeters covering the forward hemisphere

The downstream tracking system had 3 analyzing magnets, 26 PWC planes with position resolution $\sim 0.6 - 1mm$, and 3 vector drift chamber stations with 24 $\sim 100\mu m$ resolution planes each.

A major innovation in E781 was the use of an **online charm filter**: a program that analyzed events online and selected those that have evidence for a secondary vertex. The filter algorithm started from downstream tracking to find high-momentum ($p > 20 GeV$) tracks in the PWCs after the second analyzing magnet. Candidate tracks were extrapolated back to vertex silicon within roads predicted by downstream tracking. If hits in silicon planes along these roads form tracks for at least two candidates, a full vertex reconstruction of all these high-momentum tracks found in the silicon detectors was performed. The filter rejected events if all tracks form only a primary vertex. This reduced the data size by a factor of 8, at a cost of about factor of 2 in charm written to tape, as a study of special sets of unfiltered data, taken simultaneously, showed.

III CHARM PHYSICS RESULTS

Results presented in this talk were obtained in the first processing of the whole data set. In total in this analysis we processed events recorded from 15.2 billion inelastic interactions. The distribution between different beam particles is summarized in Table 1. It should be mentioned that the code for the first data processing was not fully optimized, and we expect higher final yields after a second data processing pass.

Beam	Billion interactions
Σ^-	10.20
π^-	2.08
p	2.77
π^+	0.14

TABLE 1. E781 data sets

To extract charm signals we applied the following cuts

1. Good separation between primary and secondary vertices: $L/\sigma > 8$, where L was the distance between vertices and σ is an error on L
2. Secondary tracks formed a good secondary vertex $\chi^2_{sec} < 5$
3. Charm track pointed back to primary vertex
4. K, p were identified by RICH

To suppress background even more we applied additional cuts:

5. Events with slow secondary pion ($P_\pi > 8 GeV$) were excluded, as multiple scattering for those tracks was so big, that it was not possible to resolve if they came from primary or secondary vertex.
6. Charm secondary tracks had a minimum transverse miss distance at the primary vertex

Mass spectra of D^0, D^+, D_s^+ and Λ_c^+ candidates after the first four cuts were applied to our sample are shown in Fig. 2.

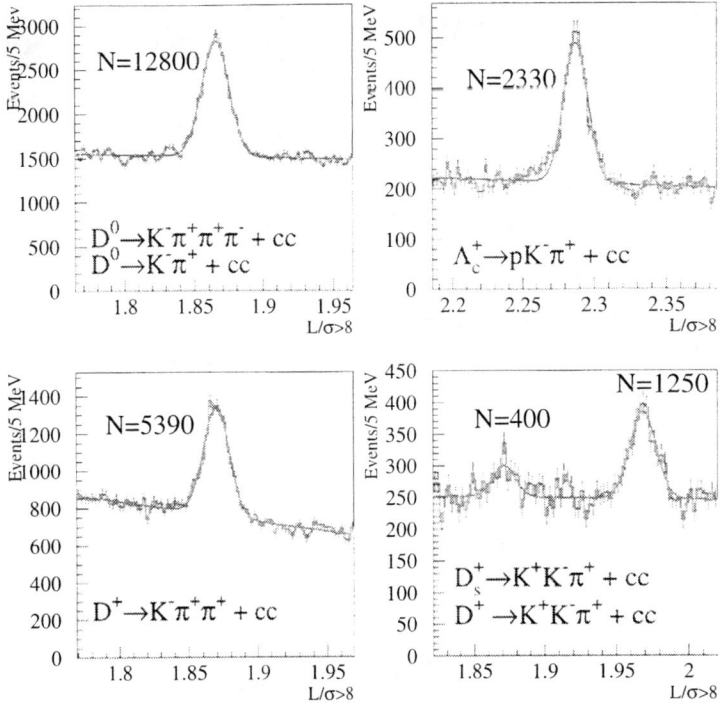

FIGURE 2. E781 charm signals for $D^0, D^+, D_s^+, \Lambda_c^+$

A Charm Hadroproduction

Production of $c\bar{c}$ pairs to the lowest order QCD is symmetric. Next to leading order (NLO) perturbative QCD calculations do not predict large asymmetries either [2,4,5]. However recent experimental data show [8,9] that charm particles which share valence quarks with the beam particle (leading particle) are produced more copiously in forward direction as compared to charm particles which do not share valence quarks with the beam particle (non-leading particle). Leading particle phenomenology also predicts higher asymmetry for leading/non-leading particles at large x_F. Progress is being made in developing both intrinsic charm models [6] and string-fragmentation models [7] to explain the data.

E781 preliminary results on the integrated charm production asymmetry for $x_F > 0.3$ are summarized in Fig. 3. The asymmetry for charm baryons $\Lambda_c^+/\overline{\Lambda_c^-}$ is much stronger than for charm mesons D^-/D^+, especially when the charm baryons are produced by a baryon (proton or Σ^-) beam. The large $\Lambda_c^+/\overline{\Lambda_c^-}$ asymmetry for Σ^- is consistent with WA89 results [9]. Also results for D^-/D^+ asymmetry in π^- beam are consistent with E791 results [8].

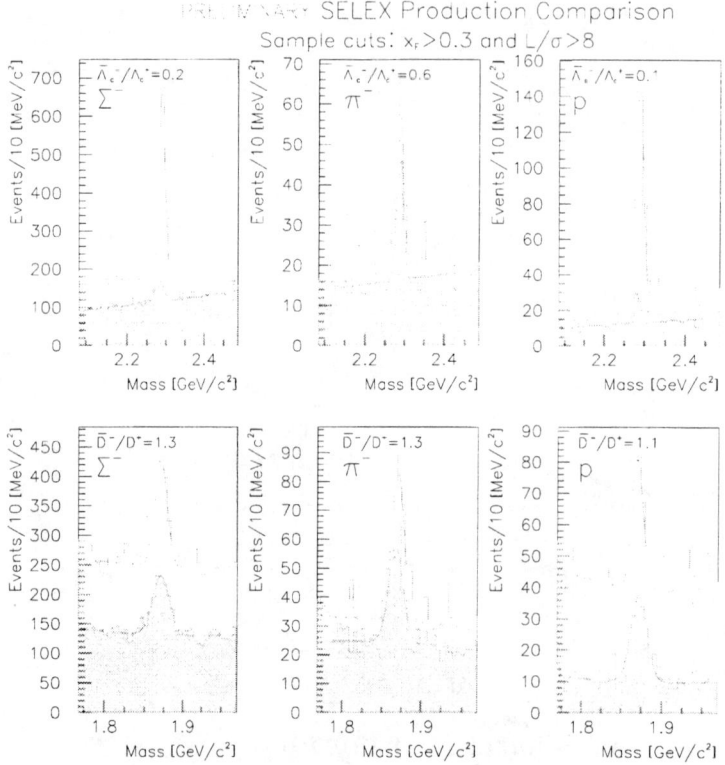

FIGURE 3. Charm and and Anticharm signals for $x_F > 0.3$ for Σ^-, π^- and p beams. $\Lambda_c^+ \to pK^-\pi^+$ and $D^+ \to K^-\pi^+\pi^+$ decay modes were used from charm particles. Charge conjugate modes were used for anticharm particles.

TABLE 2. Relative Charmed Particle Yields for $x_F \geq 0.3$ versus beam type

Relative Yields	p	π^-	Σ^-
$\overline{\Lambda_c}$	0.25	1.0	1.1
Λ_c^+	0.8	1.0	1.2
D^-	0.4	1.0	0.8
D^+	0.2	1.0	0.4

The relative yields of charm particles normalized to production in the π^- beam are summarised in Table 2. This study required careful comparison of data sets for different beams and is still at a very preliminary stage. Thus, no errors were included in the table. A surprising result from this table is that baryon beams, especially Σ^- are very good charm baryon producers, at least for $x_F > 0.3$. It will be interesting to extend this table to cover charm-strange baryons when we extend our analysis. For those states we expect higher yields for Σ^- beams.

A preliminary result on the x_F dependence of Λ_c^+ production for three different beams is shown in Fig. 4. For all three beams Λ_c^+ is a leading particle, but the spectrum is much softer for the pion than the baryon beams. Another interesting result is that hadroproduction by Σ^- seems to be different from that by protons.

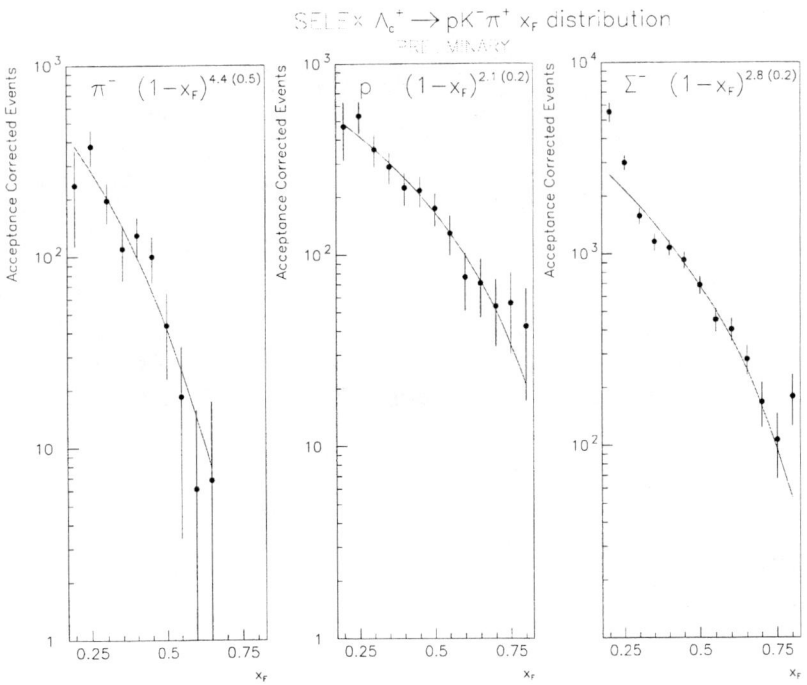

FIGURE 4. Λ_c^+ production as a function of Xf for three different beam particles

B Λ_c^+ Lifetime measurement

To select clean sample of Λ_c^+ events we required a vertex-separation significance $L/\sigma > 10$. We used the conventional reduced proper time analysis [10], histogramming $ct_r = \frac{L-10\sigma}{\gamma}$. The signal distribution was extracted using the sideband-subtraction method. The efficiency of the apparatus as a function of ct_r was found using D^0 samples from these data and checked using K_s decays which occur in the first few centimeters after the primary interaction. This experimentally-derived acceptance was applied to correct the Λ_c^+ ct_r distribution. Results of this study are summarised in Fig. 5.

Our preliminary result for the Λ_c^+ lifetime is $177 \pm 10 fs$. Only the statistical error is included. This result is close to the PDG value $206 \pm 12 fs$ [11]. One of the checks of systematic error is to compare the Λ_c^+ lifetime result from each of the 5 different targets. This is shown in lower-left plot in Fig. 5. The results are independent of the target within statistical errors.

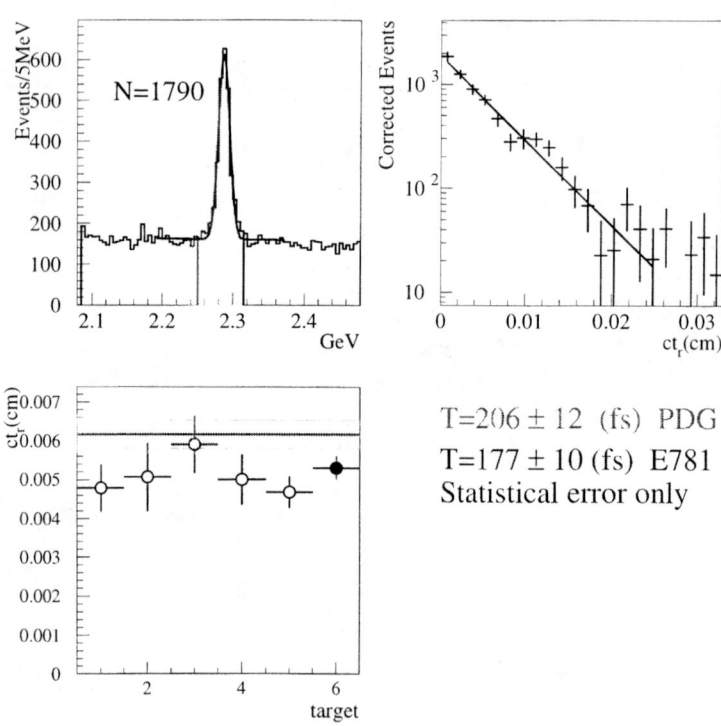

FIGURE 5. Λ_c^+ Preliminary Lifetime measurement

C Reconstruction of Charm-Strange Baryons

The majority of charm-strange baryon decays have hyperons in their final state: $\Sigma^+, \Sigma^-, \Lambda^0, \Xi^-, \Xi^0, \Omega^-$. In our apparatus a large fraction (80-90%) of those hyperons decayed inside magnetic field of the first analyzing magnet (M1), or before it. Those hyperon decays are hard to reconstruct, and we did not reconstruct them in the first processing of the data. We do plan to do it in the next pass through the data. Instead we reconstructed those charged hyperons that decayed after magnet M1 (10-20%), so that the hyperon track was momentum-analyzed. The consequence is that in this analysis we have only limited acceptance for charm-strange baryons.

SELEX results for reconstruction of charm baryon decays with hyperons in the final state are shown in Fig. 6 on the two left plots. The bump on the right side of $\Lambda_c^+ \to \Sigma^- \pi^+ \pi^+$ is the reflection on this plot of $\Xi_c \to \Xi^- \pi^+ \pi^+$. In our hyperon reconstruction method we cannot distinguish between Σ^- and Ξ^-.

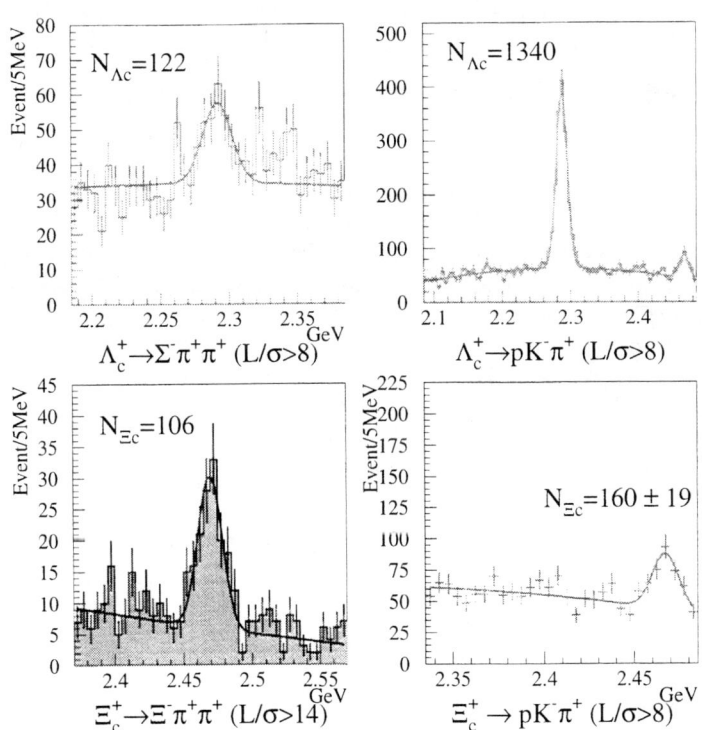

FIGURE 6. Observed charm baryon states

D First observation of Cabibbo suppressed $\Xi_c^+ \to pK^-\pi^+$ decay

The upper-right plot in Fig. 6 shows our preliminary result for the mass reconstruction of the $pK^-\pi^+$ final state. The big peak in the middle represents the expected $\Lambda_c^+ \to pK^-\pi^+$ decay. There is a little bump in the right corner of the mass plot, corresponding to the Cabibbo-suppressed decay $\Xi_c^+ \to pK^-\pi^+$. An enlarged plot of this region is shown in the lower-right figure. The observed signal corresponds to a $\sim 7\sigma$ effect. In this preliminary analysis there is no right shoulder of the background on the plot because we did not expect any signal in that region and saved only $pK^-\pi^+$ events which had mass inside 200 MeV window around Λ_c mass. That will be fixed in the next data reprocessing pass.

Another confirmation that the observed decay is indeed Cabibbo suppressed comes from an estimate of the branching ratio of this mode to the analogous Cabibbo-allowed mode $\Xi_c^+ \to \Sigma^+ K^-\pi^+$. We have only a small signal in $\Xi_c^+ \to \Sigma^+ K^-\pi^+$ mode, but a stronger signal in the companion mode $\Xi_c^+ \to \Xi^-\pi^+\pi^+$ signal (Fig. 6). The branching ratio for the two favored modes has been measured by CLEO to be $1.18 \pm 0.26 \pm 0.17$ [12]. Using Monte-Carlo simulation to calculate the apparatus efficiency, we estimate a branching ratio:

$$Br \frac{\Xi_c^+ \to pK^-\pi^+}{\Xi_c^+ \to \Sigma^+ K^-\pi^+} = 0.16 \pm 0.07 \simeq 3 \times \tan^2 \vartheta_c$$

This is a very preliminary result with only the statistical error included. Many more systematic studies are under way. Approximately one could expect branching ratio of Cabibbo suppressed to Cabibbo allowed modes to be of the order of $\tan^2 \vartheta_c$, and indeed our observed ratio was of that order. One can also compare the Cabibbo-suppressed Ξ_c^+ branching ratio to a different charm baryon Cabibbo-suppressed decay observed by E687 [13]:

$$BR \frac{\Lambda_c^+ \to pK^-K^+}{\Lambda_c^+ \to pK^-\pi^+} = 0.096 \pm 0.029 \pm 0.010$$

The scale of the two decays is clearly similar.

IV SUMMARY

SELEX processed data from 15.2 billion inelastic interactions and has **preliminary** results on:

- the observation of main decay modes for all charm mesons and Λ_c^+, Ξ_c^+ baryons

- $\overline{\Lambda_c^-}/\Lambda_c^+, D^-/D^+$ integrated production asymmetry for $x_F > 0.3$ in p, Σ^- and π^- beams

- Λ_c^+ production as a function of x_F for all three beams
- Λ_c^+ lifetime
- First observation of the Cabibbo-suppressed $\Xi_c^+ \to pK^-\pi^+$ decay

The first processing of the data produced new interesting results and also was a very valuable lesson in offline algorithm studies, showing several places for code improvement. Extending algorithms for hyperon reconstruction will allow us to increase the charm-strange baryon sample. SELEX is going to process the whole data set once again with more efficient code. Our current efforts also include tuning the Monte-Carlo simulation for accurate efficiency calculations.

ACKNOWLEDGEMENTS

We are indebted to B. C. LaVoy, D. Northacker, F. Pearsall, and J. Zimmer for invaluable technical support. This project was supported in part by Bundesministerium für Bildung, Wissenschaft, Forschung und Technologie, Consejo Nacional de Ciencia y Tecnología (CONACyT), Conselho Nacional de Desenvolvimento Científico e Tecnológico, Fondo de Apoyo a la Investigación (UASLP), Fundação de Amparo à Pesquisa do Estado de São Paulo (FAPESP), the Israel Science Foundation founded by the Israel Academy of Sciences and Humanities, Istituto Nazionale de Fisica Nucleare (INFN), the International Science Foundation (ISF), the National Science Foundation (Phy #9602178), NATO (grant CR6.941058-1360/94), the Russian Academy of Science, the Russian Ministry of Science and Technology, the Turkish Scientific and Technological Research Board (TÜBİTAK), the U.S. Department of Energy (DOE grant DE-FG02-91ER40664 and DOE contract number DE-AC02-76CHO3000), and the U.S.-Israel Binational Science Foundation (BSF).

REFERENCES

1. Carnegie Mellon University, Fermilab, University of Iowa, University of Rochester, University of Hawaii, University of Michigan-Flint, Petersburg Nuclear Physics Institute (Russia), Institute for Theoretical and Experimental Physics (Moscow), Institute for High Energy Physics (Protvino), Moscow State University, University of Sao Paulo, Centro Brasileiro de Pesquisas Fisicas, Universidade Federal de Paraiba, Insitute of High Energy Physics (Beijing), University of Bristol, Tel Aviv University, Max Planck Institut fuer Kernphysik (Heidelberg), University of Trieste and INFN, University of Rome and INFN, Universidad Autonoma de San Luis Potosi, Bogazici University
2. S. Frixione, M. L. Mangano, P. Nason, G. Ridolfi, "Heavy Flavours II", A.J. Buras and M. Lindner, eds., World Scientific Publishing Co, Singapore(1997); see also preprint hep-ph/9702287
3. J. Engelfried et al.: The SELEX Phototube RICH Detector; hep-ex/9811001. Submitted to Nuclear Instruments and Methods A.

4. P. Nason, S. Dawson, R.K. Ellis. *Nucl. Phys.* **B327** (1989) 49.
5. S. Frixione, M. L. Mangano, P. Nason, G. Ridolfi, *Nucl. Phys.* **B431** (1994) 453.
6. T. Gutierrez, R. Vogt, preprint hep-ph/9808213
7. E. Norrbin, T. Sjostrand, submitted to *Phys.Lett.* **B**; preprint hep-ph/9809266. See also these proceedings.
8. E791 collaboration, E.M. Aitala et al., *Phys.Lett.* **B371** (1996) 157.
9. WA89 collaboration, M.I.Adamovich et al., submitted to Eur. Phys. J. C; preprint hep-ex/9803021
10. E687 Collaboration, P.L. Frabetti et al.,*Phys.Rev.Lett.* **70** (1993)1755
11. Particle Data Group, *Phys. Rev.* **D54** (1996) 1.
12. CLEO collaboration, T. Bergfeld et al.. *Phys. Lett.* **B365** (1996) 431.
13. E687 Collaboration, P.L. Frabetti et al., *Phys. Lett.* **B314** (1993) 477.

Charm and Beauty Production in Experiment WA92

D. Barberis*, C. Gemme* and L. Malferrari[†]

*University of Genova and INFN, Via Dodecaneso 33, I-16146 Genova, Italy
[†]University of Bologna and INFN, Via Irnerio 46, I-40126 Bologna, Italy

Abstract. Using a sample of $1.4 \cdot 10^8$ triggered events, produced in π–Cu and π–W interactions at 350 GeV/c, we have studied charm and beauty production. J/ψ and ψ' mesons are identified through their decays into muon pairs and are selected in the Feynman-x range $x_F > 0$. Inclusive J/ψ and ψ' cross-sections and J/ψ differential cross-sections as functions of x_F and p_T are measured. A fit to the data for different targets gives $\alpha(J/\psi)\{Si, Cu, W\} = 0.87 \pm 0.05$ (stat.) ± 0.04 (syst.), $\sigma_0(J/\psi) = (216.2 \pm 5.4$ (stat.) ± 13.3 (syst.)) nb/nucleon and $\sigma_0(\psi') = (28.3 \pm 8.1$ (stat.) ± 11.3 (syst.)) nb/nucleon. For the study of beauty production, we have identified 26 beauty events. The estimated background in this sample is 0.6±0.6 events. From these data, assuming a linear A-dependence, we measure a beauty production cross-section integrated over all x_F of $5.7^{+1.3}_{-1.1}$ (stat.) $^{+0.6}_{-0.5}$ (syst.) nb/N. Differential distributions with respect to x_F and p_T^2 have been determined as well as some two-particle kinematic variables. Our results are compared with previous experiments and with a next-to-leading-order QCD calculation.

I INTRODUCTION

Heavy quark hadroproduction can in principle be described by perturbative QCD; complete calculations are nowadays available at next-to-leading order (NLO) [1]. These predictions are affected by theoretical uncertainties but, because of the large b and c-quark masses, the measurements of the charm and beauty hadroproduction cross-sections and other production characteristics are good tests of perturbative QCD.

The first theoretical descriptions of charmonium hadroproduction were the Colour-Evaporation Model [2] and the Colour-Singlet Model [3]. According to these models, the expected dominant contribution to charmonium production comes from quark-antiquark annihilation or gluon-gluon fusion, giving origin to a bound $c\bar{c}$ pair. In a recent extension of the Colour-Singlet Model, the so called Factorization Approach (FA) [4], a quark-antiquark pair can be produced on a short distance scale in both a colour-singlet or a colour-octet state. In the case of the colour-octet state the pair evolves non-perturbatively at long distance into a charmonium state by emitting soft gluons. On the other side, in the description of the Colour-Evaporation Model (CEM) there is no treatment of colour, that is assumed to be readjusted by the emission of soft gluons before the $c\bar{c}$ pair appears as an asymptotic ψ. The model assumes that charmonium states come from any $c\bar{c}$ pair created below the threshold for open charm production. Different charmonium states are predicted to be produced, independently of the center of mass energy, as a constant fraction of the total charmonium cross-section.

Data taking for the WA92 experiment was performed at the CERN Ω' spectrometer in 1992 and 1993, with a 350 GeV/c π^- beam incident on a 2 mm copper or tungsten target. The number of events recorded was 10^8 on a Cu target and $4 \cdot 10^7$ on a W target, corresponding to integrated luminosities of 8.1 (nb per Cu nucleus)$^{-1}$ and 1.3 (nb per W nucleus)$^{-1}$.

Full details of the experimental apparatus and trigger are given elsewhere [5]. The trigger acceptance was $\sim 30\%$ for beauty events, $\sim 25\%$ for J/ψ and ψ' events and $\sim 2\%$ for inelastic interactions.

II J/ψ AND ψ' PRODUCTION

A Event selection

In this analysis events containing at least two reconstructed muons with opposite charge are selected. In order to reduce the background from muons that originate in hadron showers, at least one muon of a $\mu^+\mu^-$ pair is required to be associated with the primary vertex.

Among selected events, interactions in the target material, in the In-Target counter and in the Decay detector are all considered. A distinction between them is achievable by measuring the x-coordinate of the primary event vertex, that can be reconstructed with a resolution σ_x of about 115 μm. Interactions in the target material are selected by requiring the signed x-distance of the primary vertex from the upstream and downstream target edges, Δx_U and Δx_U respectively, to satisfy $\Delta x_U/\sigma_x > -3$ and $\Delta x_D/\sigma_x < -0.5$. Conversely, interactions in the silicon detector planes are selected by requiring $\Delta x_D/\sigma_x > 0.5$, i.e. the primary vertex downstream of the target edge. The above conditions allow good separation between interactions in the target and downstream interactions: ψ's produced in the target are selected with a negligible contamination from ψ's produced in silicon, while about 9% of the sample of ψ's selected in silicon is due to ψ's produced in the target.

In order to estimate the total background, a fit to the $\mu^+\mu^-$ invariant mass distribution is performed in the range 2.2-4.2 GeV/c^2. The fitting function with best χ^2 is found to be a sum of two Gaussians centred around the J/ψ region, plus two Gaussians centred around the ψ' region, plus two exponentials representing the background. After background subtraction, we have $320 \pm 19 \pm 3$ J/ψ produced on

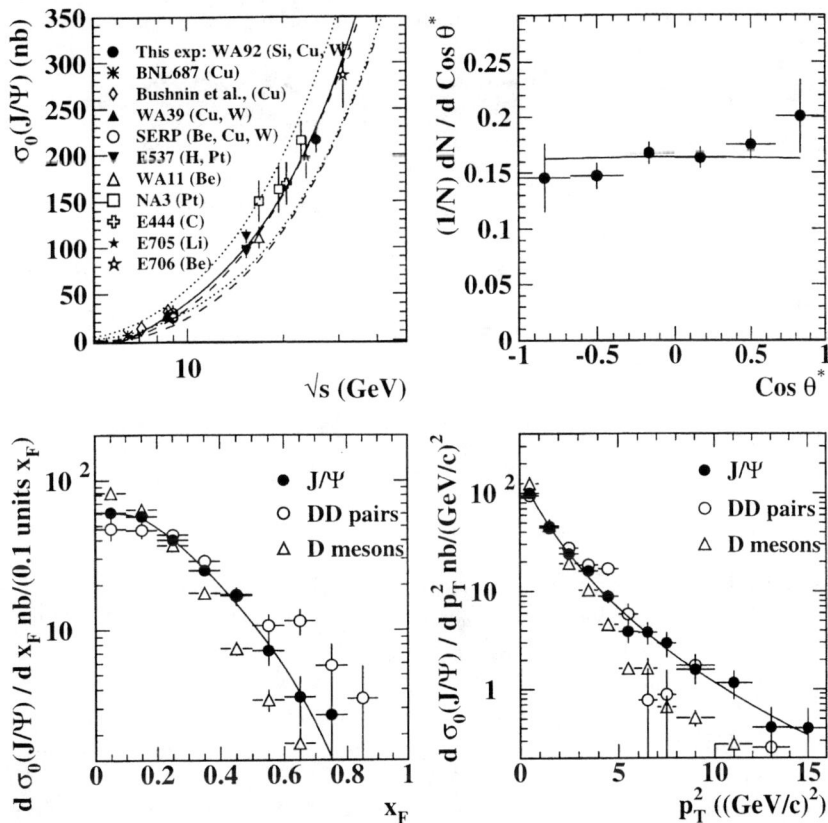

FIGURE 1. *Top left: cross-section per nucleon for J/ψ production by π^- as measured by different experiments. Top right: μ^+ emission angle in the J/ψ reference frame with respect to the J/ψ direction in the laboratory frame, after background subtraction and acceptance correction. Bottom: differential cross-section for J/ψ, $D\bar{D}$ pairs and events containing a reconstructed charmed meson as a function of x_F (left) and p_T^2 (right).*

tungsten, $719 \pm 28 \pm 4$ J/ψ on copper, $57 \pm 9 \pm 2$ J/ψ on silicon, $6.3 \pm 3.6 \pm 0.7$ ψ' produced on tungsten and $13.3 \pm 5.5 \pm 3.3$ ψ' on copper.

B Total and differential cross-sections

The nuclear cross-section is usually parameterized as $\sigma = \sigma_0 \, A^\alpha$, where σ_0 is the cross-section per nucleon and α is a parameter which is nearly independent of the centre of mass energy of the reaction. In this analysis the α parameter for inclusive J/ψ production can be determined from a fit to the three J/ψ production cross-sections measured for silicon, copper and tungsten. $\alpha(J/\psi)$ is found to be

$$\alpha(J/\psi)\{Si, Cu, W\} = 0.87 \pm 0.05 \text{ (stat.)} \pm 0.04 \text{ (syst.)}.$$

If the processes involved in J/ψ and ψ' production are the same, then similar values are expected for $\alpha(\psi')$ and $\alpha(J/\psi)$. Despite the large statistical uncertainties on ψ' cross-sections, as a cross-check, $\alpha(\psi')$ has been measured from copper and tungsten data, giving $\alpha(\psi')\{Cu, W\} = 0.86 \pm 0.66$ (stat.) ± 0.26 (syst.). For the remainder of the section the same value, $\alpha = 0.87$, for both J/ψ and ψ' is used to parameterize the A-dependence.

Using the value of $\alpha = 0.87$, the weighted averages of the inclusive J/ψ and ψ' cross-sections at $x_F > 0$ and their ratio result

$$\sigma_0(J/\psi) = (216.2 \pm 5.4 \text{ (stat.)} \pm 13.3 \text{ (syst.)}) \text{ nb/nucleon},$$

$$\sigma_0(\psi') = (28.3 \pm 8.1 \text{ (stat.)} \pm 11.3 \text{ (syst.)}) \text{ nb/nucleon, and}$$

$$\sigma_0(\psi')/\sigma_0(J/\psi) = 0.13 \pm 0.05 \text{ (stat.)} \pm 0.01 \text{ (syst.)},$$

where correlated systematic uncertainties have been taken into account. Inclusive J/ψ cross-sections per nucleon from various experiments are displayed in Fig. 1a), where errors include both statistical and systematic contributions, when available. All data have been adjusted for the assumption of $\alpha = 0.87$, for the branching fraction $B(J/\psi \to \mu^+\mu^-) = 6.01\%$ and for the kinematical region $x_F > 0$.

Experimental data can also be compared to the theoretical predictions provided by some of the currently proposed models. Calculations performed using the Factorization Approach [6] predict at next-to-leading order a J/ψ cross-section energy dependence as shown in Fig. 1a) for two choices of the strong coupling constant, $\mu = m_c$ (lower dashed curve) and $\mu = 0.8\, m_c$ (upper dashed curve), where m_c is fixed at 1.5 GeV/c^2. The prediction of the total J/ψ cross-section has been evaluated also at next-to-leading order within the Color Evaporation Model framework [7,8]. The cross-section energy dependence is shown in Fig. 1a) for two values of the charm mass, $m_c = 1.28$ and 1.38 GeV/c^2 (upper and lower dotted curves). Both models describe rather well the energy dependence of the J/ψ cross-section, however the absolute normalization of the predictions is affected by large uncertainties because of the strong dependence on the choices of renormalization and factorization scale, parton-density functions and charm quark mass.

Cross-section distributions as function of x_F and p_T^2 have been determined for J/ψ events (Fig. 1c-d)). The present study offers the opportunity to compare, within the same experiment, data for charmonium production and open-charm production, the latter having recently been reported elsewhere [9].

The comparison of charmonium with open-charm production is also of particular interest as a test of the Colour-Evaporation Model. According to this model the inclusive J/ψ cross-section is a constant fraction of the open-charm cross-section, independently of the centre of mass energy. Taking the WA92 measurement of

$\sigma_0(D\bar{D}) = 12.37 \pm 0.19 \pm 0.86$ μb/nucleon (given for positive x_F and assuming $\alpha = 1$), we calculate the ratio:

$$\frac{\sigma_0(J/\psi)}{\sigma_0(D\bar{D})} = (1.75 \pm 0.05(\text{stat.}) \pm 0.15(\text{syst.})) \times 10^{-2},$$

where correlated systematics have been cancelled. This ratio is close to the value of 2×10^{-2}, estimated from data of experiments at different energies [10]. Within the CEM framework charmonium production is described by the same dynamics as $D\bar{D}$ production [10,11]. Therefore stringent similarities, in the limit $m_c \approx m_D$, between kinematic distributions of the measured cross-sections of charmonium and $D\bar{D}$ pairs are expected. WA92 already presented an analysis [12] based on a sample of 475 events, in which two charmed-particle decays are observed. The kinematical features observed in the $D\bar{D}$ pair distributions with respect to the J/ψ distributions are not fully consistent with the CEM expectations. However, corrections due to the approximation $m_c \approx m_D$ and to the energy carried by the soft gluon emission, involved in the final state interactions, could account for such differences.

C J/ψ polarization

Since the polarization magnitude is characteristic of the charmonium production process, its measurement is of particular importance in the comparison with predictions of theoretical models: in the Factorization Approach J/ψ and ψ' are produced with a large transverse polarization, estimated within the ranges $0.31 < \lambda(J/\psi) < 0.63$ and $0.15 < \lambda(\psi') < 0.44$ [6]. The CEM, on the contrary, predicts absence of polarization for J/ψ and ψ' production [10], as the initial polarization of $c\bar{c}$ pairs is destroyed by multiple exchanges of soft gluons.

The angular distribution of J/ψ decay muons, after background subtraction and acceptance correction, is shown in Fig. 1b). The μ^+ emission angle θ^* is calculated in the J/ψ rest frame with respect to the J/ψ direction in the laboratory frame. Assuming the angular distribution to be proportional to $(1 + \lambda \cdot \cos^2 \theta^*)$, where λ is the polarization magnitude, the measured value is $\lambda(J/\psi) = -0.01 \pm 0.19$.

III BEAUTY PRODUCTION

A Event selection

The procedure for extracting beauty candidates from our data sample of $\sim 10^8$ π–Cu events involves several selection steps. A complete account of the analysis strategy can be found elsewhere [13]. Secondary vertices are looked for in the region between 0.3 cm and 6 cm from the primary vertex, disregarding those close to large energy releases, usually due to hadronic interactions in the silicon planes. We then select three classes of events, consistent with beauty-decay topologies: μ events,

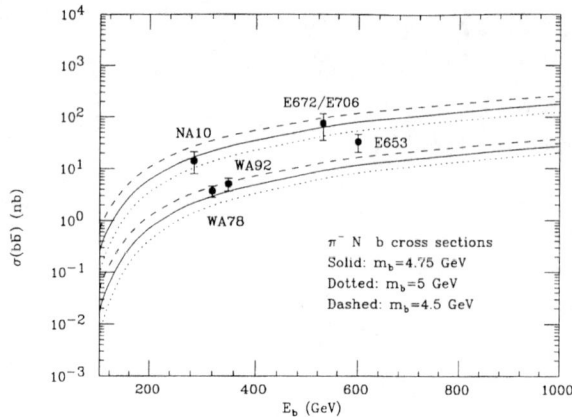

FIGURE 2. Beauty cross-section measurements in $\pi^- N$ interactions compared with theoretical predictions. Theoretical uncertainty bands are obtained by varying the b-quark mass (m_b), the factorization scale (μ_F), and the renormalization scale (μ_R). The band widths correspond to a variation of μ_F and μ_R between $m_b/2$ and $2m_b$.

characterized by a high p_T muon not associated with the primary vertex and at least one secondary vertex in the fiducial volume; *multivertex* events, events with at least three secondary vertices in the fiducial volume and satisfying a high-p_T-track requirement; *non-pointing D* events, with a fully-reconstructed Cabibbo-favoured D decay, where the D does not point back to the primary vertex, and where there is at least one other secondary vertex.

Beauty candidates have been identified in each of the three classes: 12 in the μ sample, 12 in the *multivertex* sample, 5 in the *non-pointing D* sample. In agreement with expectations from the simulation, one event is common to the three classes and one event is common to the *multivertex* and μ classes. Our total sample then consists of 26 events.

To study the background from primary charm, we simulated $c\bar{c}$ events which went through the analysis and selection chain as the experimental data; the number of simulated $c\bar{c}$ events surviving after scanning, normalized to the number of charm events in the data, is 0.6±0.6. Nuclear interactions and decays of pions and kaons into muons contribute a negligible fraction of events to the final data sample.

B Beauty cross–section

We measure an overall cross-section using all 26 beauty candidates, independent of the class to which they belong. The following beauty production cross-section, integrated over all x_F, is obtained:

$$\sigma_{b\bar{b}} = 5.7 \, {}^{+1.3}_{-1.1} \text{ (stat.) } {}^{+0.6}_{-0.5} \text{ (syst.)} = 5.7 \, {}^{+1.5}_{-1.2} \text{ nb/N}$$

Statistical errors include the fluctuations on the number of signal and background events; systematic errors take into account the uncertainties on the luminosity and efficiency.

In Fig. 2 we compare the present $\sigma_{b\bar{b}}$ measurement with results obtained in previous experiments. In the low energy region our result is in good agreement with that obtained in the WA78 experiment. The theoretical predictions [1], based on a next-to-leading-order QCD calculation, are also shown in Fig. 2. They are affected by large uncertainties, depending on the values assumed for the b-quark mass and for the factorization and renormalization scales. Within these uncertainties, the QCD predictions are in good agreement with our result.

C Beauty pair selection and momentum estimator

In our 26 $b\bar{b}$ events, nearly all of the secondary vertices are due to decays of charm and beauty. As a result, a secondary vertex to be considered as a candidate for an event's second beauty-particle decay is initially required to satisfy only minimal selection criteria, *i.e.* it must not be compatible with the decay of the charmed hadron from the beauty-particle decay already identified or from the other beauty particle. A vertex is wrongly identified as a beauty decay with a probability of only 3.6%, and the effect on our measurement is negligible.

Having applied all selection criteria, we identify the decay vertex of the second beauty particle in 13 of our 26 $b\bar{b}$ events, giving us a total of 39 beauty decays. The background, already low (0.6 ± 0.6) for the 26 $b\bar{b}$ events, is expected to be negligible for the 13 events with two identified beauty-decay vertices. The charmed-hadron daughter is identified for half of our 39 beauty particles. The basic composition of our beauty sample is well reproduced by a sample of simulated $b\bar{b}$ events passing the same selection criteria.

To obtain kinematic information from the 39 identified beauty decays, it is necessary to have an estimator for the momentum of the partially reconstructed vertex which takes into account the unseen decay products. We adopt a combination of two methods based on different physical considerations. The first method consists in closing the beauty decay by adding a missing particle, assumed to have a pion mass, requiring the vertex to have the beauty mass and that its total momentum vector point to the primary vertex. The second method [14] is based on the assumption that in the rest frame of the beauty particle the unseen momentum is emitted on average in a direction perpendicular to the beauty laboratory momentum. A comparison between these two methods and the simulation indicates that both methods give useful and independent information; the best estimator for the beauty momentum is a weighted average of the results of the two methods.

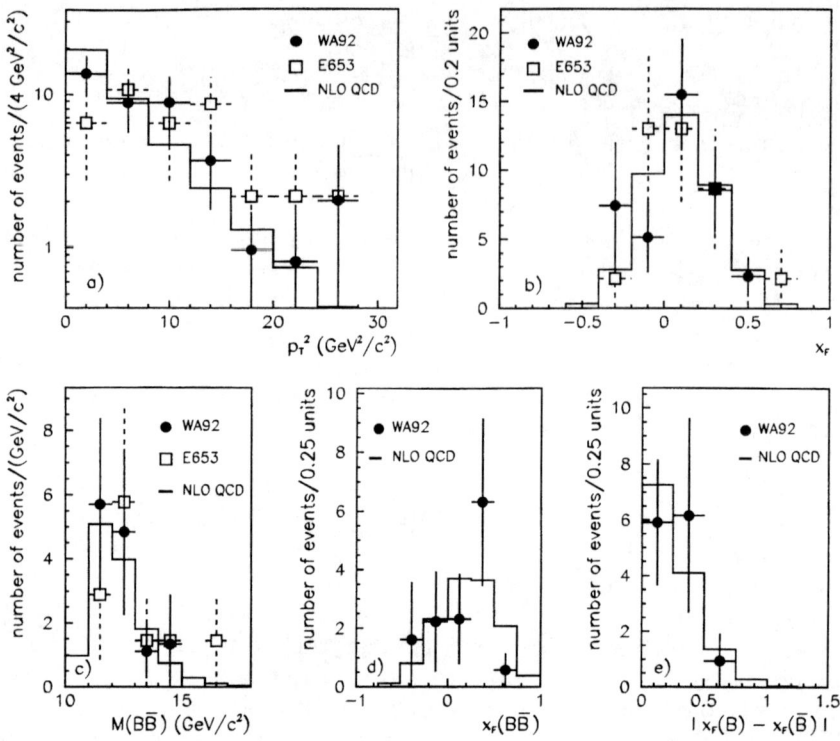

FIGURE 3. *Distributions of a) p_T^2 and b) x_F for the beauty particles; distributions of c) $B\overline{B}$ invariant mass, d) $|x_F(B) - x_F(\overline{B})|$ and e) Feynman-x of the $B\overline{B}$ system. WA92 measurements are compared with NLO QCD predictions and E653 measurements.*

D Single and double-differential distributions

The p_T^2 and x_F distributions of the beauty particles, corrected for acceptance, are shown in Fig. 3a)-b) along with the results of experiment E653 [14] and the predictions based on NLO QCD calculations [1]. Within the large statistical uncertainties, our data are consistent both with the result of experiment E653 and with the theoretical predictions. This is particularly interesting since, thanks to the large b-quark mass, the NLO QCD calculations are very reliable and corrections due to fragmentation, intrinsic transverse momentum of the incoming partons or higher order contributions are negligible.

Using the sample of 13 events in which both beauty decays have been identified, we have measured the azimuthal angle $\Delta\phi$ between the beauty particles [15] and other kinematic correlation variables: the invariant mass $M(B\overline{B})$, the Feynman-x difference $|\Delta x_F| = |x_F(B) - x_F(\overline{B})|$, and the Feynman x of the $B\overline{B}$ system. These

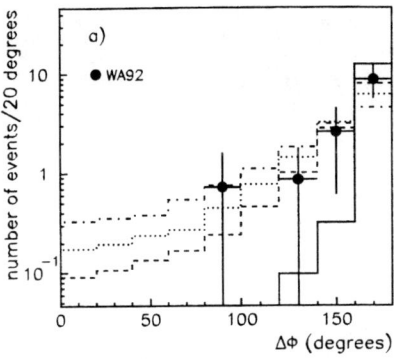 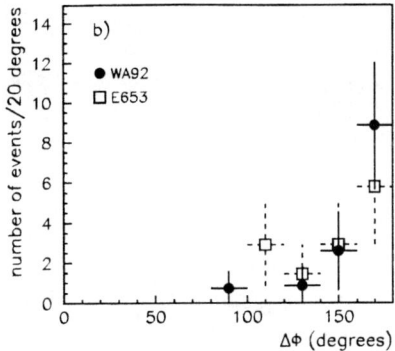

FIGURE 4. *WA92 measurement of the $\Delta\phi$ distribution compared with: a) distributions given by NLO QCD for $\langle k_T^2\rangle = 0\,GeV^2/c^2$ (solid line), $\langle k_T^2\rangle = 0.5\,GeV^2/c^2$ (dashed line), $\langle k_T^2\rangle = 1\,GeV^2/c^2$ (dotted line) and $\langle k_T^2\rangle = 2\,GeV^2/c^2$ (dashed-dotted line); b) E653 measurement.*

kinematic correlation variables are shown in Fig. 3c)-e) along with experiment E653 results, when available, and the predictions based on NLO QCD calculations. Again, within the large statistical uncertainties, our data are consistent both with the result of experiment E653 and with the theoretical predictions.

Fig. 4 shows the $\Delta\phi$ distribution for the 13 beauty pairs. This correlation variable is interesting since in leading-order production processes the B and $\overline{B}$ particles should have opposite directions in the plane perpendicular to the beam; inclusion of NLO processes results in a broadening of the distribution. Further broadening may result from perturbative corrections beyond NLO and from non-perturbative effects. However, in the QCD model [1], the $\Delta\phi$ distribution is supposed not to be modified by the fragmentation of quarks into hadrons but it is altered significantly when allowance is made for the interacting partons' momentum, k_T, transverse to the beam direction. A comparison between the measured $\Delta\phi$ distribution and the theoretical distributions for $\langle k_T^2\rangle = 0\,\text{GeV}^2/c^2$ (corresponding to bare NLO QCD), $\langle k_T^2\rangle = 0.5\,\text{GeV}^2/c^2$, $\langle k_T^2\rangle = 1\,\text{GeV}^2/c^2$ and $\langle k_T^2\rangle = 2\,\text{GeV}^2/c^2$ is shown in Fig. 4a). Our data favour a $\langle k_T^2\rangle$ value of between $0.5\,\text{GeV}^2/c^2$ and $1\,\text{GeV}^2/c^2$. Within the large statistical uncertainties, our data are also consistent with the result of experiment E653, as shown in Fig. 4b).

IV CONCLUSIONS

A study of heavy quark hadroproduction has been performed in the WA92 experiment, in which a beam of 350 GeV/c π^- particles was directed onto a 2-mm-thick copper or tungsten target.

For J/ψ events the cross-section dependence on the nuclear mass has been measured from a fit to the data on different targets, obtaining $\alpha(J/\psi)\{Si, Cu, W\} = 0.87 \pm 0.05$ (stat) ± 0.04 (syst) and the corresponding cross-section per nucleon, for α fixed at 0.87, $\sigma_0(J/\psi) = (216.2 \pm 5.4$ (stat) ± 13.3 (syst)) nb/nucleon. The ratio of the inclusive ψ' to J/ψ cross-section is measured to be $\sigma_0(\psi')/\sigma_0(J/\psi) = 0.13 \pm 0.05$ (stat) ± 0.01 (syst). From a comparison of kinematical properties obtained for J/ψ and open-charm events in which $D\bar{D}$ pairs were reconstructed, similarities have been observed, but not completely consistent with the expectations of the Colour-Evaporation Model.

The WA92 experiment has recorded 26 events featuring production of beauty particles, with a background contamination estimated to be 0.6 ± 0.6 events. Assuming a linear A-dependence, we measure a $b\bar{b}$ cross-section, integrated over all x_F, of $5.7^{+1.3}_{-1.1}$ (stat.) $^{+0.6}_{-0.5}$ (syst.) nb/N. Using this sample of beauty decays and optimizing an estimator for the momentum of the partially reconstructed vertices, we measure the distributions with respect to x_F and p_T^2 for the beauty particles as well as some correlation variables. Our results, which are consistent with the only previous measurements for 9 $b\bar{b}$ events, are compared to NLO QCD predictions. The $\Delta\phi$ distribution observed is well described by NLO QCD if the interacting partons are assumed to have a $\langle k_T^2 \rangle$ between $0.5\,\text{GeV}^2/c^2$ and $1\,\text{GeV}^2/c^2$.

REFERENCES

1. S. Frixione et al., CERN-TH/97-16 (1997), also in *Heavy Flavours II*, Advanced Series on Directions in High-Energy Physics, World Sci., Singapore, and references therein.
2. H. Fritzsch, *Phys. Lett.* **B 67** (1977) 217;
   F. Halzen and S. Matsuda *Phys. Rev.* **D 17** (1978) 1344;
   M. Gluck et al. *Phys. Rev.* **D 17** (1978) 2324.
3. E.L. Berger and D. Jones, *Phys. Rev.* **D 23** (1981) 1521;
   R. Baier and R. Ruckl, *Z. Phys.* **C 19** (1983) 251.
4. E. Braaten and S. Fleming, *Phys. Rev. Lett.*, **74** (1995) 3327;
   M. Cacciari, M. Greco, M. Mangano and A. Petrelli, *Phys. Lett.* **B 356** (1995) 560.
5. M. Adamovich et al., *Nucl. Instr. and Meth.* **A 379** (1996) 252.
6. M. Beneke and I.Z. Rothstein, *Phys. Rev.* **D 54** (1996) 2005 and **D 54** (1996) 7082.
7. E. Gregores, private communication.
8. R.K. Ellis et al., *Nucl. Phys.* **B 312** (1989) 551;
   P. Nason et al., *Nucl. Phys.* **B 327** (1989) 49.
9. M. Adamovich et al., *Nucl. Phys.* **B 495** (1997) 3.
10. J.F. Amundson, O.J.P. Éboli, E.M. Gregores and F. Halzen, *Phys. Lett.* **B 372** (1996) 127, and *Phys. Lett.* **B 390** (1997) 323.
11. R. Gavai et al., *Int. J. Mod. Phys.* **A 10** (1995) 3043.
12. M. Adamovich et al., *Phys. Lett.* **B 385** (1996) 487.
13. M. Adamovich et al., *Nucl. Phys.* **B 519** (1998) 19.
14. K. Kodama et al., *Phys. Lett.* **B 303** (1993) 359.
15. Y. Alexandrov et al., *Phys. Lett.* **B 433** (1998) 217.

Charm Detection at HERMES

Elke-Caroline Aschenauer

DESY Zeuthen, Platanenallee 6, D-15738 Zeuthen, Germany

(on behalf of the HERMES-Collaboration)

Abstract. Charm production in polarized lepton nucleon scattering is believed to be a clean probe of the nucleon spin contribution carried by gluons. An extensive search for charm signals has been carried out in the HERMES data of the years 1995 to 1997. Clean signals for both hidden and open charm could be extracted. A preliminary measurement of the J/ψ cross section near threshold is presented. Recently major detector upgrades have been implemented that will significantly improve the physics potential of HERMES in the charm sector.

I INTRODUCTION

Charm production in Deep Inelastic Scattering (DIS) is believed to be a clean probe of the distribution of the gluons $G(x)$ inside the nucleon, with x being interpreted as the fraction of the nucleon light-cone momentum carried by the gluon. It is expected to proceed mainly via photon-gluon fusion. The virtual photon emitted by the scattered lepton is absorbed by a gluon inside the nucleon, yielding a pair of $c\bar{c}$ quarks. Since the hard scale of the process is ensured by the high mass of the produced $c\bar{c}$-pair, the cross section can rather reliably be calculated from elementary diagrams (see fig. 1) in the framework of perturbative Quantum Chromo Dynamics (QCD). This allows to extract $G(x)$ from the measured production rate.

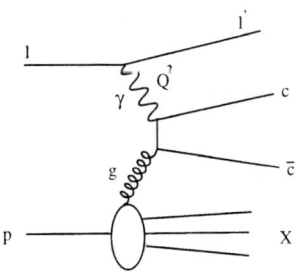

FIGURE 1. Feynman graph for charm production by photon gluon fusion in leading order QCD.

While this process is most cleanly reflected in open charm production, for the hidden charm production models [1,2] the produced c– and $\bar{c}$–quark are bound together into a color singlet object by emission of a hard gluon. In the scattering of longitudinally polarized leptons off a longitudinally polarized target the asymmetry A_{LL}^c for the production of charm states with parallel or anti-parallel lepton and target spins is, in analogy to the unpolarized case, defined as

$$A_{LL}^c = \frac{N_c^{\uparrow\downarrow} - N_c^{\uparrow\uparrow}}{N_c^{\uparrow\downarrow} + N_c^{\uparrow\uparrow}}. \qquad (1)$$

For photon gluon fusion this double spin asymmetry is related to the ratio of the polarized and unpolarized gluon distribution $\frac{\Delta G(x)}{G(x)}$ in the nucleon by

$$A_{LL}^c = \frac{1}{D f P_b P_t} \hat{a}_{QCD} \frac{\Delta G(x)}{G(x)}. \qquad (2)$$

Here D denotes the depolarization factor for the virtual photon, f the dilution by unpolarized target material, P_b and P_t the polarization of the beam and the target and $\hat{a}_{QCD}$ the analyzing power of the elementary QCD process (see fig. 1). A measurement of A_{LL}^c in polarized deep inelastic scattering should thus allow one to extract $\Delta G(x)/G(x)$, which has up to now never been directly measured and is at present rather poorly constrained by QCD fits to the g1 world data. A direct measurement might eventually help to resolve the mismatch found between the sum of the spins of the constituent quarks compared to the spin of the entire nucleon [3].

The major aim of the HERMES experiment [4] at DESY in Hamburg is to determine the different contributions of the spin of the nucleon from the combination of inclusive and semi-inclusive deep-inelastic polarized scattering data. The scattering products of the longitudinally polarized 27.5 GeV/c momentum $e^\pm$ beam from the longitudinally polarized atoms of an internal gas target are detected in an open forward spectrometer [5]. Since the apparatus has been designed to detect and identify also the hadrons produced in the deep inelastic scattering process it should also allow the detection of some open charm states. By this means HERMES may be able to reveal, for the first time, information on $\frac{\Delta G(x)}{G(x)}$ at a mean $x_{gluon} \sim 0.3$.

II PRELIMINARY RESULTS

HERMES has been commissioned in 1995. Since then, data have been collected on a variety of gaseous targets, i.e. polarized H and ^3He, but also on unpolarized H_2, D_2, ^3He and N_2 at higher density. The following section presents the hidden and open charm signals extracted from the 1995-97 data, to illustrate the experiment's charm detection capability. For open charm production the two c-quarks fragment into $D(c\bar{q}, q\bar{c})$ mesons. In case of the $J/\psi(c\bar{c})$ several processes have to be distinguished. The inelastic production is described by the Color Singlet Model [6], where a color neutral state is formed by the emission of a hard gluon. Additionally, a significant contribution from elastic J/ψ vector meson production has to be considered. From s-channel helicity conservation one would expect no dependence on ΔG for this process. Small asymmetries are predicted by models describing the hard elastic process as exchange of two gluons [7].

A J/ψ signals

J/ψ's decaying to e^+e^- are readily reconstructed in HERMES, since for this decay the excellent hadron–electron separation of the detector can be fully exploited and the trigger efficiency is close to 100 % for single electrons and positrons. Accordingly a clear signal has been observed in all data taking periods. J/ψ decaying to

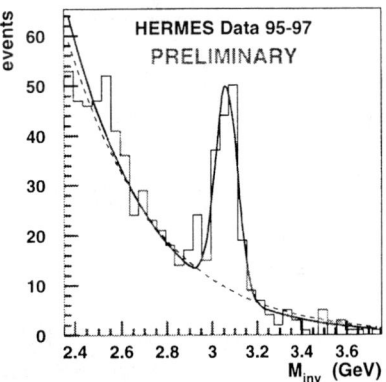

FIGURE 2. Invariant mass spectrum for $J/\psi \to e^+e^-$ & $J/\psi \to \mu^+\mu^-$ together with the renormalized spectrum of wrong charge sign combinations (shaded) from the HERMES 1995–97 data. The curves are fits of a Gaussian over an exponential l^+l^- background and of a pure exponential background from $l^{\pm}l^{\pm}$ combinations.

$\mu^+\mu^-$ has been exploited by HERMES as well, even though the experiment had no dedicated muon detection system till 1998. Particle identification was achieved in this case by requiring minimum ionizing signals in the electromagnetic calorimeter and the pre–shower detector. To provide sufficient trigger sensitivity an additional 2-track trigger has been implemented in 1996 and upgraded to full coverage in 1997.

The signals of both decay modes from the 1995–97 data on all targets are combined in figure 2: The invariant mass spectrum of l^+l^- combinations is shown together with the scaled spectrum (shaded) from l^+l^+ and l^-l^- combinations. The figure shows that there is a clear peak of about 140 J/ψ events over an exponential background of ~ 40, the shape of which is rather well described by the spectrum of $l^{\pm}l^{\pm}$ combinations. The signal is still small, since HERMES is restricted to J/ψ production near threshold by the beam energy of 27.5 GeV and was not designed for the detection of charmed final states where the opening angle between the decay products is rather large. The data have been collected on various target gases; two thirds of the statistics are from unpolarized targets. As the J/ψ's are dominantly produced at low Q^2 the scattered positron remains mostly inside the beam pipe, thus preventing the full reconstruction of the event kinematics and the separation of different production mechanisms.

B J/ψ production cross section

The behavior of the J/ψ production cross section near threshold is an interesting subject. Figure 3 shows the total electroproduction cross section at HERMES separated by year and decay channel. Despite the different trigger and particle identification, the two decay channels agree very well within errors. The preliminary average is $\sigma_{ep}^{J/\psi}[E_e = 27.5 \text{ GeV}] = (0.112 \pm 0.012)$ nb [8]. This value can be converted [9] into a photoproduction cross section $\sigma_{\gamma p}^{J/\psi}[E_\gamma = 15 \text{ GeV}] = (4.3 \pm 0.5)$ nb. The preliminary systematic uncertainty is 20%. This result is compared to existing data on the J/ψ photoproduction cross section by previous γ– and μ–beam experiments in figure 4. There is satisfactory agreement taking into account that for some of the older data only the elastic part of the cross section was taken into account.

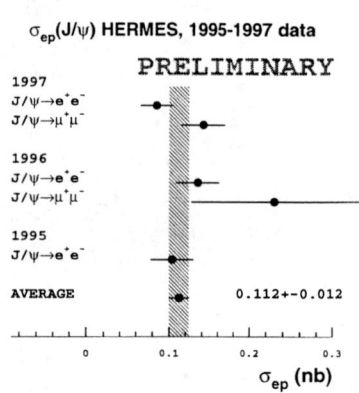

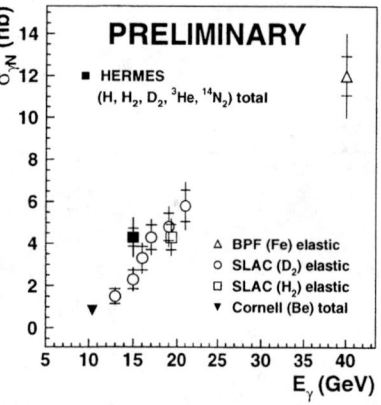

FIGURE 3. J/ψ electroproduction cross section separated by decay channel and year.

FIGURE 4. Total J/ψ photoproduction cross section extracted from the 1995–97 HERMES e^+e^- and $\mu^+\mu^-$ signals. The preliminary HERMES measurement is shown as the solid point at E_γ=15 GeV together with previous measurements using real and virtual photon beams [10].

C Open Charm

Measuring open charm production at HERMES is quite different in many respects from its measurement at dedicated charm experiments. There are a number of basic limitations for such a measurement in the HERA electron beam, which limit the number of charm events seen by HERMES and the statistics for a precise measurement of A_{LL}^c.

1. The most serious limitation comes, as in the case of the J/ψ, from the low charm production rate available to the experiment. It is mainly due to

   - The low energy of the HERA $e^\pm$ beam of 27.5 GeV, which results in a measurement near the production threshold for $c\bar{c}$ ($\sqrt{s} = 7.2$ GeV) and thereby a photo–production cross section for charm of the order of only 60 nb (see fig. 5).

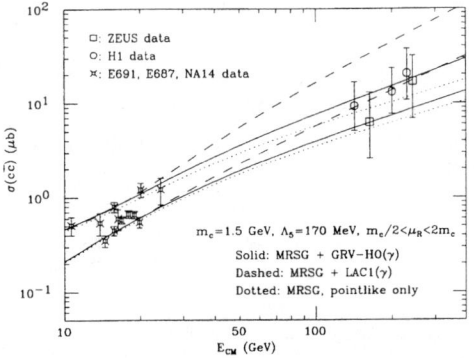

FIGURE 5. The total cross section for $c\bar{c}$ photo–production as a function of $\sqrt{s}$. Different lines correspond to different values of the parameters entering the theoretical calculation. The points are from experimental measurements with real and virtual γ's. The figure is taken from [11].

   - At present polarized internal gas targets are limited to areal densities of about 10^{14} atoms/cm^2. In conjunction with the HERA $e^\pm$ beam this implies a rather modest interaction rate and thereby luminosities of not more than 60 pb^{-1} can be achieved per data taking period.

2. Due to its high mass charm typically decays to high multiplicity final states with low branching ratios. The acceptance of the present HERMES apparatus is rather small for high multiplicity, large angle decays. While the typical efficiency for the inclusive measurement of one track (as the scattered lepton in DIS events) is close to 100 %, the acceptance for 2 body decays is more like 10 %. This represents a serious constraint on the number of charm final states accessible to HERMES.

3. The capabilities for the separation of the charm signal from the high combinatorial background due to minimum bias photo production are rather restricted. The HERMES particle identification (PID) has been originally optimized for the (inclusive) separation of electrons from hadrons and was (un-

til 1998) rather limited concerning the identification of the different hadron species. The threshold Čerenkov counter used until now can separate $\pi^\pm$ from $K^\pm$ only in the range 4 GeV/c $< p <$ 13.5 GeV/c.

The above listed experimental conditions have resulted in the following strategy adopted for the search for charm events:
The analysis has to focus on charm final states with sizeable branching ratio [14], ensuring reasonable rates and decaying via 2 body decays, like $D^0 \to K^-\pi^+$ and $\bar{D}^0 \to K^+\pi^-$ (branching ratio: 3.85%). The formation of a $D^0/\bar{D}^0$ is favored by most hadronization scenarios [12,13]. This is specially true for the production of $\bar{D}^0$, which can be produced also in conjunction with a Λ_c^+, a process which requires less energy than the production of two charm mesons and which should therefore dominate at low virtual photon energy ν.

The $K^\pm\pi^\mp$ decay mode has intensively been studied at HERMES. It turned out that the hadron identification of the HERMES experiment is not sufficient to allow the extraction of a signal out of the severe h^+h^- combinatorial background. This has been confirmed by a thorough Monte Carlo study comparing the expected $D^0/\bar{D}^0$ signal from photon gluon fusion (simulated using the AROMA generator [15]) with the hadronic background from minimum bias γp-interactions (from PYTHIA [16]) and standard deep inelastic scattering (from LEPTO [17]). It clearly demonstrates that the $D^0/\bar{D}^0 \to K^\mp\pi^\pm$ signal expected from photon gluon fusion is completely hidden under the huge h^+h^- background.

This situation makes kinematic criteria, i.e. cuts on the momenta and invariant masses of the decay products mandatory. It has been known since a long time that the $D^{*\pm}$ decay proceeding to about 68 % [14] via an intermediate $D^0/\bar{D}^0\,\pi^\pm$ state, by its emission of a soft pion, allows a very strong kinematical selection, as their decay products have to satisfy two conditions at a time:

- The q-value of the $D^{*\pm} \to D^0/\bar{D}^0\pi^\pm$ decay is only 39 MeV, which means that the momentum of the $\pi^\pm$ from the $D^{*\pm}$ decay is on the very low edge of the phase space. Since typically the amount of slow pions from other sources is small in this region, again strong background suppression is achieved.

- A very strong reduction of all allowed track combinations is achieved by the fact, that the momentum vectors of the K and fast π have to sum up to an invariant mass near the $D^0/\bar{D}^0$ mass, which is dominated by the experimental resolution.

Since the acceptance would be negligible for any decay mode involving more than 3 decay particles only the decay chain $D^{*\pm} \to K^\mp\pi^\pm\pi^\pm_{slow}$ has been analyzed. This analysis [18] led to a small but significant signal for the decay $D^* \to D^0\pi$ by tagging the missing mass (fig. 6). As most of the events do not have the scattered $e^\pm$ in the acceptance, only data from the 1997 and part of 1996 data taking period can be used, when the already above mentioned photo production trigger has been operational. For this reason, at present only a global analysis (summed over all targets) of this data from 1996 and 1997 was possible. Still, this allowed to show, that Open Charm is accessible in HERMES via this decay mode at all.

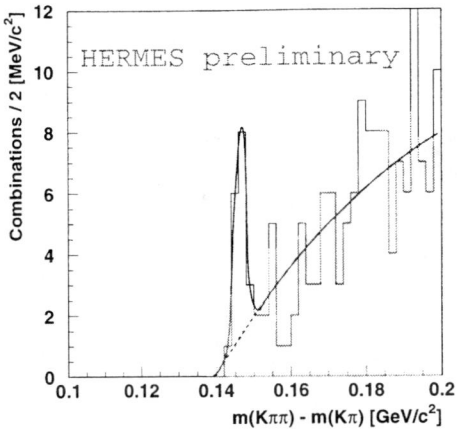

FIGURE 6. Missing mass spectrum $M(K\pi\pi) - M(K\pi)$ for the process $D^\star \to D^0(\to K\pi)\,\pi$ together with a background estimate (shaded) from the D^0 side bands.

III FUTURE PROSPECTS

To reconstruct open charm in the modes $D^0 \to X l^+ \nu_l$ and $D^0/\bar{D}^0 \to K^\mp \pi^\pm$ (direct decay) at HERMES requires either unambiguous muon identification (semi-leptonic electron decays being impractical as due to the use of an electron beam the electron can not be unambiguously identified) or clean hadron identification over the full Hermes momentum range.

To meet these requirements the HERMES detector was upgraded by a series of additional detector components [20,19], which are depicted as shaded elements in figure 7. They comprise

- A ring imaging Čerenkov counter equipped with aerogel and gaseous C_4F_{10} radiators. It will allow the separation of $\pi/K/p$ over the full momentum range accepted by the spectrometer. Monte Carlo studies showed that this $\pi/K/p$ separation reduces the combinatorial background by a factor 15 for the open charm channels $D^0 \to K^-\pi^+$ and $\bar{D}^0 \to K^+\pi^-$.

- A muon detection system, consisting of an iron wall equipped with an array of hodoscopes. It will enable the clean identification of semi-leptonic $X\mu^+\nu_\mu$ decays and improve the detection of muonic J/ψ decays.

- A low angle electron tagger, consisting of a set of small vertical drift chambers and scintillating counters mounted in two beam quadrupole magnets downstream of the experiment. This will enable HERMES to double the precentage

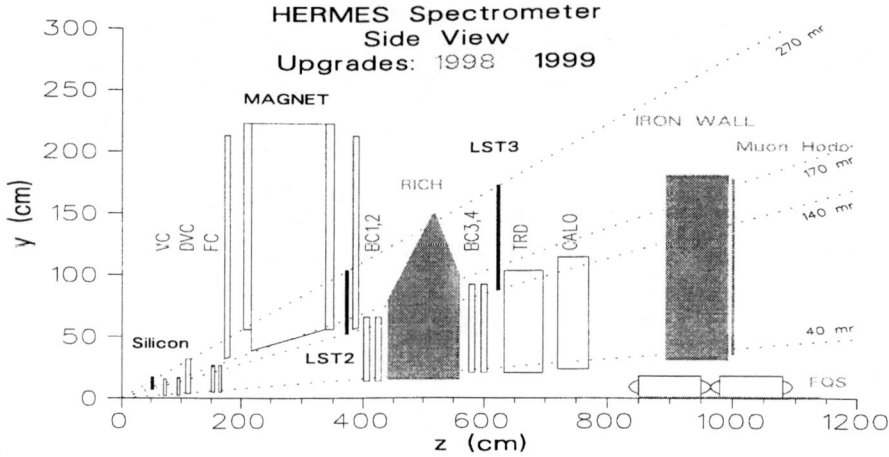

FIGURE 7. Overview over the HERMES detector upgrades to enhance charm detection.

of fully reconstructed events at low Q^2, where most of the charm events are being produced.

These detectors have been installed in the 1997–98 shutdown and successfully been tested in spring 1998. The program will be completed by the installation of large acceptance tracking detectors (labeled *Silicon*, *LST2* and *LST3* in figure 7) in spring 1999, which will enlarge the muon acceptance by a factor 2.

Having implemented these new components, HERMES will detect more statistics on open charm which will possibly allow to obtain a first result of $\Delta G(x)/G(x)$.

IV CONCLUSION

In summary, the present HERMES charm data samples are small, but well understood. Several detector upgrades have been introduced for the 1998 running. Clean $\pi/K/p$ separation by a RICH will allow direct $D^0 \to K\pi$ reconstruction. Semileptonic $D^0 \to \mu\nu K$ decays may be accessible by the new muon detector. A small angle detector enhances the acceptance of the scattered electron at low Q^2. With the help of these upgrades, charm production at HERMES makes a determination of $\Delta G(x)/G(x)$ an attainable goal.

REFERENCES

1. E.L. Berger, D. Jones, *Phys. Rev.* **D 23** (1981) 1521.
2. G.T. Bodwin, E. Braaten, G.P. Lepage, *Phys. Rev.* **D 51** (1995) 1125.

3. J. Ashman et al., *Phys. Lett.* **B 206** (1988) 364.
   J. Ashman et al., *Nucl. Phys.* **B 328** (1989) 1.
4. K. Coulter et al., Proposal to the PRC, **DESY-PRC 90/01**.
   HERMES Collaboration, Technical Design Report, **DESY-PRC 93/06**.
5. K. Ackerstaff et al., **DESY-98-057**, *Nucl. Instr. Meth.* **A** (in press)
6. E.L. Berger, D. Jones, *Phys. Rev.* D 23 (1981) 1521.
7. M. Vänttinen, L. Mankiewicz, hep-ph/9805338 and /9807287 (1998).
8. F. Meissner *Hermes* internal note 98-051.
9. C.F. Weizsäcker, *Z. Phys.* 88 (1924) 612.
10. S.D. Holmes, W. Lee, J.E. Wiss *Ann. Rev. Nucl. Part. Sci.* **35** (1985) 397.
11. S. Frixione, M.L Mangano, P. Nason, G. Ridolfi, **hep-ph/9702287**.
12. S. Frixione et al., *Nucl. Phys.* **B 431** (1994) 453.
13. B. Andersson et al., *Phys. Rep.* **97** (1983) 31.
14. R.M. Barnett et al., *Phys. Rev.* **D 54** (1996) 1.
15. G. Ingelman, J. Rathsman, G.A. Schuler, **DESY 96-058**.
16. T. Sjöstrand, *Comp. Phys. Comm.* **82** (1994) 74.
17. G. Ingelman, A. Edin, J. Rathsman, **DESY 96-057**.
18. S. Brons *Hermes* internal note, 98-057.
19. E. Cisbani et al., Proposal to the PRC, **DESY-PRC 97/04**.
20. M. Amarian et al., Proposal to the PRC, **DESY-PRC 97/03**.

Heavy Quark Production in Neutrino Deep-Inelastic Scattering

T. Adams[4], A. Alton[4], C. G. Arroyo[2], S. Avvakumov[7],
L. de Barbaro[5], P. de Barbaro[7], A. O. Bazarko[2], R. H. Bernstein[3],
A. Bodek[7], T. Bolton[4], J. Brau[6], D. Buchholz[5], H. Budd[7],
L. Bugel[3], J. M. Conrad[2], R. B. Drucker[6], J. A. Formaggio[2],
R. Frey[6], J. Goldman[4], M. Goncharov[4], D. A. Harris[6],
R. A. Johnson[1], J. H. Kim[2], B. J. King[2], T. Kinnel[8],
S. Koutsoliotas[2], M. J. Lamm[3], W. Marsh[3], D. Mason[6],
K. S. McFarland[7], C. McNulty[2], S. R. Mishra[2], D. Naples[4],
P. Nienaber[3], A. Romosan[2], W. K. Sakumoto[7], H. M. Schellman[5],
F. J. Sculli[2], W. G. Seligman[2], M. H. Shaevitz[2], W. H. Smith[8],
P. Spentzouris[2], E. G. Stern[2], B. M. Tamminga[2], M. Vakili[1],
A. Vaitaitis[2], V. Wu[1], U. K. Yang[7], J. Yu[3], G .P. Zeller[5]

Presented by T. Adams
[1] *University of Cincinnati, Cincinnati, OH 45221*
[2] *Columbia University, New York, NY 10027*
[3] *Fermilab, Batavia, IL 60510*
[4] *Kansas State University, Manhattan, KS 66506*
[5] *Northwestern University, Evanston, IL 97403*
[6] *University of Oregon, Eugene, OR 97403*
[7] *University of Rochester, Rochester, NY 14627*
[8] *University of Wisconsin, Madison, WI 45207*

Abstract. Charm production by neutrino charged-current interactions produces two muon (dimuon) events which are easily identified. This signal provides an important method to measure the strange sea and the mass of the charm quark. Several experiments, including CCFR, CDHS and CHARM II, have performed analyses of such events. The results of these analyses are summarized with emphasis on CCFR and improvements made by NuTeV.

INTRODUCTION

Lepton-nucleon deep-inelastic scattering (DIS) is an excellent method of studying nucleon structure with weak interactions reducing complications due to Quantum ChromoDynamics (QCD). In particular it becomes possible to measure the strange sea content of the nucleon. This measurement is of significant interest to many other experiments.

There are three primary methods for measuring the strange sea in scattering experiments. Firstly, the ratio of F_2 in charged lepton and neutrino scattering is sensitive to the strange sea:

$$\frac{F_2^{\ell^\pm N}(x,Q^2)}{F_2^{\nu N}(x,Q^2)} = \frac{5}{18} - \frac{3}{18}\left[\frac{xs(x,Q^2) + x\bar{s}(x,Q^2) - xc(x,Q^2) - x\bar{c}(x,Q^2)}{2xF_1^{\nu N}(x,Q^2)}\right], \quad (1)$$

commonly known as the "5/18ths rule". This technique has difficulties with normalizing data from separate experiments.

Secondly, the difference between xF_3 for neutrino and anti-neutrino scattering can measure the strange sea:

$$xF_3^{\nu N} - xF_3^{\bar{\nu} N} = 4(xs - xc) \quad (2)$$

These analyses suffer from low statistics.

The best technique is charm production in neutrino DIS. The reactions are shown in Eq. 3. The neutrino (anti-neutrino) interacts with a $d(\bar{d})$ or $s(\bar{s})$ in the nucleon producing a $c(\bar{c})$ quark. The charmed meson decays semi-muonically producing a final state which includes two oppositely charged muons (dimuon) and additional hadrons. The importance of this particular final state is discussed below.

$$\begin{aligned}
\nu_\mu + N &\longrightarrow \mu^- + c + X \\
&\hookrightarrow s + \mu^+ + \nu_\mu \\
\bar{\nu}_\mu + N &\longrightarrow \mu^+ + \bar{c} + X \\
&\hookrightarrow \bar{s} + \mu^- + \bar{\nu}_\mu
\end{aligned} \quad (3)$$

Charm production is $\sim 10\%$ of the total charged current interaction rate. Because neutrino charm production off $d(\bar{d})$ quarks is Cabibbo suppressed production off strange sea quarks is dominant. For neutrino interactions $\sim 50\%$ of charm production is from the strange sea while in anti-neutrino interactions it rises to $\sim 90\%$. The high rate of charm production and importance of strange quarks allows for measurements of the nucleon strange sea.

THEORETICAL ISSUES

There are a number of theoretical issues associated with neutrino DIS production of heavy quarks. Production of charm quarks includes complications due to quark mass. If the charm quark were massless, it could easily be treated as a "light" quark such as u and d. If the charm quark were very heavy ($m_c \ll Q^2$) it could be treated with gluon-fusion. However, the charm quark mass is intermediate and effects from both limits must be included. It affects momentum fraction (x), inelasticity (y) and energy dependence of the production rate, causing a significant suppression in cross-section for $E_\nu < 100$ GeV. One technique for handling the charm quark mass is the method of "slow rescaling" [1] which involves a change of variables from Bjorken x to ξ,

$$\xi = x(1 + \frac{m_c^2}{Q^2}). \tag{4}$$

At leading order (LO) the differential cross-section for reactions in Eq. 3 can be factorized in terms of three major components

$$\frac{d^3\sigma(\nu_\mu N \to \mu^-\mu^+ X)}{d\xi\, dy\, dz} = \frac{d^2\sigma(\nu_\mu N \to cX)}{d\xi\, dy} D(z) B_c(c \to \mu^+ X), \tag{5}$$

where the first term is the differential cross-section for charm quark production, $D(z)$ is the fragmentation function and $B_c(c \to \mu^+ X)$ is the semi-muonic branching fraction. The charm quark differential cross-section can be written as

$$\left\{\frac{d^2\sigma(\nu_\mu N \to cX)}{d\xi\, dy}\right\}_{LO} = \frac{G_F^2 M E_\nu}{\pi(1 + Q^2/M_W^2)^2}\{\,[\xi u(\xi, \mu^2) + \xi d(\xi, \mu^2)]\,|V_{cd}|^2 \tag{6}$$
$$+ 2\xi s(\xi, \mu^2)\,|V_{cs}|^2\,\}\left(1 - \frac{m_c^2}{2ME_\nu\xi}\right)$$

where ξ is the slow rescaling variable.

There are a number of additional diagrams in next-to-leading order (NLO). Figure 1 shows the diagrams which contribute up to $\mathcal{O}(\alpha_S)$. The dominant diagrams are the leading-order quark initiated diagram and the $u-$ and $t-$ channel gluon fusion diagrams (Fig 1(a)). The size of the gluon PDF's compared to the sea distributions partially compensates for the additional order of $\alpha - S$. There are several radiative gluon diagrams which are less significant (Fig. 1(b)). Also, at next-to-leading order, the differential cross-section does not completely factorize.

Calculation of neutrino charm production has several theoretical uncertainties. There are several factorization schemes available which handle s and c quarks differently. Choice of factorization scale can produce a dependence especially at low scales. The renormalization scale is usually chosen to be the same as the factorization scale, but can be different.

EXPERIMENTS

Several experiments have studied neutrino charm production (see Table 1). CDHS [2], CHARM II [3] and CCFR [4,5] have the largest data samples and

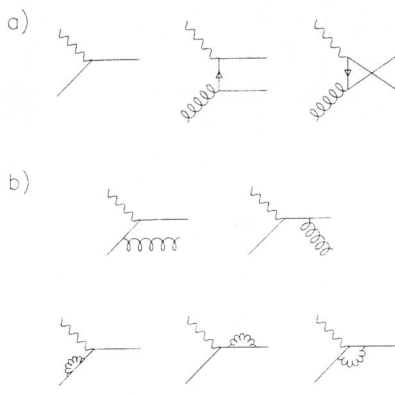

FIGURE 1. Mechanisms that contribute to neutrino production of charm up to $\mathcal{O}(\alpha_S)$. **a)** The dominant diagrams: the leading-order quark-initiated diagram, and the t channel and u channel gluon-initiated diagrams, respectively. **b)** The radiative-gluon and self-energy diagrams.

will be the focus of experimental discussion. Each of these has produced a LO analysis [2–4], while CCFR has also performed a NLO analysis [5].

CCFR used the Fermilab Lab E detector during the 1985 and 1987-88 fixed target runs collecting data on charged- and neutral- current neutrino interactions. The detector consists of two major parts, a target/calorimeter and a toroid spectrometer. The target/calorimeter is a 690-ton steel sampling calorimeter consisting of 42 segments which each contain 10-cm of iron, two scintillation counters and one drift chamber. The transverse dimensions of each segment are 3 × 3 meters. Energy resolution of the calorimeter is $\sigma/E_{had} = 0.89/\sqrt{E_{had}}$. The toroid spectrometer is located immediately downstream of the calorimeter and contains five sets of drift chambers in an iron toroid magnet. Momentum measurement is limited by multiple scattering to $\sigma_p/p = 0.11$. CCFR ran with a mixed beam of neutrinos and antineutrinos with an energy range from 30 GeV to 500 GeV. The toroid spectrometer field was alternated between focusing μ^+ and μ^-.

The mass of the calorimeter creates a heavy target for incident neutrinos and restricts shower development in interactions. A result is that muons are the only charged particles which deeply penetrate the detector. Forward-going muons entering the toroid spectrometer are momentum analyzed. For these reasons, dimuon

TABLE 1. Summary of data samples from neutrino charm production experiments. CDHS and CCFR detect dimuon events, the 15-ft BC measures $\mu - e$ events, and E531 detects inclusive decays of charmed particles in an emulsion target.

Experiment	E_ν GeV	$\mu^- l^+$ events	$\mu^+ l^-$ events	Background %
CCFR	30 – 600	5030	1060	15
$p_{\mu_2} > 5 GeV$	(> 100)	(3721)	(493)	
CDHS	30 – 250	11041	3684	13
$p_{\mu_2} > 5 GeV$	(> 100)	(3589)	(452)	
CHARM II	30 – 250	4111	871	22
$p_{\mu_2} > 6 GeV$				
15-ft BC	1 – 200	461	–	18
$p_{e^+} > 0.3 GeV$				
E531	1 – 250	122	–	3
$\nu - Emulsion$		(Charm Events)		

events have a distinctive and easily measurable signal. Table 1 lists major features of several dimuon experiments including neutrino energy range and minimum energy of the second muon.

Dimuon events are analyzed by comparing measured distributions of E_{vis}, x_{vis}, y_{vis} and z_{vis} to those generated with Monte Carlo using certain parameters. The parameters are varied until the Monte Carlo matches data. The "vis" subscript refers to measured quantities which may be missing part of the event due to final state neutrinos. E_{vis} is the sum of the energy of both muons and hadronic shower. Bjorken x_{vis} is measured via $Q^2/2ME_{hadvis}$ where $Q^2 = 2E_{vis}E_{\mu 1}(1 - cos\theta_{\mu 1})$, M is the mass of the target proton and E_{hadvis} is the sum of the energy of the hadronic shower ($E_{showervis}$) and the energy of the second muon ($E_{\mu 2}$). Inelasticity, y, is given by E_{hadvis}/E_{vis}. z_{vis} describes the fragmentation by measuring $E_{\mu 2}/E_{hadvis}$.

Five parameters are used to fit the data (κ, α, m_c, B_c, and ϵ). The first two describe the strange sea relative to the non-strange (u, d) sea. κ gives the size relative to the non-strange sea while α relates the shape of the distribution by $(1 - x)^\alpha$. m_c is the mass of the charm quark. B_c is the semi-muonic branching fraction of all produced charmed mesons. ϵ is used in the parameterization of the fragmentation.

Monte Carlo simulation is an important part of dimuon measurements. The most important pieces are the cross-section model, fragmentation/hadronization model, decay kinematics, background sources and detector simulation. The cross-section model can be either leading order or next-to-leading order and include different factorization/renormalization schemes and parton distribution functions (PDF's). The primary source of background is π/K decays in hadronic showers from normal charged-current interactions. CCFR has tuned the detector simulation using data from two weeks of calibration beam.

RESULTS

CCFR has performed both a LO [4] and NLO [5] analysis of dimuon events. The LO analysis used a simple parton scattering model, slow rescaling and a parameterization of Buras-Gaemers PDF's [6]. The NLO analysis used the scheme of Aivazis et. al. [7], included the gluon fusion diagrams (but not the gluon radiation diagrams) and PDF's extracted from CCFR data using Duke and Owens' [8] implementation of the DGLAP evolution equation.

Table 2 summarizes results from the three largest data samples, CCFR [4,5], CDHS [2] and CHARM II [3]. Figure 2 gives a graphical comparison of results for κ, α and m_c. All three experiments performed a LO analysis while CCFR also performed a NLO analysis with two different fragmentation parameterizations, Peterson [9] and Collins-Spiller [10]. Their results were independent of the choice of fragmentation.

TABLE 2. Summary of experimental results from CCFR, CDHS and CHARM II.

Exp. χ^2/dof	Fragmentation	κ	α	B_c	m_c (GeV)
CCFR NLO fit 52.2/65	Collins-Spiller $\epsilon = 0.81$	0.477 ±0.045 ±0.14	−0.02 ±0.57 ± 0.27 ±0.024	0.1091 ±0.0078 ± 0.0057	1.70 ±0.17 ±0.09
CCFR NLO fit 41.2/46	Peterson $\epsilon_P = 0.20$	0.468 ±0.053 ±0.04	−0.05 ±0.47 ± 0.27 ±0.025	0.1047 ±0.0076 ± 0.0057	1.69 ±0.16 ±0.11
CCFR LO fit 42.5/46	Peterson $\epsilon_P = 0.20$	0.373 ±0.045 ±0.04	2.50 ±0.58 ± 0.31 ±0.018	0.1050 ±0.007 ± 0.005	1.31 ±0.21 ±0.12
CDHS LO fit	δ-function ($z = 0.68$)	0.478 ±0.094	Not fit Set 0.0	0.0839 ±0.0147	Not fit 1.2 − 1.8
CHARM II LO fit 96/80	Peterson $\epsilon_P = 0.073$ ±0.017	0.388 ±0.074 ±0.067	1.12 ±0.78 ± 1.03		1.79 ±0.28 ±0.27

The results yield several interesting insights. The most useful conclusion is that κ has a value near 0.4. This shows the sea is not SU(3) symmetric. A SU(3) symmetric sea would yield a κ of 1.0. The u and d contributions to the sea are larger than the s contribution. All three experiments agree on this point. This has significant implications for theoretical work and should be taken into account.

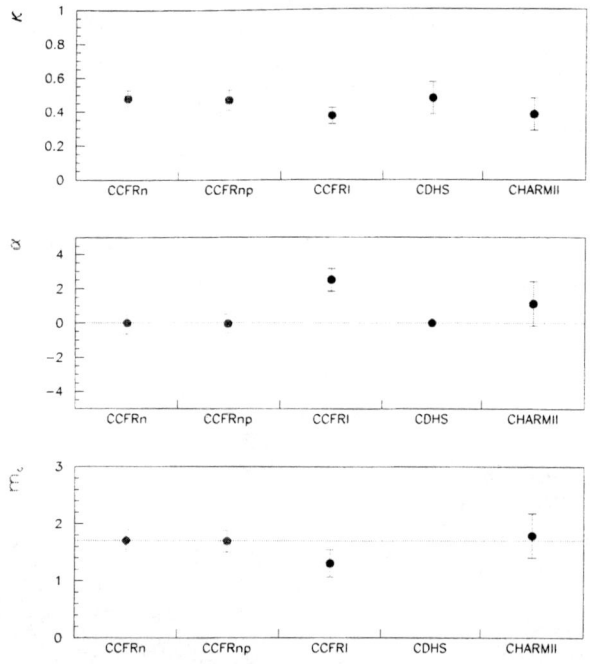

FIGURE 2. The parameters κ, α and m_c from (left to right) the two CCFR NLO, CCFR LO, CDHS LO and CHARM II LO analyses.

CCFR NLO and CHARM II LO results find the shape of the strange sea to be consistent with the shape of the non-strange sea. CDHS LO did not allow the strange sea to have a different shape. CCFR LO needed a softer $(1-x)^{2.5}$ strange sea. The semi-muonic branching fraction was found to be consistent between CCFR and CDHS.

There is a difference in m_c between analyses. CCFR NLO and CHARM II LO analyses found a charm quark mass around 1.7 GeV/c^2. CDHS did not fit m_c, but varied it between 1.2-1.8 GeV/c^2. CCFR LO found m_c to be around 1.3 GeV/c^2. This difference is not understood.

CCFR has performed a check for an asymmetry between momentum distributions of the s and $\bar{s}$ seas. An additional parameter was fit $\Delta\alpha = \alpha - \alpha'$ where $(1-x)^\alpha$ parameterizes the shape of the s sea contribution relative to the u and d sea and $(1-x)^{\alpha'}$ parameterizes the $\bar{s}$ sea. The result was $\Delta\alpha = -0.46 \pm 0.42 \pm 0.36$. This shows no definite asymmetry and limits it (at 90% confidence level) to $-1.9 < \Delta\alpha < 1.0$.

NUTEV EXPERIMENT

Fermilab E815 (NuTeV) is the successor to CCFR. It took data with the Lab E detector during the 1996-1997 fixed target run with more than 2.5×10^{18} protons on target recorded. NuTeV had several improvements over CCFR. A continuous calibration beam was available with a range of energies and particle composition (electrons, muons, or hadrons). This data allows an improved calibration and additional tuning of the detector simulation.

However, the major improvement of NuTeV is the beam which used the newly installed Sign-Selected Quadrapole Train (SSQT). Secondary particles from the primary target were selected to send only particles of one particular sign into the decay region. The resulting neutrino beam was pure to 1×10^{-3} for neutrinos and 2×10^{-3} for anti-neutrinos. Sign-selection was alternated to take approximately equal amounts of beam in each mode. The toroid spectrometer was set to focus muons from the beam-selected charged-current interactions.

Knowledge of the incident neutrino type should improve the strange sea measurement. In the CCFR analyses, neutrino type (ν or $\bar{\nu}$) was determined by an algorithm based on transverse momentum (p_t) of muons in relation to the hadron shower. The muon from the primary vertex was assumed to have a larger p_t. The error on this identification was up to 20% for low-x in anti-neutrino interactions. This part of the distribution is most sensitive to the strange sea. NuTeV does not have this level of contamination and should result in an improved measurement of the strange sea.

Figure 3 shows a display of a dimuon event from NuTeV. This event is neutrino induced with two low energy muons present. The x's show hits in drift chambers, while vertical bars at the top indicate energy deposit in the scintillators. The focused muon is the primary while the defocused muon is from the charm decay. The sign-selection of the neutrino beam allows this identification without use of a p_t algorithm. In CCFR this event may have been misidentified, especially if the toroid polarity had been set to defocusing. NuTeV should be able to increase its acceptance while decreasing its background.

SUMMARY

In conclusion, a number of experiments have analyzed neutrino dimuon production. CCFR has performed the only next-to-leading order analysis to date. NuTeV has an improved data sample which will allow for significant reduction in errors. We will perform a full next-to-leading order analysis (including gluon radiative diagrams) and study the scheme and scale dependences. There remain some open questions about the strange sea and the mass of the charm quark which NuTeV should be able to shed some light upon.

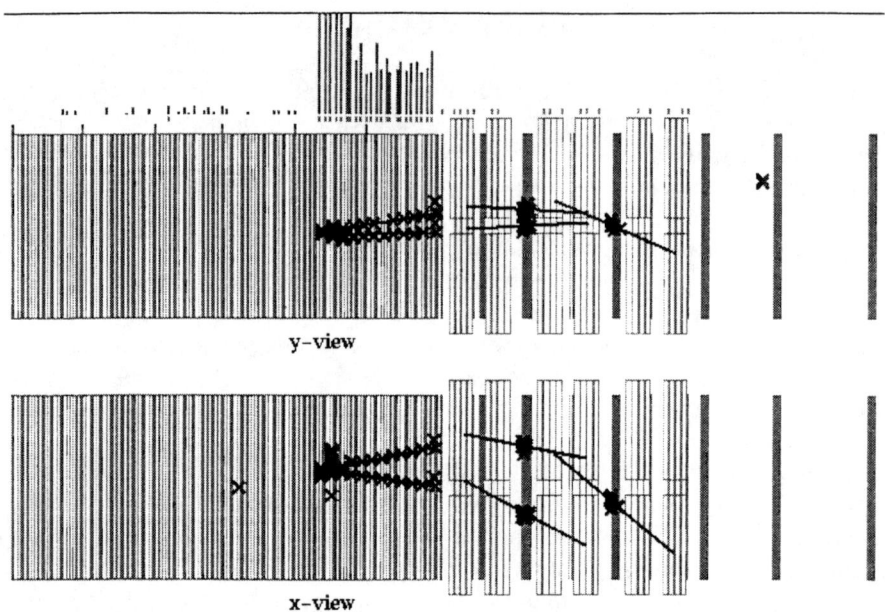

FIGURE 3. Dimuon event from NuTeV.

REFERENCES

1. Georgi, H. and Politzer, H.D¿, *Phys. Rev.* **D 14**, 1829 (1976).
2. Abramowitz, H., *Z. Phys.*, **C 15**, 19 (1982).
3. Vilain, P., *CERN-EP/98-128* (1998).
4. Rabinowitz, S.A., *Phys. Rev. Let.* **70**, 134 (1993).
5. Bazarko, A.O., *et. al.*, *Z. Phys.* **C65**, 189 (1995).
6. Buras, A.J. and Gaemers, K.J.F, *Nucl. Phys.* **B132**, 249 (1978).
7. Aivazis, M.A.G., *et. al.*, *Phys. Rev.* **D 50**, 3085 (1994); **50**, 3102 (1994).
8. Duke, D. and Owens, J., *Phys. Rev.* **D 30**, 49 (1984).
9. Peterson, C, *Phys. Rev.* **D 27**, 105 (1983).
10. Collins, P. and Spiller, T., *J. Phys.* **G 11**, 1289 (1985).

Heavy Flavour Physics at HERA

Beate Naroska

University of Hamburg
II. Institut für Experimentalphysik
Luruper Chaussee 149
D 22761 Hamburg
E-mail:naroska@mail.desy.de

Abstract. New results with increased statistics are presented for heavy flavour production at $Q^2 \lesssim 150\,\text{GeV}^2$ and in the photoproduction limit $Q^2 \to 0$. Cross sections for D^* production, F_2^c, the gluon density in the proton, and inelastic J/ψ production are discussed and compared to theoretical calculations. A first measurement of the $b\bar{b}$ cross section is reported.

Introduction

At HERA positrons of 27.5 GeV collide with 820 GeV protons yielding a center of mass energy of 300 GeV. Heavy flavours are predominantly produced in pairs by photon gluon fusion. Charm quark production is expected to be a factor 200 more abundant than bottom quarks at this energy. Heavy flavour processes give new opportunities of studying perturbative QCD at center of mass energies roughly a factor 10 higher than in fixed target experiments.

There are several methods to tag heavy flavours: "Open" charm production is tagged via reconstruction of D^* (H1 and ZEUS) or via semi-leptonic decays to

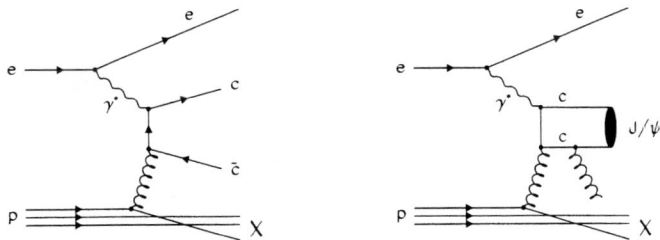

FIGURE 1. Charm production via photon gluon fusion (left) and J/ψ production in the color singlet model.

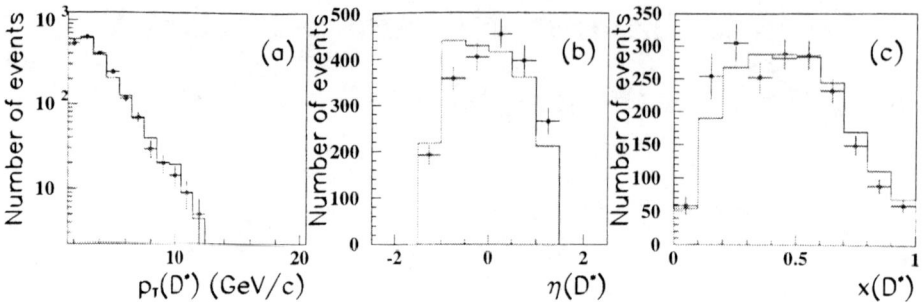

FIGURE 2. ZEUS: Reconstructed D^* related quantities compared to Monte Carlo simulation RAPGAP [2].

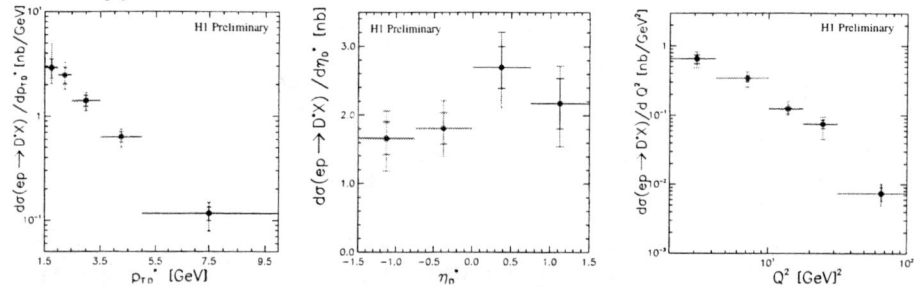

FIGURE 3. H1: Cross sections ($2 < Q^2 < 100\,\text{GeV}^2$, $0.05 < y < 0.7$, $p_T^{D^*} > 1.5\,\text{GeV}$, $|\eta^{D^*}| < 1.5$) compared to NLO calculations [3]. The shaded band represents the uncertainty in m_c.

electrons (ZEUS). b quarks have been measured via semi–muonic decays by H1. Finally "hidden" charm is studied via reconstruction of J/ψ (see fig. 1). $\psi(2s)$ and Υ Mesons have as yet been reconstructed only in diffractive processes [1a] at HERA and will not be reported here.

The integrated luminosity delivered by HERA has steadily increased over the years. This review will cover data from 1995 ($\sim 6\,\text{pb}^{-1}$), 1996 ($\sim 10\,\text{pb}^{-1}$) and 1997 ($\sim 26\,\text{pb}^{-1}$). Most results are preliminary.

The usual kinematic variables for deep inelastic scattering are used:
$s = (k+P)^2$; $Q^2 = -(k-k')^2$; $x = \frac{Q^2}{2P \cdot q}$; $y = \frac{q \cdot P}{k \cdot P}$; $W_{\gamma p}^2 = (q+P)^2 = sy - Q^2$

where k and P are the four vectors of incoming electron and proton, and q of the exchanged photon.

Determination of F_2^c

The inclusive cross section for production of charm in deep inelastic scattering (DIS) can be written as

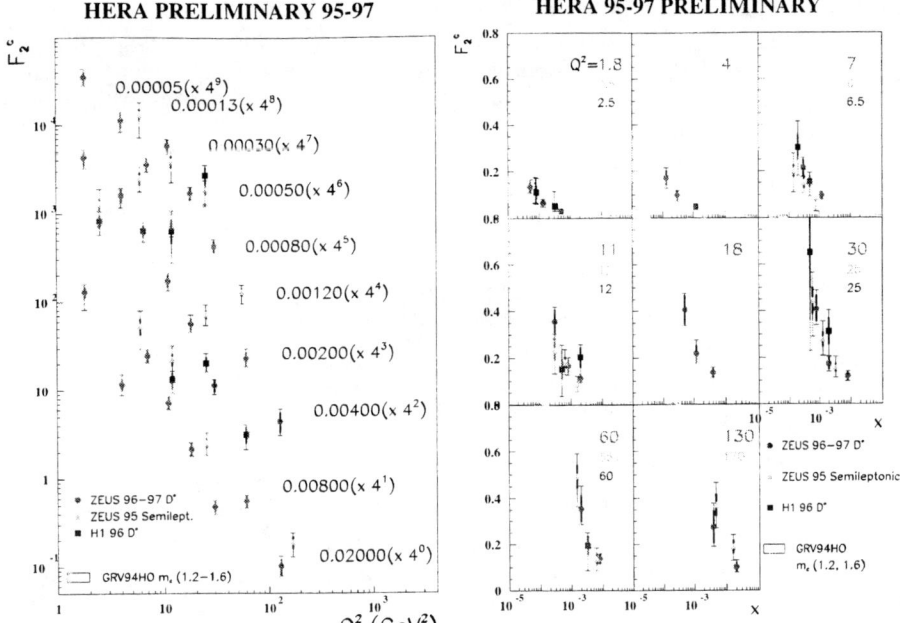

FIGURE 4. F_2^c as function of Q^2 at fixed x (scale factors applied for clarity); F_2^c as function of x in bins of Q^2. For comparison a NLO prediction using the GRV94HO parton densities is shown with the uncertainty due to the charm mass.

$$\frac{d^2\sigma^{ep\to ec\bar{c}X}}{dx dQ^2} = \frac{2\pi\alpha^2}{xQ^4}(1+(1-y)^2) \cdot F_2^c(x,Q^2)$$

where the contribution due to F_L has been neglected since it is expected to be small.

Charm is tagged through reconstruction of $D^{*+} \to D^0\pi^+$ with subsequent decay $D^0 \to K^-\pi^+$ and also the charge conjugate decay. ZEUS has presented a new analysis of semileptonic charm decays $c \to e^+ + X$. The electron was identified using the electromagnetic calorimeter and the specific energy loss dE/dx in the driftchamber. Details of the analysis from H1 and ZEUS can be found in [1b].

For D^* production a comparison to the RAPGAP Monte Carlo [2] simulation is shown in fig. 2. Reasonable agreement is found. After unfolding detector effects cross sections are obtained in a restricted kinematical region, examples from H1 data are shown in fig. 3. The data are compared to a NLO calculation by Harris and Smith [3] using the Peterson fragmentation function. The agreement is good and the extrapolation to the full kinematic region is done with this calculation.

The resulting Q^2 and x dependence of F_2^c is shown in fig. 4. The data span Q^2 values from 1.8 to 130 GeV^2 and $5 \cdot 10^{-5} \leq x \leq 0.02$. The agreement of the different data sets is reasonable within errors. Also shown is the theoretical NLO calculation using the GRV94-HO parton density functions which reproduces the data well. A strong rise of F_2^c towards low x is observed at fixed Q^2 and in Q^2 strong scaling violations are seen at fixed x. F_2^c gives a contribution of between 10% (low Q^2) and 30% (high Q^2) to the inclusive F_2 at an $x \sim 5 \cdot 10^{-4}$.

Photoproduction of D^*

When the exchanged photon is almost real, contributions due to its hadronic nature have to be taken into account ("resolved" processes). In NLO QCD calculations an unambiguous separation of the direct process (fig. 5a) and resolved processes (b and c) is no longer possible, only the sum of the two is well defined. There are two approaches to calculate the photoproduction cross sections in next to leading order:

The "massive" approach [4] where only the light quarks u, d and s and gluons are active partons in the photon (and proton). Charm is only generated in the hard subprocess (see also fig. 5b). This approach is valid for $m_c \gg \Lambda_{QCD}$. The *"massless" approach* [5,6] where also charm is an active flavour. This approach is valid at $p_t \gg m_c$.

The high statistics data from ZEUS [7] are shown in fig. 6. They are found to be above the massive and massless calulations. The comparison of H1 data [1b] with massive calculations shown in fig. 7 is satisfactory.

ZEUS has presented an analysis of D^* events which contain two jets [7]. In these events the *observed* momentum fraction x_γ^{obs} can be calculated, which describes the fraction of the photon energy contributing to the production of the two jets. A significant tail at low x_γ^{obs} is found in the data. In the generator HERWIG this tail can be described by charm excitation in the photon, while considering only light flavours leads to discrepancies.

Gluon density from D^* events

H1 extracted the proton's gluon density function in DIS ($2 < Q^2 < 100\,\text{GeV}^2$, $0.05 < y < 0.7$) and in photoproduction ($0.02 < y < 0.32$; $0.29 < y < 0.62$; $Q^2 \sim 0$) [1b].

The observed momentum fraction x_g^{obs} of the gluon is reconstructed from the kinematics of the final state and a differential cross section $d\sigma/dx_g^{obs}$ is determined, which for the DIS data is shown in fig. 8. The correlation of x_g^{obs} with the true x_g as given by the NLO QCD calculations [3] – also shown in fig. 8 – is used in an iterative unfolding procedure to obtain $d\sigma/dx_g$. The gluon density is then obtained by reweighting the calculation with the measured cross section. The result is shown in fig. 8 as a momentum distribution $x_g \cdot g(x_g)$. The range $10^{-3} < x_g < 0.02$ is covered. The data from photoproduction and DIS agree well within the large errors.

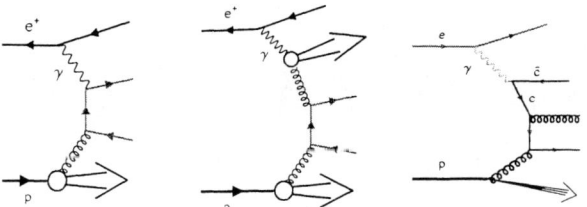

FIGURE 5. Generic diagrams for a) direct, b) resolved charm production. In c) a NLO diagram is shown (flavour excitation).

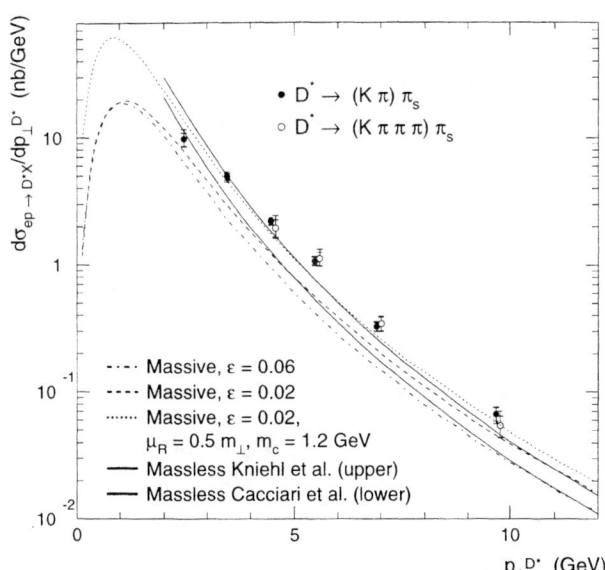

FIGURE 6. $d\sigma/dp_\perp$ for D^* photoproduction from ZEUS for $130 < W_{\gamma p} < 280\,\text{GeV}$, $Q^2 < 1\,\text{GeV}^2$ and $|\eta^{D^*}| < 1.5$ (η^{D^*} is the pseudorapidity of the D^*) for the $(K\pi)\pi$ and $(K\pi\pi\pi)\pi$ channels. The curves represent "massless" and "massive" calculations as indicated.

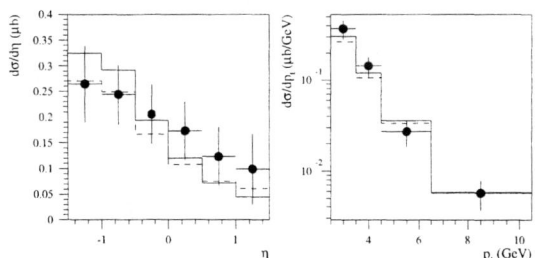

FIGURE 7. Photoproduction of D^* from H1: $d\sigma^{\gamma p}/d\eta$ for $p_\perp > 2.5\,\text{GeV}$ and $d\sigma^{\gamma p}/dp_\perp$ for $|y^{D^*}| < 1.5$, where y^{D^*} is the rapidity. The histograms show the NLO QCD predictions calculated according to [4] with the MRSG (solid) and MRSA' (dashed) proton parton density parametrizations.

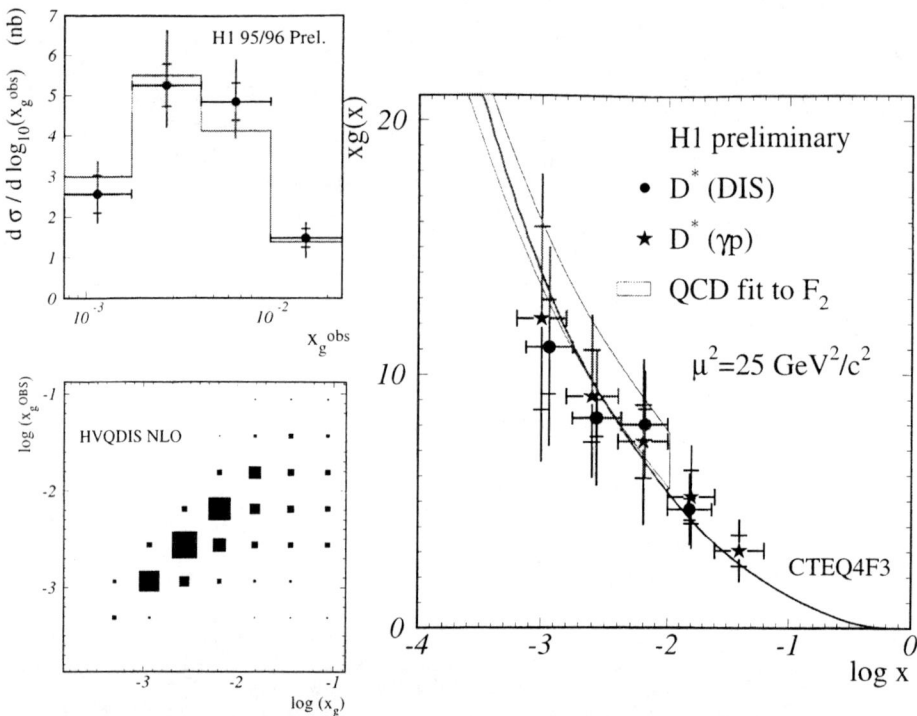

FIGURE 8. (a) Differential cross section for $ep \to eD^*X$ as function of x_g^{obs} in the visible range compared to the NLO QCD prediction using the CTEQ4F3 parton distribution. The shaded band reflects the uncertainty due to the charm mass $1.3 < m_c < 1.7$ GeV. (b) Correlation between x_g^{obs} and true momentum fraction x_g. (c) Comparison of the gluon densities obtained from the two D^* analyses in DIS and photoproduction. The systematic error is dominated by the uncertainty of the charm quark mass and the fragmentation parameter. For comparison the H1 QCD analysis of the inclusive F_2 measurement (shaded band) and the CTEQ4F3 parametrization are shown.

They also agree with the result from an analysis of scaling violations in the inclusive measurement of F_2 [8].

$b\bar{b}$ Production

Due to the higher mass of the b quark the total cross section for $b\bar{b}$ production is expected to be 200 times smaller than that for $c\bar{c}$ production. The theoretical uncertainties in calculating the next to leading order predictions are, however, smaller [9]. H1 determined the cross section for the first time in the HERA energy range using semi–muonic b decays [1c].

A photoproduction event sample was selected containing two jets of transverse energy $E_T > 6$ GeV and a muon of transverse momentum (relative to the beam

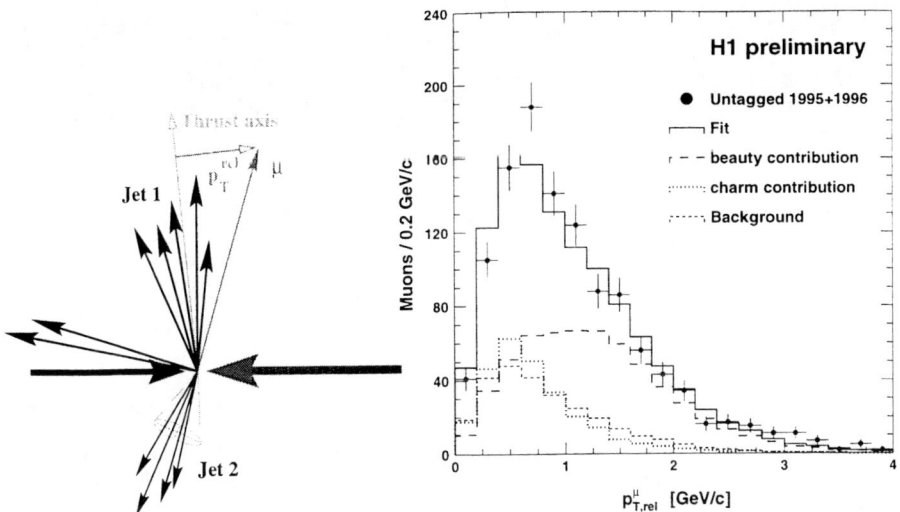

FIGURE 9. *Left:* Definition of $p_{T,rel}^{\mu}$. *Right:* The measured $p_{T,rel}^{\mu}$ distributions and the result of the fit (solid line); the contributions from beauty, charm and background are shown separately.

direction) $p_T^{\mu} > 2$ GeV in the central detector region $35° < \theta^{\mu} < 130°$.

The thrust axis[1] was determined for each jet in order to approximate the b flight direction. The transverse momentum of the muon $p_{t,rel}^{\mu}$ with respect to the jet is used as a discriminating variable: Muons from b decay show a $p_{t,rel}^{\mu}$ spectrum extending to higher values than c decays (see fig. 9 for an illustration of the method).

The background comes from the production of the light quarks u, d and s, which is roughly a factor 2000 larger than b production. Punch through and decay in flight lead to false muon signatures. The contribution is determined from data using an independent dataset and using the muon fake probability and the hadron composition from a well tuned and checked simulation program. The resulting $p_{t,rel}^{\mu}$ spectrum is shown in fig. 9 indicating the background contribution (23.5%), which is absolutely determined. The fractions of b and c quarks are obtained from a fit to the data distribution, yielding $(51.4 \pm 4.4)\%$ and $(23.5 \pm 4.3)\%$, respectively.

The cross section in the visible kinematic range of $Q^2 < 1 \text{ GeV}^2$; $p_T^{\mu} > 2 \text{ GeV}$; $95 \leq W_{\gamma p} \leq 270 \text{ GeV}$; $35° < \theta^{\mu} < 130°$ is determined as

$$\sigma(ep \to e\,b\bar{b} + X)^{vis} = 0.93 \pm 0.08^{+0.21}_{-0.12} \text{ nb},$$

where the first error is statistical and the second systematic. Contributions to the systematic errors are the branching ratio $b \to X\mu\nu$, the energy scale of calorimeters and detector efficiencies.

---

[1] The thrust axis is the axis which maximizes $T = \max(\frac{\Sigma p_i^L}{\Sigma |p_i|})$, where the sum runs over all particles belonging to the jet and p_i^L is the component of the particle momentum parallel to the thrust axis.

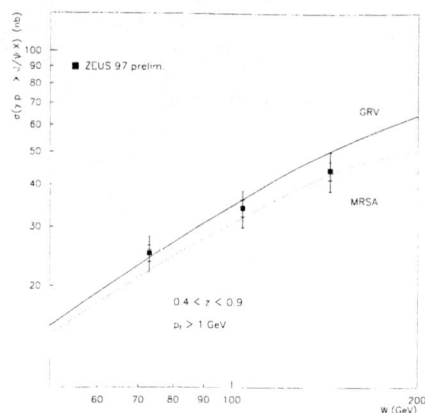 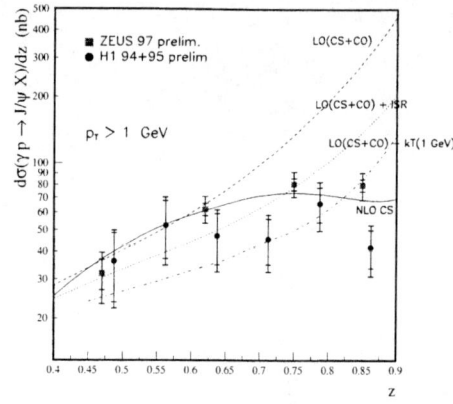

FIGURE 10. Inelastic J/ψ photoproduction: a) cross section as function of $W_{\gamma p}$ for $0.4 < z < 0.9$ and $p_T > 1\,\text{GeV}$. The lines correspond to the NLO predictions from [13] with GRV or MRSA' [20] parton density functions ($m_c = 1.4$ GeV and $\Lambda_{QCD} = 300$ MeV). b) $d\sigma/dz$ for $50 < W_{\gamma p} < 180\,\text{GeV}$ and $p_T > 1\,\text{GeV}$. The NLO computation is shown as a solid line. The dashed and dotted lines are given by the sum of the colour–singlet and the colour–octet leading order calculations performed in [15] and in [16]. The latter include estimates of higher order QCD corrections due to initial state radiation.

The corresponding direct LO cross section from the AROMA [10] simulation is 0.19 nb, roughly a factor 5 lower. The fraction of c quarks determined from this analysis leads to the same cross section for $ep \to ec\bar{c}X$ as previously determined from analysis of D^* production [11].

Inelastic J/ψ production

New data on charmonium (J/ψ, ψ') and Υ production have been presented by H1 and ZEUS [1a]. Here we will concentrate on "inelastic" J/ψ production as opposed to the diffractive processes which dominate the cross section at low Q^2. Inelastic J/ψ production could at lower $W_{\gamma p}$ (fixed target regime) be well described by the Colour Singlet Model (CSM). For HERA the CSM cross section calculations are available in NLO [13].

As is well known the CSM fails to describe charmonium production in $p\bar{p}$ collisions at high p_T [12]. Colour octet contributions have been proposed for an adequate description. The NRQCD factorisation approach (NRQCD = Non Relativistic QCD) describes any process $A + B \to J/\psi + X$ as a sum over colour singlet and colour octet contributions.

Whereas the transformation of a colour singlet $^3S_1$ state into a J/ψ can be calculated using the measured semileptonic decay width, the transition of a colour octet

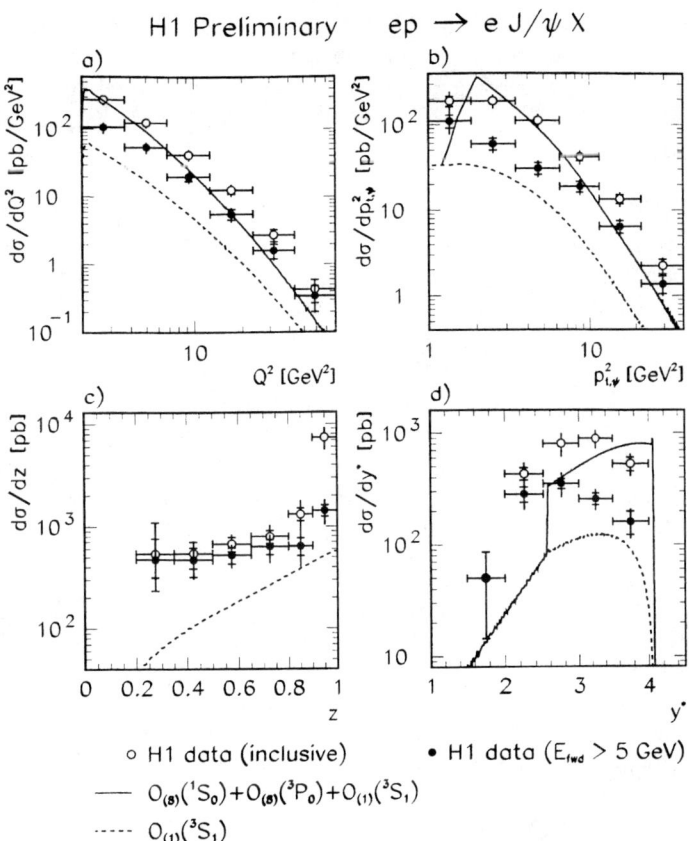

FIGURE 11. Differential cross sections for the inclusive (open points) and inelastic ($E_{fwd} > 5$ GeV, black points) $ep \to e J/\psi X$ process. a) $d\sigma/dQ^2$, b) $d\sigma/dp_{t,\psi}^2$, c) $d\sigma/dz$ and d) $d\sigma/dy^*$. The curves are predictions [17] within the NRQCD factorization approach for the colour singlet contribution (dashed) and the sum of singlet and octet contributions (full line).

state to J/ψ is non–perturbative and at present not calculable. Therefore predictions for the cross section at HERA use the non perturbative transition matrix elements extracted from the CDF data.

The ZEUS collaboration has updated their photoproduction data [1d]. The results for the γp cross section as function of $W_{\gamma p}$ and of z are shown in fig. 10. The data agree well with the next to leading order pQCD calculation in the color singlet model [13]. The variable z is defined as $z = \frac{P_\psi \cdot P}{P \cdot q} \approx \frac{E_\psi}{E_\gamma}$ where the latter approximation holds in the proton rest frame. In fig. 10 b in addition to the CSM in NLO calulations using the NRQCD/factorisation approach [14–16] are shown. The upper curve was calculated in LO using the transition matrix elements extracted from CDF data in LO and shows a strong rise towards high z values. The lower curves also take into account higher orders *approximately* as explained in refs. [14–16]. Doing so leads to modifications in the non perturbative matrix elements and/or in the cross sections themselves. The net effect is a decrease of the predicted rise at high z.

H1 has for the first time determined the cross sections for inelastic J/ψ production at $Q^2 > 2$ GeV^2 [1d]. The results are shown in fig. 11. Two data sets are shown, a completely inclusive one (open points) and one where the diffractive contributions have been removed by a cut on the energy in the forward region of the detector as suggested by Fleming and Mehen [17], whose LO calculations are shown for comparison. The data are seen to be far above the CS contributions. The magnitude of the data is reproduced better taking into account colour octet contributions. The shape of the latter leaves, however, much room for improvement, in particular in the rapidity y^* in the $\gamma^* p$ center of mass system. Note that the NRQCD calculations are performed at the parton level, no smearing due to the transition into J/ψ is taken into account.

Summary

Due to increased statistics detailed analyses of heavy flavour production in ep collisions are performed in a variety of channels and kinematic regions. $b\bar{b}$ production was observed for the first time in photoproduction via semi–muonic decay of the b–quark. The cross section was found to be considerably larger than the leading order predictions for the direct process.

In the range $2 \lesssim Q^2 \lesssim 130$ GeV^2 cross sections and the charm contribution to F_2^c are determined and found to agree with next to leading order predictions. In photoproduction the validity of different approaches to calculate next to leading order corrections is being studied in various kinematic regions.

Since photon gluon fusion is the dominant process a direct determination of the gluon density in the proton was carried out in DIS and in the photoproduction regime. The result agrees with the indirect determinations from scaling violations.

Inelastic J/ψ production is studied in photoproduction and DIS and is well described in photoproduction by the color singlet model alone in next to leading order. In DIS the data have been compared to LO color singlet and color octet predictions (at parton level). In the latter rough agreement in absolute normalisation is found, while the color singlet model reproduces the shape of the data slightly better.

Acknowledgement I wish to thank the organisers for a very pleasant and fruitful meeting and my colleagues at ZEUS and H1 for supplying their data and for discussions.

REFERENCES

1. Contributed papers to International Conference on High Energy Physics, ICHEP98, Vancouver, July 1998.
   http://www-h1.desy.de/h1/www/publications/conf/list.vancouver98.html
   and http://www-zeus.desy.de/conferences98/ichep98papers/index.html
   a) ZEUS Coll., contr. paper 791 (Υ),792 (J/ψ) , H1 Coll., contr. paper 572 ($J/\psi,\psi(2s)$), 574 (Υ)
   b) H1 Coll., contr. paper 538, 540; ZEUS Coll., contr. papers 768, 772
   c) H1 Coll., contr. paper 575
   d) ZEUS Coll., contr. paper 814; H1 Coll., contr. paper 573
2. Jung H., *Comp. Phys. Commun.* **86**, 147 (1995)
3. Harris B.W., hep-ph/9608379; DESY 97-111, FSU-HEP-970527, May 1997; *Nucl. Phys.* **B452**, 109 (1995); *Phys. Lett.* **B353**, 535 (1995)
4. Frixione S. et al., *Phys. Lett.* **B348**, 633 (1995); *Nucl. Phys.* **B454**, 3 (1995)
5. Kniehl B.A. et al., *Z. Phys.* **C76**, 689 (1997);
   Binnewies J. et al., *DESY* 97-241, hep-ph/9712482,*Z. Phys.* **C76**, 677 (1997)
6. Cacciari M. et al., *Phys. Rev.* **D55**, 2736 (1997); ibid. 7134
7. ZEUS Coll., J. Breitweg et al., DESY 98-085, hep-ex/9807008
8. Chekelian V., DESY 97-248 (1997)
9. Nason P., Dawson S., and Ellis R.K., *Nucl. Phys.* **B303**, 607 (1988);
   Ellis R.K. and Nason P., *Nucl. Phys.* **B312**, 551 (1989);
   Smith J. and Van Neerven W.L., *Nucl. Phys.* **B374**, 36 (1992)
10. Ingelman G., Rathsman J., Schuler G.A., *Comput. Phys; Commun.* **101**, 135 (1997)
11. H1 Coll., *Nucl. Phys.* **B472**, 32 (1996)
12. CDF Coll., Abe F. et al., *Phys. Rev. Lett.*, **69**, 3704 (1992); ibid. **79**, 572 (1997);
    DØ Coll., Abachi S. et al., *Phys. Lett.* **B370**, 239 (1996)
13. Krämer M., *Phys. Lett.* **B348**, 657 (1995); *Nucl. Phys.* **B459**, 3 (1996)
14. Beneke M. and Krämer M., *Phys. Rev.* **D55**, 5269 (1997);
    Kniehl B.A. and Kramer G., *Phys. Rev.* **D56**, 5820 (1997); **B413**, 416 (1997);
    Beneke M., Rothstein I.Z. and Wise M.B., *Phys. Lett.* **B408**, 373 (1997)
15. Cano–Coloma B. and Sanchis–Lozano M.A., *Nucl. Phys.* **B508**, 753 (1997)
16. Sridhar K., Martin A.D., Stirling W.J. DTP-98-30 (1998), hep-ph 9806253
17. Fleming S. and Mehen T., *Phys. Rev.* **D57**, 1846 (1998)

B PRODUCTION and ONIUM PRODUCTION AT THE TEVATRON

Andrzej Zieminski
Representing the CDF and DØ Collaborations

Indiana University, Bloomington, IN 47405, USA
email: zieminski@ind.physics.indiana.edu

Abstract. We review recent results obtained by the DØ and CDF collaborations on the rapidity dependence of b-quark and onia production in $p\bar{p}$ collisions at $\sqrt{s} =$ 1.8 TeV. The DØ experiment has measured the inclusive production cross section for muons originating from b-quark decays in the forward rapidity region of $2.4 <| y^\mu |< 3.2$ and compared it with its published results for muons in the central region. There is an indication of an increasing discrepancy between the data and the next-to-leading order QCD calculations at larger rapidities. The CDF experiment has measured a ratio of the $b\bar{b}$ correlated cross sections for a central-central and central-forward production. Reduced experimental and theoretical uncertaintities allow to treat this measurement as a direct probe of the high-x gluon content of the proton. Finally, the DØ collaboration extended studies of the inclusive J/ψ production into the forward pseudorapidity region $2.5 <| \eta^{J/\psi} |< 3.7$ and transverse momenta between 1 and 16 GeV/c. The new results are consistent with the predictions of theoretical models for charmonium production, whose parameters were adjusted to describe the direct J/ψ production in the central rapidity region.

INTRODUCTION

The production of b quarks in hadronic collisions is a subject of persistent theoretical and experimental interest. The comparison of the measured b quark production cross section with QCD calculations reveals the underlying dynamics. The next-to-leading order (NLO) calculations have been available for a decade [1]. In addition to the lowest order $b\bar{b}$ production processes (s-channel gluon diagrams or t-channel quark exchange), sizable contributions come from $\mathcal{O}(\alpha_s^3)$ diagrams including the t-channel gluon exchanges with subsequent gluon splitting, and flavor excitation in which a gluon–$b\bar{b}$ virtual fluctuation is put on-mass shell by an interaction.

The NLO QCD predictions show a large dependence on the choice of the renormalization and factorization scale, μ. This scale is usually chosen as $\mu = \mu_o \equiv \sqrt{m_b^2 + (p_T^b)^2}$, where p_T^b is the transverse momentum of the b-quark and m_b is its mass. The theoretical uncertainties are obtained by varying μ between $\mu_0/2$ and

CP459, *Heavy Quarks at Fixed Target*
edited by Harry W. K. Cheung and Joel N. Butler
© 1999 The American Institute of Physics 1-56396-864-9/99/$15.00

$2\mu_0$. The theoretical predictions are less affected by the choice of m_b, with a central value usually set to 4.75 GeV/c^2, and varied between 4.5 and 5.0 GeV/c^2 for the uncertainties.

Both CDF and DØ have extracted the inclusive b quark cross sections for rapidity $|y| < 1$ using several different final state channels [2,3]. These measurements were found to agree in shape with the NLO QCD predictions over the studied transverse momentum range, $6 < p_T^b < 40$ GeV/c, but to be larger by a factor ~ 2.5 (DØ) to ~ 2.8 (CDF) than the central value [4,5]. In section 2 we present the preliminary DØ measurement of the differential muon cross section due to b quark decays in the forward region, $2.4 < |y^\mu| < 3.2$, as a function of the muon transverse momentum, p_T^μ.

Studies of $b\bar{b}$ correlations provide an even more stringent test of the NLO QCD calculations. Several previous studies of the $b\bar{b}$ correlations performed by CDF [6] and DØ [7] focused on the difference in azimuthal angle between the two muons coming from $b\bar{b}$ decays. The shape of the opening angle between the two muons is found to be consistent with the theory within the uncertainties. However, another CDF analysis [8] of the correlations between the b and $\bar{b}$ quarks in μ-jet events, exploring production of b and $\bar{b}$ quarks with relatively large transverse momenta, shows discrepancies between the predicted and observed shape of the azimuthal opening angle between the two quarks.

Recently, CDF has extended its correlation studies to the forward region [9], measuring the $b\bar{b}$ production cross section when one quark is produced in the forward region ($1.8 < |\eta^b| < 2.6$) and the other in the central pseudorapidity range ($|\eta^b| < 1.5$). This measured cross section was found to be higher than the NLO QCD prediction by a factor $2.4^{+0.6}_{-0.5}$, with the uncertainty reflecting the experimental error only. The latest CDF analysis, described in section 3, studies a ratio of $b\bar{b}$ production cross sections with either both quarks in the central region or one of the quarks produced centrally and the other in the forward region. This approach significantly reduces the systematic uncertainties affecting both the data and the theory.

Finally, in section 4, we discuss a recent DØ measurement [10] of the J/ψ production in previously unexplored kinematical regions of small J/ψ transverse momenta and large rapidities.

FORWARD μ AND B-QUARK PRODUCTION

The DØ analysis of the b-quark production in the forward region is based on the small angle muon spectrometer [11], SAMUS, consisting of magnetized iron toroids and drift tube stations (18 planes on each side of the interaction region). The data were collected in special runs during the 1994-95 Tevatron run and correspond to an integrated luminosity of 104 ± 6 nb^{-1}.

Muons are selected in the rapidity range $2.4 < |y^\mu| < 3.2$, with momentum $p^\mu < 150$ GeV/c and transverse momentum $p_T^\mu > 2$ GeV/c. Additional cuts are applied to ensure a good momentum measurement and an energy deposition in the calorimeter consistent with that from a minimum ionizing particle, leading to a final sample of 5106 events.

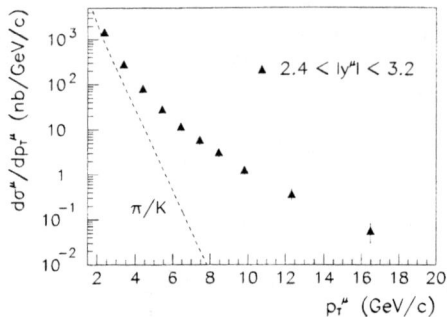

FIGURE 1. The inclusive muon cross section, per unit of rapidity, in the forward region as a function of p_T^μ. Only the statistical errors are shown. The p_T independent systematic uncertainty is 14%. The remaining systematic uncertainties vary between 15 and 45%, the major error sources being the momentum resolution, momentum scale and momentum smearing unfolding. The dashed line represents the expected contributions from π/K decays.

The measured inclusive muon cross section as a function of p_T^μ is shown in Fig. 1. The three major contributions to the inclusive muon cross section consist of in-flight decays of pions and kaons, and, b- and c-quark decays. The expected π/K contribution, obtained using ISAJET [12], dominates the inclusive muon cross section for low p_T^μ but falls much more rapidly than the data, the excess being attributed to the heavy flavors decays. After subtraction of the π/K contribution, the muon cross section is multiplied by the b-quark fraction, f_b, defined as the ratio of the muon yield due to b-quark decays to the total muon yield from b- and c-quark decays. This ratio is determined using NLO QCD predictions for b- and c-quark production cross sections. DØ uses events containing a muon associated with a jet ($E_T^{jet} > 10$ GeV) to compare the b-quark fraction obtained from the data against Monte Carlo predictions. The f_b fraction is determined by fitting distributions of the muon transverse momentum with respect to the jet direction, p_T^{rel}, to the expected shapes from $b, c,$ and π/K decay. Unfortunately only 7% of events in the cross section sample have an associated jet. Therefore, another sample of 31,000 SAMUS muon plus jet events, collected during the entire 1994-95 Tevatron run, is used for the verification. A good agreement between data and Monte Carlo simulation (not shown) is found over the entire p_T^μ range studied.

The differential muon cross section from b-quark production and decay is presented in Fig. 2. The systematic uncertainties on this measurement include those of

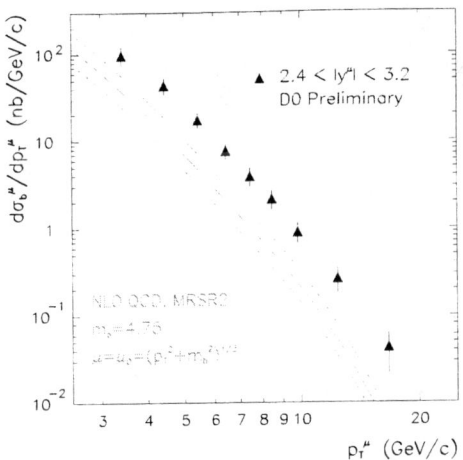

FIGURE 2. p_T spectrum of forward muons from b decays compared to NLO QCD prediction, generated with $m_b = 4.75$ GeV/c, $\mu = \mu_0$, and the MRSR2 [13] parton distribution functions. The theoretical uncertainty is determined by varying the parameters m_b and μ from 4.5 and 5.0 GeV/c^2 and from $\mu_0/2$ to $2\mu_0$, respectively.

the inclusive muon cross section together with uncertainties due to the f_b estimation (17 – 8)% and the π/K subtraction (14 – 0)%. In Fig. 2, the data are compared to the NLO QCD prediction obtained using the Monte Carlo simulation that relies on MNR calculations [1] for the production of the $b/\bar{b}$ partons and ISAJET for the fragmentation to b hadrons and the decay of these hadrons to muons. The predictions of this model match the shape of the measured cross section fairly well, but are approximately a factor of four lower than the data.

These measurements, combined with previous DØ results [3] obtained in the central region ($|y^\mu| < 0.8$) permit to study the rapidity dependence of the b-produced muon cross section. The differential muon cross section from b decay as a function of rapidity is compared to the NLO QCD predictions in Fig. 3 for two p_T^μ ranges, $p_T^\mu > 5$ GeV/c and $p_T^\mu > 8$ GeV/c. For $p_T^\mu > 5$ GeV/c, the ratio data/theory is equal to 2.5 ± 0.5 in the central region and increases to 3.6 ± 0.8 in the forward region. The uncertainty on these ratios only reflects the experimental errors. For $p_T^\mu > 8$ GeV/c, these numbers become 3.0 ± 0.6 and 4.6 ± 1.3 in the central and forward regions, respectively.

There have been several theoretical attempts to account for the observed increase in the data/theory discrepancy with increasing b quark rapidity. New calculations based on a variable flavor number scheme [14] predict an increase of the forward b-quark production cross section by 30 to 50% with respect to the standard NLO

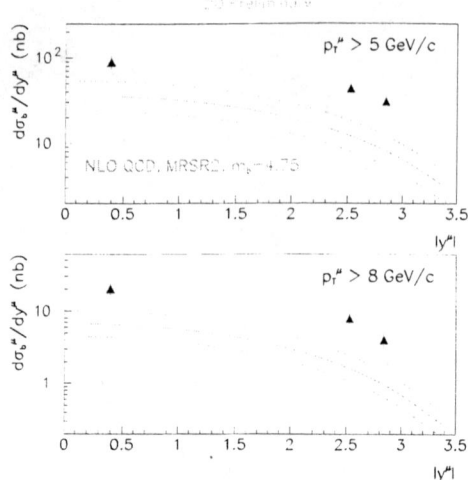

FIGURE 3. b-produced muon cross section $vs.$ rapidity compared to the NLO QCD prediction.

QCD calculations. An increase in the B-meson production cross section of 50% in the forward region and 30% in the central region can also be obtained by using a stiffer b-quark fragmentation function than the usual Peterson one [15].

$B\bar{B}$ RAPIDITY CORRELATIONS

This analysis represents an extension of the CDF absolute forward-central $b\bar{b}$ cross section measurement mentioned in the introduction. In order to reduce the systematic uncertainties on both the data and the theory, CDF now measures [9] a ratio of correlated cross sections for a central-central and central-forward $b\bar{b}$ production.

The event selection requires a central b-quark jet with $E_T > 26$ GeV and $\mid \eta^{jet} \mid <$ 1.5, accompanied by a second b-quark decaying into a muon and a jet. This jet must have an $E_T > 15$ GeV and the events are classified as central-central or forward-central depending upon the muon pseudorapidity range, $\mid \eta^\mu \mid < 0.6$ or $2.0 < \mid \eta^\mu \mid < 2.6$, respectively. For the μ-jet combination, the b-quark decay is identified on the basis of the muon momentum transverse to the μ-jet direction, p_T^{rel}. The central b-quark jet is identified through a secondary vertex tag. The final data sample, corresponding to an integrated luminosity of 80 pb^{-1}, contains 382 forward-central and 7544 central-central events.

To extract the signal fraction in each sample, the p_T^{rel} of the muon and the pseudo-$c\tau$ of the b-jet are fitted simultaneously using a binned maximum likelihood method. The template histograms for the signal and the dominant background sources ($c\bar{c}$

production and $b\bar{b}$ production where one b tag is a fake) are obtained either from Monte Carlo simulation or directly from the data. The $b\bar{b}$ signal fraction is about 75% in the forward-central events and 60% in the central-central events.

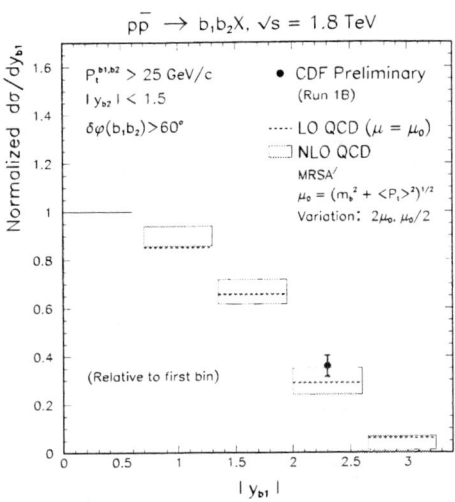

FIGURE 4. Evolution with rapidity of the cross section ratio.

The ratio of the cross sections is given by:

$$R_{\text{data}} = \frac{\sigma(p\bar{p} \to b_1 b_2 X; \; 2.0 < |y_{b_1}| < 2.6)}{\sigma(p\bar{p} \to b_1 b_2 X; \; |y_{b_1}| < 0.6)} \quad (1)$$

where $p_T(b_1, b_2) > 25$ GeV/c, $|y_{b_2}| < 1.5$, and, $\delta\phi(b\bar{b}) > 60°$ for both cross sections. It is calculated using the number of signal events in both samples and the ratio of the total efficiency for central-central and forward-central events. Due to the much smaller kinematic acceptance of the forward toroids relative to the central detector, this ratio is equal to 5.4. The result,

$$R_{\text{data}} = 0.361 \pm 0.041(\text{stat})^{+0.011}_{-0.023}(\text{syst}),$$

is in good agreement with the NLO QCD prediction obtained with MRSA' [16]:

$$R_{\text{theory}} = 0.338^{+0.014}_{-0.097}.$$

The systematic errors on R_{data} are due to the uncertainty in the energy scale (3%), acceptance (3%), fragmentation (3%), and background assumption (5%). The uncertainty on R_{theory} results from changing the μ scale between $2\mu_0$ and $\mu_0/2$. The predicted evolution of the cross section ratio with the rapidity range

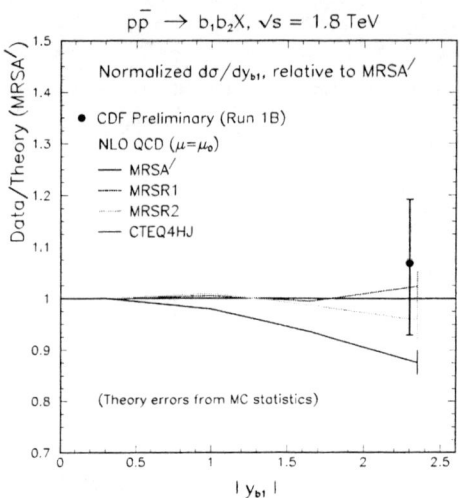

FIGURE 5. Comparison of the cross section ratio with various parton distribution functions.

of the muon coming from the b-quark semileptonic decay is shown in Fig. 4. The measurement is in agreement with both the LO and NLO QCD predictions.

With the cross section definition given in (1), the forward-central events arise from collisions between one parton with an average $x \approx 0.025$ and a second parton at $x \approx 0.25$. The cross section ratio should thus be sensitive to the gluon distribution in the proton at high x values, i.e. in a region where this gluon distribution is not very well known [17].

Figure 5 shows a comparison of the R_{data} measurement with R_{theory} obtained using the parton distribution sets MRSR1(2) [13] and CTEQ4HJ [18]. The data point and theory curves are normalized to MRSA' [16] and are presented as a function of the rapidity of the b-quark decaying to μ-jet. The data point is in good agreement with the MRS sets and slightly disfavors the CTEQ4HJ distribution. Since the CDF result is dominated by the statistical uncertainty, this comparison illustrates the potential resolving power of a high statistics forward b production measurement.

SMALL ANGLE J/ψ PRODUCTION

In high energy $p\bar{p}$ collisions J/ψ's are produced directly, from decays of higher mass charmonium states [χ and $\psi(2S)$], and from b-quark decays. Existing experimental results in the central rapidity region from UA1 [19] at $\sqrt{s} = 0.63$ TeV, and from CDF [20] and DØ [21] at $\sqrt{s} = 1.8$ TeV demonstrate that the measured inclusive J/ψ transverse momentum distribution cannot be described solely by contributions from b quark decays and prompt production predicted by the color singlet model [22]. In the color singlet model the charmonium meson retains the quantum numbers of the produced $c\bar{c}$ pair and thus each J/ψ state can only be directly produced via the corresponding hard scattering color singlet subprocess. The model predicts direct J/ψ and $\psi(2S)$ production rates fifty times smaller than those observed by CDF [20]. To explain this discrepancy, a color octet model was introduced [23,24]. The color octet mechanism extends the color singlet approach by taking into account the production of $c\bar{c}$ pairs in a color octet configuration accompanied by a gluon. The color octet state evolves into a color singlet state via emission of a soft gluon. The parameters of the model were derived from a fit to CDF data for direct J/ψ and $\psi(2S)$ production at central rapidity.

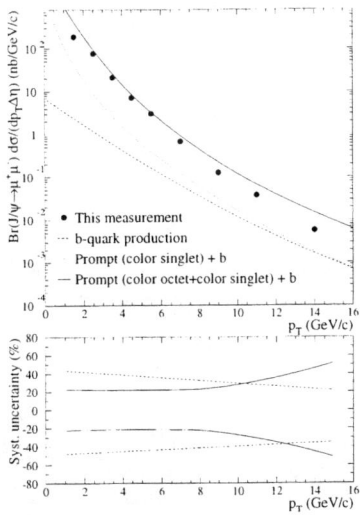

FIGURE 6. The p_T dependence of the J/ψ differential cross section and its theoretical predictions (upper figure). Only the statistical errors are shown. The lower figure presents systematic uncertainties; the solid curves are the sum of all systematic errors, the dashed curves represent the uncertainty band due to J/ψ polarization. The upper (lower) dashed curve corresponds to 100% transverse (longitudinal) polarization.

Recently, the DØ collaboration utilized the large rapidity coverage of the DØ muon system to study the process $p\bar{p} \to J/\psi + X \to \mu^+\mu^- + X$ in previously unexplored kinematical regions of small J/ψ transverse momenta and large rapidities [10]. In this analysis, events are selected requiring one interaction vertex, a single muon or dimuon trigger, and at least two reconstructed muon tracks. The muon track selection criteria are identical to the single muon analysis described in section 2. In total, 1779 events with opposite sign muon pairs and 281 events with same sign muon pairs are selected from the data sample with integrated luminosity of

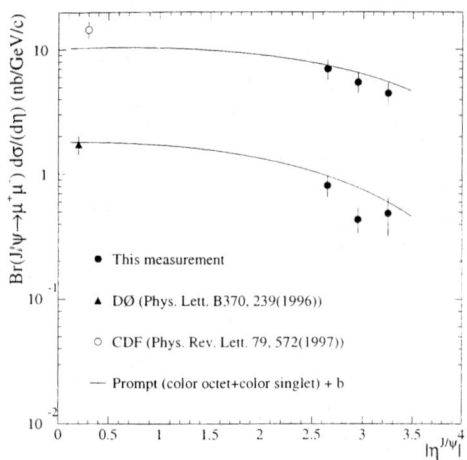

FIGURE 7. The pseudorapidity dependence of the J/ψ production cross section with $p_T > 5$ GeV/c (upper points and curve) and $p_T > 8$ GeV/c (lower points and curve). The error bars are statistical and systematic errors (polarization uncertainties not included) summed in quadrature.

9.8 ± 0.5 pb^{-1}. The fit to the opposite sign dimuon mass spectrum (not shown) yields 691 ± 41 J/ψ events. The fit is also performed in five $\eta^{\mu\mu}$ and nine $p_T^{\mu\mu}$ intervals.

In Fig. 6 we compare the resulting J/ψ cross section with current models of charmonium production. For J/ψ from b-quarks DØ uses the NLO QCD predictions [1] with the renormalization/factorization scale $\mu = \frac{1}{3}\mu_0$. The scale is chosen to match theory predictions to the published DØ b-quark cross sections in the central rapidity region [21]. ISAJET is used to fragment b quarks into J/ψ. The color octet and color singlet contributions to the direct J/ψ production and radiative χ decays are taken from Ref. [24]. The term representing the direct J/ψ production is increased by 12% to account for the contribution from $\psi(2S)$ decays [20].

Figure 7 shows the pseudorapidity dependence of the measured J/ψ cross section for $p_T^{J/\psi} > 5$ and 8 GeV/c along with the corresponding central rapidity measurements of DØ [21] and CDF [20]. Within uncertainties, the color octet model plus b quark decays describe the η dependence of the inclusive J/ψ production in the full rapidity region. Therefore, the new results are consistent with the published results for the central J/ψ production, once the expected effects due to the process kinematics and variations of the gluon structure functions are taken into account.

REFERENCES

1. M. Mangano, P.Nason and G. Ridolfi, *Nucl. Phys.* B **373**, 295 (1992).
2. F. Abe *et al.*, (CDF Collaboration), *Phys. Rev. Lett.* **71**, 500, 2396, 2537 (1993); *Phys. Rev. Lett.* **75**, 1451 (1995); *Phys. Rev. Lett.* **79**, 572 (1997).
3. S. Abachi *et al.*, (DØ Collaboration), *Phys. Rev. Lett.* **74**, 3548 (1995).
4. D. Zieminska, in: Proceedings of the XXXIIIrd Rencontres de Moriond '98, QCD and Hadronic Interactions.
5. Most of the available DØ results on the inclusive b-quark production [3,4] were derived from the measured inclusive muon transverse momentum spectra. The transition from the muon spectrum to the inclusive b-quark cross sections involves b-quark decay tables, which have changed since the completion of the analyses. The quoted data/theory ratio for the DØ inclusive b-quark cross sections [4] has been corrected upwards by a factor ~ 1.2, to account for the change between ISAJET version 7.09 used in Ref. [3] and versions 7.22 and higher used in more recent analyses.
6. F. Abe *et al.*, (CDF Collaboration), *Phys. Rev.* D **55**, 2546 (1997).
7. D. Fein, (DØ Collaboration), FERMILAB-Conf-97/008-E (1997).
8. F. Abe *et al.*, (CDF Collaboration), *Phys. Rev.* D **53**, 1051 (1996).
9. F. Abe *et al.*, (CDF Collaboration), www - cdf.fnal.gov /physics /new /bottom /bottom.html, to be submitted to Phys. Rev. Lett.
10. B. Abbott *et al.* (DØ Collaboration), Fermilab-Pub-98/237-E, Phys. Rev. Lett. - in press.
11. S. Abachi *et al.*, (DØ Collaboration), *Nucl. Instrum. Methods* **A338**, 185 (1994).
12. F. Paige and S.D. Protopopescu, BNL-Report BNL-38034 (1986).
13. A.D. Martin, R.G. Roberts and W.J. Stirling, *Phys. Lett.* B **387**, 419 (1996).
14. F.I. Olness, R.J. Scalise, and Wu-Ki Tung, FERMILAB-Pub-97/428-T, hep-ph/9712494.
15. M.L. Mangano, hep-ph/9711337 ; CERN preprint TH-97-328 (unpublished).
16. A.D. Martin, R.G. Roberts and W.J. Stirling, *Phys. Rev.* D **47**, 867 (1993).
17. J. Huston *et al.*, (CTEQ Collaboration), *Phys. Rev.* D **58**, 4034 (1998).
18. H. L. Lai *et al.*, (CTEQ Collaboration), *Phys. Rev.* D **55**, 1280 (1997).
19. C. Albajar *et al.* (UA1 Collaboration), Phys. Lett. B **256**, 112 (1991).
20. F. Abe *et al.* (CDF Collaboration), Phys. Rev. Lett. **69**, 3704 (1992); *ibid* **79**, 572 (1997); *ibid* **79**, 578 (1997).
21. S. Abachi *et al.* (DØ Collaboration), Phys. Lett. B **370**, 239 (1996).
22. R. Baier and R. Ruckl, Z. Phys. C **19**, 251 (1983).
23. E. Braaten and S. Fleming, Phys. Rev. Lett. **74**, 3327 (1995); P. Cho and M. Wise, Phys. Lett. B **346**, 129 (1995).
24. P. Cho and A.K. Leibovich, Phys. Rev. D **53**, 6203 (1996).

Heavy Quark Fragmentation

Emanuel H. Norrbin

Department of Theoretical Physics, Lund University
Helgonavägen 5, S-223 62, Lund, Sweden
e-mail: emanuel@thep.lu.se

Abstract. I present the main aspects of open charm production in π^-p collisions in the context of the Lund String Fragmentation Model as implemented in the Monte Carlo program PYTHIA. The emphasis is on the transition from large to small strings and the dependence on model parameters. A modified version is presented and compared with experimental results both on asymmetries, single-charm spectra and correlations.

INTRODUCTION

Since the discovery of charm in 1974 numerous experiments on charm production at fixed target have been performed [1]. In recent years the precision has increased significantly [2–4], which enables a detailed comparison with theory. Perturbative QCD can describe some of the phenomenology of charm production, but not all [5,6]. Most notably the asymmetry between leading and non-leading particles, which is negligible in NLO QCD, has been shown to increase with x_F [2–4]. Also the momentum spectra of produced D mesons are harder than the NLO QCD c quark predictions, especially for leading particles, see e.g. [2,6]. These facts imply that nonperturbative effects are important in the production of charmed hadrons. In the string fragmentation approach both these aspects are included in the 'drag' effect, whereby a charm quark can gain momentum when it is connected to a beam remnant. The extreme case of this effect is the collapse of a small string into a single hadron, which gives rise to a dependence on the flavour contents of the beam.

BASICS OF STRING FRAGMENTATION

The Lund String Fragmentation Model [7] is best explored in e^+e^- annihilation, where the produced q and $\bar{\text{q}}$ are connected by a linear force field with a string-like topology. The $q\bar{q}$ production process is described by perturbative QCD, with a parton-shower approximation to higher orders. Radiated gluons are interpreted as 'kinks' on the string. The nonperturbative hadronization of a string proceeds via the production of $q\bar{q}$ pairs in the colour force field, which arrange themselves to produce the observed hadrons. A strongly constrained fragmentation function can be derived from very general and physically intuitive assumptions about the fragmentation process [8]. Because of the non-negligible mass of the charm quark the fragmentation function has to be modified for heavy-flavour production [9]. This model has been implemented in the Monte Carlo program PYTHIA [10], which has been tuned to e^+e^- experiments to give a good description of available data.

Hadron-hadron collisions

When we carry this model over to hadron-hadron physics, we again divide the process into a perturbative and a nonperturbative part and assume factorization between the two. In addition we assume that the fragmentation process is universal, i.e. the hadronization of a colour singlet is independent of how it was produced. In a hadron-hadron collision, such as π^-p, several ambiguities not present in e^+e^- annihilation are introduced. The main ones are presented in the following.

Structure of the incoming hadrons. The particles that participate in the collision are not point-like but have an internal structure. The longitudinal structure is parameterized by *parton distribution functions* (PDFs) which have been determined from other experiments such as deep inelastic scattering (DIS). These will not be discussed in the following, but in principle they give rise to some ambiguity, especially for small momentum fractions.

Structure of the beam remnant. When a parton has been picked out of a hadron, what is left continues in the direction of the beam and is called a beam remnant. If the beam remnant can be viewed as consisting of two or more objects its structure must be described. How this should be done is not known from first principles and has not been studied much. This aspect is therefore parameterized in *beam remnant distribution functions* (BRDFs) and several variants are considered.

Primordial transverse momentum. The partons inside the hadron are confined to a transverse dimension less than 1 fm; therefore by the uncertainty principle the spread of the transverse momentum should be of the order .2 GeV. This is modeled by adding a *primordial $k_\perp$* to the partons going into the hard scattering process. In the default version of PYTHIA, $k_\perp$ is assumed distributed as a Gaussian with a width of 0.44 GeV. Several studies [6,11,12] imply that this value is too small and a value of 1 GeV or more is needed to describe the data. This remains somewhat of a mystery and will not be resolved here.

Small strings in hadronization. Quark masses. When the colour topology of an event has been determined, every colour singlet is hadronized as a string would in e^+e^-. This works for strings with a mass of a few GeV or more. For strings with masses near the two-particle threshold, the standard string fragmentation approach can not be used and some other scheme is needed. As we will see later this introduces a large dependence on the quark masses.

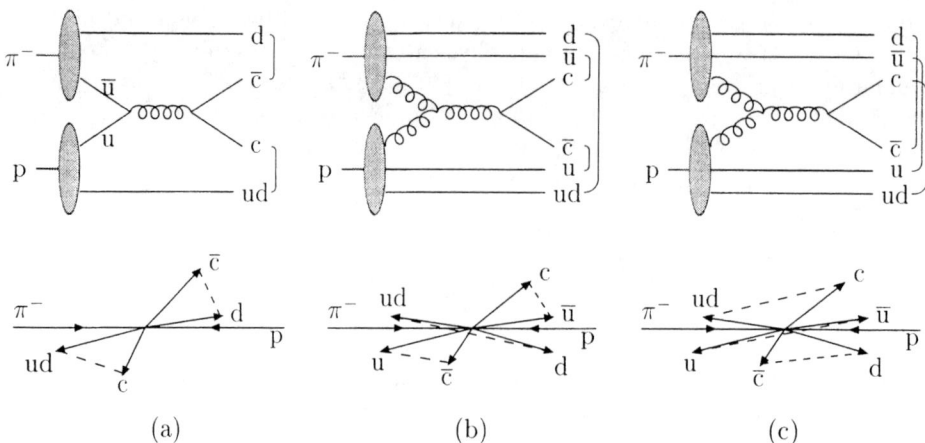

FIGURE 1. Examples of different string configurations in a π^-p collision: (a) $u\bar{u} \to c\bar{c}$ has a unique colour flow; (b,c) $gg \to c\bar{c}$ with the two possible colour flows. Dashed lines are strings.

String topologies and the 'beam drag' effect

In a π^-p collision we include charm production via the leading-order production mechanisms of quark and gluon fusion ($q\bar{q} \to c\bar{c}$ and $gg \to c\bar{c}$ respectively). The partons of the hard interaction and of the beam remnants are connected by strings, representing the confining colour field [7]. Each string contains a colour triplet endpoint, a number (possibly zero) of intermediate gluons and a colour anti-triplet end. The string topology can usually be derived from the colour flow of the hard process. For instance, consider the process $u\bar{u} \to c\bar{c}$ in a π^-p collision. The colour of the incoming u is inherited by the outgoing c, so it will form a colour-singlet together with the proton remnant, here represented by a colour anti-triplet ud diquark. In total, the event will thus contain two strings, one c–ud and one $\bar{c}$–d (Fig. 1a). In $gg \to c\bar{c}$ a similar inspection shows that two distinct colour topologies are possible. Representing the proton remnant by a u quark and a ud diquark (alternatively d plus uu), one possibility is to have three strings c–$\bar{u}$, $\bar{c}$–u and d–ud (Fig. 1b), and the other is the three strings c–ud, $\bar{c}$–d and u–$\bar{u}$ (Fig. 1c).

Other production mechanisms such as charm excitation and charm from parton showers (i.e. higher order effects) are not included in this study. Because of the relatively low virtuality of the hard process at fixed target energies the second mechanism is negligible, but it will become increasingly important at higher energies; at the LHC, e.g., this mechanism will dominate over the fusion mechanisms. Charm excited from the sea could give a non-negligible contribution, but we will not include it here.

Consider a colour singlet in Fig. 1 containing a charm endpoint. The hadronization of this string is performed in the CM system of the string. In that frame hadronization always results in a deceleration of the quark. After the rotation and boost to the hadron-hadron CM system, on the other hand, the net result of hadronization can be either an acceleration or a deceleration of the charm quark. This is interpreted as the beam remnant dragging the charm quark in the direction of the beam. This effect alone does not account for the asymmetry because of the cancellation between the diagrams in Fig 1b and c. We must therefore consider the flavour contents of the beam.

Cluster collapse of small strings

In e^+e^- annihilation the string mass is fixed by the CM energy of the process. In a hadron-hadron collision, on the other hand, charmed strings can have any mass ranging from m_q+m_c to $\sqrt{s}$. For string masses larger than some cut-off (here taken as $m_q+m_c+m_0$, with $m_0 \approx 1$ GeV and q a light quark) the Lund string fragmentation approach can be used. The model assumes a continuous final-state phase space, and an iterative scheme is used which demands that at least two particles are produced from a string. A string with a mass smaller than the cut-off we call a *cluster* and it is hadronized in the following way:

Cluster decay. A $q\bar{q}$ pair is created, using standard flavour selection, from the force-field connecting the cluster endpoints, and two hadrons are produced. If kinematically possible the cluster will decay isotropically into these two hadrons.

Cluster collapse. If it is not kinematically possible for the cluster to decay into two hadrons, it will be forced to collapse into a single hadron under conservation of the flavour quantum numbers. Since the mass of the cluster most likely will not correspond to any physical state (e.g. D or D*), the energy and momentum of the cluster must be slightly modified in order to put it on the hadronic mass shell.

There are two main ambiguities in this scheme. The first is that there is no clear separation between the two hadronization mechanisms, and the second is energy/momentum conservation in the cluster collapse.

To justify the cluster collapse approach we use an argument based on local duality, which has been shown to hold in e^+e^- annihilation, DIS [13] and τ decay [14]. In e^+e^- annihilation into hadrons around the $c\bar{c}$ threshold the observed cross section consists of peaks at J/ψ and ψ' and a continuum above the $D\bar{D}$ threshold. The perturbatively calculated cross section, on the other hand, is continuous from $\sqrt{s} = 2m_c$ onwards. However, if the experimental cross section is suitably smeared, it approximately agrees with the perturbative one. Another way of stating this is that the integrated cross section (over $\sqrt{s}$) should be the same provided that the integration interval is suitably large. We use the same argument in the present case, by replacing $\sqrt{s}$ with M_{string} and the J/ψ and ψ' peaks with D and D*. The duality argument could then be stated in the following way:

$$\int_{m_1}^{m_2} \frac{d\sigma_{\text{Partons}}}{dM_{\text{String}}} dM \approx \int_{m_1}^{m_2} \frac{d\sigma_{\text{Hadrons}}}{dM_{\text{String}}} dM \qquad (1)$$

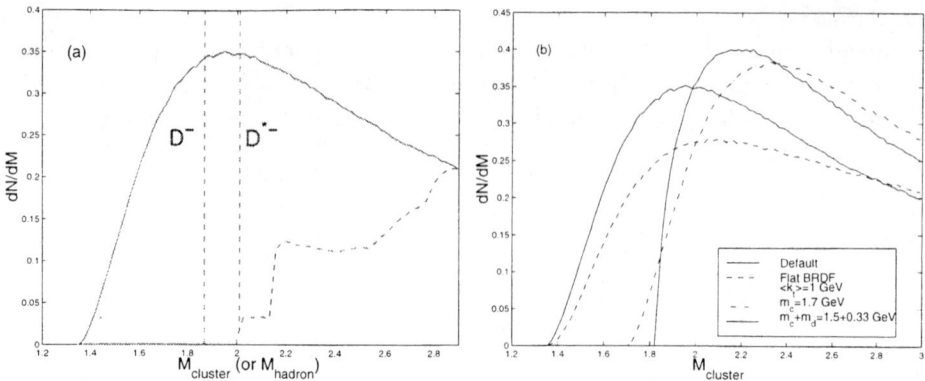

FIGURE 2. (a) Distribution of cluster (full) and meson (dashed) masses in the string model. Clusters within the gray area collapse to D^- or D^{*-}. (b) Dependence of the parton level mass distribution on some parameters of the model.

Fig. 2a shows how this looks using PYTHIA with standard parameters. The solid line is the mass distribution of produced clusters at the parton level and the dashed one is the produced hadrons. The clusters in the gray area have collapsed into single hadrons. The parton level distribution depends on many of the parameters such as the BRDF, primordial $k_\perp$, and quark masses. This is shown in Fig 2b. Consider e.g. an increase of the charm mass. The threshold will be shifted towards higher values and fewer clusters will be forced to collapse (the gray area is decreased). It is also possible to decrease the number of collapses from above by increasing the probability for a cluster above the $D\pi$ threshold to decay.

Fig. 3 shows the x_F distributions for different production channels and different parameter sets. These parameters will be discussed in the following. The explanation of the leading particle asymmetry in this model is that D^+ cannot be produced from cluster collapse because it has no quark in common with the beam.

Dependence on parameters

In this section the different parameters that have already been introduced will be discussed in more detail. We start with the BRDFs. How the energy and momentum in the beam remnant should be split between the constituents is not known from first principles, so we consider two extreme cases. In the first (default) scenario, we use BRDFs similar to those in PDFs, where one object takes a small fraction of the available energy. In the other extreme, we use naive counting rules where the energy is shared democratically between the constituents. How this effects the distributions is shown in Fig. 3. Most notably the dip around $x_F \sim 0.5$ in the cluster collapse distribution is smeared out when an even sharing is used.

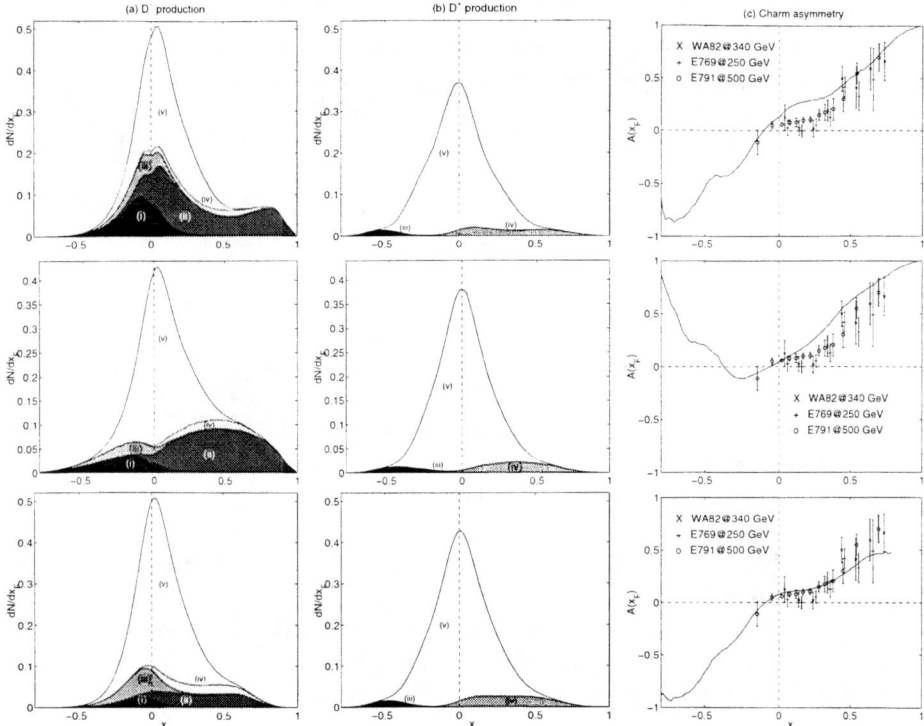

FIGURE 3. (a) D^- and (b) D^+ meson production in a π^-p collision at a π^- beam momentum of 500 GeV with different parameter sets. From top to bottom these are: default, using flat BRDFs and the new parameter set presented in this talk. The distributions are divided into different production channels: (i) Cluster collapse, light quark from p end, (ii) Cluster collapse, light quark from π^- end, (iii) Cluster decay, light quark from p end, (iv) Cluster decay, light quark from π^- end and (v) String fragmentation. (c) The resulting asymmetry.

Also note that the asymmetry is much more sensitive to the BRDF in the proton region, where one of the constituents of the beam remnant is a diquark. In the tuned version we will use an intermediate energy sharing.

We now come to the problem of energy/momentum conservation in the cluster collapse. To understand the dependence on this we consider two mechanisms that are of opposite nature:

Old method: 'far away'. In the first scheme, energy and momentum is shuffled to the parton in the event that is farthest from the cluster, in order to minimize the recoil. In this approach the D meson is often harder than the cluster.

New method: 'local'. In this new scheme we conserve energy locally by exchanging 'gluons' between the cluster and the string in the event that is closest to the cluster.

The details are presented in [11] and the conclusion is that the observables are not sensitive to the details of the energy/momentum conservation scheme, except for $x_F > 0.5$, where cluster collapse dominates but data is scarce.

Looking at Fig. 3 we see that the reason for the large asymmetry using the default parameters is that they allow many clusters to collapse for $x_F > 0$. The following set of parameters are inspired by the E791 collaboration and data from WA82 and they aim at decreasing the asymmetry by decreasing the number of clusters that collapse into one particle [15].

- Quark masses. The charm mass used in PYTHIA is by default set to the current algebra one (1.35 GeV). This is the value used in e.g. Higgs physics but it is far from obvious that this is the relevant mass in the present case. We therefore use constituent quark masses: $m_c = 1.5$ GeV; $m_d = m_u = 0.33$ GeV; $m_s = 0.5$ GeV. This will shift the threshold of the distribution of cluster masses and thus decrease the number of cluster collapses.
- Cluster decay. Increase the probability for a cluster to decay.
- Cluster collapse. Use new 'local' collapse mechanism.
- BRDF. We use an intermediate energy sharing that is more democratic, but not completely.
- Primordial $k_\perp$. The width of the Gaussian primordial $k_\perp$ is increased from 0.44 to 1.0 GeV. This allows cluster collapses between a charm quark and a beam remnant to occur also at fairly large values of $p_\perp$, thus leading to an essentially $p_\perp$-independent asymmetry. In addition, the $p_\perp$ kick added to charm quarks and beam remnants tends to increase the average invariant mass of the produced clusters, thereby reducing the number of cluster collapses.

COMPARISON WITH EXPERIMENT

In this section we compare the model (with the new parameter set [15]) to some data. Fig. 4 shows the asymmetry as a function of x_F and $p_\perp^2$ as well as single-charm spectra from WA82 for D^+ and D^- individually. In this case the data is nicely described by the new parameter set.

Next we compare the model to the new correlation data from E791 [12]. Several correlations in events where a pair of $D\overline{D}$ mesons with rapidity in $-0.5 < y < 2.5$ is fully reconstructed are studied. The distributions in Fig. 5 show that the longitudinal correlations in the string model are generally different from the data and are better described by NLO QCD [12]. The transverse correlations on the other hand are better described by the model, mostly because of the increased primordial $k_\perp$.

It is clear that the correlation between charm quark pairs should be modified by hadronization, but the string model seems to produce D mesons that are too far from each other in momentum (rapidity). To attempt a description of the data we use the independent fragmentation approach where a charmed hadron is simply a

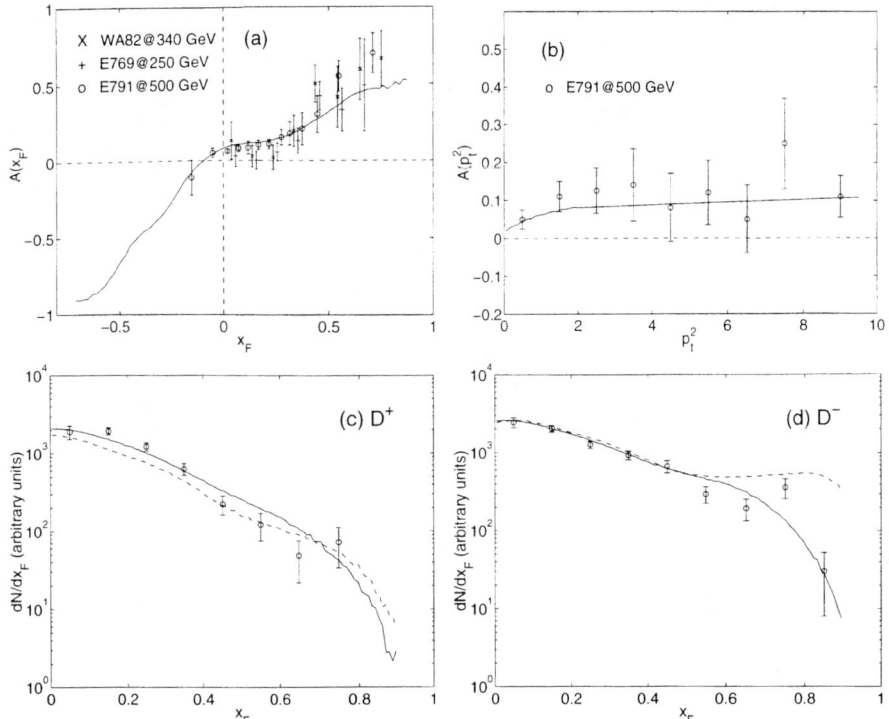

FIGURE 4. Comparison between the new PYTHIA parameter set and data. Asymmetry as a function of (a) x_F and (b) $p_\perp^2$. Single-charm spectra from the WA82 CERN experiment [2] for (c) D^+ and (d) D^- (dashed line is standard parameters and the full line is the new).

scaled-down version of the charm quark. In this way the beam remnants do not affect the charm quarks at all. We use the fragmentation function of Peterson et. al. [16] with $\epsilon_c = 0.05$. Surprisingly the longitudinal correlations are quite nicely described by this scheme. Especially the rapidity distribution is interesting because NLO QCD also fails to describe the strong peaking at small rapidities seen in the E791 data, see Fig. 6. However, independent fragmentation of course fails to describe the single-charm spectra from WA82, falling way below the data and lacking any asymmetry at large x_F. Thus there seems to be a contradiction between the data on single-charm and correlations that we cannot explain. There are several possibilities that deserve to be investigated. There could be some cut in the data that we don't understand, or problems with acceptance corrections at large x_F. Alternatively, some other mechanisms that we have not yet included (sea, intrinsic charm, parton showers) could give a sizable contribution.

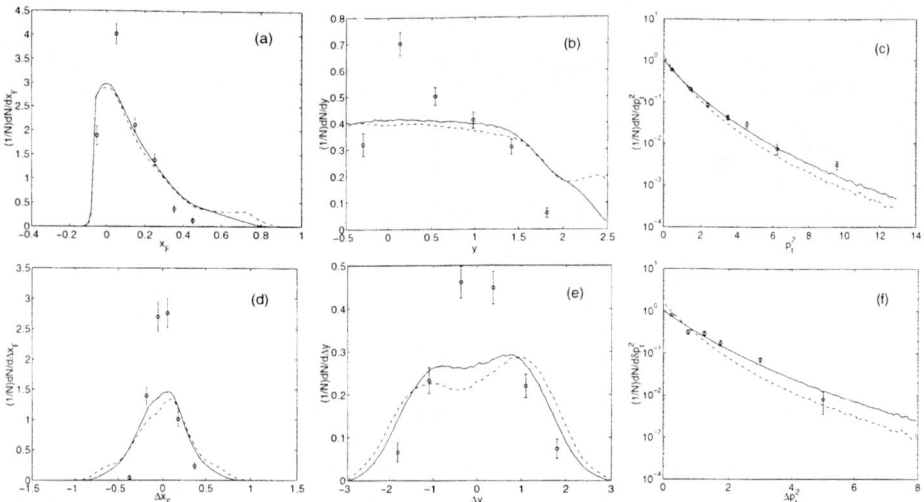

FIGURE 5. $D\bar{D}$ correlations. (a) x_F; (b) y; (c) $p_\perp^2$; (d) $\Delta x_F = x_{F,D} - x_{F,\bar{D}}$; (e) $\Delta y = y_D - y_{\bar{D}}$; (f) $\Delta p_\perp^2 = |p_{\perp,D}^2 - p_{\perp,\bar{D}}^2|$. Experimental data are from the E791 experiment [12] compared to the default (dashed) and modified (full) PYTHIA predictions.

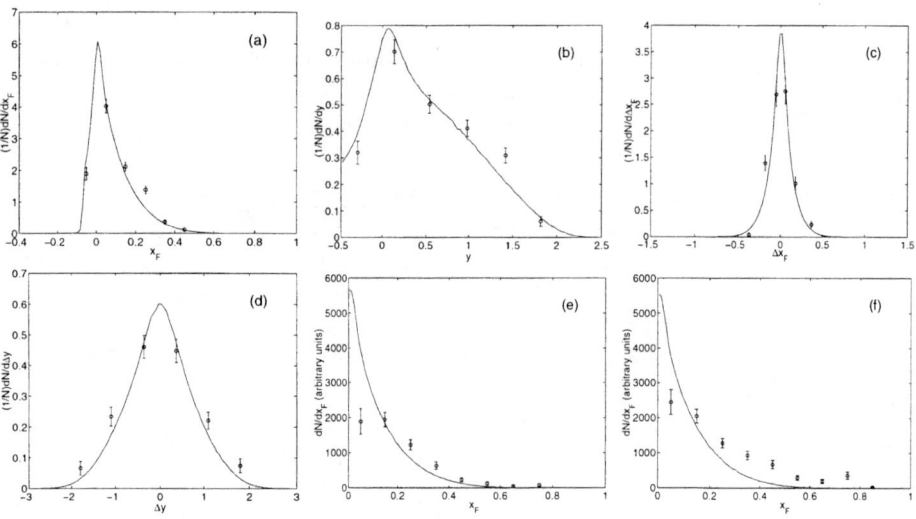

FIGURE 6. $D\bar{D}$ correlations compared to independent fragmentation. Correlations: (a) x_F; (b) y; (c) $\Delta x_F = x_{F,D} - x_{F,\bar{D}}$; (d) $\Delta y = y_D - y_{\bar{D}}$. Single-charm from WA82: (e) D^+ and (f) D^-.

SUMMARY

To summarize we have described the string fragmentation approach to charm production in hadronic collisions. A number of uncertainties have been identified and studied in detail, in particular the transition from a continuous string-mass distribution to a discrete hadron-mass one. The conclusion is that the model can describe asymmetries, single-charm spectra and transverse correlations but not longitudinal correlations. Also, these data do not fully constrain the choice of model parameters. Further data on charm production in π^-p collisions may provide further information, as may charm production e.g. in ep collisions.

REFERENCES

1. M. Basile et al., *Nuovo Cim.* **A66** (1981) 129;
   ACCMOR Collaboration, R. Bailey et al., *Phys. Lett.* **132B** (1983) 237;
   NA27 Collaboration, M. Aguilar-Benitez et al., *Phys. Lett.* **B161** (1985) 400;
   NA32 Collaboration, S. Barlag et al., *Z. Phys.* **C49** (1991) 555
2. WA82 Collaboration, M. Adamovich et al., *Phys. Lett.* **B305** (1993) 402
3. E769 Collaboration, G.A. Alves et al., *Phys. Rev. Lett.* **72** (1994) 812
4. E791 Collaboration, E.M. Aitala et al., *Phys. Lett.* **B371** (1996) 157
5. P. Nason, S. Dawson and R.K. Ellis, *Nucl. Phys.* **D327** (1989) 49;
   W. Beenakker, R. Meng, G.A. Schuler, J. Smith and W.L. Van Neerven, *Nucl. Phys.* **B351** (1991) 507
6. S. Frixione, M.L. Mangano, P. Nason and G. Ridolfi, *Nucl. Phys.* **B431** (1994) 453
7. B. Andersson, G. Gustafson, G. Ingelman and T. Sjöstrand, *Phys. Rep.* **97** (1983) 31
8. B. Andersson, G. Gustafson and S. Söderberg, *Z. Phys.* **C20** (1983) 317
9. M.G. Bowler, *Z. Phys.* **C11** (1981) 169
10. T. Sjöstrand, *Comput. Phys. Commun.* **82** (1994) 74
11. E. Norrbin and T. Sjöstrand, hep-ph/9809266 (to appear in *Phys. Lett.* **B**)
12. E791 Collaboration, E.M. Aitala et al., hep-ex/9809029 (submitted to *Phys. Rev.* **D**)
13. E.D. Bloom and F.J. Gilman, *Phys. Rev.* **D4** (1971) 2901;
    J.J. Sakurai, *Phys. Lett.* **46B** (1973) 207;
    H. Fritzsch, *Phys. Lett.* **67B** (1977) 217;
    R.A. Bertlmann, G. Launer and E. de Rafael, *Nucl.Phys.* **B250** (1985) 61;
    and references therein.
14. E. Braaten, S. Narison and A. Pich, *Nucl. Phys.* **B373** (1992) 581
15. The specific parameters of PYTHIA 6.1 that we use are: PMAS(1,1)=PMAS(2,1)=0.33D0; PMAS(3,1)=0.5D0; PMAS(4,1)=1.5D0 (quark masses), MSTP(92)=3 (intermediate BRDFs), PARP(91)=1.D0 (width of Gaussian primordial $k_\perp$), PARP(93)=5.D0 (the maximum allowed $k_\perp$). The changes in cluster decay/collapse are not implemented in the standard distribution yet. Patches can be found on the PYTHIA homepage at http://www.thep.lu.se/~torbjorn/Pythia.html.
16. C. Peterson, D. Schlatter, I. Schmitt and P. Zerwas, *Phys. Rev.* **D27** (1983) 105

Heavy Quark Production

Fredrick I. Olness

Department of Physics, Southern Methodist University
Dallas, Texas 75275-0175

Abstract. We provide a brief overview of some current experimental and theoretical issues of heavy quark production.

INTRODUCTION

The production of heavy quarks in high energy processes has become an increasingly important subject of study both theoretically and experimentally. The theory of heavy quark production in perturbative Quantum Chromodynamics (pQCD) is more challenging than that of light parton (jet) production because of the new physics issues brought about by the additional heavy quark mass scale. The correct theory must properly take into account the changing role of the heavy quark over the full kinematic range of the relevant process from the threshold region (where the quark behaves like a typical "heavy particle") to the asymptotic region (where the same quark behaves effectively like a massless parton). With steadily improving experimental data on a variety of processes sensitive to the contribution of heavy quarks (including the direct measurement of heavy flavor production), this is a very rich field for studying different aspects of the QCD theory including the problems of multiple scales, summation of large logarithms, subtleties of renormalization, and higher order corrections. We shall briefly review a limited subset of these issues.[1]

THE FACTORIZATION THEOREM

Perturbative calculations for heavy quark production are performed in the context of the factorization theorem expressed below in the commonly used form:

$$\sigma_{a \to c} = f_{a \to b}(x, \mu^2) \otimes \hat{\sigma}_{b \to c}(Q^2/\mu^2, M_H^2/\mu^2, \alpha_s(\mu)) + \mathcal{O}(\Lambda_{QCD}^2/Q^2) \tag{1}$$

---

[1] For a recent comprehensive review, see: Frixione, Mangano, Nason, and Ridolfi, Ref. [1]

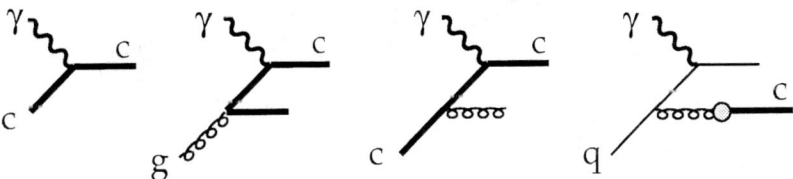

FIGURE 1. Basic processes for DIS heavy quark production. a) $\mathcal{O}(\alpha_s^0)$ flavor excitation: $\gamma + c \to c$; b) $\mathcal{O}(\alpha_s^1)$ flavor creation: $\gamma + g \to c + \bar{c}$; c) $\mathcal{O}(\alpha_s^1)$ flavor excitation: $\gamma + c \to c + g$; d) $\mathcal{O}(\alpha_s^1)$ light-quark (q) fragmentation: $(\gamma + q \to q + g) \otimes (g \to c)$.

While the factorization was originally proven for massless quarks, [2] the theorem has recently been extended by Collins [3] to incorporate quarks of any mass, including "heavy quarks." (Note, we have explicitly retained the M_H^2 dependence in $\hat{\sigma}$.) It is important to note that the corrections to the factorization are only of order Λ_{QCD}^2/Q^2, and not M_H^2/Q^2, *even for the case of general quark masses.*

The factorization theorem can also be expressed as a composition of t-channel two particle irreducible (2PI) amplitudes:[2]

$$\sigma_{a \to c} \simeq \hat{\sigma}_{b \to c} \otimes f_{a \to b} \simeq \left[C \cdot \frac{1}{1-(1-Z)K} \right] \cdot Z \cdot \left[\frac{1}{1-K} \cdot T \right] \qquad (2)$$

Here, C represents the graph for a hard scattering, K represents the graph for a rung, T represents the graph that couples to the target, and Z represents a collinear projection operator. The first term in brackets roughly corresponds to the hard scattering coefficient function $\hat{\sigma}$, and the second term to the parton distribution function (PDF), f. Note that these two terms only communicate through a collinear projection operator, Z. Part of the effort in generalizing the factorization theorem for the case of massive quarks involves constructing the proper Z, and demonstrating that terms containing (1-Z) are power suppressed. However, once Z is determined, Eq.(2) yields an *all-orders* prescription for computing for both the hard scattering coefficient ($\hat{\sigma}$) and the parton distribution function (f). A calculation using this formalism was first performed by ACOT [4] for the case of heavy quark production in deeply inelastic scattering, and we now examine this process in detail.

HEAVY QUARK PRODUCTION IN DIS

Several experimental groups [5] have studied the semi-inclusive deeply inelastic scattering (DIS) process for heavy-quark production $\ell_1 + N \to \ell_2 + Q + X$. New data from HERA investigates the DIS process in a very different kinematic range from

---

[2]) I must necessarily leave out many details here; for a precise treatment, see Collins [3].

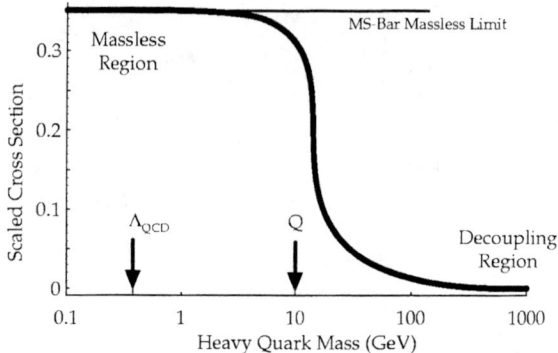

FIGURE 2. The scaled cross section for DIS heavy quark production as a function of the quark mass m_H.

that available at fixed-target experiments. This perception has changed the way that we compute the semi-inclusive DIS heavy quark production. Traditionally, the heavy quark mass was treated as a large scale, and the number of active parton flavors was fixed to be the number of quarks lighter than the heavy quark. In this scheme, the perturbation expansion begins with the $\mathcal{O}(\alpha_s^1)$ heavy quark creation fusion process $\gamma g \to c\bar{c}$, (cf., Fig. 1b). We refer to this approach as the Fixed Flavor Number (FFN) scheme since the number of flavors coming from parton distributions is fixed at three for charm production.[3]

More recently, a Variable Flavor Number (VFN) scheme (ACOT [4]) has been proposed which includes the heavy quark as an active parton flavor with non-zero heavy quark mass. In this case, the perturbation expansion begins with the $\mathcal{O}(\alpha_s^0)$ heavy quark excitation process $\gamma c \to c$, (cf., Fig. 1a). The key advantages of this scheme are: [7]

1. By incorporating the heavy quark into the parton framework, the composite scheme yields a result which is valid from threshold to asymptotic energies; in contrast, the FFN scheme contains unsubtracted mass singularities which will vitiate the perturbation expansion in the $m_c \to 0$ or $E \to \infty$ limit.

2. Because the composite scheme resums the large logarithms appearing in the FFN scheme into the parton distribution functions, it includes the numerically dominant terms of the $\mathcal{O}(\alpha_s^2)$ FFN scheme calculation in a $\mathcal{O}(\alpha_s^1)$ calculation.

In effect, the VFN scheme subsumes the FFN scheme. To illustrate this fact with a concrete calculation, in Fig. 2, we plot the cross section for "heavy" quark production as a function of the quark mass.[4] This figure clearly shows the three important

---

[3] The necessary diagrams have been computed to $\mathcal{O}(\alpha_S^2)$ by Smith, van Neerven, and collaborators, cf., Ref. [6].
[4] To be specific, we have computed single quark production for a photon exchange with $x = 0.1$, $\mu = Q = 10$ GeV, and the cross section is in arbitrary units.

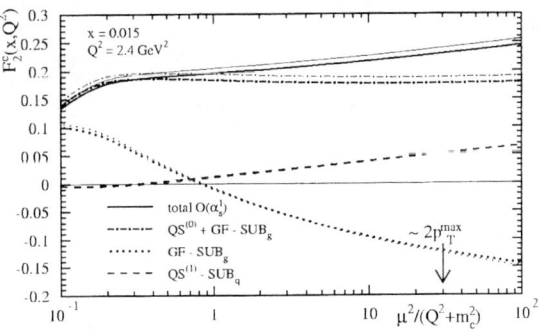

FIGURE 3. The contributions to DIS charged current inclusive F_2^{charm} vs. μ. For each separate contribution, the thick lines are the $\overline{MS}$ result ($m_s = 0$), and the thin lines are the ACOT result with $m_s = 0.5$ GeV. From Kretzer, Ref. [8].

kinematic regions. 1) In the massless region, where $m_H \ll Q$, the ACOT VFN result reduces precisely to the massless $\overline{MS}$ result. 2) In the decoupling region, where $m_H \gg Q$, this "heavy quark" decouples and its contribution vanishes. 3) In the transition region, where $m_H \sim Q$, this (not-so) "heavy quark" plays an important dynamic role. While the FFN scheme is appropriate only when $m_H \sim Q$, we see that the VFN scheme is valid throughout the full kinematic range.[5]

This point is also illustrated in a calculation by Kretzer [8] (cf., Fig. 3) which shows the partial contributions to the charged current F_2^{charm}.[6] In this figure, each line is actually a pair of lines: the thin lines represents the result for F_2^{charm} using the ACOT scheme with $m_s = 0.5$ GeV, and the thick lines regularize the strange quark with the massless $\overline{MS}$ prescription. (The charm mass is, of course, retained.) The fact that these two calculations match so closely (particularly in comparison to the μ-variation) indicates: 1) the ACOT scheme smoothly reduces to the desired massless $\overline{MS}$ limit as $m_H \to 0$, and 2) for $m_H \lesssim \Lambda_{QCD}$ we see that the quark mass no longer plays a dynamic role in the process and becomes purely a regulator.

HEAVY QUARKS AND THE GLOBAL PDF ANALYSIS

Recent precision data on F_2 and on F_2^{charm} from HERA indicate that the charm contribution can rise to 25% of the total F_2 at small-x. These results clearly imply the need to perform new global analyses to account for the correct physics behind

---

[5] Buza et al., have determined the asymptotic form of the heavy quark coefficient functions which are then used to determine the threshold matching conditions between the three- and four-flavor shemes, Ref. [6]. Thorne and Roberts have a similar scheme with slightly different matching conditions, Ref. [9].
[6] Kretzer and Schienbein have performed the first calculation of the $\mathcal{O}(\alpha_S)$ quark initiated process for general masses and general couplings, Ref. [8].

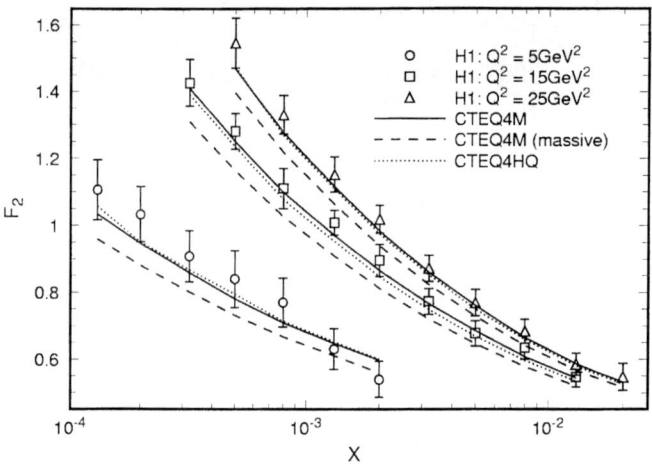

FIGURE 4. Comparison of H1 data in the small-x region. From Lai and Tung, Ref. [10].

these measurements. Tung and Lai [10] have repeated the CTEQ4M global analysis, [11] but this time implementing the heavy quark leptoproduction within the ACOT formalism to obtain a CTEQ4HQ set of PDF's. The deviation of CTEQ4HQ distributions from CTEQ4M are minimal, and are most noticeable at small-x; interestingly, the differences are larger for the light quarks than for the gluon and charm.

The effect of these new PDF's and the comparison with data are shown in Fig. 4. The solid curves show the CTEQ4M distributions convoluted with massless matrix elements. The dashed curves show the CTEQ4M distributions convoluted with massive matrix element; while technically this is a mismatch of schemes, this comparison is useful to gauge the magnitude of the heavy quark effects, (which we observe are comparable to the experimental uncertainties). Finally, the dotted curves show the CTEQ4HQ distributions convoluted with massive matrix element. When a consistent scheme is used for both the matrix elements and the PDF's, the agreement with data is excellent. (This is as expected since this data was included in the fit.) It is interesting to note that overall χ^2 for CTEQ4HQ (χ^2=1293) was slightly improved compared to the previous best fit CTEQ4M (χ^2=1320) for 1297 data points. While this difference is small, we find it reassuring that the proper treatment of the heavy quark mass resulted in an improved fit; particularly when compared with a 4-flavor FFN fit (χ^2=1349) or a 3-flavor FFN fit (χ^2=1380).

A recent re-analysis of the EMC data [12] concluded that there could be an intrinsic charm component in the proton of $0.86 \pm 0.60\%$. It would be interesting to repeat this calculation in the context of a global analysis using the VFN ACOT formalism to see if more recent data favor an intrinsic charm component.

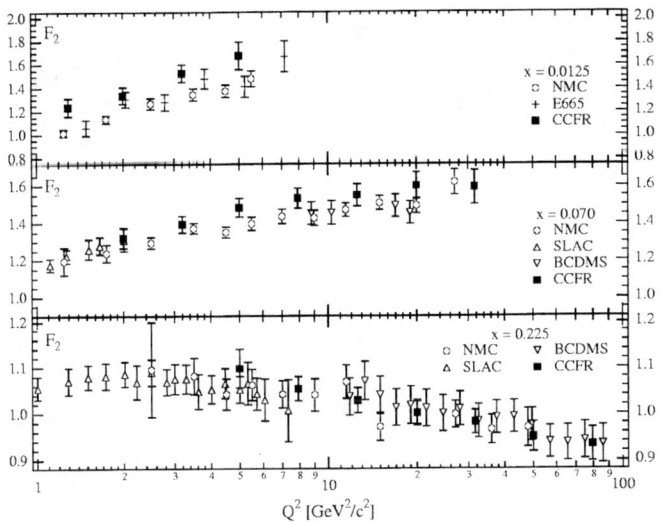

FIGURE 5. Comparison of F_2 from charged and neutral current DIS. From Seligman, et al., Ref. [13].

HEAVY QUARKS AND EXTRACTION OF $S(X)$

A topic closely related to DIS charm production is the extraction of the strange quark distribution.[7] In principle, we can extract $s(x)$ by comparing DIS neutral and charged current data. To leading order, we have:

$$\frac{F_2^{NC}}{F_2^{CC}} \simeq \frac{5}{18} \left\{ 1 - \frac{3}{5} \frac{(s+\bar{s}) - (c+\bar{c}) + ...}{q+\bar{q}} \right\} \quad . \tag{3}$$

While the individual F_2 structure functions are measured precisely (cf., Fig. 5), [13] this approach is indirect in the sense that small uncertainties in the larger valence distributions will magnify the uncertainty on the extracted $s(x)$.

A direct method of obtaining $s(x)$ is to use the neutrino induced di-muon process: $\nu_\mu N \to \mu^- c X$ with the subsequent decay $c \to s \mu^+ \nu_\mu$. Here, the di-muon signal is directly related to the charm production rate, which goes via the process $W^+ s \to c$ at leading order. The method has the advantage that the signal from the s-quark is not a small effect beneath the valence process.

A complete NLO experimental analysis was performed using the CCFR data set. [15] The recently collected data from the NuTeV experiment will provide an opportunity to extend the precision of these investigations still further. [16] Their high intensity sign-selected neutrino beam and the new calibration beam allows for large improvement in the systematic uncertainty while minimizing statistical errors. (See the paper by T. Adams, this meeting. [17])

---

[7] For a comprehensive review, see Conrad, Shaevitz, and Bolton, Ref. [14].

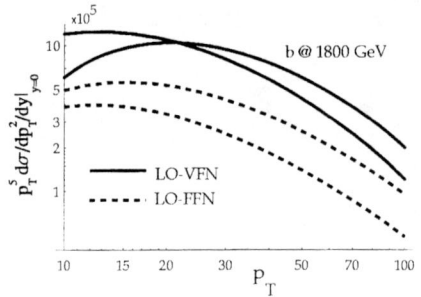
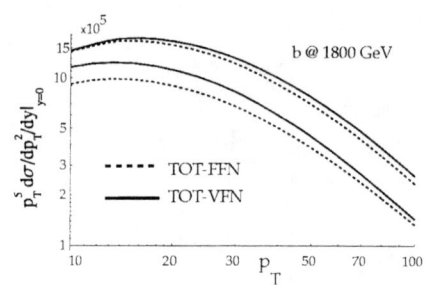

FIGURE 6. The scaled cross section (nb-GeV^3) vs. p_T for the a) LO-FFN and LO-VFN contributions, and b) TOT-FFN and TOT-VFN contributions. For each contribution, we choose $\mu = [M_T/2, 2M_T]$, with $M_T = \sqrt{m_H^2 + p_T^2}$, to gauge the μ-variation. From Ref. [18].

HADROPRODUTION OF HEAVY QUARKS

We now turn to the hadroproduction of heavy quarks, and discuss how the method of ACOT [4,18] is used to provide a dynamic role for the heavy quark parton. We concentrate mostly on b-production at the Tevatron for definiteness, and present typical results for b quark production. [1,19,20] (See the paper by A. Zieminski, this meeting. [21]) Fig. 6a shows the scaled differential cross section vs. p_T for b production at 1800 GeV for the leading order (LO) calculations. The heavy creation (HC) process[8] ($gg \to b\bar{b}$) represents the LO contribution to the fixed-flavor-number (FFN) scheme result. The heavy excitation (HE) process ($gb \to gb$) plus the HC term represents the LO contribution to the variable-flavor-number (VFN) scheme result. The pair of lines for each result shows the effect of varying μ. In a similar manner, Fig. 6b shows the total FFN and VFN results.[9]

Two interesting features are worth noting. 1) Examining Fig. 6a we observe the HE contribution is comparable to the HC one, in spite of the smaller b-quark PDF compared to the gluon distribution. Closer examination reveals that two effects contribute to this unexpected result: a larger color factor and the presence of t-channel gluon exchange diagrams for the HE process, as compared to the HC process. 2) The LO-VFN (=HC+HE) contributions (Fig. 6a) (tree processes) give a reasonable approximation to the full cross section TOT-VFN (Fig. 6b); thus, the NLO-VFN correction is relatively small. This is in sharp contrast to the familiar FFN scheme where the TOT-FFN term is more than twice as large as the LO-FFN (=HC). This is, of course, an encouraging result, suggesting that the VFN scheme heavy quark parton picture represents an efficient way to organize the perturbative

---

[8]) In this section we let g represent both gluons and light quarks, where applicable. Therefore, the HC process described as $gg \to b\bar{b}$ also includes $q\bar{q} \to b\bar{b}$.
[9]) The formidable calculations of the NLO $gg \to b\bar{b}$ process were computed by Nason, Dawson, and Ellis (Ref. [22]), and also by Beenakker et al., (Ref. [23]). These calculations were implemented in a Monte Carlo framework (including correlations) by Mangano, Nason, and Ridolfi, (Ref. [24]).

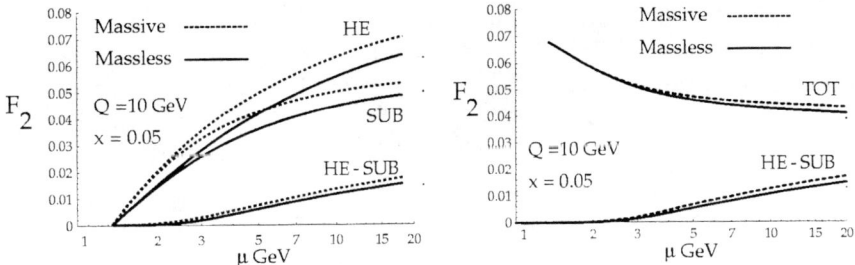

FIGURE 7. F_2 vs. μ for DIS c-production. a) F_2^{HE}, F_2^{SUB} and the difference ($F_2^{HE} - F_2^{SUB}$). The solid curves are for the mass-independent evolution scheme, and the dashed curves are for the mass-dependent evolution scheme. b) F_2^{TOT} and $F_2^{HE} - F_2^{SUB}$. The difference between the mass-independent evolution and mass-dependent evolution for F_2^{TOT} is higher order and comparable or less than the μ-variation. From Ref. [27].

QCD series.

In Fig. 6a, we also observe that while the TOT-VFN result provides minimal μ-variation for low p_T, the improvement is decreased for large p_T. This may be, in part, due to that fact that the TOT-VFN result shown here is missing the NLO-HE process $gb \to ggb$ since this calculation, with masses retained, does not exist. In a separate effort, Cacciari and Greco [25] have used a NLO fragmentation formalism to resum the heavy quark contributions in the limit of large p_T. This calculation effectively includes the massless limit of the $gb \to ggb$ contribution (omitted above); the result is a decreased μ-variation in the large p_T region. Recently, this calculation has been merged with the massive FFN calculation by Cacciari, Greco, and Nason, (Ref. [26]); the result is a calculation which matches the FFN calculation at low p_T, and takes advantage of the NLO fragmentation formalism in the high p_T region.

MASSIVE VS. MASSLESS EVOLUTION

In a consistently formulated pQCD framework incorporating non-zero mass heavy quark partons, there is still the freedom to define parton distributions obeying either mass-independent or mass-dependent evolution equations. With properly matched hard cross-sections, different choices merely correspond to different factorization schemes, and they yield the same physical cross-sections. We demonstrate this principle in a concrete order α_s calculation of the DIS charm structure function. [27] In Fig. 7 we display the separate contributions to F_2^{charm} for both mass-independent and mass-dependent evolution. The matching properties are best examined by comparing the (scheme-dependent) heavy excitation F_2^{HE} and the subtraction F_2^{SUB} contributions of Fig. 7a.

We observe the following. 1) Within each scheme, F_2^{HE} and F_2^{SUB} are well matched near threshold, cf., Fig. 7a. Above threshold, they begin to diverge, but the difference ($F_2^{HE} - F_2^{SUB}$), which contributes to F_2^{TOT}, is insensitive to the different schemes. 2) It is precisely this matching of F_2^{HE} and F_2^{SUB} which ensures the scheme dependence of F_2^{TOT} is properly of higher-order in α_s, (cf., Fig. 7b).

This matching is not accidental, but simply a result of using a consistent renormalization scheme for both F_2^{HE} and F_2^{SUB}. To understand this we expand these terms near threshold ($\mu \sim m_H$) where the m_H/Q terms are relevant:

$$\sigma_{SUB} = {}^R f_{g/P} \otimes {}^R \hat{\sigma}^{(1)}_{g\gamma^* \to c\bar{c}} = {}^R f_{g/P} \otimes \frac{\alpha_s}{2\pi} \int_{m_H^2}^{\mu^2} \frac{d\mu^2}{\mu^2} \, {}^R P^{(1)}_{g \to c} \otimes \sigma^{(0)}_{c\gamma^* \to c} + 0$$

$$\sigma_{HE} \simeq {}^R f_{c/P} \otimes {}^R \hat{\sigma}^{(0)}_{c\gamma^* \to c} \simeq {}^R f_{g/P} \otimes \frac{\alpha_s}{2\pi} \int_{m_H^2}^{\mu^2} \frac{d\mu^2}{\mu^2} \, {}^R P^{(1)}_{g \to c} \otimes \sigma^{(0)}_{c\gamma^* \to c} + \mathcal{O}(\alpha_s^2)$$

Here, the prescript R specifies the renormalization scheme. From these relations, it is evident that F_2^{HE} and F_2^{SUB} will match to $\mathcal{O}(\alpha_s^2)$ so long as a consistent choice or renormalization scheme R is made for the splitting kernels, ${}^R P^{(1)}_{g \to c}$. This is the key mechanism that compensates the different effects of the mass-independent vs. mass-dependent evolution, and yields a σ_{TOT} which is identical up to higher-order terms. The lesson is clear: the choice of a mass-independent $\overline{MS}$ or a mass-dependent (non-$\overline{MS}$) evolution is purely a choice of scheme, and becomes simply a matter of convenience–*there is no physically new information gained from the mass-dependent evolution.*

CONCLUSIONS

We have provided a brief overview of some current experimental and theoretical issues of heavy quark production. The wealth of recent heavy quark production data from both fixed-target and collider experiments will allow us to to extract a precision measurement of structure functions which can provide important constraints for searches of new physics at the highest energy scales. As an important physical process involving the interplay of several large scales, heavy quark production poses a significant challenge for further development of QCD theory.

We thank J.C. Collins, R.J. Scalise, and W.-K. Tung for valuable discussions, and the Fermilab Theory Group for their kind hospitality during the period in which part of this research was carried out. This work is supported by the U.S. Department of Energy, and the Lightner-Sams Foundation.

REFERENCES

1. S. Frixione, M. L. Mangano, P. Nason, and G. Ridolfi, hep-ph/9702287; M. L. Mangano, hep-ph/9711337.
2. J. Collins, D. Soper, and G. Sterman, Nucl. Phys. **B250**, 199 (1985).
3. J. C. Collins, Phys. Rev. **D58**, 094002 (1998).
4. M. A. G. Aivazis, J. C. Collins, F. I. Olness, and W.-K. Tung, Phys. Rev. D **50**, 3102 (1994).

5. H1 Collaboration (C. Adloff *et al.*). Z. Phys. C72, 593 (1996).
   ZEUS Collaboration (J. Breitweg *et al.*). Talk given at International Europhysics Conference on High-Energy Physics (HEP 97), Jerusalem, Israel, 19-26 Aug 1997, N-645.
6. E. Laenen, S. Riemersma, J. Smith, W.L. van Neerven. Phys. Rev. **D49**, 5753 (1994); M. Buza, Y. Matiounine, J. Smith, and W. L. van Neerven, hep-ph/9707263; hep-ph/9612398; M. Buza and W. L. van Neerven, Nucl. Phys. **B500**, 301 (1997).
7. C. Schmidt, hep-ph/9706496; J. Amundson, C. Schmidt, W. K. Tung, X. Wang, MSU preprint, in preparation.
8. S. Kretzer, e-Print hep-ph/9808464, S. Kretzer, I. Schienbein, Phys. Rev. **D58**, 094035 (1998).
9. R.S. Thorne, R.G. Roberts, Phys. Lett. **B421**, 303 (1998); R.S. Thorne, R.G. Roberts, Phys. Rev. **D57**, 6871 (1998).
10. H. L. Lai and W.-K. Tung, Z. Phys. **C74**, 463 (1997). The displayed figure is a reproduction of Fig. 2 from Lai and Tung, is copyrighted by Springer-Verlag, and used by permission.
11. H. L. Lai *et al.*, Phys. Rev. D **55**, 1280 (1997).
12. B.W. Harris, J. Smith, R. Vogt Nucl. Phys. **B461**, 181 (1996).
13. CCFR Collaboration (W.G. Seligman et al.), Phys. Rev. Lett. **79**, 1213 (1997).
14. Janet M. Conrad, Michael H. Shaevitz, and Tim Bolton. hep-ex/9707015
15. A. O. Bazarko *et al.*, Z. Phys. **C65**, 189 (1995).
16. NuTeV Collaboration: Jaehoon Yu *et al.*, hep-ex/9806030; K.S. McFarland *et al.*, hep-ex/9806013.
17. T. Adams,*Heavy Quark Production in Neutrino Deep-Inelastic Scattering.* HQ'98, Fermilab, October 10-12, 1998.
18. F.I. Olness, R.J. Scalise, Wu-Ki Tung, hep-ph/9712494. Phys. Rev. **D59**, 014506 (1999).
19. CDF Collaboration (F. Abe *et al.*), Phys. Rev. D **50**, 4252 (1994); Phys. Rev. Lett. **75**, 1451 (1995).
20. D0 Collaboration (S. Abachi *et al.*), Phys. Rev. Lett. **74**, 3548 (1995).
21. A. Zieminski,*B Production and Onium production at the Tevatron.* HQ'98, Fermilab, October 10-12, 1998.
22. P. Nason, S. Dawson, and R. K. Ellis, Nucl. Phys. **B303**, 607 (1988); **B327**, 49 (1989); **B335**, 260(E) (1990).
23. W. Beenakker, H. Kuijf, W. L. van Neerven, and J. Smith, Phys. Rev. D **40**, 54 (1989); W. Beenakker, W. L. van Neerven, R. Meng, G. A. Schuler, and J. Smith, Nucl. Phys. **B351**, 507 (1991).
24. M. L. Mangano, P. Nason, and G. Ridolfi, Nucl. Phys. **B373**, 295 (1992).
25. M. Cacciari and M. Greco, Nucl. Phys. **B421**, 530 (1994).
26. M. Cacciari, M. Greco, and P. Nason, hep-ph/9803400, J. High Energy Phys. **05**, 007 (1998).
27. F. I. Olness and R. J. Scalise, Phys. Rev. D **57**, 241 (1998).

PART IV
HADRONIC DECAYS

Hadronic Charm Decays and Lifetimes from E791

Nader Copty[1]

Department of Physics and Astronomy
University of South Carolina
Columbia, SC 29208
U.S.A.

Abstract. Fermilab E791 is a high statistics charm experiment with over 2×10^{10} events collected to study the production and decay of charm. We report results of a precise measurement of the D_s lifetime, preliminary results on the measurement of the D^0 lifetime and a limit on the D^0-$\overline{D}^0$ mixing rate due to lifetime difference, results of the study of the singly Cabibbo-suppressed decay of $D^0 \to K^-K^+\pi^-\pi^+$, and results of the study of the doubly Cabibbo-suppressed decay of $D^+ \to K^+\pi^+\pi^-$.

INTRODUCTION

E791 [1] is a high statistics charm experiment which acquired data at Fermilab during the 1991-1992 fixed-target run. The experiment combined a fast data acquisition system with an open trigger. Over 2×10^{10} events were collected with the Tagged Photon Spectrometer using a 500 GeV π^- beam. There were five target foils with 15 mm center-to-center separations: one 0.5 mm thick platinum foil followed by four 1.6 mm thick diamond foils. The spectrometer included 23 planes of silicon microstrip detectors (6 upstream and 17 downstream of the target), 2 dipole magnets, 10 planes of proportional wire chambers (8 upstream and 2 downstream of the target), 35 drift chamber planes, 2 multi-cell threshold Čerenkov counters that provided π/K separation in the 6–60 GeV/c momentum range [2], electromagnetic and hadronic calorimeters, and a muon detector.

We present the results of four topics: (1) a precise measurement of the D_s lifetime; (2) preliminary results on the measurement of the D^0 lifetime and a limit on the D^0-$\overline{D}^0$ mixing rate due to lifetime difference; (3) results of the study of the singly Cabibbo-suppressed decay of $D^0 \to K^-K^+\pi^-\pi^+$; and (4) results of the study of the doubly Cabibbo-suppressed decay of $D^+ \to K^+\pi^+\pi^-$. Unless otherwise specified, charge conjugate states are implicitly included in this paper.

---

[1] Representing the Fermilab E791 Collaboration.

CP459, *Heavy Quarks at Fixed Target*
edited by Harry W. K. Cheung and Joel N. Butler
© 1999 The American Institute of Physics 1-56396-864-9/99/$15.00

MEASUREMENT OF THE D_S LIFETIME

Precise measurements of the lifetimes of the weakly decaying charm mesons are useful for understanding the contributions of various weak decay mechanisms. The least well measured lifetime of the ground state charm mesons is that of the D_s, which is known only to about the 4% level [3]. The current measurements are only suggestive of a difference between the lifetimes of the D^0 and the D_s. A more precise measurement of the lifetime of the D_s could more clearly establish whether the lifetimes of the D^0 and D_s are indeed different. A difference between the D^0 and D_s lifetimes could be due to phase space differences, color factors, W-exchange, W-annihilation, decay via τ lepton modes, or relativistic effects [4].

We report results of a high-statistics measurement of the D_s lifetime [5] which establishes more certainly a difference between the lifetimes of the D^0 and the D_s. We use only the decay mode $D_s \to \phi\pi$ since the ϕ mass constraint gives us a D_s sample with a very good signal-to-noise ratio. The reconstruction efficiency as a function of lifetime was determined from data for the Cabibbo-favored (CF) decay $D^+ \to K^-\pi^+\pi^+$ and from the ratio of the efficiency for $D_s \to \phi\pi$ to that for $D^+ \to K^-\pi^+\pi^+$ obtained from Monte Carlo (MC) simulations. With this technique, most uncertainties that arise from MC simulation should cancel.

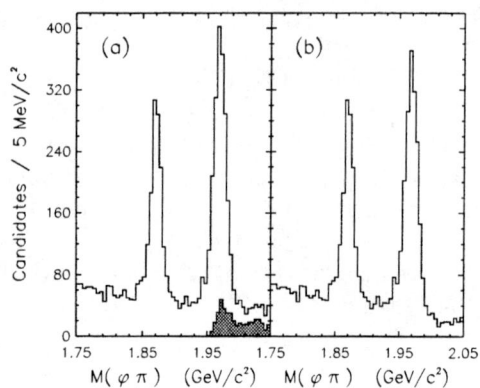

FIGURE 1. The $\phi\pi$ mass distributions: (a) before excluding candidates with $M(K^-\pi^+\pi^+)$ within ± 30 MeV/c^2 of the D^+ mass, which are shown as the hatched histogram; (b) after excluding candidates with $M(K^-\pi^+\pi^+)$ within ± 30 MeV/c^2 of the D^+ mass, as described in the text.

To eliminate any possible background due to reflections from $D^+ \to K^-\pi^+\pi^+$ with a pion misidentified as a kaon, which appear under and near the D_s signal, we excluded all $\phi\pi$ candidate events with $M(K^-\pi^+\pi^+)$ within ± 30 MeV/c^2 of the D^+ mass. The mass distributions for all $\phi\pi$ candidates that survived our selection criteria are shown in Fig. 1. The $\phi\pi$ mass distribution before $D^+ \to K^-\pi^+\pi^+$ background subtraction is shown in Fig. 1(a), with candidates consistent with $D^+ \to K^-\pi^+\pi^+$ background shown as the hatched region. The $\phi\pi$ mass distribution after the reflection subtraction is shown in Fig. 1(b). The hatched region in Fig. 1(a) appears to be a small D_s signal and a linear background that turns on around 1.95 GeV/c^2. The background in Fig. 1(b) is therefore modeled as a piecewise linear function with a discontinuity fixed at 1.95 GeV/c^2. We have studied the sensitivity of our measured lifetime to the position of this discontinuity and have found very little effect.

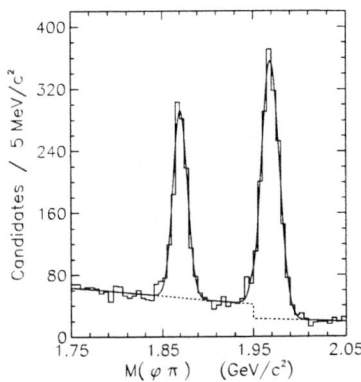

FIGURE 2. The projection from the unbinned maximum-likelihood fit on the $M(\phi\pi)$ distribution, with the discontinuity in the linear background fixed at 1.95 GeV/c^2. The two peaks are the SCS D^+ signal with a fitted yield of 997 ± 39 events, and the D_s signal with a fitted yield of 1662 ± 56 events, as described in the text.

To extract the D_s lifetime we performed a simultaneous unbinned maximum-likelihood fit to the distribution of mass and the reduced proper decay time (t^R) of all $\phi\pi$ candidates that pass the selection criteria. The t^R is defined as ($L - L_{min}$)/$c\beta\gamma$, where L is the distance between the secondary and primary vertices, and L_{min} is the minimum L value allowed by the cut on the separation between the secondary and primary vertices for this event. The projections of the data and the best fit function from the unbinned maximum-likelihood fit on the $M(\phi\pi)$ and t^R distributions for all events that pass the selection criteria are shown in Figs. 2 and 3(a), respectively. For the subset of events within ± 25 MeV/c^2 of the D_s mass, the projections of these fit results on the t^R distribution, scaled for the new mass range, are shown in Fig. 3(b).

The fit yields 1662 ± 56 $D_s \rightarrow \phi\pi$ signal events with a D_s lifetime $\tau(D_s) = 0.518 \pm 0.014 \pm 0.007$ ps. The first error reported is statistical and the second is systematic. Since ratios of MC efficiencies are used in the fit, most systematic errors cancel in the final D_s lifetime measurement. However, some uncertainties remain. In general, the estimates of these uncertainties come from refitting after varying data selection criteria, parameters, and mass or t^R regions used. Using our result and the world average D^0 lifetime [6], we find the ratio of our D_s lifetime to

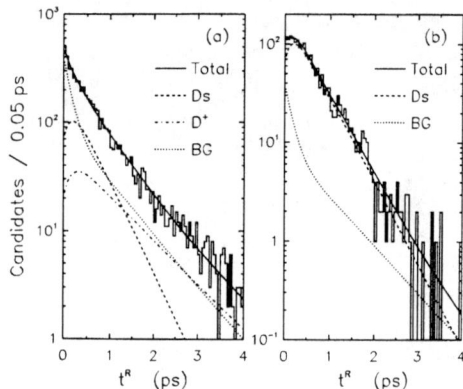

FIGURE 3. The projection of the data and the best fit function from the unbinned maximum-likelihood fit, on the reduced proper decay time distribution. (a) All events that pass the selection criteria (histogram). Also shown are the total fit function (solid curve), and the contribution to the fit from $D_s \to \phi\pi$ (dashed curve), $D^+ \to \phi\pi^+$ (dashed-dotted curve), and $\phi\pi$ background (dotted curve). (b) Events within ± 25 MeV/c^2 of the D_s mass (histogram). Also shown are the total fit function (solid curve), and the contribution to the fit from $D_s \to \phi\pi$ (dashed curve) and from $\phi\pi$ background (dotted curve), as described in the text.

the D^0 lifetime to be

$$\frac{\tau(D_s)}{\tau(D^0)} = 1.25 \pm 0.04 \quad (6\sigma \text{ difference from unity})$$

showing significantly different lifetimes for the D_s and D^0.

MEASUREMENT OF THE D^0 LIFETIME AND SEARCH FOR D^0-$\overline{D}^0$ MIXING

To date, particle-antiparticle mixing has been observed only in K^0 and B^0 particles. In these cases, the Standard Model is thought to explain the observed mixing. In the charm sector, Standard Model contributions to D^0-$\overline{D}^0$ mixing are expected to be too small to be measured [7]. Thus, observation of such mixing would be considered evidence of new physics.

The rate of mixing is usually characterized by the parameter r_{mix}, which may have contributions from both the mass difference ΔM and the width difference $\Delta \Gamma$ between the two neutral D mass eigenstates according to:

$$r_{mix} = \frac{\Gamma(D^0 \to \overline{D}^0 \to \overline{f})}{\Gamma(D^0 \to f)} = \frac{(\Delta M)^2}{2\Gamma^2} + \frac{(\Delta \Gamma)^2}{8\Gamma^2}$$

where Γ is the mean width of the two states.

We report here preliminary results on the first direct measurement of the decay width difference $\Delta\Gamma$ for the D^0-$\overline{D}^0$ system via the lifetime of the decays $D^0 \to K^-\pi^+$ and $D^0 \to K^+K^-$. In looking for decay time differences for different final states, it is convenient to think in terms of odd and even CP eigenstates D_1^0 and D_2^0. The Cabibbo-suppressed decay $D^0 \to K^+K^-$ represents a CP-even final state while the Cabibbo-allowed decay $D^0 \to K^+\pi^-$ represents a CP-mixed final state. If the physical D^0 mesons are CP eigenstates, in the limit of small $\Delta\Gamma \equiv \Gamma_2 - \Gamma_1$, we have:

$$\frac{1}{\tau_{K^-K^+}} - \frac{1}{\tau_{K^-\pi^+}} = \Gamma_{K^-K^+} - \Gamma_{K^-\pi^+} \approx \frac{\Delta\Gamma}{2}$$

The measurement of $\Delta\Gamma$ is based on the observation of $D^0 \to K^+K^-$ and $K^+\pi^-$ decay modes in the same apparatus with essentially the same detector and topological biases and uncertainties, leading to a precise comparison of the lifetimes of the two modes. Čerenkov identification is used to improve the statistical significance of the $D^0 \to K^+K^-$ sample. We obtained the necessary Čerenkov track identification efficiencies from a study of the $D^0 \to K^+\pi^-$ data. We also used MC to correct for small acceptance differences between the $K^+\pi^-$ and K^+K^- final states. The fits to the invariant mass distributions, shown in Fig. 4, yielded 60581 ± 353 $D^0 \to K^-\pi^+$ and 6683 ± 161 $D^0 \to K^-K^+$ signal events.

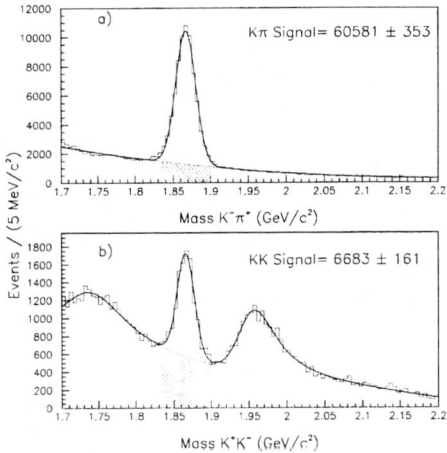

FIGURE 4. Invariant mass distributions of Čerenkov efficiency corrected candidates for: (a) $D^0 \to K^-\pi^+$ and (b) $D^0 \to K^-K^+$ decays. The solid lines shows the fits to the data.

The reduced proper decay time distributions after acceptance corrections are shown in Fig. 5. The analysis is still in progress and the fits yield lifetimes τ_{KK} and $\tau_{K\pi}$ for $D^0 \to K^-\pi^+$ and K^-K^+ with sensitivity of the order 0.7% and 2.7%, respectively. Using the above equation, we calculate $\Delta\Gamma$ to be consistent with zero and with errors of the order of 0.15 ps^{-1}, which implies that the contribution of

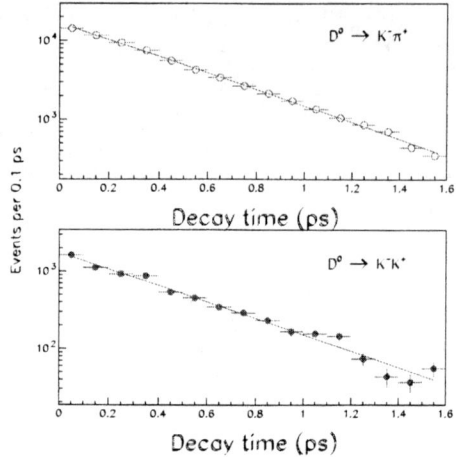

FIGURE 5. The reduced proper decay time distributions and fits for: (top) $D^0 \to K^-\pi^+$ and (bottom) $D^0 \to K^-K^+$ decay distributions after acceptance corrections.

$\Delta\Gamma$ to r_{mix} would have sensitivity less than 0.03%. We therefore observe no D^0-$\overline{D}^0$ mixing at our level of sensitivity, and thus no evidence of physics beyond the Standard Model.

STUDY OF THE DECAY $D^0 \to K^-K^+\pi^-\pi^+$

The Cabibbo-suppressed decay $D^0 \to K^-K^+\pi^-\pi^+$ can be the result of intermediate two-body decays, three-body decays, or four-body non-resonant decays. For most decays, there are two tree-level amplitudes with Cabibbo factors having opposite sign, leading to GIM cancellation. Among the exceptions are decays with a ϕ as an intermediate state.

We report branching ratio and resonant substructure measurements of $D^0 \to K^-K^+\pi^-\pi^+$ [9]. While several measurements are already available on this decay mode [10], all previous measurements ignored the interference among the various decay amplitudes. We have used a coherent amplitude analysis, which does allow for such interference, to estimate the resonant substructure.

The signal level in each mode was estimated using unbinned maximum likelihood fits to the observed invariant-mass distributions. Fig. 6 shows the final observed $K^-K^+\pi^-\pi^+$ and $K^-\pi^+\pi^-\pi^+$ signals. The fits found 136 ± 15 $D^0 \to K^-K^+\pi^-\pi^+$ and 8245 ± 101 $D^0 \to K^-\pi^+\pi^-\pi^+$. Although the selection criteria for the two decay modes were identical, the geometric acceptance and reconstruction efficiencies differed because of the different kinematics and resonant substructure of the decay modes. MC events were used to study these effects. The efficiency of the Čerenkov selection criteria used to identify kaons was estimated using E791

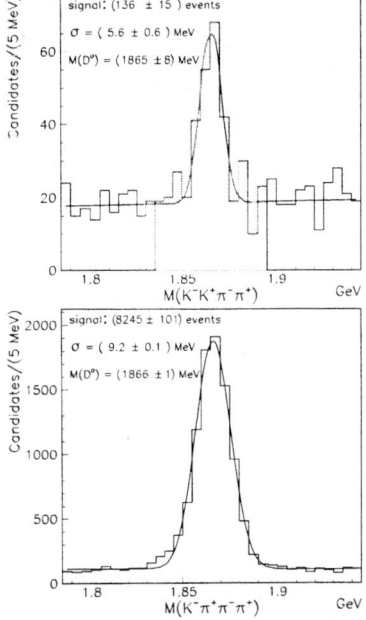

FIGURE 6. Invariant masses of $D^0 \to K^-K^+\pi^-\pi^+$ and $K^-\pi^+\pi^-\pi^+$ candidates. The events in the shaded region were used in the resonant substructure analysis.

$D^0 \to K^-\pi^+\pi^-\pi^+$ events.

With these relative acceptances and efficiencies, we measured the ratio of decay widths to be

$$\frac{\Gamma(D^0 \to K^-K^+\pi^-\pi^+)}{\Gamma(D^0 \to K^-\pi^+\pi^-\pi^+)} = (3.13 \pm 0.37 \pm 0.36)\%,$$

somewhat lower than the naive expectation of $\tan^2\theta_C = 5\%$. This may be evidence of the tree-level GIM suppression expected for most of the amplitudes. Using the $D^0 \to K^-\pi^+\pi^-\pi^+$ branching fraction from the PDG [11], we obtain $B(D^0 \to K^-K^+\pi^-\pi^+) = (0.254 \pm 0.030 \pm 0.033)\%$. The first error reported is statistical and the second is systematic. Sources of systematic uncertainties include the detector simulation, the background shapes, and the acceptance and efficiency.

The resonant substructure of $D^0 \to K^-K^+\pi^-\pi^+$ was studied using events from the branching fraction sample with $1835 < M(KK\pi\pi) < 1895$ MeV. Contributions from resonant modes to the observed signal were estimated using a series of unbinned maximum likelihood fits. For each model of the data, we performed a coherent amplitude analysis to extract relative magnitudes and phases of amplitudes. From these, we extracted the decay fractions $\Gamma(D^0 \to f)/\Gamma(D^0 \to K^-\pi^+\pi^-\pi^+)$ to be $(2.0 \pm 0.9 \pm 0.8)\%$ for $f = \phi\rho^0$, $(0.9 \pm 0.4 \pm 0.5)\%$ for $f = \phi\pi^+\pi^-$, $< 2.0\%$ at the 90% CL for $f = K^{*0}(892)\overline{K}^{*0}(892)$, and $< 1.0\%$ at the 90% CL for f = the sum of $\overline{K}^{*0}(892)K^+\pi^-$ and $K^{*0}(892)K^-\pi^+$. The phase angles among the different modes were found to be significant indicating that interference between the decay amplitudes is strong.

STUDY OF THE DECAY $D^+ \to K^+\pi^-\pi^+$

The origin of the differences between the charm meson lifetimes is the difference in their hadronic decay widths. The Cabibbo-favored (CF) hadronic decay rate of the D^0 is 3.2 times that of the D^+, while in the simplest picture, ignoring non-spectator amplitudes, the doubly Cabibbo-suppressed (DCS) decay rate of the D^+ is $\tan^4\theta_C$ times that of the D^0. Therefore we expect that $\Gamma_{DCS}(D^+)/\Gamma_{CF}(D^+) \approx 3.2 \times \tan^4\theta_C$. Thus, measurements of the DCS decays of D^+ can provide important insights into the D meson lifetime pattern.

We report a measurement of the branching fraction for the DCS decay $D^+ \to K^+\pi^-\pi^+$ and results of a Dalitz plot analysis to search for resonant structures [12]. Due to particle misidentification and reconstruction errors, several charm decays contribute to the background in the $K^+\pi^-\pi^+$ sample. The major sources of charm background are: (a) D^+ and D_s with missing neutrals, (b) D^0 2-prong decays with a spurious added track and D^0 4-prong decays with a missing track, and (c) D^+ and D_s 3-prong hadronic decays which produce structures in the $K^+\pi^-\pi^+$ mass distribution. The range of the $K^+\pi^-\pi^+$ masses over which we could reliably model the background was 1.76 to 2.06 GeV/c^2. Within this interval, backgrounds of type (a) and (b) did not produce peaks, while the parameters for the reflection shapes resulting from background (c) were modeled using real data and MC simulations.

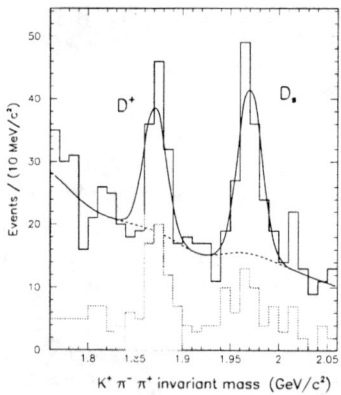

FIGURE 7. Mass spectrum of candidate $K^+\pi^-\pi^+$ events, as discussed in the text. The events in the shaded area were used in the Dalitz plot analysis

The $K^+\pi^-\pi^+$ mass distribution for the final sample is shown in Fig. 7. The spectrum was fit to the sum of reflections described above, a smooth background function, and two Gaussian functions representing the D^+ and $D_s \to K^+\pi^-\pi^+$ signals. The number of $D^+ \to K^+\pi^-\pi^+$ signal events determined by the fit was 59 ± 13. CF $D^+ \to K^-\pi^+\pi^+$ events used as the normalization signal were selected with the same criteria as the DCS signal. The number of $D^+ \to K^-\pi^+\pi^+$ signal events was found to be 7688 ± 90.

Using MC simulations, we found that the product of acceptance and efficiency was the same for the CF and DCS samples within the $\pm 2\%$ statistical errors in the simulations. The ratio of the branching fractions was thus found to be

$$\frac{B(D^+ \to K^+\pi^-\pi^+)}{B(D^+ \to K^-\pi^+\pi^+)} = (7.7 \pm 1.7 \pm 0.8) \times 10^{-3}.$$

Using the PDG [11] value for the CF branching fraction, we find $B(D^+ \to K^+\pi^-\pi^+) = (7.0 \pm 1.5 \pm 0.9) \times 10^{-4}$. The first error reported is statistical and the second is systematic, which was dominated by uncertainties in the background shape and particle identification. The value we have measured for the ratio of DCS to CF branching fractions is $(3.0 \pm 0.8) \times \tan^4 \theta_C$, which agrees well with the simple spectator picture discussed above, and with the result from E687 [13].

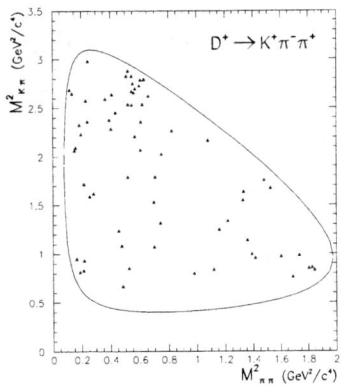

FIGURE 8. Dalitz plot of events represented by the shaded region in Fig. 7.

To study the DCS amplitudes which lead to the decay $D^+ \to K^+\pi^-\pi^+$, we have analyzed the Dalitz plot of a smaller but cleaner sample of events. This cleaner sample contains 42 ± 9 signal events which correspond to the shaded area at the bottom of Fig. 7. The Dalitz plot of these events is shown in Fig. 8 and the Dalitz plot was fit using an unbinned maximum-likelihood method. The $D^+ \to K^+\pi^-\pi^+$ decay amplitude was represented by a uniform non-resonant component plus relative amplitudes corresponding to the decays $D^+ \to K^+\rho^0(770)$ and $K^{*0}(890)\pi^+$. Using the decay fractions from the fit, we obtain $B(D^+ \to K^{*0}(890)\pi^+) = (2.5 \pm 1.2) \times 10^{-4}$, $B(D^+ \to K^+\rho^0(770)) = (2.6 \pm 1.3) \times 10^{-4}$, and $B(D^+ \to \text{non-resonant } K^+\pi^-\pi^+) =$

$(2.5 \pm 1.3) \times 10^{-4}$. The two resonant modes are approximately in phase with each other and roughly $90°$ out of phase with the non-resonant part.

CONCLUSION

E791 has a new precise measurement of the D_s lifetime, showing significantly different lifetimes for the D_s and D^0. We have preliminary measurements of lifetimes of $D^0 \to K^-\pi^+$ and K^-K^+, and the first directly measured limit on the decay width difference for the D^0-$\overline{D}^0$ system. We also measure the branching fraction of the SCS decay $D^0 \to K^-K^+\pi^-\pi^+$ and perform a coherent amplitude analysis to study resonant substructure. In addition, we measure the branching fraction of the DCS decay $D^+ \to K^+\pi^+\pi^-$ and perform Dalitz plot analysis to search for resonant substructure. These E791 results are either new or have more precision than previous results.

REFERENCES

1. E. M. Aitala et al., submitted to Phys. Rev. D, FERMILAB-PUB/98-297-E and hep-ex/9809029 (1998).
2. D. Bartlett et al., Nucl. Instrum. Methods Phys. Res., Sect. A 260 (1987) 55.
3. S. E. Csorna et al., Phys. Lett. B 191 (1987) 318; H. Albrecht et al., Phys. Lett. B 210 (1988) 267; J. R. Raab et al., Phys. Rev. D 37 (1988) 2391; M. P. Alvarez et al., Z. Phys. C 47 (1990) 539; S. Barlag et al., Z. Phys. C 46 (1990) 563; P. L. Frabetti et al., Phys. Rev. Lett. 71 (1993) 827.
4. R. J. Morrison and M. S. Witherell, Ann. Rev. Nucl. Part. Sci. 39 (1989) 183; I. I. Bigi and N. G. Uraltsev, Z. Phys. C 62 (1994) 623; T. E. Browder, K. Honscheid, and D. Pedrini, Ann. Rev. Nucl. Part. Sci. 46 (1996) 395; G. Bellini, I. Bigi, and P. J. Dornan, Phys. Rept. 289 (1997) 1.
5. E. M. Aitala et al., submitted to Phys. Lett. B, FERMILAB-PUB/98-320-E and hep-ex/9811016 (1998).
6. Particle Data Group, C. Caso et al., Eur. Phys. J. C 3 (1998) 1.
7. Alexei A. Petrov, Phys. Rev. D 56, 1685 (1997); E. Golowich, Proceedings of the Conference on B Physics and CP Violation, Honolulu, HI (March, 1997), hep-ph/9703335; E. Golowich and A. A. Petrov, Phys. Lett. B 427, 172 (1998).
8. P. L. Frabetti et al., Phys. Lett. B 323 (1994) 459.
9. E. M. Aitala et al., Phys. Lett. B 423 (1998) 185.
10. J. C. Anjos et al., Phys. Rev. D 43, (1991) 6351; R. Ammar et al., Phys. Rev. D 44, (1991) 3383; H. Albercht et al., Z. Phys. C 64 (1994) 375; P. L. Frabetti et al., Phys. Lett. B 354 (1995) 486.
11. Particle Data Group, R. M. Barnett et al., Phys. Rev. D 54 (1996) 1.
12. E. M. Aitala et al., Phys. Lett. B 404 (1997) 187.
13. P. L. Frabetti et al., Phys. Lett. B 359 (1995) 403.

The FOCUS Spectrometer and Hadronic Decays in FOCUS

Jonathan M. Link
On behalf of the FOCUS Collaboration[1]

University of California, Davis
Davis, California 95616

Abstract. An overview of Fermilab's FOCUS experiment will be presented. The spectrometer is described with special emphasis on upgrades since the previous experiment E687. The FOCUS data set, obtained in Fermilab's 1996-97 fixed target run, will be shown to be at least 14 times E687's with significant improvements in overall signal quality. Finally, hadronic decays of charm mesons will be discussed, stressing our ability to make significant contributions in this area in the near future.

INTRODUCTION

Fermilab experiment 831, known as FOCUS, is an international collaboration of physicists from institutions in Brazil, Italy, Korea, Mexico, Puerto Rico and the United States. The experiment is a continuation of the highly successful experiment 687. Both experiments study decays of photoproduced charm.

The goal of FOCUS is to collect a sample of charm particle decays 10 times larger that the E687 sample. In this paper I will show that this goal was clearly surpassed. After a brief description of the experiment, I will discuss the size and quality of the data set. Then I will discuss the physics of charm meson hadronic decays, and I will finish with several preliminary examples of charm hadronic decay signals from FOCUS.

THE EXPERIMENTAL APPARATUS

The Photon Beam

The generation of the photon beam begins with 800 GeV protons from the Tevatron which interact in a primary target of liquid deuterium. The resulting particles are collimated and charged particles are swept out of the beam. The remaining neutral beam is passed through a lead converter in which $\sim 60\%$ of the photons are converted into e^+e^- pairs. These pairs are bent around a neutral particle dump

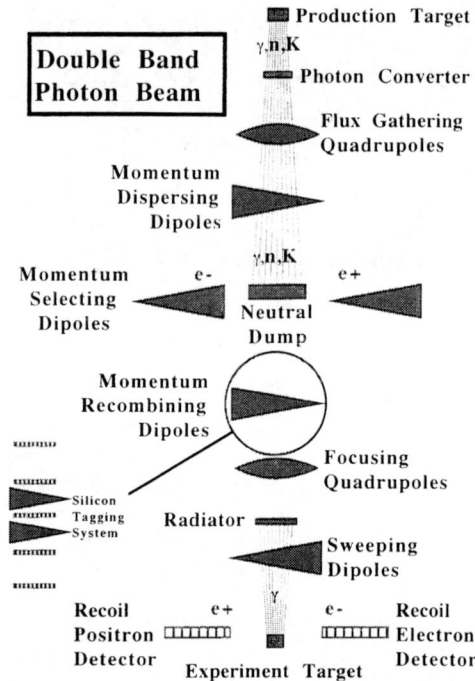

FIGURE 1. Schematic of the double band photon beam used by FOCUS.

using beam optics tuned to select particles with a median energy of 300 GeV. The two beam arms are recombined and their momenta are measured with 5 beam silicon stations. Finally, the beam passes through a lead radiator of $\sim 20\%$ radiation length and the resulting bremsstrahlung photons, with a mean energy of 170 GeV, impinge on the experimental target. The energy of the recoil $e^{\pm}$ is measured with 2 arrays of shower hodoscopes (See Figure 1).

Several changes were made to the E687 beam conditions [2] with the goal of increasing the number of photons on target. Most notably, the number of primary protons delivered per spill was increased by 80%; the median photon energy was lowered from 350 GeV – used in E687 – to 300 GeV (a 70% increase in photons on target); and the positron band was used to get another factor of 1.7 in beam to target.

The Spectrometer

At the core of the experiment are sixteen layers of silicon microstrips, which are used to reconstruct vertices and separate production vertices from decay vertices. Two large aperture dipole magnets are used to measure the momentum of the decay

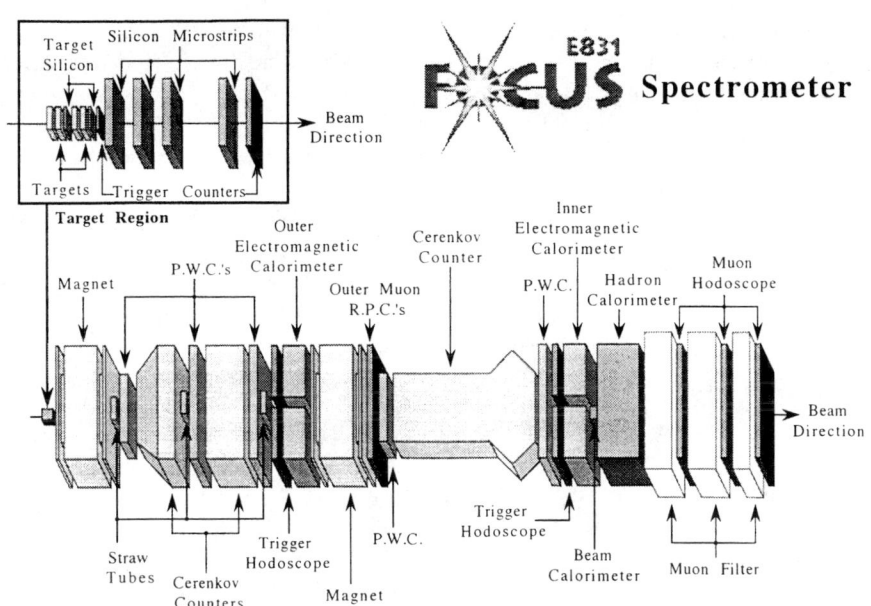

FIGURE 2. The FOCUS spectrometer

fragments. Between the magnets are three proportional wire chambers stations (PWC) and three straw tube stations for tracking, two threshold Čerenkov counters for particle identification, and the Outer Electromagnetic Calorimeter. Beyond the second magnet is the Outer Muon System, two more PWC stations, one Čerenkov counter, and the Inner Electromagnetic Calorimeter and Hadronic Calorimeter. At the downstream end of the detector is the Inner Muon System consisting of three blocks of steel filters with scintillator hodoscopes in between. This detector design provides excellent particle identification, momentum and energy resolution, and vertexing capabilities. (See Figure 2)

Targeted upgrades were made to the E687 spectrometer [3] in order to improve resolution, increase reliability, and handle the higher beam flux and data rate.

The target region (inset in Figure 2.) had several upgrades. A segmented target was introduced to increase the number of decays occurring outside of the target material. Such decays have significantly improved signal to noise with the added benefit that their primary source of background comes from charm decays which we model well. Between the target segments we added 4 new planes of silicon microstrips. The addition of these planes improved the vertex resolution, which would otherwise have degraded because the added length of the segmented target. We also changed the target material from the beryllium used in E687 to beryllium oxide which is more compact at the same interaction length.

Three narrow straw tube arrays were added to augment tracking in the high intensity e^+e^- pair region. The Outer Muon system was rebuilt, replacing proportional tubes with resistive plate chambers (RPCs). The Inner Electromagnetic Calorimeter was rebuilt using lead glass in place of plastic scintillators. Proportional tubes in the Inner Muon System were replaced with scintillating hodoscopes.

Finally, the hadronic calorimeter was rebuilt using scintillating tiles read out by optical fibers. This critical upgrade allowed the use of hadronic energy in the first level trigger. This reduced the trigger rate and significantly enhanced the charm fraction in the triggered data set. The trigger and data acquisition systems were also completely overhauled which allowed the experiment to operate with a live time greater that 80%. In comparison, E687, with only one fifth of the beam flux of FOCUS, had an average trigger live time of about 60%.

THE DATA SET

The amount of the charm in the FOCUS data set can be estimated by looking at the yield of the fully charged Cabibbo favored decays of the D^+ and D^0 mesons ($D^+ \rightarrow K^-\pi^+\pi^+$, $D^0 \rightarrow K^-\pi^+$ and $D^0 \rightarrow K^-\pi^+\pi^+\pi^-$). E687 counted 62,000 decays in these three modes at a particular set of cuts. For the same cuts, a plot combining these channels from FOCUS is shown in Figure 3a. This plot, generated by the reconstruction monitoring in the first level of processing, shows over 845,000 decays - nearly 14 times the E687 sample. It is also worth noting that the cuts used have not been optimized for FOCUS and the old E687 version of the particle identification (ID) code was used. As a result, we believe that the overall gain in

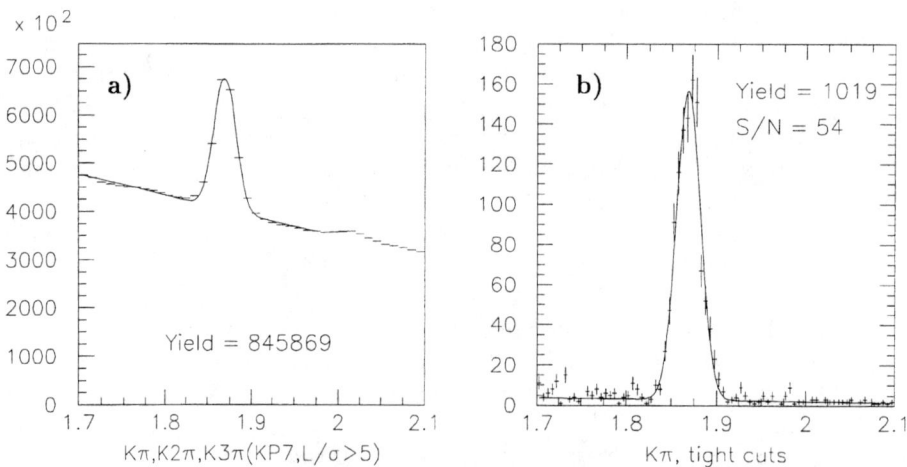

FIGURE 3. a) Combined invariant mass plot of the 3 fully charged, Cabibbo favored D decay from full FOCUS data set. **b)** Tight cuts on a sample of $D^0 \rightarrow K^-\pi^+$ from 3% of the data set.

statistics is somewhat higher.

In addition to the larger data set, there are dramatic gains in signal quality and sensitivity due to the upgraded spectrometer and improved reconstruction code. For instance, Figure 3b shows a nearly background free $D^0 \to K^-\pi^+$ mass plot based on only 3% of the data set. Our ability to obtain ultra clean signals, like Figure 3b, is the result of several factors including:

- A new Čerenkov particle ID alogithm, which assigns a probability and confidence level for each particle hypothesis.
- Overall improvement in particle ID from the upgraded calorimeter and muon systems.
- Improved vertex resolution from the new silicon planes.
- More decays occurring outside of the target material due to segmentation.

The availability of such clean high statistics charm signals, will not only improve physics results, but will also enhance our ability to study the detector and to model its response to charm decays.

HADRONIC DECAYS OF CHARM MESONS

The study of charm meson hadronic decays is important because it addresses important questions about non-spectator contributions to charm decays and the

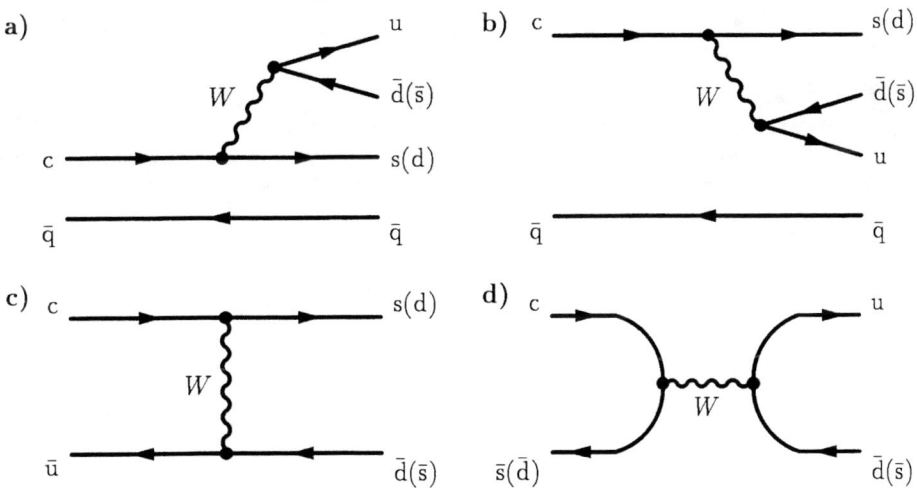

FIGURE 4. Lowest order Feynman diagrams for charm meson decays: **a)** External W emission spectator, **b)** Internal W emission spectator, **c)** W exchange, and **d)** Annihilation. Quarks associated with the Cabibbo suppressed modes are shown in parentheses.

role of final state interactions. Also, the charm system is the the only quark system in which, Cabibbo favored (CF), singly Cabibbo suppressed (SCSD) and doubly Cabibbo suppressed decays (DCSD) have all been observed.

The four lowest order weak decay Feynman diagrams are shown in Figure 4. The spectator processes (Figure 4a,b) are available to all charm mesons, and are the dominant decay mechanisms. The W exchange diagram (Figure 4c) is available only to the D^0 meson, while the annihilation diagram (Figure 4d) is only possible for the charged D mesons. Their contributions to hadronic decays is suppressed with respect to the spectator processes. The simple picture of Figure 4 is complicated by QCD corrections. In addition, once the hadrons are formed in the final state, they can continue to interact strongly. These final state interactions can distort naive predictions based on the weak processes shown in Figure 4.

One example of the importance of final state interactions in charm hadronic decays can be found in the two body, singly Cabibbo suppressed decays of $D^0 \to \pi^+\pi^-$ and $D^0 \to K^+K^-$. Both decays can occur via the external spectator or the W exchange processes. Naively one might assume that the branching ratio of K^+K^- to $\pi^+\pi^-$ would be very close to 1 (up to a small difference in the phase space factors) since their decay diagrams only difference is the vertex at which Cabibbo suppression is applied. E687 measured the branching ratio [4] and found:

$$\frac{\Gamma(D^0 \to K^+K^-)}{\Gamma(D^0 \to \pi^+\pi^-)} = 2.54 \pm 0.46 \pm 0.19. \tag{1}$$

This measurement is consistent with the current world average of 2.79 ± 0.17 [5], but not close to 1. This apparent discrepancy could be explained by a large contribution from final state interactions. (Preliminary $\pi^+\pi^-$ and K^+K^- signals from FOCUS are shown in Figure 5.)

At this early stage of the data processing we have not had the opportunity to carefully study many of the charm hadronic decay modes. Nevertheless, several representative decays were shown during the conference, but can not be included here because of space limitations. The modes shown include:

- D^+ and $D_s^+ \to K^+K^-\pi^+$ (SCSD and CF)
- $D^0 \to K^+K^-\pi^+\pi^-$ (SCSD)
- $D^0 \to \pi^+\pi^+\pi^-\pi^-$ (SCSD)
- $D^+ \to K_s\pi^+$ (CF)
- D^+ and $D_s^+ \to K_s K^+$ (SCSD and CF)
- $D^0 \to K_s K^+ K^-$ (SCSD)
- $D^0 \to K_s K^{\pm}\pi^{\mp}$ (SCSD)

All of the above listed modes, when extrapolated to the full data set, are consistent with statistics of 14 to 20 times E687. Also, we expect to have large samples of D decay modes with π^0's.

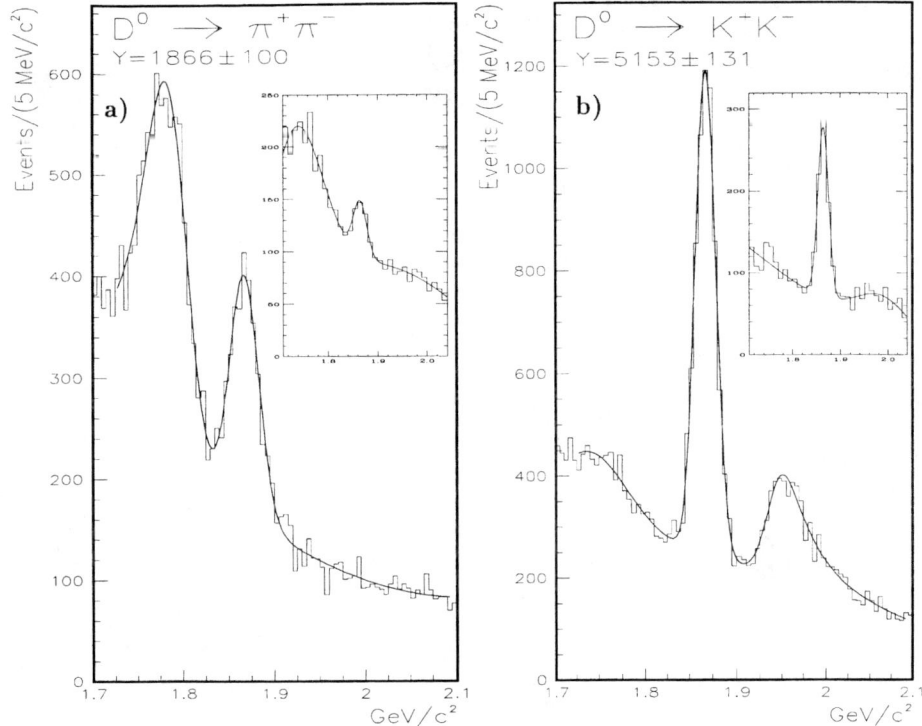

FIGURE 5. Two body singly Cabibbo suppressed decays from 40% of the FOCUS data set. **a)** $D^0 \to \pi^+\pi^-$ invariant mass plot. **b)** $D^0 \to K^+K^-$ invariant mass plot. **Insets:** The E687 $\pi^+\pi^-$ and K^+K^- samples [4] are shown for comparison.

Doubly Cabibbo Suppressed Decays

In the case of DCSD both W vertices in the lowest order Feynman diagrams involve quarks from different generations. Therefore each vertex contributes a factor of $\sin\theta_c$ to the matrix element for a total Cabibbo suppression factor of about 3×10^{-3}.

The first observation of a DCSD mode in charm was made by E687 with the observation of 21 events in the $D^+ \to K^+\pi^+\pi^-$ mode [6] (See Figure 6 inset). E687 measured the following branching ratio:

$$\frac{\Gamma(D^+ \to K^+\pi^+\pi^-)}{\Gamma(D^+ \to K^-\pi^+\pi^+)} = 0.0072 \pm 0.0023 \pm 0.0017. \tag{2}$$

A preliminary plot of the $K^+\pi^+\pi^-$ mode from FOCUS is shown in Figure 6. With the full FOCUS data set we will be able to conduct a comprehensive Dalitz analy-

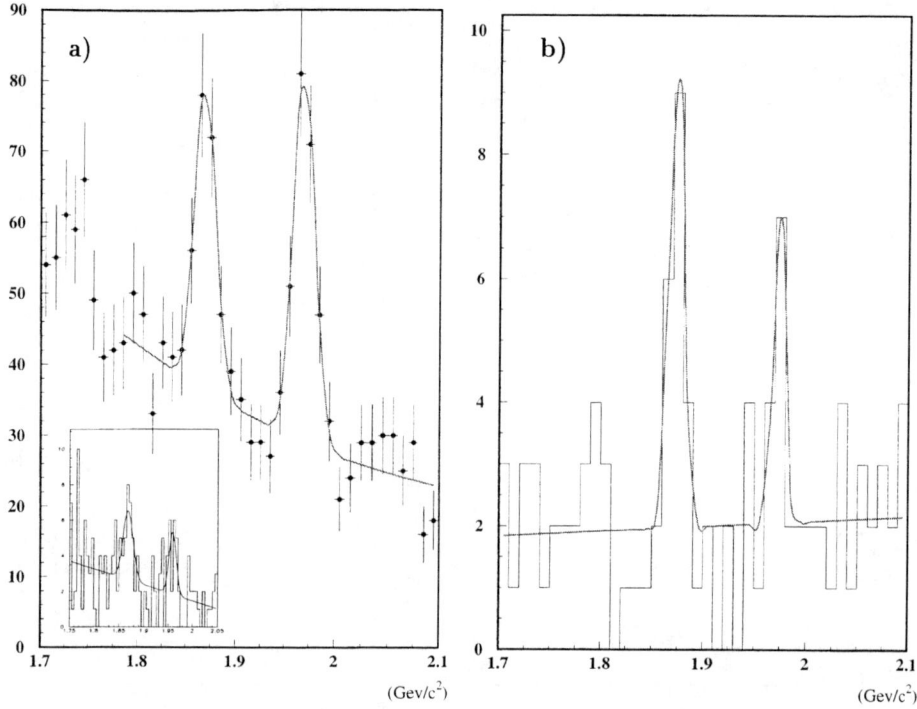

FIGURE 6. DCSD D^+ and SCSD D_s^+ from 25% of the FOCUS data set. **a)** The $K^+\pi^+\pi^-$ invariant mass plot. Inset: The E687 $K^+\pi^+\pi^-$ sample [6] is shown for comparison. **b)** The $K^+K^+K^-$ invariant mass plot (work in progress).

sis [7], which will provide detailed information about the resonant structure of this mode.

A second fully charged DCSD mode of the D^+ is $K^+K^+K^-$. In addition to double Cabibbo suppression, this mode is further suppressed by the fact that there is no simple spectator process which leads to this final state. It can occur via the quark annihilation process of Figure 4d, or it can proceed through a resonance, such as the f_0, that couples $d\bar{d}$ to $s\bar{s}$. An observation of this mode has never been reported. A very preliminary plot of possible $K^+K^+K^-$ candidates from FOCUS is shown in Figure 6b. The consistency of the signal is still under investigation.

CONCLUSION

The FOCUS data set contains a charm sample that is unprecedented in both size and data quality. With well over a million fully reconstructed charm hadronic decays we will significantly improve many existing measurements as well as make measurements that, in previous experiments, were simply not possible because of low statistics or large backgrounds. We expect to produce our first results by the summer of 1999.

ACKNOWLEDGEMENTS

We wish to acknowledge the assistance of the staffs of Fermi National Accelerator Laboratory, the INFN of Italy and the physics departments of the collaborating institutions. This research was supported in part by the National Science Foundation, the U.S. Department of Energy, the Italian Istituto Nazionale di Fisica Nucleare and Ministero dell'Università e della Ricerca Scientifica e Tecnologica, the Brazilian Conselho Nacional de Desenvolvimento Científico e Tecnológico, CONACyT-México, the Korean Ministry of Education and the Korean Science and Engineering Foundation.

REFERENCES

1. FOCUS (FNAL-E831) collaboration, in *Heavy Quarks at Fixed Target*, 1998
2. P.L. Frabetti, et al., Nucl. Instrum. Methods **A329**, 62 (1993).
3. P.L. Frabetti, et al., Nucl. Instrum. Methods **A320**, 519 (1992).
4. P.L. Frabetti, et al., Phys. Lett. **B321**, 295 (1994).
5. C. Caso, et al., Eur. Phys. J. **C3** 1 (1998).
6. P.L. Frabetti, et al., Phys. Lett. **B359**, 403 (1995).
7. E.W. Vaandering, Dalitz analisys, rare decays and charmed baryons at FOCUS and E687, in *Heavy Quarks at Fixed Target*, 1998

Dalitz Analysis, Rare Decays, and Charmed Baryons at FOCUS and E687

Eric W. Vaandering
for the FOCUS Collaboration[1]

University of Colorado
Department of Physics
Boulder Colorado 80309-0390

Abstract. Preliminary results from FOCUS demonstrate that FOCUS will set new standards in many areas of charm physics, including amplitude analysis of three body D decays, rare and forbidden decay physics, charmed baryon spectroscopy, and lifetime analyses.

THE FOCUS EXPERIMENT

The Fermilab experiment FOCUS (**P**hotoproduction **O**f Charm with an **U**pgraded **S**pectrometer) is a fixed target photoproduction experiment with a goal to accumulate 10^6 fully reconstructed charmed particles. The experiment accumulated data during the 1996–1997 Fermilab fixed target run. FOCUS (FNAL–E831) is an upgrade of the highly successful FNAL–E687. E687 is thoroughly described elsewhere [2]. The FOCUS upgrades are described elsewhere in these proceedings [3].

DALITZ ANALYSIS

Dalitz plot analyses of D mesons are an active area of study for several reasons. Primarily they give insight into the decay dynamics and provide direct information about the resonant structure. In this way, one can extend tests of the factorization models to quasi-two body decays such as pseudoscalar-vector decays as well as to vector-vector decays by analizing the resonant structure in multibody, nonleptonic, charmed meson decays. The proposed factorization models can only predict the partial decay widths of two body decays of charmed mesons in the absence of final state interactions. The amplitude analyses performed on Dalitz plots provide additional handles on final state interaction effects through interference of the amplitudes describing competing resonant channels [4].

TABLE 1. E687 Results for $D_s^+ \to \pi^-\pi^+\pi^+$ Dalitz Analysis.

Decay Mode	Decay Fraction	Phase (Degrees)	Amp. Coeff.
$f_2(1270)\pi$	0.147 ± 0.053	83 ± 16	0.42 ± 0.08
$f_0(980)\pi$	0.848 ± 0.067	0 (fixed)	1.0 (fixed)
$\mathcal{S}(1475)\pi$	0.341 ± 0.078	210 ± 10	0.63 ± 0.09

Previously E687 has published several Dalitz analyses of D^+ and D_s^+ meson decays. E687 performed a high statistics study of the decay $D^+ \to K^-\pi^+\pi^+$ [5] and lower statistics studies of the Cabibbo suppressed decays $D^+ \to \pi^-\pi^+\pi^+$ [6] and $D^+ \to K^-K^+\pi^+$ [7]. D_s^+ decays to these three channels were also studied using the Dalitz technique. The decay $D_s^+ \to \pi^-\pi^+\pi^+$ is particularly interesting since neither of the initial quarks ($c\bar{s}$) appear in the final state. The initial explanation for this decay was W-annihilation, however experiments have failed to observe a conclusive signature for this process. This channel remains one of the most promising in which to look for evidence of W-annihilation in purely hadronic decays.

With significant improvements to the FOCUS detector as well as improvements to reconstruction algorithms, FOCUS will have about 15 times the statistics of E687 in these modes. This makes a high statistics analysis of the Cabibbo suppressed D^+ and Cabibbo favored D_s^+ decay modes possible. In addition, a full phase-amplitude fit of the doubly Cabibbo suppressed D^+ and singly Cabibbo suppressed D_s^+ decay modes will be possible.

As a demonstration of the capabilities of FOCUS, we preview the Dalitz analysis of $D_s^+ \to \pi^-\pi^+\pi^+$ from FOCUS (Figure 2) and compare with the results of E687 (Figure 1). The final E687 Dalitz fit [6] had three statistically significant components, the $f_2(1270)$, the $f_0(980)$, and what was termed the $\mathcal{S}(1475)$ a scalar meson with $\Gamma = 100$ MeV/c^2. These are the approximate parameters of the $f_0(1500)$ but it should be noted that several interfering resonances in this region could mimic the contribution of this component. The improved quality of the FOCUS data is apparent in Figure 2. However, this analysis awaits a full phase-amplitude fit to determine the exact contributions to the decay process.

Other Dalitz analyses will be similarly improved.

RARE D MESON DECAYS

The E687 collaboration set the lowest limits for several decay modes of the D^+ which are either forbidden or predicted to be extremely rare by the standard model. Due to the greatly increased statistics available in FOCUS and judging by the nearly background free signals obtained by E687, FOCUS expects to improve upon the E687 sensitivity by a factor of 10–30.

The E687 group searched for decay modes of the D^+ which would indicate flavor changing neutral currents (FCNC), lepton family number violation (LFNV), or lepton number violation (LNV). A summary of those results is presented in Table 2.

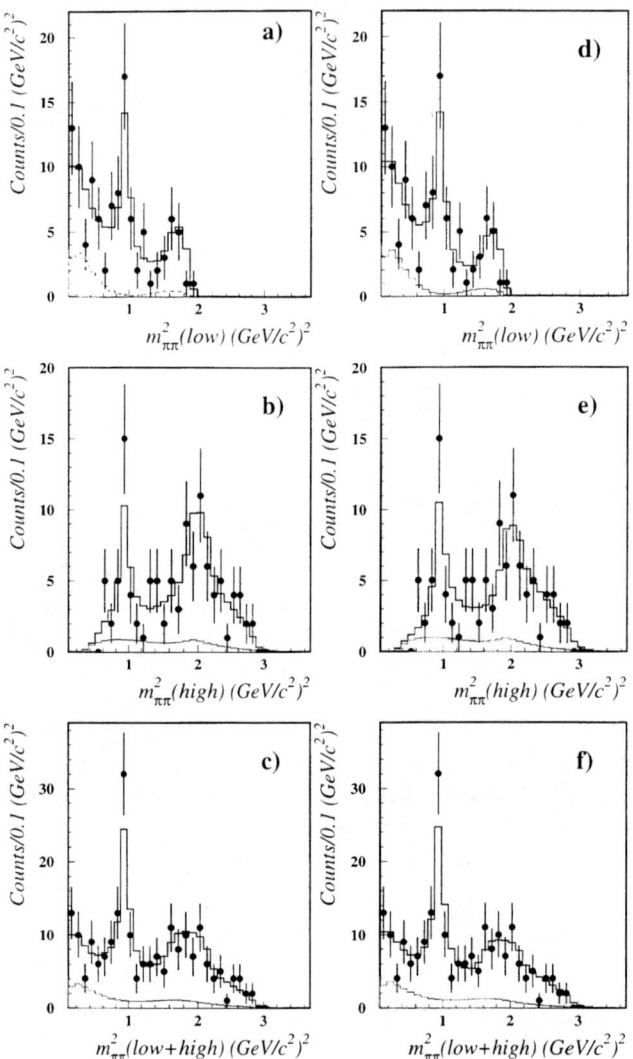

FIGURE 1. E687 $D_s^+ \to \pi^-\pi^+\pi^+$ results. The solid histogram is the fit; the points are the data. Plots a)–c) show a 5 component fit (non-resonant, $\rho(770)\pi$, $f_2(1270)\pi$, $f_0(980)\pi$, and $\mathcal{S}(1475)\pi$) plots d)–f) show the fit for the three significant components ($f_2(1270)\pi$, $f_0(980)\pi$, and $\mathcal{S}(1475)\pi$). Plots a) and d) show the low mass $\pi^+\pi^-$ combination, b) and e) show the high mass. c) and f) show the sum of two.

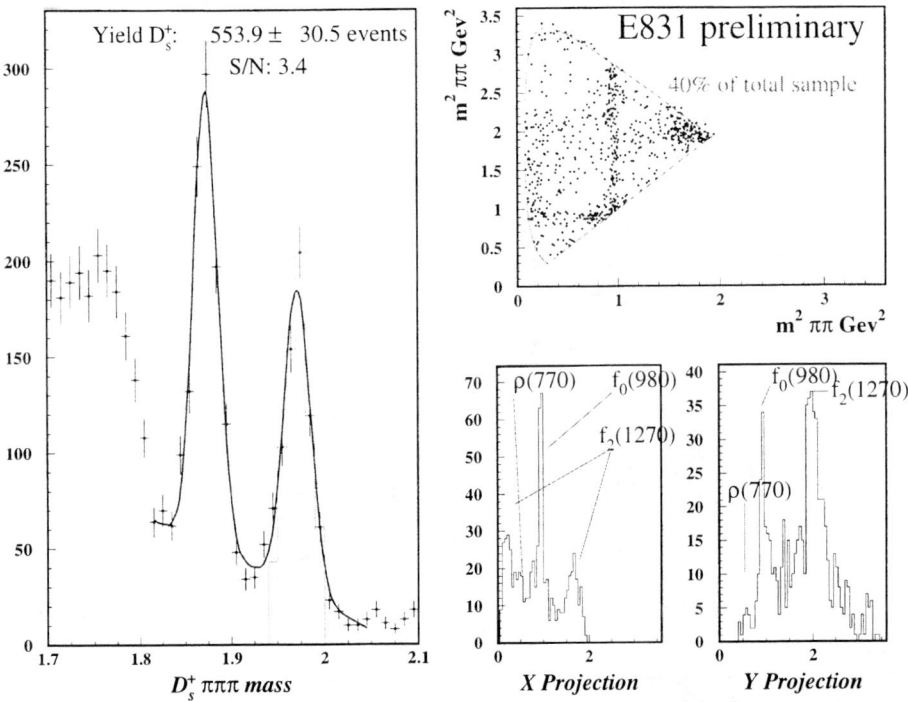

FIGURE 2. $D_s^+ \to \pi^-\pi^+\pi^+$ Dalitz plot from 40% of the FOCUS data sample. The Dalitz phase space plot is made from the outlined area in the invariant mass plot. The x & y projections are of the low and high mass $\pi^+\pi^-$ combinations respectively.

Complete results can be found in [8].

The Standard Model predicts branching ratios for the FCNC $D^+ \to \pi^+ l^+ l^-$ decay (which can occur via penguin diagrams) at or below the 10^{-8} level. $D^0 \to l^+ l^-$ can proceed via box diagrams, but these modes are helicity suppressed an *additional* factor of 10^{-3} for muons and 10^{-7} for electrons. Standard Model predictions for branching ratios of $D^0 \to \mu^+\mu^-$ are typically $< 3 \times 10^{-15}$. The LFNV and LNV decays are strictly forbidden in the Standard Model.

Any observation by FOCUS of these rare or forbidden decay modes would be an indication of new physics. FOCUS should achieve a branching ratio sensitivity of $< 10^{-5}$, well above the Standard Model prediction for FCNC decays.

TABLE 2. E687 FCNC, LFNV, and LNV Limits.

D^+ Decay Mode	E687 Limit ($\times 10^{-5}$)	Previous Limit ($\times 10^{-5}$)
$\pi^+ e^+ e^-$	11	6.6 [9]
$K^+ e^+ e^-$	20	480 [10]
$\pi^+ \mu^+ \mu^-$	8.9	1.8 [9]
$K^+ \mu^+ \mu^-$	9.7	32 [11]
$\pi^+ \mu^+ e^-$	13	330 [10]
$K^+ \mu^+ e^-$	12	340 [10]
$\pi^+ e^+ \mu^-$	11	330 [10]
$K^+ e^+ \mu^-$	13	340 [10]
$\pi^- e^+ e^+$	11	480 [10]
$K^- e^+ e^+$	12	910 [10]
$\pi^- \mu^+ \mu^+$	8.7	22 [11]
$K^- \mu^+ \mu^+$	12	32 [11]
$\pi^- \mu^+ e^+$	11	370 [10]
$K^- \mu^+ e^+$	13	400 [10]

CHARMED BARYONS

Spectroscopy

The four ground states in the singly-charmed baryon sector have all been observed. The conclusive observation of the Ω_c^0 by E687 [12,13] was the most recent. In addition, 13 excited charmed baryon states have been discovered. These excited states decay via π or γ emission to their corresponding ground states. These states, mass splittings, decay modes, and assumed $I(J)^P$ values are presented in Figure 3. With the exception of the Λ_c^+, none of the angular momentum quantum numbers have been measured experimentally.

Clearly, there are a large number of excited states which are unobserved. Most notably absent are the Σ_c^{*+}, the $J = \frac{1}{2}$ Ξ_{c1} states, and excited states of the Ω_c^0. There must also be higher L states and radial excitations. Many of these states are expected to be too broad to be observed, but some theories [14] predict that the Σ_{c2} may be narrow and observable.

Studies of charmed baryon spectroscopy allow tests to be made of the heavy quark and light flavor symmetries in the heavy quark limit. These symmetries rely on the relationship $m_c \gg \Lambda_{QCD}$ where $\Lambda_{QCD} \sim 500$ MeV. Previously these types of detailed studies have only been possible in the hyperon sector, which is problematic since $m_s \sim \Lambda_{QCD}$.

All of the observed charmed baryon states have been seen by FOCUS or E687 except the Ξ_c' and Ξ_{c1} states. Searches for these states in the FOCUS data set are under way. Prospects for observing the known and unknown excited states are promising given the large data samples of FOCUS. These samples will provide a significant increase in statistics over previous fixed target experiments. Preliminary

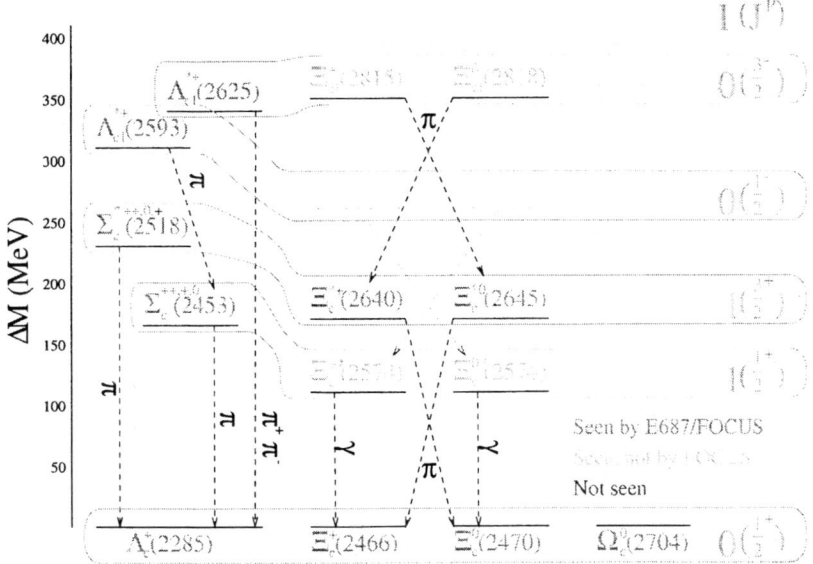

FIGURE 3. The spectrum for the observed charmed baryons, including some unobserved states.

samples of charmed baryons in typical modes are shown in Figure 4. Especially notable is our ability to obtain a Λ_c^+ signal with very good signal-to-noise while maintaining high statistics. These preliminary signals are based on 40% of the FOCUS data set.

Recently, the CLEO collaboration reported the existence of two excited states of the Σ_c baryon, decaying in the channel $\Sigma_c^* \to \Lambda_c^+ \pi^{\pm}$. CLEO observes a mass splitting of 234.5 ± 1.4 MeV and a width of 18 ± 5 MeV for the Σ_c^{*++} and a mass splitting of 232.6 ± 1.3 MeV and a width of 13 ± 5 MeV for the Σ_c^{*0} [15].

With 40% of our data set, FOCUS confirms this observation with a clear observation in the $\Lambda_c^+\pi^+$ channel and a statistically significant excess in the $\Lambda_c^+\pi^-$ channel. Our results are shown in Figure 5; the numbers are consistent with those reported by CLEO.

Lifetime measurements

The E687 collaboration is the only experiment to measure lifetimes for all four ground state charmed baryons [16–19]. The E687 measurements are presented with the corresponding PDG [20] averages in Figure 6. The recent E687 result for the Ξ_c^+ lifetime [21] is not included in the PDG average. The E687 values still dominate the world averages. With it's increased statistics and better lifetime resolution, FOCUS should make substantial improvements over the E687 measurements.

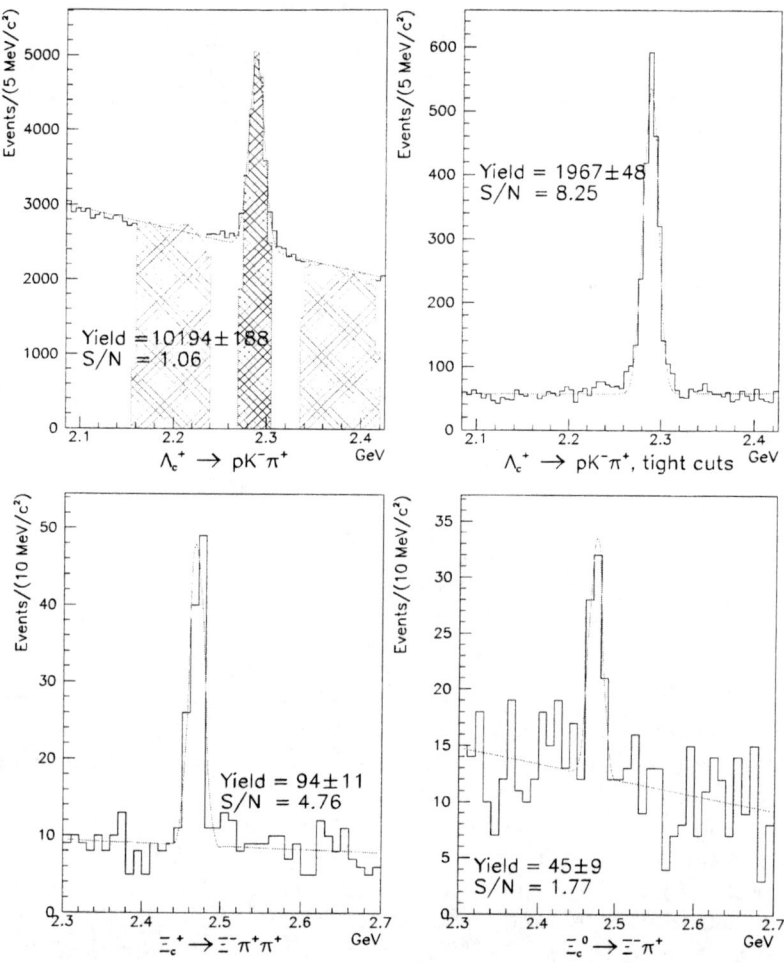

FIGURE 4. Charmed baryon samples from 40% of the FOCUS data set. In the upper left plot, the hatched sideband area is used to find the background shape in Figure 5.

An accurate determination of charmed baryon lifetimes is important from a theoretical basis. There are many significant contributions to the total widths of charmed baryons, including semi-leptonic spectator decay, internal and external hadronic decays, and W-exchange. If, as expected, W-exchange is the dominant effect for the Λ_c^+ and Ξ_c^0, then Cabibbo suppressed W-exchange in the Ξ_c^+

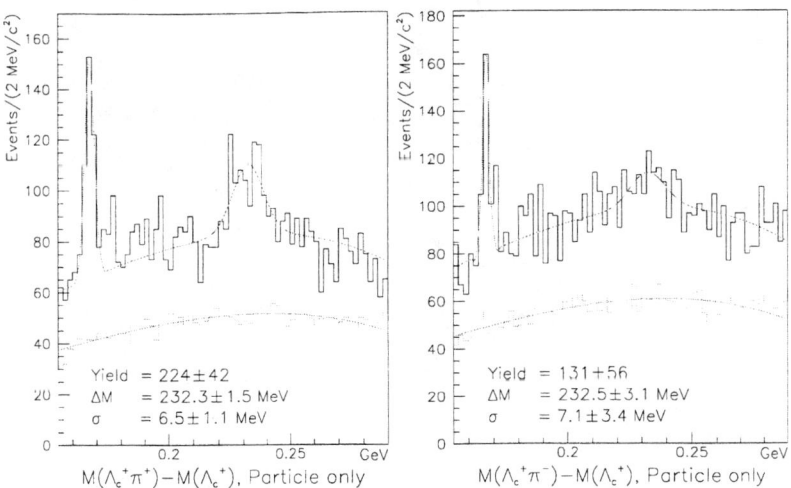

FIGURE 5. $\Sigma_c^* \to \Lambda_c \pi^\pm$ mass difference plots for 40% of the FOCUS data set. The lower plots show the background coming from the Λ_c^+ sidebands in Figure 4. Here σ is a Gaussian width and not the proper Γ from the Breit-Wigner distribution.

and Ω_c^0 will also be important. Accurate measurements are needed to distinguish among theories which predict different values for these contributions to the total width. The current data support the predictions of Guberina, et. al. who predict $\tau(\Omega_c^0) \approx \tau(\Xi_c^0) < \tau(\Lambda_c^+) < \tau(\Xi_c^+)$ [22]. However, the experimental measurements of $\tau(\Omega_c^0)$ and $\tau(\Xi_c^0)$ support nearly any theoretical prediction. A measurement of these two lifetimes to 10% or better accuracy should resolve this issue.

CONCLUSIONS

We have shown that the quantity and quality of FOCUS will advance many areas of charm physics well beyond that achieved by E687 and other experiments. Reconstruction and event selection of the FOCUS data set is still under way and should be completed in early 1999. First physics results will be available in mid-1999.

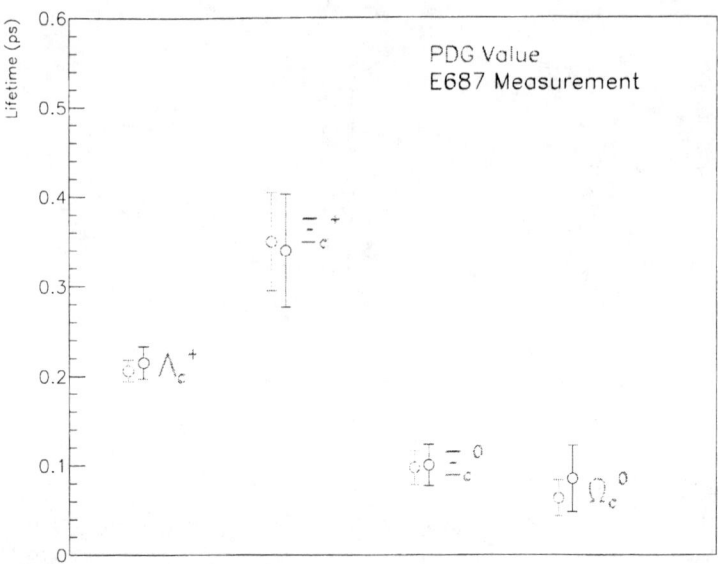

FIGURE 6. A comparison of E687 and PDG values for the lifetimes of the charmed baryons. The PDG values are on the left.

ACKNOWLEDGEMENTS

We wish to acknowledge the assistance of the staffs of Fermi National Accelerator Laboratory, the INFN of Italy, and the physics departments of the collaborating institutions. This research was supported in part by the U.S. National Science Foundation, the U.S. Department of Energy, the Italian Istituto Nazionale di Fisica Nucleare and Ministero dell'Università e della Ricerca Scientifica e Tecnologica, the Brazilian Conselho Nacional de Desenvolvimento Científico e Tecnológico, CONACyT-México, the Korean Ministry of Education, and the Korean Science and Engineering Foundation.

REFERENCES

1. FOCUS (FNAL-E831) collaboration, in *Heavy Quarks at Fixed Target*, 1998.
2. P. L. Frabetti et al., Nucl. Instrum. Methods **A320**, 519 (1992).
3. J. M. Link, The FOCUS spectrometer and hadronic decays in FOCUS, in *Heavy Quarks at Fixed Target*, 1998.
4. J. E. Wiss, in *Proceedings of the International Summer School of Physics, Varenna 1997*, edited by I.Bigi and L.Moroni, pages 39–94, IOS Press, Amsterdam, 1998.
5. P. L. Frabetti et al., Phys. Lett. **B331**, 217 (1994).
6. P. L. Frabetti et al., Phys. Lett. **B407**, 79 (1997).
7. P. L. Frabetti et al., Phys. Lett. **B351**, 591 (1995).
8. P. L. Frabetti et al., Phys. Lett. **B398**, 239 (1997).
9. E. M. Aitala et al., Phys. Rev. Lett. **76**, 364 (1996).
10. A. J. Weir et al., Phys. Rev. **D41**, 1384 (1990).
11. K. Kodama et al., Phys. Lett. **B345**, 85 (1995).
12. P. L. Frabetti et al., Phys. Lett. **B338**, 106 (1994).
13. P. L. Frabetti et al., Phys. Lett. **B300**, 190 (1993).
14. T.-M. Yan and D. Pirjol, Phys. Rev. **D56**, 5483 (1997).
15. G. Brandenburg et al., Phys. Rev. Lett. **78**, 2304 (1997).
16. P. L. Frabetti et al., Phys. Rev. Lett. **70**, 1755 (1993).
17. P. L. Frabetti et al., Phys. Rev. Lett. **70**, 1381 (1993).
18. P. L. Frabetti et al., Phys. Rev. Lett. **70**, 2058 (1993).
19. P. L. Frabetti et al., Phys. Lett. **B357**, 678 (1995).
20. C. Caso et al., Eur. Phys. J. **C3**, 1 (1998).
21. P. L. Frabetti et al., Phys. Lett. **B427**, 211 (1998).
22. B. Guberina, R. Ruckl, and J. Trampetic, Z. Phys. **C33**, 297 (1986).

Hadronic Decays of Beauty and Charm from CLEO

Jorge L. Rodriguez
(CLEO Collaboration)

*Department of Physics and Astronomy University of Hawaii
2505 Correa Road Watanabe Hall 227
Honolulu, Hawaii 96822*

Abstract. A selection of recent results on hadronic charm and beauty decays from the CLEO experiment are presented. We report preliminary evidence for the existence of final state interactions in B decays and the first observation of the decay $B^0 \to D^{*+}D^{*-}$ with a branching fraction of $(7.8^{+5.4}_{-3.8} \pm 1.5) \times 10^{-4}$. We also present preliminary results on the first observation of the broad, $J^P = 1^+$, charmed meson resonance with a mass of $m_{D_1(j=1/2)^0} = 2.461^{+0.41}_{-0.34} \pm 0.010 \pm 0.032$ GeV and a width of $\Gamma = 290^{+101}_{-79} \pm 26 \pm 36$ MeV and branching fraction measurements of the $B^- \to D_J^0 \pi^{-1}$ decay. Finally, we report on our search for the radial excitation of a spin 1 charmed meson, the $D^{*\prime +}$, and on an improved measurement of the ratio of decay rates $\Gamma(D^0 \to K^+\pi^-)/\Gamma(D^0 \to K^-\pi^+)$.

INTRODUCTION

The CLEO experiment has provided important contributions to our understanding of hadronic decays of the beauty and charm systems since it began taking data in the early 1980s. The wealth of results is due primarily to the large data samples collected over the years and the excellent tracking, energy resolution and reasonably good particle ID of the CLEO series of detectors. In this paper we present five analyses on hadronic decays of charm and bottom mesons.

CLEO is currently analyzing data from two separate runs taken with different detector configurations. The first run ended in the Summer of 1995 and includes a total luminosity of 3.1 fb^{-1} on and 1.6 fb^{-1} taken 60 MeV below the $\Upsilon(4S)$. Given the $B\bar{B}$ cross section this sample corresponds to 3.1 million $B\bar{B}$ pairs. The configuration of the detector during the first run is described in detail in Ref. [1]. This dataset will be referred to as the CLEOII dataset hereafter. At the end of the CLEOII run the detector was significantly improved with the replacement of the inner straw tube drift chambers by a 3 layer silicon vertex (SVX) detector [2]. In addition, the argon-ethane gas in the drift chambers was replaced with

---

[1] Unless otherwise indicated complex conjugate states are implied throughout this paper

CP459, *Heavy Quarks at Fixed Target*
edited by Harry W. K. Cheung and Joel N. Butler
© 1999 The American Institute of Physics 1-56396-864-9/99/$15.00

a helium-propane mixture which improved both particle ID and the momentum resolution in the drift chambers. Finally, the track fitting software was updated to one based on the Kalman filtering algorithm. These improvements in tracking and particle ID, while featured in only two of the analysis presented here will be become increasingly important in future analyses. The data collected and reconstructed, with the CLEOII upgrade, (CLEOII/SVX), consists of 2.5 fb^{-1} on and 1.3 fb^{-1} off resonance. The data run for CLEOII/SVX will be completed at the end of 1998.

First Observation of $B^0 \to D^{*-}D^{*+}$

The Cabbibo suppressed decay $B^0 \to D^{*-}D^{*+}$ is a potentially interesting CP violation mode, whose rate is expected to be comparable to the gold plated CP $B^0 \to J/\Psi K_s$ decay. Since the $D^{*+}D^{*-}$ final state can be obtained from either B^0 or a $\bar{B}^0$ this decay mode mode can be used to extract $\sin 2\beta$ through $B^0 \bar{B}^0$ mixing. The amplitude for this decay is dominated by the external tree diagram and we can estimate its rate by comparison to the measured $B^0 \to D^{*-}D_s^{*+}$ rate, after taking into account the appropriate ratio of decay constants and CKM matrix elements. While the expected rate is of order 0.1 %, the rather large number of particles, six in the lowest multiplicity mode, in the decay chain significantly reduces the expected yield.

CLEO has performed a search for this mode by examining all of the currently available data collected on the $\Upsilon(4S)$ [3]. This includes the complete CLEOII (3.1 fb^{-1}) and the available portion of CLEOII/SVX data (2.5 fb^{-1}). The decay chain is fully reconstructed cutting on kinematic variables to reduce backgrounds. In this analysis only three of the possible four combinations of the $D^{*+}D^{-*}$ were used. The decay mode with two soft π^0, the $B^0 \to (D^+\pi^0)(D^-\pi^0)$ decays was not used due to background considerations. For events in the CLEOII/SVX sample an additional requirement was imposed to take advantage of the better position resolution obtained from the SVX.

The observables used to extract the signal were the beam-constrained mass (M_{BC}) and the difference between the reconstructed energy and the beam energy (ΔE). At CLEO, the Bs are produced nearly at rest so ΔE, for real events, is peaked at zero while backgrounds from other decays peak one or more pion mass away from zero. The beam-constrained mass variable is just the usual invariant mass with the beam energy substituted for the measured energy. The resolution of M_{BC} is significantly better than the invariant mass, by about an order of magnitude, due to the small spread of the beam energy. In Figure 1 we show the scatter plots of the on-$\Upsilon(4S)$ distributions for events that pass all of event selection criteria. The solid rectangle in Figure 1 (left) is the signal region. A total of 4 events were observed.

To estimate the backgrounds that enter into the signal region, two independent methods were used. The first estimate is based on events in the ΔE vs. M_{BC} sideband indicated by the region outside dashed rectangle in Figure 1 (left).

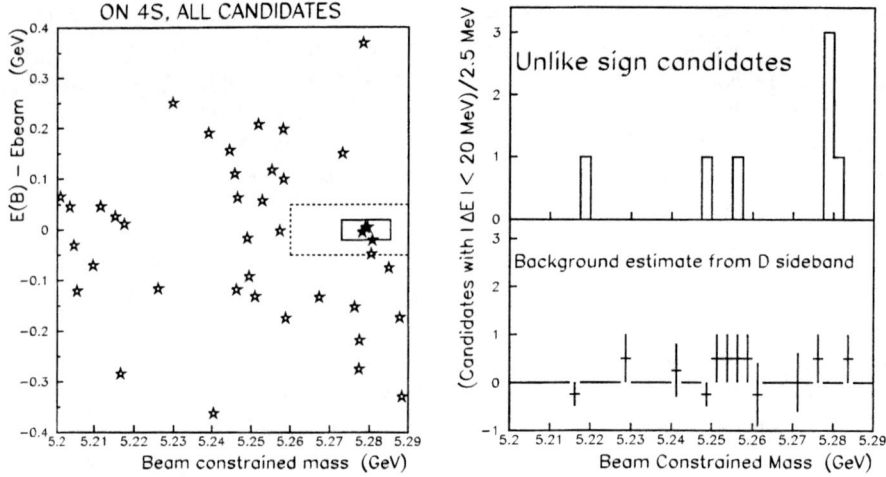

FIGURE 1. Scatter plot of on 4S events in ΔE and M_{BC} (left). Beam-constrained mass distribution for events that lie within 20 MeV of 0.0 in ΔE (right-top). The plot (right-bottom) shows the backgrounds from the D mass sidebands.

An estimate of 0.26 ± 0.04 events is determined from this sideband estimate. A second estimate is obtained by adding contributions from continuum, $B\bar{B}$ (other kinematically similar B decays that can fake the signal) and random combination that are reconstructed as signal. Each of these contributions were modeled by off-resonance data, Monte Carlo, and/or D mass sidebands. This estimate predicts a background of 0.37 ± 0.05 events.

The branching fraction measured given the four observed events is:

$$\mathcal{B}(B^0 \to D^{*-}D_s^{*+}) = (7.8^{+5.4}_{-3.8} \pm 1.5) \times 10^{-4}. \tag{1}$$

This value is determined from an unbinned likelihood fit using the larger of the two background estimates. This value is consistent with the expected rate of 0.1% given the measured branching fraction of the $B^0 \to D^{*-}D_s^{*+}$, our knowledge of decay constants and the CKM matrix elements [3].

Angular Distributions in $B \to D^*\rho$

A full partial wave analysis of the decays $B^- \to D^{*0}\rho^-$ and $\bar{B}^0 \to D^{*+}\rho^-$ has been performed using the entire CLEOII data sample. These decays proceed primarily through tree-level $b \to cW$ emission and, to first order, the amplitude for the decays are independent of a CKM phase. The absence of a weak phase suggests a clean model to study the effects of final state interactions (FSI) in hadronic B decays.

The full partial wave decomposition, with its own phases, provides us with a way to determine the strong phases through an analysis of the angular distribution of the final states.

In order to extract information on the strong phases we first need to express the differential decay rate in terms of complex amplitudes and helicity angles. In this analysis we use the helicity basis expressed in three components; two, the $H_\pm$, represent the transverse components and one, the H_0 describes the longitudinal component. Squaring and factoring the amplitude gives the differential decay rate in terms of the helicity amplitudes and the helicity angles $\theta_\rho, \theta_{D^*}$ and χ. The form of the expression is,

$$\frac{d\Gamma}{d\cos\theta_{D^*} \cdot d\cos\theta_\rho d\chi} = 4|H_0|^2 \cos^2\theta_{D^*} \cos^2\theta_\rho + \left(|H_-|^2 + |H_+|^2\right)\sin^2\theta_{D^*}\sin^2\theta_\rho$$
$$+ 2\left[Re(H_+ H_-^*)\cos 2\chi - Im(H_+ H_-^*)\sin 2\chi\right]\sin^2\theta_{D^*}\sin^2\theta_\rho$$
$$+ [Re(H_+ H_0^* + H_- H_0^*)\cos\chi - Im(H_+ H_0^* - H_- H_0^*)\sin\chi]\sin 2\theta_{D^*}\sin 2\theta_\rho. \quad (2)$$

The two helicity angles are defined in the rest frame of the decay as the angle between one of the daughters and the direction of the parent in the rest frame of the B. The angle χ is the angle between the two decay planes and is related to the azimuthal direction of the helicity axis by $\chi = \phi_{D^*} - \phi_\rho$. In the amplitude, the strong phase information is contained in the terms with the imaginary parts; in Equation (2), no FSI implies that either $Im(H_+ H_0^* - H_- H_0^*)$ and $Im(H_+ H_-^*)$ are zero, or conversely, that all amplitudes are relatively real [4].

All events are required to pass a series of selection criteria to fully reconstruct the decay chain of the B using three decay modes of the D^0, the $K\pi, K\pi\pi^0$ and $K3\pi$, and the dominant decay modes of the D^* [6]. Two methods are used to extract the phase information in Equation (2): a moments analysis [5] in which the components of each term in Equation (2) are extracted and a direct determination of the magnitude and phases of the helicity amplitudes from a three dimensional (3D) unbinned maximum likelihood fit of the data to the functional form in Equation (2).

In the unbinned likelihood fit and moments analysis the likelihood function includes terms for the signal and background contributions, factorizing each term into an angular part and a mass part. The mass part characterizes the ρ invariant mass with a relativistic Breit-Wigner and the beam-constrained mass with a Gaussian function. The angular part is modeled by Equation (2). To minimize the number of free parameters the fit is first performed ignoring the angular part and the parameters of the mass function are extracted. In the second step, the fit is redone including the angular function and fixing the mass parameters to values extracted from the first fit. The parameters in the angular part of the likelihood function are the phases and magnitudes of the transverse helicity amplitudes relative to the longitudinal component which is set to $H_0 = 1$ and $\delta_0 = 0$ in the fit. The amplitudes are then rescaled to the satisfy the normalization condition $|H_0|^2 + |H_-|^2|H_+|^2 = 1$ The results of the likelihood fit for the strong phases and amplitudes are given in

Table 1. The coefficients of Equation (2) have also been determined from the fit and a comparison made with the values obtained from the moments analysis. We find that the results are consistent with each other within statistical errors (see Table 2). Our values for the phases in Table 1 suggest non-trivial strong phases in both the $B^- \to D^{*0}\rho^-$ and $\bar{B}^0 \to D^{*+}\rho^-$ modes, however the statistical size of our sample does not provide us with an independent confirmation of the results in a 1D fit of the data to the $\sin\chi$ or $\sin 2\chi$ distributions.

The results of the fit are also used to test the factorization hypothesis by comparing the polarization of the $\bar{B}^0 \to D^{*+}\rho^-$ decay with the polarization in the semi-leptonic $\bar{B}^0 \to D^{*+}l^-\bar{\nu}$ at the appropriate q^2 scale. The predicted values for the longitudinal polarization in the semi-leptonic decay at $q^2 = m_\rho^2$, range from 85% to 88% and the recent CLEO results is $91.4 \pm 15.2 \pm 8.9\%$ [6]. The longitudinal polarization from the fit is $87.8 \pm 3.4 \pm 4.0$ % consistent, within errors, with both the theoretical predictions and the semi-leptonic measurement.

TABLE 1. Phases extracted from the unbinned likelihood fit

Parameter	$\bar{B}^0 \to D^{*+}\rho^-$	$B^- \to D^{*0}\rho^-$
δ_-	$0.19 \pm 0.23 \pm 0.14$	$1.13 \pm 0.27 \pm 0.17$
δ_+	$1.47 \pm 0.37 \pm 0.32$	$0.95 \pm 0.31 \pm 0.19$
$|H_-|$	$0.317 \pm 0.052 \pm 0.013$	$0.283 \pm 0.068 \pm 0.039$
$|H_+|$	$0.152 \pm 0.058 \pm 0.037$	$0.228 \pm 0.069 \pm 0.036$

Measurements in $B^- \to D^0_J \pi^-$

The $L = 1, n = 1$ charmed mesons are the P wave orbital excitations in which four spin-orbit configurations are possible. In the heavy quark limit, these combinations can be identified by the j_l quantum number which couples the spin of the light quark with the orbital angular momentum. These states form two doublets: a $j_l = 3/2$ and a $j_l = 1/2$ which, from conservation of parity and angular momentum, decay via either S or D waves. In the heavy quark limit, the $j_l = 1/2$ decays only via S wave while the $j_l = 3/2$ can decay only via D wave. The only $L = 1$ states so far observed have been the narrow $L = 1, n = 1$ resonances, the $D_1(2420)$ and the $D_2^*(2460)$ with widths of order 20 MeV. These states have been assigned

TABLE 2. Coefficients in Equation (3) extracted from the likelihood fit

	$\bar{B}^0 \to D^{*+}\rho^-$		$\bar{B}^0 \to D^{*+}\rho^-$	
Coefficient	From L.L. fit	From moments	From L.L. fit	From moments
H_0^2	0.856	0.751 ± 0.073	0.859	0.626 ± 0.074
$H_+^2 + H_-^2$	0.140 ± 0.040	0.159 ± 0.034	0.143 ± 0.060	0.168 ± 0.036
$Im(H_-H_0^* - H_+H_0^*)$	0.110 ± 0.074	0.042 ± 0.103	-0.071 ± 0.109	-0.145 ± 0.101
$Re(H_-H_0^* + H_+H_0^*)$	0.341 ± 0.088	0.352 ± 0.104	0.250 ± 0.105	0.193 ± 0.109
$Im(H_+H_-^*)$	0.053 ± 0.021	0.057 ± 0.024	-0.011 ± 0.032	0.002 ± 0.027
$Re(H_+H_-^*)$	0.023 ± 0.024	0.018 ± 0.023	0.068 ± 0.029	0.043 ± 0.025

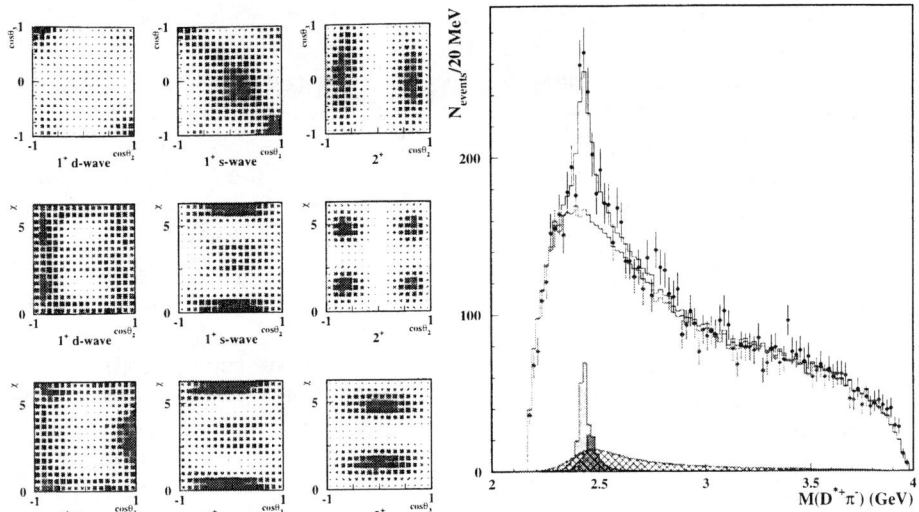

FIGURE 2. The reconstructed 2D angular distributions for the decays of the B^- meson to each of the three D_J^0 resonances (left). The $D^{*+}\pi^-$ invariant mass distribution for on 4S data with the 4D-MLF projections from each of the three D_J^0 candidates, the two narrow 1^+ (light) and the 2^+ (darker) resonance and the broad 1^+ (cross-hatched) plus the total background are superimposed on the plot (right).

to the $j_l = 3/2$ by observing the angular distributions of the decay products and measurements of the ratio of $D^*\pi$ to $D\pi$ decays in continuum production where the D_J are unpolarized.

A full partial wave analysis of the decay $B^- \to D_J^0 \pi_1^-, D_J^0 \to D^{*+}\pi_2^-, D^{*+} \to D^0 \pi_3^+$ has been performed to measure the product branching fraction of the $\mathcal{B}(B) \times \mathcal{B}(D_J)$ decays and the properties, the mass and width, of the broad $D_1(j = 1/2)$ resonance. The measurements are extracted from a 4D unbinned maximum likelihood fit (4D-MLF) to the data, where the independent variables are the three helicity angles and the invariant mass of the D_J^0 resonance.

An important point in this analysis is the fact that the D_J is completely polarized since it is the decay product of a pseudo-scalar decay to a vector plus another pseudo-scalar. Knowing the initial polarization of the D_J and the fact that angular momentum and parity are both conserved provides us with a clear picture, in the heavy quark limit, of the angular decay distribution in final states that first decay through one of the three intermediate D_J^0 resonances. In other words, we can distinguish among the three possible $L = 1$ states, those that decay to a $D^{*+}\pi^-$, not only by using the invariant mass but also by examining the full angular distribution of the final state.

A partial reconstruction technique selects the events from among the entire CLEOII dataset. These events are used in the fit to the 4D maximum likelihood function. In the partial reconstruction method the entire decay is reconstructed, up to a quadratic ambiguity, from the 4-momenta of the three pions (π_1, π_2, π_3) in the decay chain and imposing energy-momentum conservation at each decay vertex [9]. This method improves statistics by about an order of magnitude over the usual full reconstruction technique since it eliminates the explicit reconstruction of the charmed meson. A trade off to the gain in statistics comes from the increased complexity of the analysis and higher levels of backgrounds. These backgrounds are however, modeled by using the 1.6 fb^{-1} of off-resonance data and Monte Carlo simulations. In Figure 2 (left) we show the $D^{*+}\pi^-$ invariant mass distribution taken from events that pass the selection criteria described in Ref. [10] in the on-resonance CLEOII dataset. Superimposed on the plot are the $D^{*+}\pi^-$ invariant mass projection from the 4D-MLF for each of D_J^0 candidates plus the total background contribution from various sources; continuum and other $B\bar{B}$ decays with similar kinematics. The ability of the angular information to distinguish between the three possible resonances is illustrated in Figure 2 (right) where Monte Carlo simulations of the B^- decays to each of the three $L = 1$ D_J resonances are shown.

The 4D likelihood function used in the fit includes terms for the angular distribution, mass amplitudes of the resonances, strong phase shifts and parameters that allow for mixing between the two 1^+ states. It also allows for detector smearing and acceptance. The functional form of the amplitude is

$$\mathcal{A}_{B \to D^{*+}\pi^-\pi^-} = \alpha_{nr} e^{i\delta_{nr}} + \alpha_2 A_2 a_2 e^{i\delta_2} + \alpha_{1n} A_{1n} \left(a_{1d} \cos\beta + a_{1s} \cos\beta e^{i\phi} \right)$$
$$+ \alpha_{1b} A_{1b} \left(a_{1s} \cos\beta - a_{1d} \cos\beta e^{i\phi} \right) e^{i\delta_1} \quad (3)$$

where the α_i allows for different contributions from the various resonant and non-resonant $D^{*+}\pi^-\pi^-$ components, the A_i are Breit-Wigner amplitudes, and the a_i are the angular ($D_{m,m'}^j$) amplitudes. The mixing between the narrow and broad 1^+ resonances is currently described by the mixing angles β and ϕ and the strong phases for the resonant and non-resonant components are included via the δ parameters. The 1n,2,1b, and nr subscripts refer to the narrow $1^+, j_l = 3/2$ and $2^+, j_l = 3/2$ resonances, the broad $1^+, j_l = 1/2$ resonance and the non-resonant component respectively. This parameterization is not unique and an alternative parameterization has been used as a systematic check. The variation in the results due to the alternative parameterization is quoted as an additional systematic error. The fit is performed with the mass and width of the relativistic Breit-Wigners for the narrow 2^+ and the 1^+ states fixed to their measured values [7]. The normalization of each of the three resonant and the non-resonant components plus the mass and width of the broad 1^+ state are allowed to float in the fit. From the fit we extract the invariant mass and width of the broad 1^+ resonance to be:

$$M_{D_1(j=1/2)^0} = 2.461^{+0.41}_{-0.34} \pm 0.010 \pm 0.032 \quad \text{GeV}$$
$$\Gamma_{D_1(j=1/2)^0} = 290^{+101}_{-79} \pm 26 \pm 36 \quad \text{MeV} \quad (4)$$

where the first error is the statistical uncertainty, the third is the uncertainty from the parameterization of the amplitude and the second is the systematic uncertainty from all other sources. From the 4D fit we also extract the product branching ratios of the B^- to each of the three D_J^0 plus a single pion. These results are given in Table 3 together with the yields and the B^- branching fractions using the D_J^0 branching fractions from isospin symmetry. The second systematic error in Table 3 represents the uncertainty in the parameterization of the grand-amplitude. These results are consistent with the values obtained earlier using a simpler 2D-MLF where not all of the angular information was used [10]. These branching fraction measurements, however, disagree with theoretical expectations from heavy quark effective theory which predict the rates to be about three times smaller than the our results [11]. Our preliminary results on the mass and width of the broad 1^+ charmed meson are in agreement with the quark model [12].

Search for First Radial Excitations in Charmed Meson

In 1997 the DELPHI collaboration claimed evidence for the 1st radially excited charmed meson $D^{*\prime+}$ [8]. They found an excess of 66 ± 14 events in their sample of reconstructed $D^{*+}\pi^-\pi^+$ with a mass of $2637 \pm 2 \pm 6$ MeV and a small width. The assignment of the quantum numbers was based primarily on the mass measurement which is consistent with theoretical expectations for a $D^{*\prime+}$ [13]. The width of the bump is approximately equal to the detector resolution so DELPHI sets an upper bound on the decay width of the $D^{*\prime+}$ to be < 15 MeV at the 95% confidence level. The OPAL experiment has also performed a search for the $D^{*\prime+}$ in the same final state and in the DELPHI mass window using a similar analysis procedure. They however, found no excess and set an upper limit on $D^{*\prime+}$ production of $f_{Z^0 \to D^{*\prime+}} \times \mathcal{B}(D^{*\prime+} \to D^{*+}\pi^-\pi^+) < 2.1 \times 10^{-3}$ 95% C.L. [14]. Both experiments collect data at the Z^0 mass so $D^{*\prime+}$ production is from the $c\bar{c}$ and/or $b\bar{b}$ continuum. Both experiments also estimate that about half of their candidates are from $c\bar{c}$ production.

The analysis procedures used at CLEO are similar to that employed by both DELPHI and OPAL. First pion and kaon candidates are selected from tracks originating at the IP. These are then combined to form D^0 and D^{*+} candidates requiring consistency with particle ID and that the invariant masses be within the nominal values. To test the reconstruction procedure and reduce the systematic errors, the

TABLE 3. Results of the 4D maximum likelihood fit to the decay $B^- \to D_J^0 \pi^-$

B Decay Mode	Event Yield	$\mathcal{B}(B^- \to D_J^0 \pi^-)^a \times 10^{-3}$	$\mathcal{B}(B^-) \cdot \mathcal{B}(D_J^0) \times 10^{-4}$
$D_1^0(j_l = 1/2)\pi^-$	237.1 ± 42	$1.59 \pm 0.29 \pm 0.26 \pm 0.03 \pm 0.035$	$10.6 \pm 1.9 \pm 1.7 \pm 2.3$
$D_1(2420)^0\pi^-$	420.0 ± 41	$1.04^{+0.27}_{-0.21} \pm 0.17 \pm 0.02 \pm 0.07$	$6.9^{+1.8}_{-1.4} \pm 1.1 \pm 0.4$
$D_2^*(2460)^0\pi^-$	109.5 ± 26	$1.55 \pm 0.42 \pm 0.23 \pm 0.03 \pm 0.14$	$3.1 \pm 0.84 \pm 0.45 \pm 0.28$
$D^{*+}\pi^-\pi^-$	160.0 ± 61	$9.7 \pm 3.6 \pm 1.5 \pm 1.9$	
Total		$29.2 \pm 4.5 \pm 3.7 \pm 3.1$	

[a] We use the D_J branching fraction assumed in much of the literature $\mathcal{B}(D_1^0 \to D^{*+}\pi^-) = 2/3$ and $\mathcal{B}(D_2^{*0} \to D^{*+}\pi^-) = 0.2$

$D^{*\prime +}$ yields are compared to D_J^0 yields since the reconstruction procedures differ only by a single charged pion. Also, the $D^{*\prime +}$ to D_J^0 production ratio allows for a more direct comparison between the LEP and the CLEO results.

Using the entire CLEOII data set we have searched for the $D^{*\prime +}$ in the mass region suggested by the DELPHI results. We find no evidence of an excess the region between 2590 MeV and 2670 MeV and set a preliminary upper limit of:

$$\frac{N_{D^{*\prime +}} \cdot \mathcal{B}(D^{*\prime +} \to D^{*+}\pi^-\pi^+)}{N_{D_2^{*0}} \cdot \mathcal{B}(D_2^{*0} \to D^{*+}\pi^-) + N_{D_1^0} \cdot \mathcal{B}(D_1^0 \to D^{*+}\pi^-)} < 0.16 \text{ @ } 90\% \text{ C.L.} \quad (5)$$

This may be compared with the DELPHI measurement of $0.49 \pm 0.18 \pm 0.10$ for this rate [8]. The DELPHI result includes both $b\bar{b}$ and $c\bar{c}$ production.

Measurement of $D^0 \to K^+\pi^-$ Decays

The decay $D^0 \to K^+\pi^-$ can proceed either a through doubly Cabbibo suppressed decay (DCSD) channel or through $D^0 - \bar{D}^0$ mixing. In the standard model, the rate is expected at 0.3% level so this decay can be used to search for exotic or beyond-standard model decay mechanisms. Standard model predictions for R, defined as $R = \Gamma(D^0 \to K^+\pi^-)/\Gamma(D^0 \to K^-\pi^+)$, from mixing vary considerably from about 10^{-3} to 10^{-10}. The contributions to R from DCSD is of order $\tan^4\theta_C \sim 3 \times 10^{-3}$ [15]. To separate the mixing and DCSD contributions a measurement of the decay time distribution is required. With the new silicon vertex detector, CLEO can now perform this measurement, however, in the discussion that follows we present only an R measurement that includes contributions from both mixing and DCSD. We plan to eventually add the time-dependent measurement to this analysis.

To measure the ratio R we have to determine the decay rates $\Gamma(D^0 \to K^+\pi^-)$ and $\Gamma(D^0 \to K^-\pi^+)$. These rates are extracted from analyzing high momentum continuum $D^{*+} \to D^0\pi_s^+ \to (K\pi)\pi_s^+$ events, where the sign of the slow pion (π_s^+) tags whether the $K\pi$ combination comes from a D^0 or a $\bar{D}^0$. The combination where the sign of the K and the slow pion (π_s^+) are the same is referred to as the "wrong sign" combination while events where the K and the π_s^+ have opposite signs is referred to as the "right sign" combination. The wrong/right sign signal yields are determined by fitting the distribution of mass differences (δM) between the D^{*+} and the D^0 and requiring the D^0 invariant mass (M_D) to be within ± 13 MeV (2σ) of the nominal D^0 mass. An important feature in this analysis is the small width of both the M_D and δM mass distributions as compared to other experiments. This is due primarily to the improvements in the tracking algorithm and the vertex resolution of the SVX detector. For example, in the CLEOII/SVX data the resolution of M_D is now 6.5 MeV, while the δM resolution is 200 keV compared with the CLEOII pre re-processed (data reconstructed prior to the application of the Kalman algorithm) values of about 12 MeV and 750 keV respectively. The improved mass resolution allows for a greater separation of signal from backgrounds. This improvement is evident in the low levels of backgrounds in Figure 3.

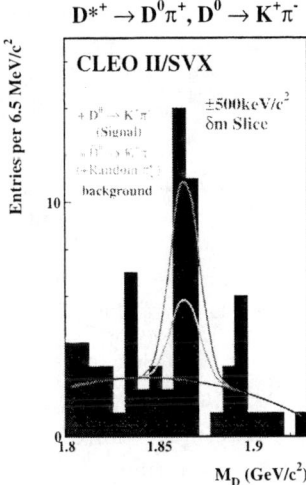 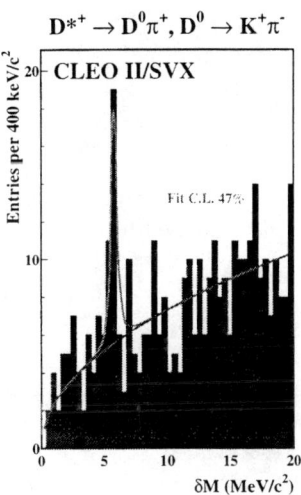

FIGURE 3. The wrong sign M_{D^0} and δM distributions with the results of the 1D-ML fit with the signal and background contributions superimposed on the data.

Because of the rarity of the $D^0 \to K^+\pi^-$ events, a significant amount of work has been done to both reject and understand the backgrounds which enter the M_D and δM distributions. The most significant background components are due to real $\bar{D}^0 \to K^+\pi^-$ combined with a random slow π_s^+ and backgrounds from $D^{*+} \to (K^+\pi^-)\pi_s^+$ where the kaon and the pion are mis-identified. The random slow π_s^+ background tends to peak in the D^0 mass region but not in δM, while the doubly mis-identified background tends to peak in δM but not in the D^0 mass. The latter is reduced by forming the $m_{\text{flp}}(D^0)$, where the mass assignments to the kaon and pion are switched, vetoing the event if it's "mass-flip" mass is within 4σ of the nominal D^0 mass. The remaining backgrounds are modeled by Monte Carlo simulations. The background distributions are used in the likelihood function for the 1D maximum likelihood fit.

The preliminary result of the 1D ML fit for the ratio of decay rates is:

$$\frac{\Gamma(D^0 \to K^+\pi^-)}{\Gamma(D^0 \to K^-\pi^+)} = 0.0032 \pm 0.0012 \pm 0.0015 \quad (6)$$

This result was obtained from an analysis of the current CLEOII/SVX dataset which includes both the off and on-resonance samples for a total of 3.8 fb^{-1}. The new result is already more statistically significant, by itself, than the current world average of 0.0072 ± 0.0025 [7]. There is currently about a factor of two more CLEOII/SVX data yet to be analyzed so we expect the statistical significance to improve. We are also working on improving the estimate of the systematic error and expected it to be much smaller than the value quoted in Equation (6). Finally, it is worth noting that while compatible, within errors, with the world average, the new R measurement is lower by about a factor of two and therefore more consistent with theoretical expectations.

Summary and Conclusions

We have presented preliminary results from five CLEO analysis in hadronic decays of bottom and charmed mesons. From the B hadronic analyses we have shown results which provide us with the first observation and measurements of the decay $B^0 \to D^{*-}D^{*+}$, a first hint of final state interactions in B decays and a measurement of the decay rates of the B^- meson decaying into three of the $L = 1$ charmed mesons plus a single pion. From the $B^- \to D_J^0 \pi^-$ analysis we have the first observation of the broad $L = 1, j_l = 1/2$ charmed meson and have determined its mass and decay width. Our preliminary search for the first radial excitation of the charmed meson (the $D^{*\prime +}$) has been unable to confirm the observation by the DELPHI collaboration. Finally, our preliminary results on a measurement of the ratio $\Gamma(D^0 \to K^+\pi^-)/\Gamma(D^0 \to K^-\pi^+)$ gives a results lower than previous measurements. This new measurement is an improvement over the previous CLEO results with a substantial increase in data and improved detector.

REFERENCES

1. Y. Kubota, et al., (CLEO Collaboration), Nucl. Inst. Meth., **A320**, 66 (1992)
2. J. Alexander, et al., Nucl. Inst. Meth., **A326**, 243 (1993)
3. D. E. Jaffe, et al., (CLEO Collaboration), CLEO conference report CLEO CONF98-07/ICHEP98-849 (1998). Submitted to Phys. Rev. Lett.
4. G. Kramer, T. Mannel, and W.F. Palmer, Z. Phys., **C55**, 497 (1992)
5. A. Dighe, I. Dunietz and R. Fleischer, CERN preprint CERN-TH-98-85/hep-ph/9804253
6. G. Bonvicini et al., (CLEO Collaboration), CLEO conference report CLEO CONF98-23/ICHEP9-852 (1998)
7. C. Caso et al., (Particle Data Group), Eur. Phys. J., **C3** (1998)
8. P. Abreu et al., (DELPHI Collaboration), Phys. Lett. B, **426**, 231 (1998)
9. A description of the partial reconstruction technique as applied to a similar mode can be found in G. Brandenburg et al., (CLEO Collaboration), Phys. Rev. Lett., **80**, 2762 (1998)
10. J. Gronberg et al., (CLEO Collaboration),CLEO conference report CLEO CONF96-25/ICHEP96 PA05-69 (1996)
11. M. Neubert, Phys. Lett. B, **418**, 173 (1998): hep-ph/9709327
12. S. Godfrey and R. Kokoski, Phys. Rev. D, **43**, 1679 (1991)
13. S. Godfrey and N. Isgur, Phys. Rev. D, **32**, 189 (1985)
14. OPAL Collaboration, OPAL conference report OPAL PN352/ICHEP98-1037 (1998)
15. D. Cinabro et al., (CLEO Collaboration) Phys. Rev. Lett., **72**, 1406 (1994)

Review of Charm and Beauty Lifetimes

Harry W. K. Cheung[1]

Fermi National Accelerator Laboratory. P. O. Box 500, Batavia, IL 60510-0500, U.S.A.

Abstract. A review of the latest experimental results on charm and beauty particle lifetimes is presented together with a brief summary of measurement methods used for beauty particle lifetime measurements. There have been significant updates to the D_s^+/D^0, B^+/B_d^0 and Λ_b^0/B_d^0 lifetime ratios which have some theoretical implications. However more precise measurements are still needed before one can make conclusive statements about the theory used to calculate the particle lifetimes.

INTRODUCTION

Motivation

The study of the charm and beauty particle lifetimes is broadly motivated by two main goals. The first is to convert relative branching fractions to partial decay rates and the second is to learn about the strong interaction.

Experimental data on decays are normally obtained by measuring decay fractions, e.g. $\Gamma(D^0 \to K^-\pi^+)/\Gamma(D^0 \to X)$, whereas theory calculates the partial decay rate, e.g. $\Gamma(D^0 \to K^-\pi^+)$. The lifetime of the particle, $\tau = \hbar/\Gamma(D^0 \to X)$, is needed in order to convert the experimentally measured decay fractions into decay rates. Not only does this allow tests of theoretical predictions but it also enables the extraction of Standard Model parameters if the theoretical calculations are reliable, e.g. a comparison of B semileptonic decay rates may allow the extraction of $|V_{cb}|$ and $|V_{ub}|$.

The second motivation for the study of lifetimes is that they are interesting in their own right. They allow us to learn more about the "Theoretically-Challenged" part of the Standard Model, *i.e.* non-perturbative QCD. This is one of the few areas of the Standard Model where experimental data and theoretical ideas closely interact. Although this has been touted as an area of the Standard Model not worth pursuing since a lot more theoretical understanding is needed before tests of the Standard Model can be made, it is also one area that is intellectually interesting.

---

[1] This work was supported by the Fermi National Accelerator Laboratory, which is operated by the Universities Research Association, Inc., under contract DE-AC02-76CHO3000 with the U.S. Department of Energy

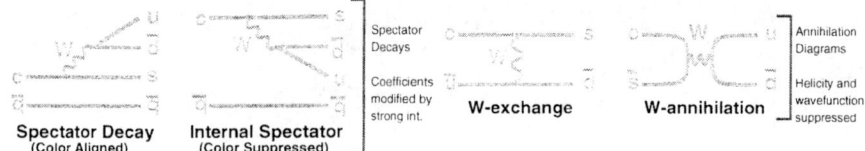

FIGURE 1. Hadronic decay diagrams for charm meson decays.

For example, even though we have some models, we have little idea about exactly *how* quarks turn into hadrons. In my view this *is* new physics since it is beyond what the Standard Model can do right now. Calculations using Lattice QCD are only just now being used to study the *dynamics* of decays and may start producing reliable results [1].

Decay Diagrams

The lifetime of a particle is given by the following expression:

$$\tau = \frac{\hbar}{\Gamma_{SL} + \Gamma_{NL} + \Gamma_{PL}} \quad (1)$$

where Γ_{SL} is the semileptonic decay rate, (*e.g.* $\Gamma(D^+ \to \ell^+ \nu_\ell X)$), Γ_{NL} is the non-leptonic or hadronic decay rate, (*e.g.* $\Gamma(D^+ \to \text{hadrons})$), and Γ_{PL} is the purely leptonic decay rate, (*e.g.* $\Gamma(D^+ \to \ell^+ \nu_\ell)$). Compared to the total rate, the purely leptonic decay rate is normally very small due to helicity suppression.[2] In addition current data for D meson decays indicates that the semileptonic rate for D^+ and D^0 are equal within at least about 10% if not better.[3] This means that the large difference between the observed D^+ and D^0 lifetimes ($\tau(D^+)/\tau(D^0) = 2.55 \pm 0.04$) is due to a large difference in the hadronic decay rates for the D^+ and the D^0. Thus in contrast to the spectator model [4] which has only the free charm quark decay diagram and predicts equal D^+ and D^0 lifetimes, we need to take into account spectator quark effects. This entails taking into account other decay diagrams like those in Figure 1 and any interferences between them.

The conventional wisdom used to explain the smaller hadronic width of the D^+ relative to the D^0 is that in the D^+ Cabibbo allowed decays ($c\bar{d} \to s(u\bar{d})\bar{d}$), there

---

[2] The B and D mesons both have spin 0 so that in the decay, the resulting lepton (anti-lepton) and anti-neutrino (neutrino) must *both* be either left-handed or *both* right-handed in order to conserve angular momentum. However the $V - A$ nature of the weak interaction requires left-handed particles and right-handed anti-particles [2].

[3] The semileptonic decay rate is given by the ratio of the semileptonic branching ratio to the lifetime. Using the world average values for these compiled by the Particle Data Group [3], $\Gamma_{SL}(D^+) = (1.071 \pm 0.119) \times 10^{-13}$ GeV and $\Gamma_{SL}(D^0) = (1.067 \pm 0.041) \times 10^{-13}$ GeV.

exist identical quarks in the final state unlike for D^0, so there are additional (destructive) interference contributions for the D^+. Or, if we are talking about exclusive rather than inclusive decays, one can view the interference as that between the external spectator and internal spectator decay diagrams of Figure 1 which can lead to the same exclusive final state. It is relatively easy to show that the additional interference for inclusive hadronic decays for D^+ is destructive and can lead to a lifetime ratio of $\tau(D^+)/\tau(D^0) \sim 2.0$. However it is difficult to determine exactly how large a ratio of $\tau(D^+)/\tau(D^0)$ interference effects can accomodate and therefore how large is the additional contribution of Cabibbo allowed W-exchange decays in the D^0. Cabibbo allowed W-exchange decay is expected to contribute to lowering the D^0 lifetime but this contribution is wavefunction and helicity suppressed ($\sim \frac{|f_d|^2 m_s^2}{m_c^4}$) and is difficult to calculate reliably.

Clearly a better understanding of both charm and beauty inclusive decays is necessary. Experimental data on lifetimes from all the charm and beauty particles will allow us to learn more about how they decay and in turn use the data to extract standard model parameters like quark masses and the CKM matrix elements $|V_{cs}|$, $|V_{cd}|$, $|V_{cb}|$ and $|V_{ub}|$.

Theoretical Overview for Inclusive Decays

A systematic approach now exists for the treatment of inclusive decays that is based on QCD and consists of an Operator Product Expansion in the Heavy Quark Mass [5]. In this approach the decay rate is given by:

$$\Gamma_{H_Q} = \frac{G_F^2 m_Q^5}{192\pi^3} \Sigma f_i |V_{Qq_i}|^2 \left[A_1 + \frac{A_2}{\Delta^2} + \frac{A_3}{\Delta^3} + \frac{A_4}{\Delta^4} + ... \right] \quad (2)$$

where the expansion parameter Δ is often taken as the heavy quark mass and f_i is a phase space factor. $A_1 = 1$ gives the spectator model term and the A_2 term produces differences between the baryon and meson lifetimes. The A_3 term includes the non-spectator W-annihilation and Pauli interference effects. For mesons this term can be related to certain observables whereas for baryons particular quark models or QCD sum rules are needed to determine the parameters fully. Though scaling formally as $1/\Delta^3$, these non-spectator terms actually scale like f_Q^2/M_Q^2 and thus predict that the lifetime differences for the beauty particles should be about 10% of those seen in charm.

A theoretical review is outside the scope of this article and the reader is referred to other reviews [5].

REVIEW OF EXPERIMENTAL RESULTS

There have been new measurements of charm and beauty lifetimes since the 1998 review performed by the PDG [3]. Some are results published in journals while others were presented at conferences this year. Given the title of this Workshop, it is interesting to note that all the measurements of the B particle lifetimes come from collider experiments whereas all the ones for charm come essentially only from fixed target experiments.

In the past and even today, as well as lifetime measurements of the different species of B particles, measurements are often given for admixtures of B hadrons, either B^+/B_d^0, $B^+/B_d^0/B_s^0/b$-baryon or b-baryons. Given how well the lifetimes of the different species of B particles are now measured I think reviewing the measurements of the admixtures no longer makes sense from a physics standpoint. Even the measurement for b-baryons is not so much more precise than that for Λ_b^0, and the b-baryons sample contains a large contamination of Ξ_b with unknown lifetime. In the same vein I shall not include lifetime ratios derived from ratios of branching fractions as these involve additional assumptions and also because they do not significantly change the world average values.

Measurements of Beauty Particle Lifetimes

Methods used for the measurement of lifetimes of the B mesons can be broadly divided into three classes. The B meson can be completely reconstructed ("*Exclusive*"); the B meson decay vertex is reconstructed but its momentum is only partially reconstructed ("*Inclusive 1*"); and neither the B meson vertex not its momentum is completely reconstructed ("*Inclusive 2*"). It is interesting to separate the measurements into these three different classes as they have different systematic uncertainties and any differences may reflect effects which are not sufficiently understood.

Measurements of B^+ and B_d^0 Lifetimes

Exclusive reconstruction is performed by the CDF and ALEPH collaborations. CDF uses B meson decays to $J/\psi K$, $J/\psi K^*$, $\psi' K$ and $\psi' K^*$ [6], whereas ALEPH looks for B decays to $J/\psi K$, $J/\psi K^*$ and D, D^* plus charged pions, (either direct pions or pions from the decay of the ρ or a_1) [7]. Measurements are still very much statistics limited since the B meson is completely reconstructed. The dominant systematics in the measurements are from non-Gaussian tails, possibly from misreconstruction, and uncertainties in the background lifetime distribution.

The *Inclusive 1* reconstruction involves reconstruction of semileptonic decays $B_d^0 \to D^{*+}\ell^- X$, $B_d^0 \to D^+\ell^- X$ and $B^+ \to D^0\ell^- X$, where the D^0 is explicitly tested to make sure it is not compatible with being from $D^{*+} \to D^0\pi^+$. However there is still contamination or dilution of the B^+ sample from B_d^0 decays and vice-versa. For example, the $D^{(*)+}\ell^- X$ sample can contain 10–25% B^+ decays, due to $B^+ \to D^{**0}$, $D^{**0} \to D^{(*)+}\pi^-$. The B^+ and B_d^0 lifetimes are obtained by a simultaneous fit to the $D^{(*)+}\ell^- X$ and $D^0\ell^- X$ samples. Since the lifetimes of the B^+ and B_d^0 are so close together the B compositions of the samples are constrained

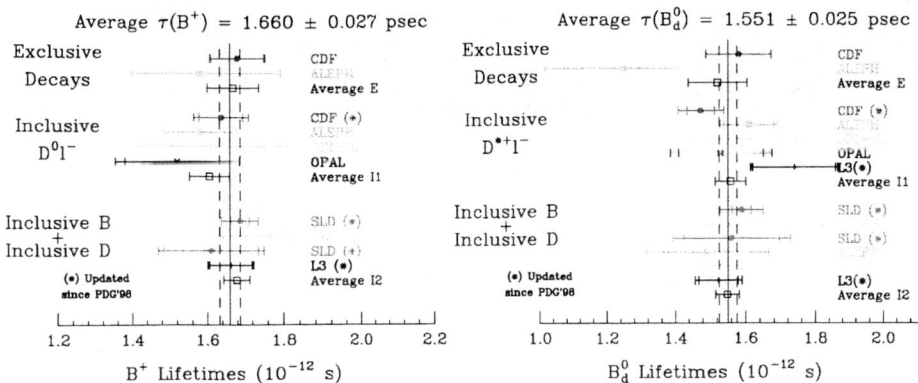

FIGURE 2. Lifetime measurements for B^+ and B_d^0 split into the three measurement method classes described in the text.

to be within the uncertainties of measurements for the various branching fractions needed in the calculation of the compositions. In additional, although the B vertex is obtained from the intersection of the D momentum vector and the lepton, the B momentum is not completely determined because of the unobserved neutrino in the B semileptonic decay. A Monte Carlo simulation is used to correct for this effect. The dominant systematic uncertainties for this type of measurement are the uncertainties in the composition and the corrections for the B momentum as well as uncertainties in the background. Measurements using this method have been done by the DELPHI [8] and OPAL [9] collaborations including using the $D^+\ell^-X$ sample, whereas the CDF [10] and ALEPH [7] collaborations use only the $D^{*+}\ell^-X$ and $D^0\ell^-X$ samples. L3 only uses their $D^{*+}\ell^-X$ sample in a (single) fit to obtain the B_d^0 lifetime [11].

In the *"Inclusive 2"* method the B meson decay vertex is not fully reconstructed. This includes a topological/vertex-charge method as well as methods based on the correlations of the B and subsequent charm decay products. The SLD [12], DELPHI [13] and L3 [14] collaborations use a topologcal method where a well separated (smeared B-C) secondary vertex is reconstructed from charged tracks to help select out a B particle decay and the vertex charge is used to tag whether the decay is from the B^+ or from the B_d^0. The SLD group also includes an event sample where a lepton is part of the secondary vertex to help further discriminate between B particle decays and background. Of the B particle decays the vertex charge selects out only about 55-77% of the correct B meson type. The fit for the B_d^0 and B^+ lifetimes are obtained by simultaneosuly fitting the charged and neutral vertex samples and with the help of Monte Carlo simulations. Since the composition has to be modeled as well as the smearing for both the B decay length and momentum, the measurement of the B lifetimes are already systematics limited for the measurements with the smallest statistical errors [12]. However the measurement of the B^+/B_d^0 lifetime ratio is not yet systematics limited. The other measurements

TABLE 1. Summary of B lifetime measurements split by experiment

Experiment	$\tau(B^+)$ ps	$\tau(B_d^0)$ ps	$\tau(B^+)/\tau(B_d^0)$
CDF[a]	1.661 ± 0.052	1.513 ± 0.053	1.091 ± 0.050
SLD[a]	1.678 ± 0.046	1.586 ± 0.057	1.059 ± 0.039
L3[a]	1.662 ± 0.061	1.571 ± 0.058	1.090 ± 0.076
ALEPH	1.580 ± 0.095	1.550 ± 0.067	1.030 ± 0.082
DELPHI	1.700 ± 0.090	1.548 ± 0.051	1.050 ± 0.100
OPAL	1.520 ± 0.166	1.530 ± 0.144	0.990 ± 0.147
World Average	1.660 ± 0.027	1.551 ± 0.025	1.065 ± 0.026

[a] Updated since PDG98

included in this class are from ALEPH [7] and DELPHI [13]. ALEPH used $\pi_B^- \pi_D^+$ correlations in $\overline{B_d^0} \to \pi_B^- D^{*+} X$ ($D^{*+} \to D^0 \pi_D^+$), and DELPHI used $\pi_D^+ \ell^-$ correlations in $\overline{B_d^0} \to D^{*+} \ell^- X$ ($D^{*+} \to D^0 \pi_D^+$) to select out a signal and to fit for the B_d^0 lifetime.

The measurements for $\tau(B^+)$, $\tau(B_d^0)$ and $\tau(B^+)/\tau(B_d^0)$ split into the three measurement method classes are shown in Figure 2. One can conclude that no systematic differences between the measurement methods exist to within the current uncertainties. The measurements split by collaboration are summarized in Table 1. Note that each collaboration already did an excellent job of calculating their own averages taking into account the correlations of their measurements using different methods. The LEP B Lifetime Working Group also calculates a world average as well as a LEP average. Using almost the same measurements as input they obtain the following world averages: $\tau(B^+) = 1.67 \pm 0.03$, $\tau(B_d^0) = 1.57 \pm 0.03$ ps and $\tau(B^+)/\tau(B_d^0) = 1.07 \pm 0.03$ [15].

The lifetime for the B^+ is now measured to be 2.6σ higher than for the B_d^0, which is becoming more significant.

Measurements of the B_s^0, Λ_b^0 and B_c^+ Lifetimes

Just as for the B^+ and B_d^0 lifetime measurements, the methods for measuring the B_s^0 lifetime can be split into the three measurement method classes described previously. Only CDF uses a fully reconstructed mode, $B_s^0 \to J/\psi \phi^0$ [6]. For the *Inclusive 1* method, OPAL [16], ALEPH [17], CDF [18] and DELPHI [19] uses $\bar{B}_s^0 \to D_s^+ \ell^- X$ and ALEPH [20] and DELPHI [19] also use $\bar{B}_s^0 \to D_s^+ h^-$ where h is one (or more) hadrons. For the *Inclusive 2* where the B_s^0 vertex is not reconstructed we have measurements from OPAL [21] and DELPHI [19] using $\bar{B}_s^0 \to D_s^+ X$ and DELPHI includes $\bar{B}_s^0 \to \phi^0 \ell^-$. These measurements are shown in Figure 3(a) and summarized in table 2.

The Λ_b^0 lifetime has been measured using $\Lambda_b^0 \to \Lambda_c^+ \ell^- X$ where the Λ_c^+ decay is fully reconstructed via various decay modes. The sample typically contains 70-90% Λ_b^0 with a 1% contamination from Ξ_b. There has only been one minor update

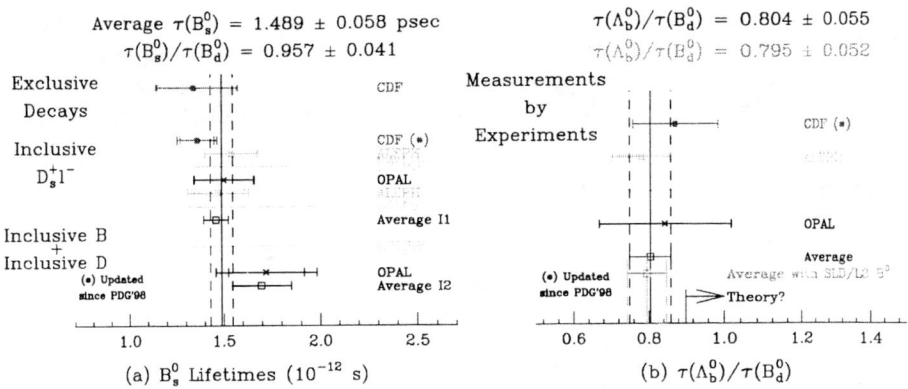

FIGURE 3. Lifetime measurements for (a) B_s^0 split by measurement methods as described in the text; and (b) $\tau(\Lambda_b^0)/\tau(B_d^0)$ split by experiment.

TABLE 2. Summary of B_s^0 and Λ_b^0 lifetime measurements split by experiment

Experiment	$\tau(B_s^0)$ ps	$\tau(\Lambda_b^0)$ ps	$\tau(\Lambda_b^0)/\tau(B_d^0)$
CDF[a]	1.356 ± 0.094	1.320 ± 0.166	0.872 ± 0.114
OPAL	1.570 ± 0.140	1.290 ± 0.238	0.843 ± 0.175
ALEPH	1.510 ± 0.110	1.210 ± 0.110	0.781 ± 0.079
DELPHI[b]	1.670 ± 0.140	1.170 ± 0.195	0.756 ± 0.128
World Average	1.489 ± 0.058	1.237 ± 0.078	0.804 ± 0.055[c]

[a] Updated since PDG98 for $\tau(B_s^0)$
[b] Updated since PDG98 for $\tau(\Lambda_b^0)$
[c] 0.795 ± 0.052 including other measurements for $\tau(B_d^0)$.

from DELPHI [22] since PDG98. The results are summarized in Table 2 and the results for $\tau(\Lambda_b^0)/\tau(B_d^0)$ are shown in Figure 3(b) compared to the theoretical limit. Although $\tau(\Lambda_b^0)/\tau(B_d^0)$ is small compared to theory it is still within 2σ of the theoretical limit.

CDF has observed the B_c^+ meson with a lifetime of $0.46^{+0.18}_{-0.16} \pm 0.03$ ps [23].

Measurements of Charm Particle Lifetimes

The World average lifetimes for the weakly decaying charm particles are dominated by measurements from Fermilab E687 published in 1993-1995. However there have been a few updates this year. The Fermilab E687 collaboration has published an new value for the lifetime of $\Xi_c^+ = 0.34^{+0.07}_{-0.05} \pm 0.02$ ps using an additional decay mode [24]. This supercedes their earlier published result. Direct measurements of charm particle lifetimes using the CLEO 2.5 experiment has been reported at conferences this year: $\tau(D^+) = 1.034 \pm 0.033^{+0.033}_{-0.038}$ ps [25], $\tau(D_s^+) =$

$0.475 \pm 0.024 \pm 0.025$ ps [25] and a more updated $\tau(D^0) = 0.410 \pm 0.006 \pm 0.005$ ps [26]. Although their new silicon tracker enables them to measure lifetimes to a precision rivaling the fixed target E687 results, the next generation fixed target experiment FOCUS will be overwhelming with a huge sample of fully reconstructed charm decays [27]. Results have also been presented by Fermilab E791 for $\tau(D^0) = 0.413 \pm 0.003(stat)$ ps [28] and $\tau(D_s^+) = 0.518 \pm 0.014 \pm 0.007$ ps [29].

The most significant consequence of these new measurements is that $\tau(D_s^+)$ is conclusively larger than $\tau(D^0)$, since the world average is now $\tau(D_s^+)/\tau(D^0) = 1.193 \pm 0.027$ compared to the earlier PDG98 value of 1.125 ± 0.042.

LIFETIMES AND THEORY

Λ_b^0, B^+ and B_d^0 Lifetimes

The smallest of the Λ_b^0 lifetime compared to the B_d^0 is often cited as a problem. In fact the measured ratio of $\tau(\Lambda_b^0)/\tau(B_d^0)$ is only 2σ from the theoretical limit of 0.9 given by Bigi [30]. Furthermore it was pointed out recently by Neubert and Sachrajda that if they use the same theoretical approach but without model dependent constraints on the parameters of the mass expansion, they can obtain ratios as low as 0.8 for $\tau(\Lambda_b^0)/\tau(B_d^0)$ and also the sign of $\tau(B^+) - \tau(B_d^0)$ is not determined [31]. Given that the measurements now indicate $\tau(B^+) > \tau(B_d^0)$ their results would point to a larger theoretical limit than 0.8. However more precise lifetime measurements are still needed.

In particular measurements to convincingly show that $\tau(B^+) > \tau(B_d^0)$ is interesting since studies of B^+ exclusive decay modes give the same sign for the external and internal spectator diagrams [32] which would suggest constructive interference in B^+ decays which could lead to a shorter lifetime than for the B_d^0.

Other theoretical approaches have been reported to explain the shortness of the Λ_b^0 lifetime. Datta, Lipkin and O'Donnell have shown that the Λ_b^0 lifetime can be shortened by phase space enhancement through isospin conservation [33]. Other authors have used the mass expansion approach but with the hadron mass instead of the heavy quark mass [34], or with the energy release instead of the quark mass as the expansion parameter [35].

D^0 and D_s^+ Lifetimes

The D_s^+ lifetime is now conclusively measured to be above the D^0 lifetime. However exactly how much larger is still not measured that precisely. A more accurate measurement is needed and may tell us much more about the nature and size of the W-annihilation contribution to the inclusive decay. Bigi and Uraltsev have used the mass expansion approach to analyze this lifetime difference and have concluded that a difference of $< 7\%$ is possible without WA contributions. Their estimation of the WA contribution puts an upper limit of $\tau(D_s^+)/\tau(D^0) < 1.20$ [36].

CONCLUSIONS

The most significant updates are that the D_s^+ lifetime is now conclusively measured to be above the D^0 lifetime, $\tau(D_s^+)/\tau(D^0) = 1.193 \pm 0.027$, however more precise measurements are needed to really study the size of weak annihilation effects. The B_c^+ meson has been observed with a lifetime of $0.46^{+0.18}_{-0.16} \pm 0.03$. The measured values of $\tau(B^+)/\tau(B_d^0) = 1.068 \pm 0.026$ and $\tau(\Lambda_b^0)/\tau(B_d^0) = 0.795 \pm 0.052$ are becoming more precise but still more accurate measurements are needed for both of these ratios. These ratios taken together, if more precisely measured, can indicate in a model independent fashion whether there is really a contradiction between theory and the measurements.

We can look forward to much more precise charm particle lifetimes from FOCUS [27] within a year and the future for new beauty particle lifetime measurements is also very bright with several new experiments ready soon or within a few years to take data (BELLE, BABAR, HERA-B, CDF and D0).

The study of charm and beauty lifetimes should continue to be an exciting field.

REFERENCES

1. Kronfeld, A., these proceedings.
2. Aitchison, I. J. R., and Hey, A. J. G., *Gauge Theories in Particle Physics*, 2nd ed., London: IOP Publishing Ltd., 1989, ch. 10, pp. 347-356.
3. Caso, C., et al., PDG, *Eur. Phys. J.* **C 3**, 1-794 (1998).
4. Gaillard, M. K., Lee B. W., and Rosner, J. L., *Rev. Mod. Phys.* **47** 277-310 (1975); Altarelli, G., Cabibbo, N., and Maiani, L., *Phys. Rev. Lett.* **35**, 635-638 (1975); Cabibbo, N., and Maiani, L., *Phys. Lett.* **B 73**, 418-422 (1978) and references therein.
5. Bigi, I. I., et al., "Non-leptonic Decays of Beauty Hadrons - From Phenomenology to Theory", in Stone, S. (ed.), *B Decays*, revised 2nd ed., Singapore: World Scientific, 1994, pp. 132-157, and references therein; Cheung, H. W. K., "The Physics of Charm Lifetimes", in *Proceedings of the 8th Meeting of the Division of Particles and Fields of the American Physical Society, Albuquerque, New Mexico*, Aug. 2-6, 1994, pp. 517-520, and references therein; Bellini, G., Bigi, I., and Dornan, P. J., *Phys. Rept.* **289**, 1-155 (1997).
6. Abe., F., et al., CDF Collaboration, *Phys. Rev.* **D 57**, 5382-5401 (1998).
7. Buskulic, D., et al., ALEPH Collaboration, *Z. Phys.* **C 71**, 31-44 (1998).
8. Abreu, P., et al., DELPHI Collaboration, *Z. Phys.* **C 68**, 13-23 (1995).
9. Akers, R., et al., OPAL Collaboration, *Z. Phys.* **C 67**, 379-388 (1995).
10. Abe, F., et al., CDF Collaboration, *Phys. Rev.* **D 58**, 092002, 12 pages (1998).
11. L3 Collaboration, "Measurement of the B_d^0 Meson Lifetime using the decay $\bar{B}_d^0 \to D^{*+}\ell^-\bar{\nu}$", presented at the 29th Int. Conf. on High Energy Phys., Vancouver, Canada, July 23-29, 1998. L3 Note 2142.
12. Abe, K., et al., SLD Collaboration, "Measurement of the B^+ and B_d^0 Lifetimes using Topological Vertexing at SLD", presented at the 29th Int. Conf. on High Energy Phys., Vancouver, Canada, July 23-29, 1998. SLAC Preprint SLAC-PUB-7868.

13. Adam, W., et al., DELPHI Collaboration, Z. Phys. **C 68**, 363-374 (1995).
14. L3 Collaboration, "Upper Limit on the Lifetime Difference of Short- and Long-lived B_s^0 Mesons", presented at the 29th Int. Conf. on High Energy Phys., Vancouver, Canada, July 23-29, 1998. L3 Note 2281.
15. Alcaraz, J., et al., LEP B lifetimes working group, See the latest writeup at the following URL http://home.cern.ch/~claires/lepblife.html.
16. Ackerstaff, K., et al., OPAL Collaboration, Phys. Lett. **B 426** 161-179 (1998).
17. Buskulic, D., et al., ALEPH Collaboration, Phys. Lett. **B 377** 205-221 (1996).
18. Abe, F., CDF Collaboration, "Measurement of the B_s^0 Meson Lifetime Using Semileptonic Decays", Fermilab preprint FERMILAB-Pub-98/172-E, August 1998.
19. Abreu, P., et al., DELPHI Collaboration, Z. Phys. **C 71**, 11-30 (1996).
20. Barate, R., et al., ALEPH Collaboration, Eur. Phys. J. **C 4** 367-385 (1998).
21. Ackerstaff, K., et al., OPAL Collaboration, Eur. Phys. J. **C 2** 407-416 (1998).
22. Asman, B., et al., DELPHI Collaboration, "Measurement of the lifetime of b-baryons", presented at the 29th Int. Conf. on High Energy Phys., Vancouver, Canada, July 23-29, 1998. DELPHI 98-72 CONF 140.
23. Abe, F. et al., CDF Collaboration, Phys. Rev. **D 58**, 112004, 29 pages (1998); Phys. Rev. Lett. **81**, 2432-2437 (1998).
24. Frabetti, P. L., et al., E687 Collaboration, Phys. Lett. **B 427**, 211-216 (1998).
25. Drell, S., et al., CLEO Collaboration, "Precise Measurements of Charm Meson Lifetimes Using the New CLEOII Silicon Vertex Detector", presented at the 32nd Rencontres de Moriond, Les Arcs, France, Mar. 21-28, 1998.
26. Richichi, S. J. et al., CLEO Collaboration, "Precise Measurements of the D^0 Meson Lifetime with the CLEOII.V Silicon Vertex Detector", presented at the 29th Int. Conf. on High Energy Phys., Vancouver, Canada, July 23-29, 1998.
27. FOCUS Collaboration results presented by Johns, W., Link, J. and Vaandering, E. in these proceedings.
28. Kwan, S. et al., E791 Collaboration, "Measurement of a Limit on the $D^0 \bar{D}^0$ Mixing Rates due to Lifetime Difference", presented at the 29th Int. Conf. on High Energy Phys., Vancouver, Canada, July 23-29, 1998.
29. E791 Collaboration results presented by Copty, N. in these proceedings.
30. Bigi, I. I., "Heavy Quark Expansions for Inclusive Heavy Flavor Decays and the Lifetimes of Charm and Beauty Hadrons," in *Proceedings of the 1996 Workshop on Heavy Quarks at Fixed Target, St. Goar, Germany*, Oct. 3-6, 1996, pp. 131-157.
31. Neubert, M., "Theory of Beauty Lifetimes", presented at the 2nd Int. Conf. on B Physics and CP Violation, Honolulu, Hawaii, Mar. 24-27, 1997. Neubert, M. and Sachrajda, C. T., Nucl. Phys. **B 483**, 339-370 (1997).
32. Honscheid, K., in Caso, C., et al., PDG, Eur. Phys. J. **C 3**, pp. 522-533 (1998).
33. Datta, A., Lipkin, H. J. and O'Donnell, J. O., "Implications of Isospin Conservation in Λ_b^0 Decays and Lifetime", Univ. of Toronto preprint, UTPT-98-15, Sept. 1998.
34. Cheng, H. Y., Phys. Rev. **D 56**, 2783-2798 (1997). Cheng, H. Y., and Yang, K. C., "Nonspectator Effects and B Meson Lifetimes from a Field-theoretical Calculation", Taiwan Inst. Phys. preprint IP-ASTP-03-98, May 1998.
35. Ito, T., Matsuda, M. and Matsui, Y. Prog. Theor. Phys. **99**, 271-280 (1998).
36. Bigi, I. I., and Uraltsev, N. G., Z. Phys. **C 62**, 623-632 (1994).

PART V
SEMILEPTONIC AND LEPTONIC DECAYS

Hyperon Physics Results from SELEX

Ivo Eschrich

Max Planck Institute for Nuclear Physics, Heidelberg, Germany

on behalf of the SELEX Collaboration

Abstract. In parallel to charm hadroproduction the experiment SELEX (E781) at Fermilab is pursuing a rich hyperon physics program. SELEX employs a 600 GeV/c beam consisting of 50 % Σ^- and π^- each. The three-stage magnetic spectrometer covering $0.1 \leq x_F \leq 1.0$ features a high-precision silicon vertex system, broad-coverage particle identification using TRD and RICH, and a three-stage lead glass photon calorimeter. First results for the Σ^- charge radius, total Σ^-- nucleon cross sections, and a new upper limit for the radiative width of the $\Sigma(1385)^-$ are presented.

I INTRODUCTION

In parallel to charm hadroproduction the hyperon beam experiment SELEX (E781) at Fermilab is pursuing a rich hyperon physics program. One year after the end of the 1997 fixed target run, first results are available.

The apparatus has been described already elsewhere in these proceedings [1]. Ongoing projects include the measurement of hyperon charge radii by hyperon-electron elastic scattering, the Σ^--proton total cross section, Primakoff production of hyperon resonances, Σ^+ production polarization, and the search for exotic particles produced by hyperon-induced reactions to name a few examples. The hyperon physics program is complemented by analysis of the data taken with π^- and proton beams, which has also yielded first results [2].

II CHARGE RADII

Hadrons as we understand them today are composite systems which we characterize by their static properties. One static property which reflects the phenomenon unique to hadrons – quark confinement – is the size of the particle.

Elastic scattering of an electron off a charged hadron is modified from a point interaction by the form factor $F(Q^2)$ where Q^2 is the four-momentum transfer squared. At zero momentum transfer the mean squared charge radius is related to the slope of the form factor by

$$\langle r^2 \rangle = -6\hbar^2 \frac{dF(Q^2)}{dQ^2}\bigg|_{Q^2=0}.$$

Charge radii are known only for five different hadrons so far. The fact that the K^- radius has been found to be smaller than that of the π^- by $\sim 0.1\,\text{fm}^2$ suggests that the size of a hadron is related to the flavor composition of its constituent quarks. There is supporting evidence from a study of strong interaction radii [3] which finds that replacing an *up* or *down* quark by a *strange* quark in a baryon decreases its radius by approximately $0.08\,\text{fm}^2$. Consequently the Σ^- radius should be smaller than the proton radius, and larger than the Ξ^-. The definition of a strong-interaction radius, however, is model-dependent. The significance of the above observation is therefore limited unless validated by a systematic study of hyperon charge radii.

For hadron-electron elastic scattering, two hits in the negative and none in the positive half of a hodoscope downstream of the second magnet in coincidence with a multiplicity of two in a set of scintillation counters 3 cm downstream of the target constituted a valid trigger condition. The typical trigger rate at this level was 3000 per 20-second spill at a beam rate of 10^7 particles per spill. An online filter performed a preliminary track reconstruction in the M2 spectrometer. Requiring at least one track with negative and none with positive slope together with other conditions crucial to a complete reconstruction reduced this sample by a factor of 1:1.7.

In the 1997 run SELEX has recorded 215 million candidates for hadron-electron scattering with the Σ^-/π^--beam. In preparation for a first analysis with the software tools available at that time the negative-beam sample was stripped to 10 % of its original size by cutting on an electron signature, unambiguous identification of the beam particle, and a two-negative-track event topology. A second-stage strip required a two-prong vertex, again reducing the sample by a factor of 10.

Out of the stripped data sample described above, 12,000 Σ^-- electron and 26,000 π^-- electron elastic scattering events were extracted. For each event, the incoming and outgoing tracks in the vertex were required to be coplanar. Particle identification for the two outgoing tracks was performed by combining information from he transition radiation detectors with kinematic constraints. Events with ambiguous particle identification were discarded. For Σ^-, decays upstream of the M2 chambers were rejected by requiring the scattered beam particle to have at least 60 % of the incoming beam particle's momentum. Finally, electron momentum and scattering angle had to match their expected kinematic relation to better than 10 %.

The charge radii were determined by fitting the differential cross section with an assumed radius as the single parameter to the observed distribution of the four-momentum transfer squared Q^2 (Fig. 1). Since the shape of the Q^2-distribution yields the radius no absolute normaliziation is needed. In this first analysis, Q^2 was calculated from the beam momentum and the scattering angle of the electron.

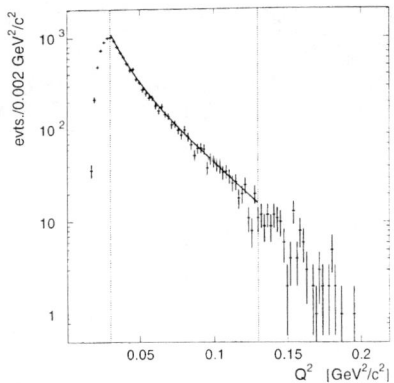

FIGURE 1. Q^2 distribution of Σ^--electron scattering events. Vertical lines indicate the region accepted for fitting.

From Monte Carlo studies the Q^2 resolution was estimated to be 1.5 %. Preliminary acceptance studies were performed using generated elastic scattering events embedded in real data. The geometrical and reconstruction-dependent acceptance was modeled and a preliminary evaluation of the trigger efficiency performed.

For the Σ^- data, a Q^2 region with flat acceptance was chosen for fitting the radius. For the π^- data, an acceptance correction was applied. Each event was normalized to its individual beam momentum to eliminate effects of the beam momentum spread. An unbinned maximum likelihood fit using dipole electric and magnetic form factors for the Σ^- yields a mean squared charge radius of

$$\langle r^2 \rangle_{\Sigma^-} = 0.60 \pm 0.08\,(stat.) \pm 0.08\,(syst.)\,\text{fm}^2$$

in the Q^2 region of $0.03 \leq Q^2 \leq 0.16$ GeV$^2/c^2$ (7,800 events) [4]. This result is well inside the limits determined by the WA89 collaboration [5], 0.4 fm$^2 \leq \langle r^2 \rangle_{\Sigma^-} \leq 1.4$ fm^2 (Fig. 2).

For the negative pion, a monopole electric form factor is used. We find

$$\langle r^2 \rangle_\pi = 0.45 \pm 0.03\,(stat.) \pm 0.07\,(syst.)\,\text{fm}^2,$$

where $0.03 \leq Q^2 \leq 0.20$ GeV$^2/c^2$. (12,000 events) [6]. This result is in excellent agreement with the so far best direct measurement [7] of $\langle r^2 \rangle_\pi = 0.44 \pm 0.01$ fm^2 as well as a recent calculation which takes into account form factor measurements in both space-like and time-like regions [8]: $\langle r^2 \rangle_\pi = 0.463 \pm 0.005$ fm^2.

Major contributions to the systematic error come from the Q^2 resolution, uncertainties in the corrections for trigger efficiency, and beam contamination by other particles, particularly Ξ^-. Significant improvement is expected for all of these when

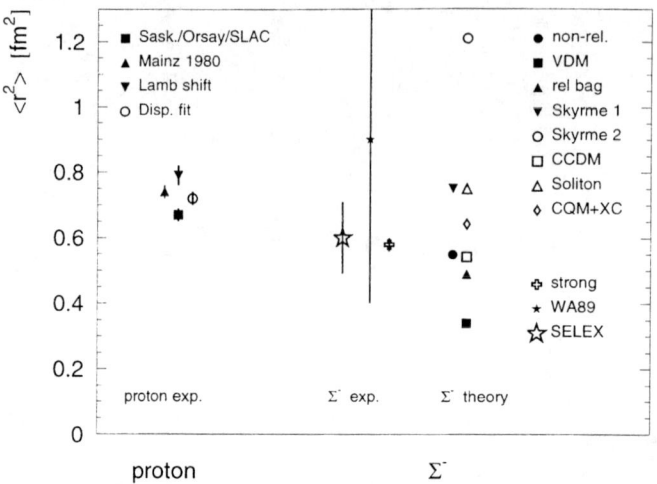

FIGURE 2. The Σ^- charge radius compared to various results for the proton radius (left): *Sask./Orsay/SLAC* [10,11], *Mainz* [9], *dispersion-theoretical fit* to all of above [12], and *Lamb shift* [13]. – Experimental results (center): *SELEX*: this measurement, *WA89*: WA89 result [5], *strong*: strong interaction radius [3]. – The predictions for Σ^- refer to the following models: *non-rel.*: non-relativistic quark model, *VDM*: vector dominance model, *rel bag*: relativistic bag model (all three values from [3]), *Skyrme 1*: Skyrme model [14], *Skyrme 2*: Skyrme model [15], *CCDM*: Chiral color dielectric model [16], *Soliton*: Soliton model [17], *CQM+XC*: Chiral constituent quark model including exchange currents [18].

advanced reconstruction and simulation software is used to refine the data sample. Q^2 will be determined from all kinematic variables and events with identified Σ^- decays accepted as well. We anticipate a statistical error of less than 10 % in the final analysis of the Σ^- radius.

III Σ^-p TOTAL CROSS SECTION

In general, the difference of total hadronic cross sections is ascribed to the difference in Regge residue functions, which are connected to hadronic radii, rather than to the Pomeron propagator. In the Landshoff-Donnachie [19] version of Regge theory the effective Pomeron intercept $\epsilon \approx (\alpha_P(0) - 1)$ and the effective Reggeon intercept $\eta \approx (\alpha_R(0) - 1)$ are assumed to be universal, i.e. the same for all hadrons, with $\epsilon \approx 0.08$ and $\eta \approx 0.47$.

Recent data from H1 and ZEUS on the proton structure function at small x and high Q^2 demonstrate, however, that the effective Pomeron intercept is higher for hadrons with smaller radii, up to $\epsilon = 0.4$ for high Q^2 [20]. There is further evidence for $\epsilon > 0.08$ from real exclusive photoproduction of heavy flavors. Data from HERA also show that the cross section of J/Ψ photoproduction rises by a factor of 6 from $\sqrt{s} = 10$ to 100 GeV [21,24].

This Q^2 dependence of the ϵ should have its counterpart in the ϵ dependence on the radii of the stable hyperons. The higher the quark mass, the smaller the interquark distance corresponding to the effective high Q^2 hadronic interaction.

The only available beams with flavors heavier than *up* or *down* are $K^{\pm}$ and hyperon beams. Total cross sections of Σ^- and Ξ^- on protons and deuterons have been measured at beam energies between 19–137 GeV at CERN [22,27]. Analogous data at 600 GeV would provide a sensitive test of the Pomeron universality [23].

In the 1997 fixed target run SELEX has recorded 58 million events with Σ^-/π^- and 18 million with proton/π^+ beams using a beam-only trigger. Beam particles were identified with a transition radiation detector. Corrections were applied to account for effects of Coulomb and Coulomb-nuclear interference as well as beam rate and contamination.

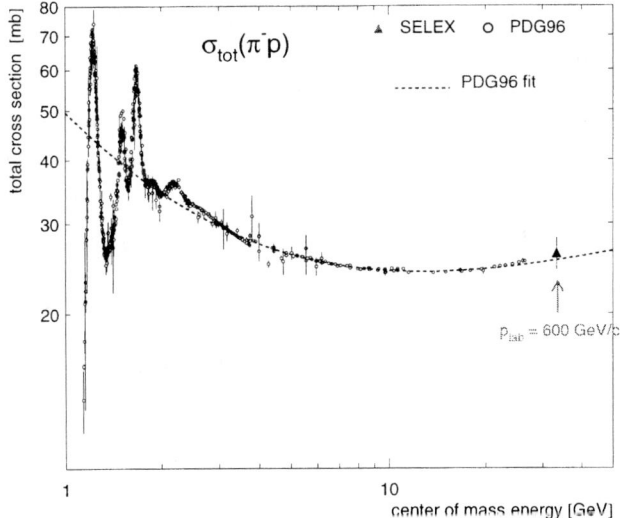

FIGURE 3. Compilation of world data on π^-p total cross sections with the preliminary SELEX result at 600 GeV/c beam momentum.

The total hadronic Σ^-p cross section at 600 GeV/c beam momentum has been determined by the transmission method:

$$\sigma_{tot}(\Omega) := \frac{1}{\rho L} \lim_{\Omega \to 0} \log \left[\frac{F_0}{F_{tr}(\Omega)} \cdot \frac{E_{tr}(\Omega)}{E_0} \right].$$

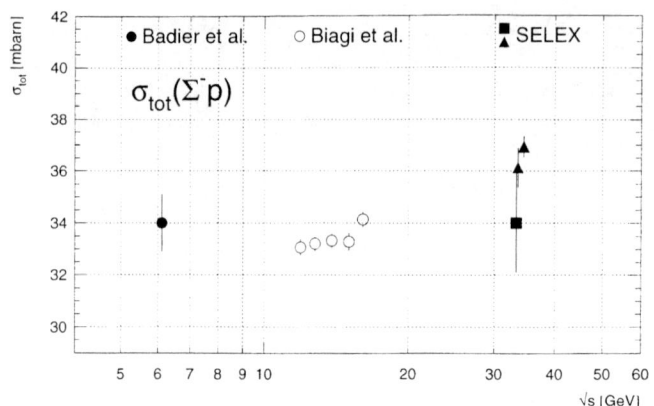

FIGURE 4. Total Σ^-p cross section at different center-of-mass energies. Low-energy data from Badier et al. [27] and WA42 (Biagi et al. [22]). SELEX results from C–CH$_2$ difference (square) and C and Be target Glauber calculations (triangles).

Here, ρ and L are density and length of the target, and F_{tr}/F_0 and E_{tr}/E_0 the transmission ratios with and without target, respectively. Instead of a liquid hydrogen target SELEX had two alternative approaches:

(1) C–CH$_2$ subtraction method. From the data sample taken with carbon and polyethylene targets the total Σ^--proton cross section was calculated from

$$\sigma_{tot}(\Sigma^-p) = \frac{1}{2}\left[\sigma_{tot}(\Sigma^-CH_2) - \sigma_{tot}(\Sigma^-C)\right].$$

This yields
$$\sigma_{tot}(\Sigma^-p) = 34.0 \pm 1.9 \text{ mb},$$

where statistical and systematic errors have been combined. As a cross check, an identical analysis was performed on the pion-proton data taken with these targets:
$$\sigma_{tot}(\pi^-p) = 26.2 \pm 1.9 \text{ mb}$$

A comparison to the current world data sample which only covers beam momenta up to 370 GeV/c [26] finds our results to be following the general trend (Fig. 3).

(2) $\sigma_{tot}(\Sigma^-p)$ was calculated from the ratio of proton-nucleus and Σ^--nucleus cross sections. Here, we used Be and C targets. The Σ^--nucleus cross sections for Be and C targets follow nicely the expected A-dependence of $\sigma_{tot}(XA) = \sigma_0 A^\alpha$, where $\alpha \simeq 0.77$ [25]. The Σ^--proton cross section was calculated from the ratio $\sigma_{tot}(\Sigma^-A)/\sigma_{tot}(pA)$. Nuclear effects were accounted for by a model based

on Glauber theory which included corrections for inelastic nuclear shadowing. This leads to the result of $\sigma_{tot}(\Sigma^-p) = 36.39 \pm 0.76$ mb (Be target, 635 GeV/c average beam momentum) and $\sigma_{tot}(\Sigma^-p) = 36.13 \pm 0.42$ mb (C target, 595 GeV/c).

The results are shown along with total Σ^-p cross sections at lower energies in Fig. 4. A fit to these data points using the Donnachie-Landshoff parametrization [26] yields $\epsilon = 0.098 \pm 0.019$ [25].

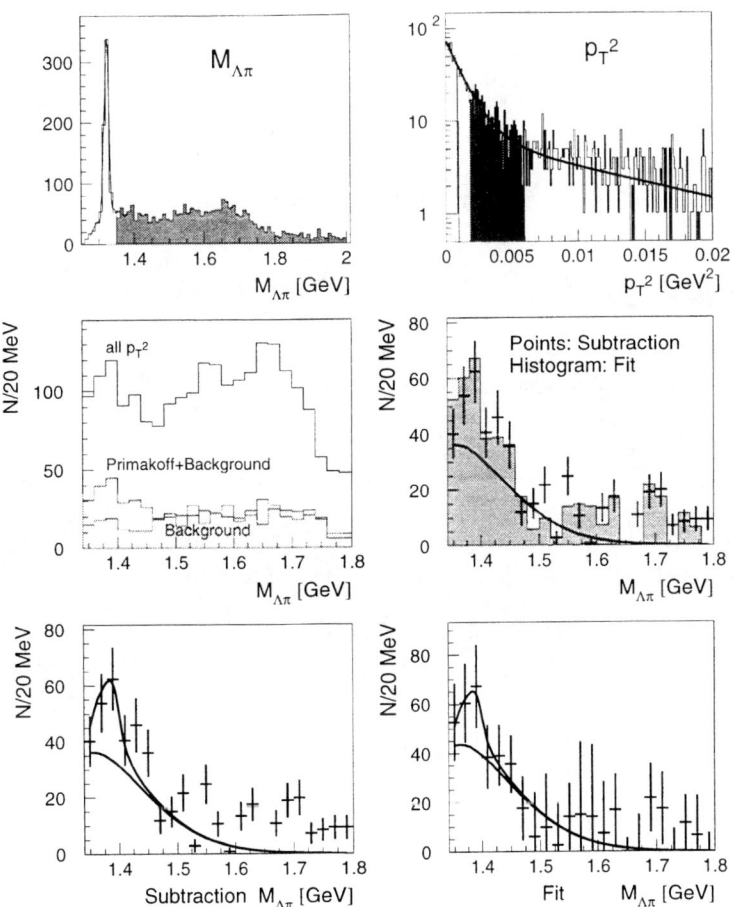

FIGURE 5. Transverse momentum squared and mass spectrum of the final state of $\Sigma^- + \text{Pb} \to \text{Pb} + (\pi^- + \Lambda)$, $\Lambda \to p + \pi^-$. Explanations in text (section IV).

IV RADIATIVE WIDTH OF $\Sigma(1385)^-$

Radiative decay widths of hyperons constitute powerful tests for dynamical theories of hadronic systems. The expected value of the SU(3)-suppressed radiative width $\Gamma(\Sigma^{*-} \to \Sigma^- \gamma)$ in different models is predicted in the region of 1–10 keV and the SU(3)-allowed width $\Gamma(\Sigma^{*+} \to \Sigma^+ \gamma)$ in the range of 100–300 keV [28]. Unfortunately, the experimental situation is difficult due to small branching ratios on one hand and large background from hadronic decays on the other.

The production of a hadron resonance state in the nuclear Coulomb field (the Primakoff formalism), on the other hand, provides a relatively clean method for the determination of radiative widths.

At SELEX, the $\Sigma(1385)^-$ was produced from Σ^- using a lead target. The differential cross section for the Primakoff reaction

$$\Sigma^- + Z \to Z + \Sigma(1385)^-, \quad \Sigma(1385)^- \to \Lambda + \pi^-$$

can be written as a function of the four-momentum transfer squared t,

$$\frac{d\sigma}{dt} = 8\pi\alpha Z^2 \frac{2J_{\Sigma^*}+1}{2J_\Sigma+1} \Gamma(\Sigma(1385)^- \to \Sigma^-\gamma) \left(\frac{m_{\Sigma^*}}{m_{\Sigma^*}^2 - m_\Sigma^2}\right)^3 \frac{t - t_{min}}{t^2} |F(t)|^2,$$

where α is the fine structure constant, t_{min} the minimal momentum transfer squared, and m_{Σ^*} the mass of the final state. Z is the charge and $F(t)$ the electromagnetic form factor of the nucleus. The t-distribution for the Primakoff reaction has a pronounced forward peak at $t = 2t_{min}$.

The beam Σ^- was identified with a TRD. Elasticity of the reaction and less than 2 GeV of energy deposition in the first lead glass calorimeter were required. The observed $\Lambda\pi^-$ mass distribution is shown in Fig. 5 (upper left), with the peak from Ξ^- decays in the lead target clearly visible. The observed p_T^2 distribution (Fig. 5, upper right) was assumed to be the sum of coherent $\Lambda\pi$ and Primakoff production. From Monte Carlo simulation we found that both coherent and Primakoff p_T^2 distributions smeared by the experimental resolution can be described by double-exponential fits [30].

Two methods were used to estimate the $\Lambda\pi$ mass distribution for the Primakoff reaction:

1. Two p_T^2 regions were defined (Fig. 5, upper right), one at $p_T^2 < 0.001$ GeV$^2/c^2$ where the Primakoff effect dominates, and one at $0.002 < p_T^2 < 0.006$ GeV$^2/c^2$ for estimation of the background from coherent production of the $\Lambda\pi$ system. The corresponding mass spectra are shown in Fig. 5, center left. Subtraction yields the data points in Fig. 5, center right and lower left.

2. All events were subdivided into 20 MeV bins of the mass spectrum of the final $\Lambda\pi$ state, and the p_T^2 distribution analyzed for each bin. The result is shown as the shaded histogram in Fig. 5, center right and lower right.

Both methods are in reasonable agreement (Fig. 5, center right). The total cross section for the process is given by the equation

$$\sigma_{tot} = \int_0^\infty \frac{d\sigma}{dt} dt = A \cdot \Gamma(\Sigma(1385)^- \to \Sigma^- \gamma).$$

A was obtained by integrating numerically over the differential cross section $d\sigma/dt$. The radiative width $\Gamma(\Sigma(1385)^- \to \Sigma^- \gamma)$ was estimated using the expression

$$\Gamma(\Sigma(1385)^- \to \Sigma^- \gamma) = \frac{N_{\Sigma^*}}{A \cdot L \cdot \epsilon \cdot \mathrm{BR}(\Sigma^* \to \Lambda\pi) \cdot \mathrm{BR}(\Lambda \to p\pi)},$$

where N_{Σ^*} is the number of observed events, L the luminosity of the experiment, ϵ the combined reconstruction efficiency where efficiency of the applied cuts and finite decay volume have been accounted for. The luminosity was determined on the basis of coherent production of $(\pi^- \pi^- \pi^+)$ by pions [29].

The upper limit for the Primakoff production of $\Sigma(1385)^-$ is: $N_{\Sigma^*} < 205$ events at 95 % confidence limit. With the above equation this yields

$$\Gamma(\Sigma(1385)^- \to \Sigma^- \gamma) < 12 \text{ keV } (95\% \text{ CL}),$$

thus improving the upper limit of 24 keV established by an experiment at Brookhaven (1977) [31].

V CONCLUSIONS

A measurement of the Σ^- mean squared charge radius has been performed by elastic Σ^--electron scattering. A preliminary analysis yields a Σ^- radius of $\langle r^2 \rangle_{\Sigma^-} = 0.60 \pm 0.08 \, (stat.) \pm 0.08 \, (syst.) \text{ fm}^2$. The π^- radius was determined in parallel and is found to be in excellent agreement with previous experiments.

The $\Sigma^- p$ total hadronic cross section has been measured at 600 GeV/c. We obtain $\sigma_{tot}(\Sigma^- p) = 34.0 \pm 1.9$ mb from the difference of results for CH_2 and C targets, and $\sigma_{tot}(\Sigma^- p) = 36.6 \pm 0.9$ mb from a Glauber model calculation using the ratios of $\Sigma^- A$ to pA cross sections on C and Be targets.

A new upper limit for the radiative width of the $\Sigma(1385)^-$ has been established at 12 keV (95 % CL) from a study of Primakoff production on lead nuclei.

Improved statistics and smaller systematic errors for these results as well as other hyperon physics results are expected as the analysis of SELEX data proceeds.

ACKNOWLEDGEMENTS

We are indebted to B. C. LaVoy, D. Northacker, F. Pearsall, and J. Zimmer for invaluable technical support. This project was supported in part by the U.S. Department of Energy (DOE grant DE-FG02-91ER40664), the National Science Foundation (Phy #9602178), Consejo Nacional de Ciencia y Tecnología (CONACyT), Fondo de Apoyo a la Investigación (UASLP), the Turkish Scientific and Technological Research Board (TÜBİTAK), NATO (grant CR6.941058-1360/94), Bundesministerium für Bildung, Wissenschaft, Forschung und Technologie, Istituto Nazionale de Fisica Nucleare (INFN), the U.S.-Israel Binational Science Foundation (BSF), the Israel Science Foundation founded by the Israel Academy of Sciences and Humanities, Fundação de Amparo à Pesquisa do Estado de São Paulo (FAPESP), Conselho Nacional de Desenvolvimento Científico e Tecnológico, the International Science Foundation (ISF), the Russian Academy of Science and the Russian Ministry of Science and Technology.

Figures 1 and 2 have been reprinted with permission from Eschrich, I. "Measurement of the Σ^- Charge Radius at SELEX", to appear in the *Proceedings of the 8th International Conference on the Structure of Baryons, Bonn, Germany*, Sept. 22-26, 1998, World Scientific.

REFERENCES

1. A. Kushnirenko, these proceedings.
2. V. Kubarovsky, "Radiative width of the a_2 meson", presented at the XXIX International Conference on High-Energy Physics, Vancouver, July 23-29, 1998.
3. B. Povh and J. Hüfner, *Phys. Lett.* B **245**, 653 (1990).
4. I. Eschrich, *Measurement of the Σ^- Charge Radius at the Fermilab Hyperon Beam*, MPIH–V22–1998, Ph.D. thesis, MPI f. Kernphysik / Univ. Heidelberg, 1998.
5. M. Adamovich et al., *First observation of Σ^--e scattering in the hyperon beam experiment WA89 at CERN*, submitted to *Eur. Phys. J.* C , (1998).
6. K. Vorwalter, *Determination of the Pion Charge Radius with a Silicon Microstrip Detector System*, MPIH–V23–1998, Ph.D. thesis, MPI f. Kernphysik / Univ. Heidelberg, 1998.
7. S. Amendolia et al., *Nucl. Phys.* B **277**, 186 (1986)
8. B. V. Geshkenbein, *hep-ph/9806418* (1998).
9. G. Simon, C. Schmitt, F. Borkowski, and V. Walther, *Nucl. Phys.* A **333**, 381 (1980)
10. L. Hand, D. Miller, and R. Wilson, *Rev Mod Phys* **35**, 335 (1963)
11. J. Murphy, Y. Shin, and D. Skopik, *Phys. Rev.* C **9**, 2125 (1974)
12. P. Mergell, U. Meissner, and D. Drechsel, *Nucl. Phys.* A **596**, 367 (1996)
13. T. Udem et al., *Phys. Rev. Lett.* **79**, 2646 (1997)
14. J. Kunz, P. Mulders, and G. Miller, *Phys. Lett.* B **255**, 11 (1991)
15. N. Park and H. Weigel, *Nucl. Phys.* A **541**, 453 (1992)
16. S. Sahu, *Mod. Phys. Lett.* A **10**, 2103 (1995)
17. H. Kim, A. Blotz, M. Polyakov, and K. Goeke, *Phys. Rev.* D **53**, 4013 (1996)
18. G. Wagner, A. Buchmann, and A. Faessler, *nucl-th/9809015*, accepted for publication by *Phys. Rev.* C , (1998).
19. A. Donnachie and P.V. Landshoff, *Phys. Lett.* B **296**, 227 (1992)
20. H1 Collaboration, *Z. Phys.* C **69**, 27 (1995).
21. ZEUS Collaboration, *Phys. Lett.* B **350**, 120 (1995).
22. WA42 Collaboration (Biagi *et al.*), *Nucl. Phys.* B **186**, 1 (1981).
23. B. Kopeliovich, private communications.
24. A. Levy, *Soft Interactions at High Energy*, DESY Report 95-204.
25. U. Dersch, *Messung totaler Wirkungsquerschnitte mit Σ^-, p, π^- und π^+ bei 600 GeV/c Laborimpuls*, Ph.D. thesis *(in German)*, MPI f. Kernphysik / Univ. Heidelberg, 1998.
26. Particle Data Group, *Phys. Rev.* D **54**, 1 (1996).
27. J. Badier *et al.*, *Phys. Lett.* B **41**, No. 3, 387 (1972).
28. L.G. Landsberg and V.V. Molchanov, *Radiative Decays of Hyperons*, IHEP 97-42, Protvino 1997 and references herein;
    H.J. Lipkin and M.A. Moinester, *Phys. Lett.* B **287**, 179 (1992);
    G. Wagner *et al.*, *Phys. Rev.* C **58**, No. 3, 1745 (1998).
29. M. Zielinski *et al.*, *Z. Phys.* C **16**, 197 (1983)
30. V. Kubarovsky, *Upper Limit for the Radiative Width of $\Sigma(1385)^-$ and Photoproduction Cross Section for the Reaction $\gamma + \Sigma^- \to \pi^- + \Lambda$*, H-Note 809 (SELEX Internal Report), 1998.
31. E. Arik *et al.*, *Phys. Rev. Lett.* **38**, No. 18, 1000 (1977).

An Overview of Hyperon Physics in KTeV

D. A. Jensen

for the KTeV Collaboration

Fermi National Accelerator Laboratory
Batavia, IL 60510

Abstract.
There is a relatively large flux of hyperons in the kaon beams in KTeV. These hyperons, particularly the Ξ^0, provide a rich opportunity for the study of hyperon physics. A general overview of the broad range of hyperon physics being studied at KTeV is presented. The results are based largely on Ξ^0 decays, both the study of the Ξ^0, and using the Ξ^0 as a source of Σ^0s and of Λs.

INTRODUCTION

KTeV is primarily a pair of neutral kaon experiments: E832 - a measurement of ϵ'/ϵ; and E799-II, a study of rare decays. Included in the E799-II running is a study of hyperon decays. The possibility of studying hyperons arises because the kaon program requires high intensity, and to achieve high intensity, the production solid angle accepted must be large, and the detector is therefore not too far from the production target. The decay region starts about 94 meters from the production target. For 250 GeV/c Ξ^0s for example, this corresponds to a little over 5.7 lifetimes. Clearly only high momentum hyperons survive to the decay region.

A wide range of hyperon physics is being done in KTeV. The following list shows many of the areas currently under study. The N_{ev} is the number of events observed before KTeV (Ref 1). N_{KTeV} is roughly the number of events expected in KTeV. The number of events in KTeV is still preliminary, and the exact number of events depends on the analysis that is in progress. The number of events also depends on the particular physics analysis. Tighter cuts may be applied in for example $\Xi^0 \to \Sigma^0 + \gamma$ to do asymmetry measurements than might be applied to obtain the branching ratio.

CP459, Heavy Quarks at Fixed Target
edited by Harry W. K. Cheung and Joel N. Butler
© 1999 The American Institute of Physics 1-56396-864-9/99/$15.00

Hyperon Physics at KTeV

Cascade Beta Decay Physics

- $\Xi^0 \to \Sigma^+ e^- \bar{\nu}_e$ $N_{ev} = 0$, , $N_{KTeV} = 1200$
- $\overline{\Xi^0} \to \overline{\Sigma^+} e^+ \nu_e$ $N_{ev} = 0$, $N_{KTeV} = 60$
- $\Xi^0 \to \Sigma^+ \mu^- \bar{\nu}_\mu$ $N_{ev} = 0$, $N_{KTeV} = 10$

Radiative Cascade Decay Physics

- $\Xi^0 \to \Sigma^0 \gamma$, $N_{ev} = 85$, $N_{KTeV} = \sim 7000$
- $\Xi^0 \to \Lambda \gamma$, $N_{ev} = 116$, $N_{KTeV} = \sim 1000$

Three Body Radiative Hyperon Decays

- $\Xi^0 \to \Lambda \pi^0 \gamma$ $N_{ev} = 0$, $N_{KTeV} \sim 30$
- $\Lambda^0 \to p \pi^- \gamma$ $N_{ev} = 72$, $N_{KTeV} \sim 7000$

Sigma Hyperon Decays

- $\Sigma^0 \to \Lambda e^+ e^-$ $N_{ev} = 0$, $N_{KTeV} \sim 10$
- $\Sigma^0 \to \Lambda \gamma \gamma$ $N_{ev} = 0$, $N_{KTeV} = ?$

$\Delta S = 2$ Hyperon Physics

- $\Xi^0 \to p\pi^-$, $\overline{\Xi^0} \to \bar{p}\pi^+$ $N_{ev} = 0$, $N_{KTeV} = 0$

Miscellaneous Hyperon Physics

- $\Xi^0 \to \Lambda \pi^0$, $\bar{\Xi}^0 \to \bar{\Lambda}\pi^0$, $\Lambda^0 \to p\pi^-$, $\overline{\Lambda^0} \to \bar{p}\pi^+$

This discussion will focus on cascade decays: both the Ξ^0 decays and the use of the Ξ^0 as a source of tagged Σ^0s. Figure 1 shows the $\Xi^0 \to \Lambda + \pi^0$ mass and momentum spectra. Note that the momentum spectrum of the Ξ^0 peaks just over 250 GeV/c. This is to be contrasted with the K^0 momentum spectrum which peaks a little below 50 GeV/c and is very small by 200 GeV/c. This difference in the momentum spectra is very helpful in reducing backgrounds to hyperon decays from K^0 decays.

BEAM AND APPARATUS

Beam

The downstream end of the proton beam line, the target area, and the upstream end of the neutral beam are shown in Figure 2. The focused proton beam is brought

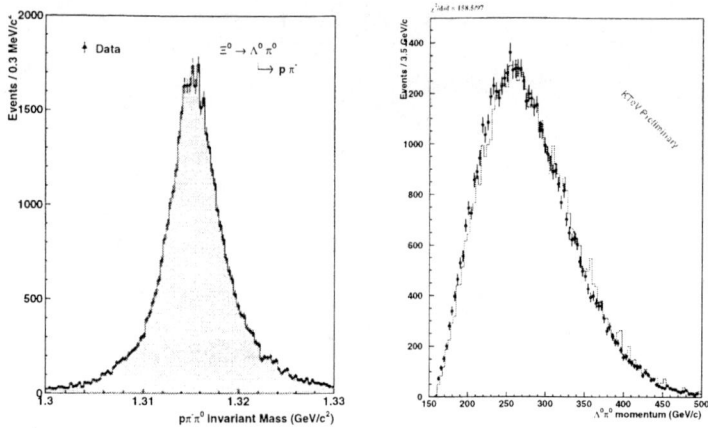

FIGURE 1. $\Xi^0 \to \Lambda + \pi^0$ Reconstructed Mass (left) Momentum Spectrum (right)

onto a BeO target at -4.5 mrad in the vertical (-y in the KTeV coordinate system direction). The neutral beam is defined by two collimators, the primary collimator shown in Figure 2, and the final or defining collimator at the entrance to the KTeV experimental halls. Charged particles are swept away by a series of muon sweeping magnets. All of the sweeping magnets bend positive particles to the left with respect to the neutral beam direction, or +x in the KTeV coordinate system. Two neutral beams are formed as that is the configuration required for E832. In the study of hyperon decays, the two beams are treated as equivalent - there being no particular advantage of multiple beams.

Hyperons produced at a finite transverse momentum are well known to be polarized. In KTeV, this polarization will be in the horizontal plane, along $\pm$ x. The muon sweeping system may be tuned, without compromising the very excellent muon sweeping, to rotate the hyperon polarization into the $\pm z$ (longitudinal) direction. A 'spin rotator' magnet in the neutral beam line, with field along the $\pm x$ direction is then used to rotate the hyperon polarization into the $\pm y$ direction. Switching between the $\pm y$ direction minimized acceptance effects as the spectrometer is very up-down symmetric. Hyperon polarization may then be measured. A preliminary measurement of the Ξ^0 polarization averaged over the observed momentum spectrum, in figure 1 yields 10%. This value has been corrected from the naive up-down/up+down type analysis to take into account the effects of daughter Λ polarization and the resulting polarization dependent acceptance effects.

The Ξ^0 flux is much less than the K^0 flux. For example, the rare branching ratio searches are based on approximately 10^{11} K^0 decays. The number of Ξ^0 decays in the same period of running is approximately a few times 10^8.

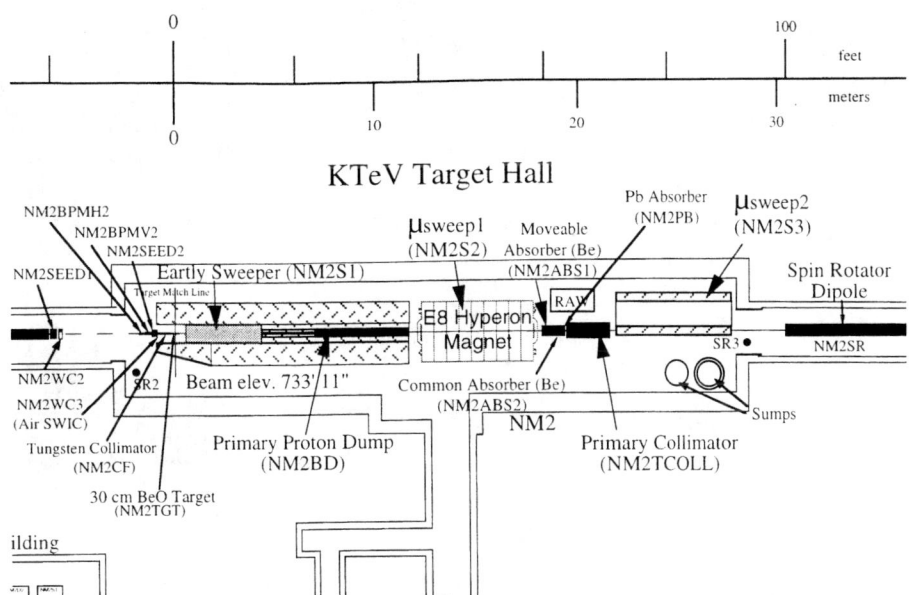

FIGURE 2. The KTeV target region showing the incoming proton beam, target station, and the sweeping magnets.

Spectrometer

The KTeV spectrometer is shown in figure 3. The apparatus consists of a charged particle spectrometer, a pure CsI electromagnetic calorimeter, a set of photon veto counters, trigger hodoscopes, TRDs for electron identification, and a muon detector.

The charged particle spectrometer consists of two drift chambers (x, x', y, y' views) upstream of a spectrometer and two more drift chambers downstream. The resolution of the drift chambers is about 100μ in both views. The spectrometer magnet was set to a field integral of 205 MeV/c. The field integrals through the spectrometer magnet are quite uniform, varying only by $\pm\ 1\%$ over the full useful aperture of the spectrometer.

Central to the KTeV spectrometer is the very excellent photon detector. The 1.9 m square electromagnetic calorimeter consists of 3100 crystals of pure CsI. The outer part of the array consists of 5 cm square crystals, whereas the inner part consists of 2.5 cm crystals. The crystals are 27 radiation lengths long. The position resolution is approximately 1mm in both the horizontal and vertical. The energy resolution may be characterized as $\sigma_E/E = (0.6 + 0.6/\sqrt{E})\%$. The average resolution is better than 1% over the full energy range of photons being detected. It is interesting to note that, with the spectrometer magnet set to 400 MeV/c, the K^0 mass resolution is nearly the same in $K^0 \rightarrow \pi^+\pi^-$ and $K^0 \rightarrow \pi^0\pi^0$. The neutral

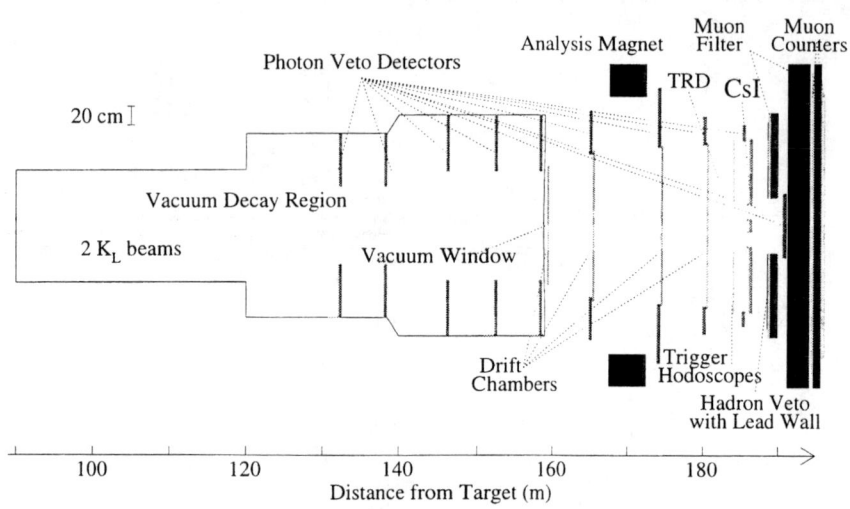

FIGURE 3. The KTeV Spectrometer as configured for E799 running.

mode is in fact slighter better.

There is also excellent particle identification. The CsI in and of itself is very good at distinguishing between pions and electrons. In addition there are the TRDs (transition radiation detectors), and the muon system.

The spectrometer aperture is assured by photon veto counters. There is the set of ring vetoes (RC) at 5 positions in the vacuum decay region, and a series of spectrometer vetoes (SA) throughout the rest of the spectrometer. A veto in the neutral beams together with the CsI detector make the spectrometer hermetic to about a hundred milliradians.

Triggers

KTeV uses three levels of triggering. The first two levels are conventional electronic levels and the third level is based on event reconstruction. Level 1 (L1) is based on hodoscopes, the veto counters, hits in drift chambers 1 and 2 and the total energy in the CsI. The hadron anti and beam counters may also be included.

At level 2 (L2) there is cluster counting based on the rather neat algorithm that looks at the hits in all the CsI blocks, the transition radiation detectors, and of particular significance for hyperon physics a stiff track trigger which requires a high momentum particle through the area of the neutral beam.

L3 is based on full online reconstruction using Silicon Graphics computers. Both charged and neutral events are fully reconstructed. Cuts may be applied to any of the parameters describing the event.

The hyperon trigger requires a high momentum track down the neutral beam line. This stiff track trigger is formed from drift chamber hits processed in a fast gate-array module. Also required is a second charged track hitting the hodoscopes, and a minimum energy and at least two clusters in the CsI.

More details regarding the overall physics program as proposed and technical details of the beam and spectrometer may be found in reference 2.

PHYSICS ANALYSIS

Results presented here are based on Ξ^0 decays. The Ξ^0s are interesting both in and of the themselves, and as a source of Σ^0s. The discussions here are, for lack of space, very brief. More details will be found in the theses and papers that will be forthcoming.

Cascade Beta Decays

One of the important pieces of physics we have gleaned from the previous run in KTeV is the first observation of and a measurement of the branching ratio for $\Xi^0 \to \Sigma^+ + e^- + \bar{\nu}_e$ with $\Sigma^+ \to p + \pi^0$. Events are selected by reconstructing a π^0 from two clusters, and forming a vertex between the stiff track and that π^0 to obtain a Σ^+. The Σ^+ and electron (identified in the CsI) are extrapolated to form a vertex - the Ξ^0 vertex. This vertex is required to be upstream of the Σ^+ vertex. Based on a modest subset of the data where the running conditions were particularly clean and stable, we obtain a branching ratio of $(2.49 \pm 0.19 \text{ (stat.)} \pm 0.31 \text{ (sys.)}) \times 10^{-4}$. The sample has been normalized to 41K $\Xi^0 \to \Lambda + \pi^0$ events. This is in good agreement with the branching ratio expected from the Cabibbo model (reference 3) of $(2.61 \pm 0.11) \times 10^{-4}$.

Plots showing the Σ^+ mass distribution with 153 events and data from the full sample are presented in figure 4. The full sample when fully analyzed will consist of over 1000 events. Additional confirmation of this Ξ^0 beta decay is presented in figure 5, which shows the reconstructed beta decay neutrino momentum spectrum for both the data and the Monte-Carlo.

This large sample of beta decays will be used to measure asymmetries and measure g_1/f_1. A first look at the e-p asymmetry in the Σ^+ frame is shown in figure

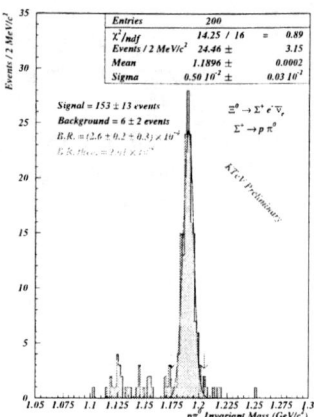 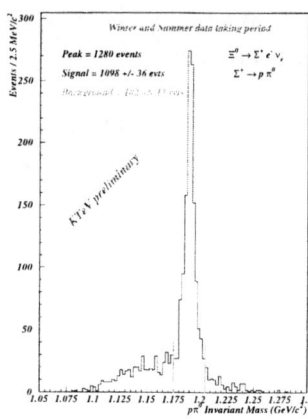

FIGURE 4. Reconstructed $\Sigma^+ \to p + \pi^0$ from Ξ^0 beta decays, from a small sample (left) the large sample (right)

5. On the same plot are shown are a number of Monte-Carlo distributions for various form factors and also pure phase space. It is clear that there is a reasonable sensitivity to the form factors. This data analysis continues.

We have also been able to observe roughly 65 ± 10 $\overline{\Xi^0}$ beta decays. The analysis that is carried out is identical to that for cascade beta decay with only the signs of the decay products changed. The absolute level of the background is comparable to that in normal cascade beta decay.

The decay $\Xi^0 \to \Sigma^+ + \mu^- + \bar{\nu}_\mu$ with $\Sigma^+ \to p + \pi^0$ has also been searched for. Six events have been observed with an expected background of less than one event. Again, the analysis is very much like the cascade beta analysis, but changing the electron identification to muon identification. The muon detection system has been well calibrated using K^0 decays. The observed branching ratio is consistent with the expected branching ratio of $(2.2 \pm 0.11) \times 10^{-6}$.

Weak Radiative Hyperon Decays

The weak radiative decays are particularly interesting in that there has been a broad spectrum of predictions for both the branching ratios and for the asymmetries - the only two parameters that are accessible in these decays. The asymmetries are the correlations between the spin of the initial state baryon and the direction of the photon.

In KTeV we have obtained large samples of both $\Xi^0 \to \Lambda + \gamma$ and $\Xi^0 \to \Sigma^0 + \gamma$. The mode $\Xi^0 \to \Sigma^0 + \gamma$ is detected through the decay $\Sigma^0 \to \Lambda + \gamma$, that is, a final state consisting of a Λ with two photons. This is exactly the same final state

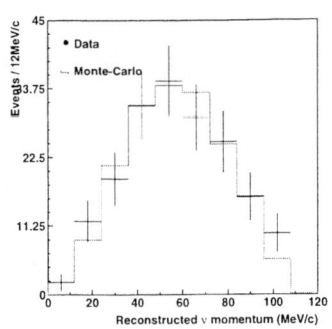 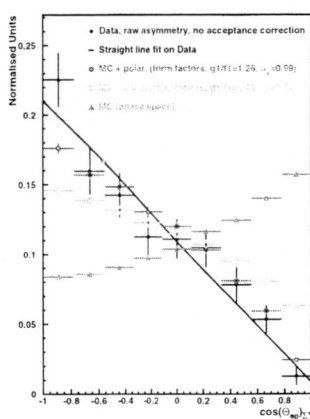

FIGURE 5. Neutrino Momentum Spectrum from Ξ^0 beta decay, data and Monte-Carlo (left), and asymmetries - data and various Monte-Carlos (right)

as $\Xi^0 \to \Lambda + \pi^0$ with $\pi^0 \to \gamma\gamma$. The only difference is the invariant mass of the two photons. Those signal events for which the two photon mass is near the π^0 mass are excluded. The final states for the signal and for the normalization process are the same, and thus the branching ratio measurement is essentially free of systematic effects. The lambda decay into p π^- is self analyzing for polarization, so the asymmetries are readily accessible.

The asymmetry in $\Xi^0 \to \Sigma^0 + \gamma$ is particularly interesting as there is the sequence of three decays, the Ξ^0, Σ^0, and the Λ decays. The whole sequence has been carefully simulated and the data compared to the Monte-Carlo simulation for various assumptions about the asymmetries. There is a clear large negative asymmetry present.

The branching ratio for $\Xi^0 \to \Sigma^0 + \gamma$ is $(3.02 \pm 0.04 \pm 0.20) \times 10^{-3}$. KTeV has about 4600 events where previous experiments have seen only about 85 events.

For $\Xi^0 \to \Lambda + \gamma$, KTeV has obtained approximately 1100 events where the previous world sample was of order 116. The branching ratio is $(0.94 \pm 0.03) \times 10^{-3}$ with no systematic errors yet determined. Again, the normalization mode is $\Xi^0 \to \Lambda + \pi^0$.

The numerous theorists who have looked at this problem have essentially filled phase space with predictions. So it makes a very interesting area of study as we attempt to project out the correct theorist.

Σ^0 and Other Radiative Processes

The Ξ^0 is a source of Σ^0s through the $\Xi^0 \to \Sigma + \gamma$. The branching ratio for $\Sigma^0 \to \Lambda + \gamma$ is nearly 100%, but other rare processes are possible. For example,

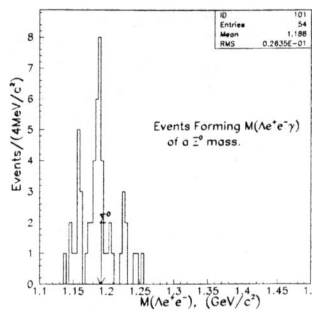
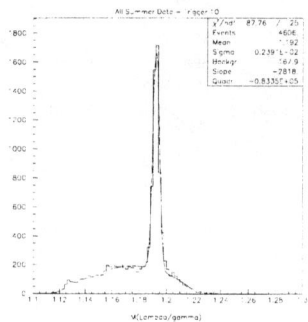

FIGURE 6. $\Sigma^0 \to \Lambda + \gamma$ Mass from $\Xi^0 \to \Sigma^0 + \gamma$ (right) and $\Sigma^0 \to \Lambda e^+ e^-$ from $\Xi^0 \to \Sigma^0 \gamma$ (left)

a search has been made to find events with two electrons and a Λ that form a Σ^0 along with a photon to reconstruct the Ξ^0. Such a sample of events has been found. The electron pairs tend generally to have high mass and are therefore well separated from conversion processes for example which might otherwise be a significant background. Figure 6 shows a plot of the mass of Λee for events which, when combined with a photon, form a Ξ^0 mass. There is a clear Σ^0 signal.

We also search for other three body hyperon radiative decays. In particular $\Lambda \to p\pi^-\gamma$ (a radiative correction type process), and $\Xi^0 \to \Lambda\pi^0\gamma$ (clearly NOT a radiative correction type process).

$\Delta S = 2$?

We have also searched for the $\Delta S = 2$ transition $\Xi^0 \to p + \pi^-$. The predicted branching ratio is of order 10^{-9} to 10^{-13} - out of reach of KTeV. An earlier experiment in 1975 placed a limit 3.6×10^{-5}. Backgrounds are $\Lambda \to p + \pi^-$ with scattered decay or mismeasured products, $K_L \to \pi e \nu$ with particle misidentification and $K_L \to \pi^+\pi^-$ again with particle misidentification. This search is normalized to $\Xi^0 \to \Lambda + \pi^0$. A limit somewhat better than the existing limit has been set. No background is observed.

CONCLUSIONS

The data taken during the previous KTeV run is indeed a rich source of hyperon physics data. A number of papers are in preparation to document these results. We are looking forward to KTeV99 - the next run in which we expect to increase the data samples by as much as a factor of two.

REFERENCES

1. Particle Data Group, C. Caso et al., Eur. Phys. J. C **3**, 1 (1998)
2. N. Cabibbo, Phys. Rev. Letters **10**, 531 (1963)
3. K. Arisaka et al., KTeV (Kaons at the Tevatron) Design Report Fermilab Report No. FN-580, 1992

Charm semileptonic decays from E791

Fermilab E791 Collaboration

Presented by Daniel Mihalcea
Department of Physics
Kansas State University
Manhattan, KS, 66506

Abstract. We report the results of a measurement of the form factor ratios $r_V = V(0)/A_1(0)$, $r_2 = A_2(0)/A_1(0)$ and $r_3 = A_3(0)/A_1(0)$ in the decays $D^+ \to \overline{K}^{*0} \ell^+ \nu_\ell$, with $\overline{K}^{*0} \to K^-\pi^+$, and $D_s^+ \to \phi \ell^+ \nu_\ell$, with $\phi \to K^-K^+$, using data from charm hadroproduction experiment E791 at Fermilab. We also report the results of an E791 measurement of the branching fraction $\mathcal{B}(D^+ \to \rho^0 \ell^+ \nu_\ell)/\mathcal{B}(D^+ \to \overline{K}^{*0} \ell^+ \nu_\ell)$.

INTRODUCTION

E791 is a fixed-target charm hadroproduction experiment [1]. Charm particles were produced in the collisions of a 500 GeV/c π^- beam with five thin targets, one platinum and four diamond. About 2×10^{10} events were recorded during the 1991-1992 Fermilab fixed-target run. We report here results from three analyses:

1. Measurement of the form factor ratios for $D^+ \to \overline{K}^{*0} \ell^+ \nu_\ell$.

2. Measurement of the form factor ratios for $D_s^+ \to \phi \ell^+ \nu_\ell$.

3. Measurement of the branching ratio $\mathcal{B}(D^+ \to \rho^0 \ell^+ \nu_\ell)/\mathcal{B}(D^+ \to \overline{K}^{*0} \ell^+ \nu_\ell)$.

Accurate particle identification is crucial for all these analyses. Hadron identification is based on the information from two multicell Čerenkov counters that provided good discrimination between kaons and pions in the momentum range $6 - 36$ GeV/c. In this momentum range, the probabability of misidentifying a pion as a kaon depends on momentum but does not exceed 5%. We identified muon candidates using a single plane of scintillator strips, oriented horizontally, located behind an equivalent of 2.4 meters of iron (comprised of the calorimeters and one meter of bulk steel shielding). The vertical position of a hit was determined from the strip's vertical position, and the horizontal position of a hit from timing information. To reduce the contamination from hadron decays in flight, only muon candidates with momenta larger than 8 GeV/c are retained. With this momentum restriction, the efficiency of muon tagging was about 85%, and the probability for a hadron to be identified as a muon was about 3%. The electron candidates were identified by electromagnetic shower shape, the match between calorimeter energy and tracking momentum, and agreement between calorimeter and tracking position measurements. Electron identification efficiency is about 70%, while the probability for a pion to be misidentified as an electron is only 1 - 2%.

MEASUREMENT OF THE FORM FACTOR RATIOS FOR $D^+ \to \overline{K}^{*0} \ell^+ \nu_\ell$

The weak decays of hadrons containing heavy quarks are substantially influenced by strong interaction effects. Semileptonic charm decays such as $D^+ \to \overline{K}^{*0} \ell^+ \nu_\ell$ are an especially clean way to study these effects because the leptonic and hadronic currents completely factorize in the decay amplitude. All information about the strong interactions can be parametrized by a few form factors. Also, according to Heavy Quark Effective Theory, the values of form factors for some semileptonic charm decays can be related to those governing certain b-quark decays. In particular, the form factors studied here can be related to those for the rare B-meson decays $B \to K^\star e^+ e^-$ and $B \to K^\star \gamma$ [2,3] which provide windows for physics beyond the Standard Model.

With a vector meson in the final state, there are four form factors, $V(q^2)$, $A_1(q^2)$, $A_2(q^2)$ and $A_3(q^2)$, which are functions of the Lorentz-invariant momentum transfer squared [3]. We assume the q^2-dependence of the form factors to be given by the nearest-pole dominance model: $F(q^2) = F(0)/(1 - q^2/m_{pole}^2)$ where $m_{pole} = m_V = 2.1$ GeV/c^2 for the vector form factor V, and $m_{pole} = m_A = 2.5$ GeV/c^2 for the three axial-vector form factors. The third form factor $A_3(q^2)$, which is unobservable in the limit of vanishing lepton mass, probes the spin-0 component of the off-shell W. Additional spin-flip amplitudes, suppressed by an overall factor of m_ℓ^2/q^2 when compared with spin no-flip amplitudes, contribute to the differential decay rate. Because $A_1(q^2)$ appears among the coefficients of every term in the differential decay rate, it is customary to factor out $A_1(0)$ and to measure the ratios $r_V = V(0)/A_1(0)$, $r_2 = A_2(0)/A_1(0)$ and $r_3 = A_3(0)/A_1(0)$ which are independent of the total decay rate.

The differential decay rate [4] is expressed in terms of four independent kinematic variables: the square of the momentum transfer (q^2), the polar angle θ_V between the π^+ and D^+ in the $\overline{K}^{*0}$ rest frame, the polar angle θ_ℓ between the ν_μ and D^+ in the W^+ rest frame, and the azimuthal angle χ between the $\overline{K}^{*0}$ and W^+ decay planes in the D^+ rest frame.

Events are selected if they contain an acceptable decay vertex determined by the intersection point of three tracks that have been identified as a muon, a kaon, and a pion. The longitudinal separation between this candidate decay vertex and the reconstructed production vertex is required to be at least 15 times the estimated error on the separation. The two hadrons must have opposite charge. If the kaon

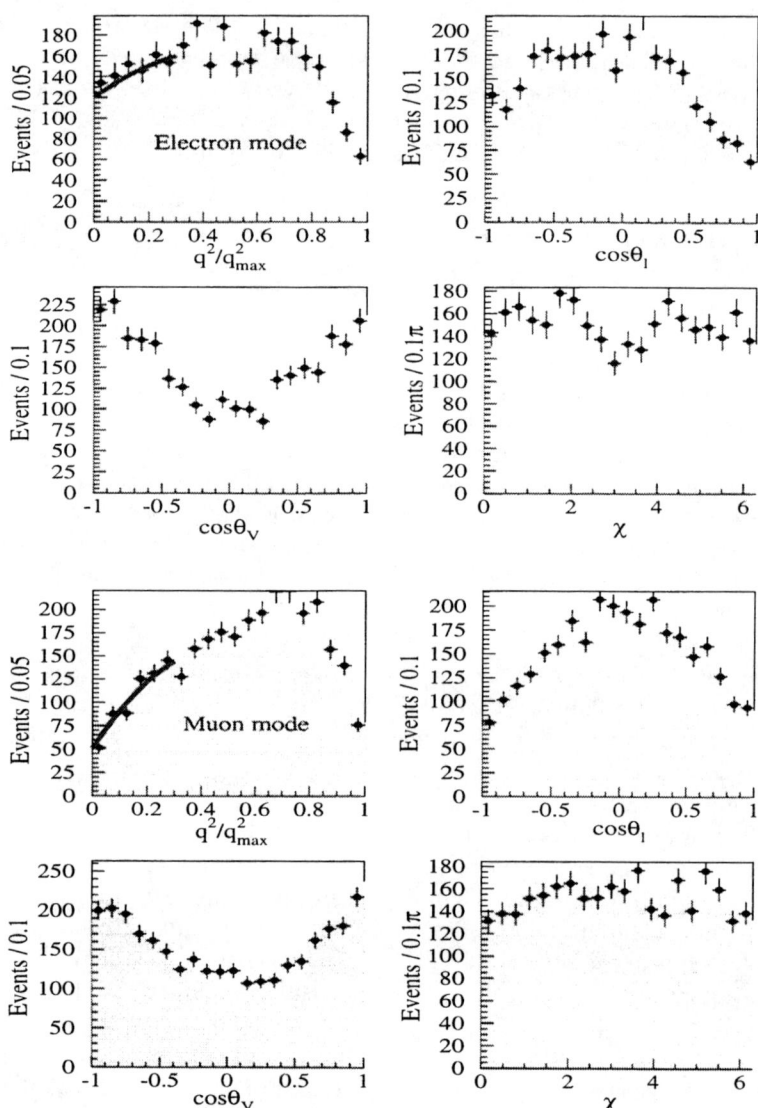

FIGURE 1. Comparison of single-variable distributions of background-subtracted data (crosses) with Monte Carlo predictions (histograms) using best-fit values for the form factor ratios. Top four plots: electron mode; bottom four plots: muon mode. Lepton mass effects contribute mostly in the region of low q^2/q^2_{max} as shown by the heavy line.

TABLE 1. Comparison of E791 results for $D^+ \to \overline{K}^{*0}\ell^+\nu_\ell$ with previous experimental results and with theoretical predictions.

	Group	Events	$r_V = V(0)/A_1(0)$	$r_2 = A_2(0)/A_1(0)$
Experiment	**E791** [18]	6000 ($e + \mu$)	$1.87 \pm 0.08 \pm 0.07$	$0.73 \pm 0.06 \pm 0.08$
	E791 [18]	3000 (μ)	$1.84 \pm 0.11 \pm 0.09$	$0.75 \pm 0.08 \pm 0.09$
	E791 [17]	3000 (e)	$1.90 \pm 0.11 \pm 0.09$	$0.71 \pm 0.08 \pm 0.09$
	E687 [22]	900 (μ)	$1.74 \pm 0.27 \pm 0.28$	$0.78 \pm 0.18 \pm 0.10$
	E653 [6]	300 (μ)	$2.00^{+0.34}_{-0.32} \pm 0.16$	$0.82^{+0.22}_{-0.23} \pm 0.11$
	E691 [7]	200 (e)	$2.0 \pm 0.6 \pm 0.3$	$0.0 \pm 0.5 \pm 0.2$
Theory	ISGW2 [8]		2.0	1.3
	WSB [9]		1.4	1.3
	AW/GS [10,11]		2.0	0.8
	Stech [12]		1.55	1.06
	LMMS [13]		1.6 ± 0.2	0.4 ± 0.4
	ELC [14]		1.3 ± 0.2	0.6 ± 0.3
	UKQCD [15]		$1.4^{+0.5}_{-0.2}$	0.9 ± 0.2
	LANL [16]		1.78 ± 0.07	0.68 ± 0.11

and the muon have opposite charge, the event is assigned to the "right-sign" sample; if they have the same charge, the event is assigned to the "wrong-sign" sample which is used to model the background. The final data samples contain 3629 right-sign and 595 wrong-sign events for the muon mode, and 3595 right-sign and 602 wrong-sign events for the electron mode.

To extract the form factor ratios, the distribution of the data points in the four-dimensional kinematic variable space is fit to the full expression for the differential decay rate by using an unbinned maximum-likelihood fitting technique. The largest systematic uncertainties are related to: (a) Monte Carlo simulation of detector effects and production mechanism; (b) fitting technique; (c) background subtraction. The fit values of the form factor ratios are $r_V = 1.84 \pm 0.11 \pm 0.09$, $r_2 = 0.75 \pm 0.08 \pm 0.09$ and $r_3 = 0.04 \pm 0.33 \pm 0.29$ for the muon mode and $r_V = 1.90 \pm 0.11 \pm 0.09$ and $r_2 = 0.71 \pm 0.08 \pm 0.09$ for the electron mode. The third form factor ratio, r_3, was not measured in the electron mode since it is not relevant in the limit of vanishing lepton mass.

Figure 1 compares the data and Monte Carlo kinematic variable distributions for both electron and muon channels. The distribution of q^2/q^2_{max} looks different for the two decay channels, mainly in the region of low q^2/q^2_{max}, because of lepton mass effects.

The consistency within errors of the results measured in the electron and muon channels supports the assumption that strong interaction effects, incorporated in the values of form factor ratios, do not depend on the particular W^+ leptonic decay. Based on this assumption, we combine the results measured for the electronic and muonic decay modes. The averaged values of the form factor ratios are $r_V = 1.87 \pm 0.08 \pm 0.07$ and $r_2 = 0.73 \pm 0.06 \pm 0.08$.

Table 1 compares the values of the form factor ratios r_V and r_2 measured by E791 in the electron, muon and combined modes with previous experimental and theoretical results. All experimental results are consistent within errors. The spread in the theoretical results is significantly larger than the E791 experimental errors.

MEASUREMENT OF THE FORM FACTOR RATIOS FOR $D_s^+ \to \Phi \ell^+ \nu_\ell$

If SU(3) symmetry is approximately valid, one might expect that replacing a spectator $\bar{d}$ quark by a spectator $\bar{s}$ quark would have relatively little effect on the form factors. We have investigated this spectator replacement experimentally in charm meson decay, and report here new measurements by Fermilab experiment E791 of form factor ratios for $D_s^+ \to \phi e^+ \nu_e$ and $D_s^+ \to \phi \mu^+ \nu_\mu$, with $\phi \to K^+ K^-$. These D_s results are then combined and compared to recent high-statistics form-factor ratio results [17,18] for $D^+ \to \overline{K}^{*0} \ell \nu_\ell$ measured in the same experiment, and also to theoretical predictions.

Events in which the lepton candidate has the same charge as the candidate decay are assigned to the "right-sign" (RS) sample, which contains both the signal and a small amount of remaining background from reconstruction errors and other charm decay channels. Otherwise, events are assigned to the "wrong-sign" (WS) channel, which is purely background; note that WS events have two hadrons of the same charge, and are thus likely to be dominated by reconstruction background. A simple counting argument suggests that the reconstruction background in RS events should be twice the number of WS events.

The final data samples contain 166 right-sign and 11 wrong-sign events for the electron channel and 161 right-sign and 17 wrong-sign events for the muon channel.

To extract the form factor ratios we use the same unbinned maximum-likelihood fitting technique as for $D^+ \to \overline{K}^{*0} \ell^+ \nu_\ell$ form factor analysis. The final form factor ratios and their statistical errors are $r_V = 2.24 \pm 0.47$ and $r_2 = 1.64 \pm 0.34$ for the electron channel, and $r_V = 2.31 \pm 0.54$ and $r_2 = 1.49 \pm 0.36$ for the muon channel.

Fig. 3 compares the background-subtracted data in the kinematic variables q^2, $\cos\theta_\ell$ and $\cos\theta_V$ for the electron and muon channels with Monte Carlo predictions using our measured form factor ratios. Note that although the q^2 distributions for the two channels are quite different, due in large part to lepton-mass effects, the form factor ratios agree well.

The largest systematic effects are associated with small densities of Monte Carlo events in kinematic variable space. We reject data events (5 total) with zero nearby Monte Carlo events in this space; the dependence of the form factor ratios on requiring more nearby Monte Carlo events was studied by varying the minimum number from 1 to 5. The final results are $r_V = 2.24 \pm 0.47 \pm 0.21$, $r_2 = 1.64 \pm 0.34 \pm 0.20$ for the electron mode and $r_V = 2.31 \pm 0.54 \pm 0.26$, $r_2 = 1.49 \pm 0.36 \pm 0.20$ for the muon mode. The weighted averages of form factor ratios for the two modes are $r_V = 2.27 \pm 0.35 \pm 0.22$ and $r_2 = 1.57 \pm 0.25 \pm 0.19$.

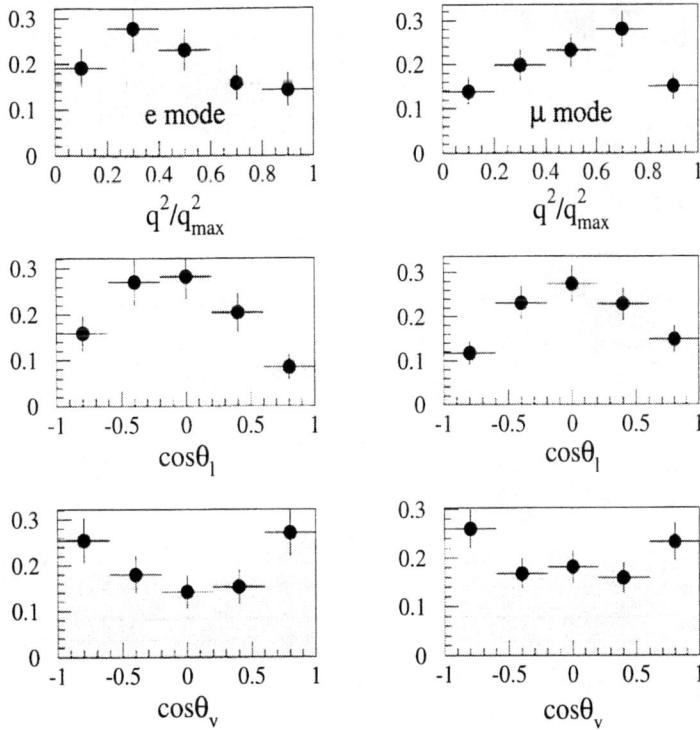

FIGURE 2. Comparison of kinematic variable distributions for electron (left) and muon (right) channels of background-subtracted data (crosses) with Monte Carlo predictions (dashed histograms), using best-fit values for the form factor ratios.

Table 2 compares the combined E791 results for r_V and r_2 to previously published experimental results (including the E791 results for $D^+ \to \overline{K}^{*0} \ell^+ \nu_\ell$) and to theoretical predictions. The E791 value of r_V is in good agreement with all quoted predictions, and is also consistent with the E791 measurement for the D^+ decay. For r_2, the theoretical predictions have a large spread compared to the E791 error; furthermore, the E791 measurement for D_s is more than two standard deviations higher than the E791 D^+ result. This discrepancy in r_2 is more significant for the new world average. Since the quoted theoretical papers predict form factor ratios equal within $\leq 10\%$ for the two decays, it is difficult for these predictions to agree well with both sets of data.

TABLE 2. Comparison of E791 results for $D_s^+ \to \phi \ell^+ \nu_\ell$ form factor ratios with previous experimental results and theoretical predictions.

	Group	Events	$r_V = V(0)/A_1(0)$	$r_2 = A_2(0)/A_1(0)$
Experiment	E791 this work	271 $(e+\mu)$	$2.27 \pm 0.35 \pm 0.22$	$1.57 \pm 0.25 \pm 0.19$
	E791 this work	144 (e)	$2.24 \pm 0.47 \pm 0.21$	$1.64 \pm 0.34 \pm 0.20$
	E791 this work	127 (μ)	$2.31 \pm 0.54 \pm 0.26$	$1.49 \pm 0.36 \pm 0.20$
	CLEO [19]	308 (e)	$0.9 \pm 0.6 \pm 0.3$	$1.4 \pm 0.5 \pm 0.3$
	E687 [22]	90 (μ)	$1.8 \pm 0.9 \pm 0.2$	$1.1 \pm 0.8 \pm 0.1$
	E653 [6]	19 (μ)	$2.3^{+1.1}_{-0.9} \pm 0.4$	$2.1^{+0.6}_{-0.5} \pm 0.2$
	E791 $D^+ \to \overline{K}^{*0} \ell^+ \nu_\ell$ [18]	6000 $(e+\mu)$	$1.87 \pm 0.08 \pm 0.07$	$0.73 \pm 0.06 \pm 0.08$
Theory	BKS [20]		$2.00 \pm 0.19^{+0.20}_{-0.25}$	$0.78 \pm 0.08^{+0.17}_{-0.13}$
	LMMS [13]		1.65 ± 0.21	0.33 ± 0.33
	ISGW2 [8]		2.1	1.3

MEASUREMENT OF THE BRANCHING RATIO
$\mathcal{B}(D^+ \to \rho^0 \ell^+ \nu_\ell)/\mathcal{B}(D^+ \to \overline{K}^{*0} \ell^+ \nu_\ell)$

Semileptonic charm decays are useful in probing the dynamics of hadronic currents since the Cabibbo-Kobayashi-Maskawa matrix elements for the charm sector are well-known from unitarity constraints. Form factors for Cabibbo-suppressed (CS) $c \to d$ semileptonic decays can be related via Heavy Quark Effective Theory (HQET) to those for $b \to u$ semileptonic decays at the same four-velocity transfer [2]. Since knowledge of the form factors in $b \to u$ transitions is vital for extracting V_{ub} from $b \to u$ semileptonic decays in a model-independent way, study of $c \to d$ semileptonic decays can improve our knowledge of V_{ub}.

Candidates for $D^+ \to \rho^0 \ell^+ \nu_\ell$, $\rho^0 \to \pi^+ \pi^-$ and $D^+ \to \overline{K}^{*0} \ell^+ \nu_\ell$, $\overline{K}^{*0} \to K^- \pi^+$ decays are selected by requiring a three-prong decay vertex of charge ± 1 with one of the decay particles being identified as a lepton. Once the lepton is identified, the other two tracks in the vertex (h_1, h_2) are assigned hadron masses. We define the right-sign (RS) sample as vertices in which the lepton and D^+ candidate have the same charge; h_1 and h_2 are then oppositely-charged. For $D^+ \to \overline{K}^{*0} \ell^+ \nu_\ell$ candidates the hadron with odd charge is assigned a kaon mass, while for $D^+ \to \rho^0 \ell^+ \nu_\ell$ candidates h_1 and h_2 are both assigned pion masses.

Figure 3 shows the signals in $D^+ \to \rho^0 \ell^+ \nu_\ell$ channel for both the electronic and muonic modes. The background contributions to the signal mainly come from real semileptonic charm decays in other channels. The amount of feedthrough from them is based on efficiencies estimated from Monte Carlo studies and Particle Data Group (PDG) [21] branching ratios.

After background subtraction, the final numbers of signal events are 49 ± 17 for $D^+ \to \rho^0 e^+ \nu_e$ and 54 ± 18 for $D^+ \to \rho^0 \mu^+ \nu_\mu$. The yields in the normalizing channels are 892 ± 52 for $D^+ \to \overline{K}^{*0} e^+ \nu_e$ and 769 ± 54 for $D^+ \to \overline{K}^{*0} \mu^+ \nu_\mu$.

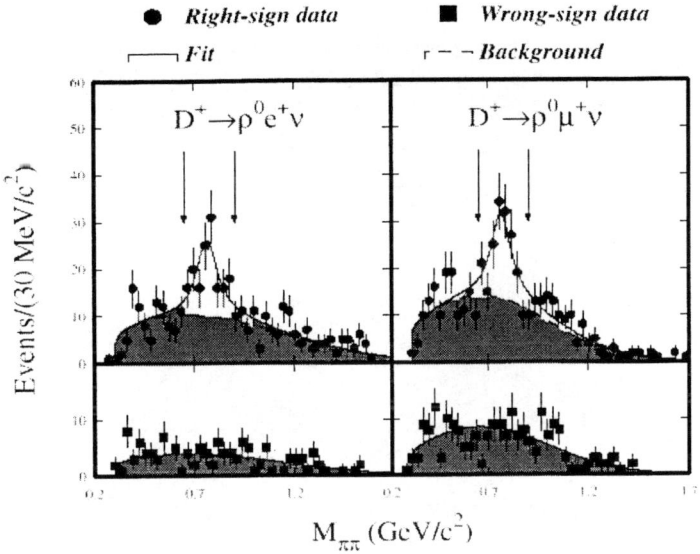

FIGURE 3. $M_{\pi\pi}$ distributions for $D^+ \to \rho^0 \ell^+ \nu_\ell$ candidates.

Systematic errors associated with lepton identification are largely cancelled in the ratio of the $D^+ \to \rho^0 \ell^+ \nu_\ell$ and $D^+ \to \overline{K}^{*0} \ell^+ \nu_\ell$ decay rates. Remaining sources of systematic error are 1) uncertainties in the branching ratios used in background subtraction; 2) uncertainty in the D_s^+ to D^+ production cross section ratio $\sigma_{D_s^+}/\sigma_{D^+}$; 3) determination of relative efficiencies and 4) the fitting procedure.

From the background-subtracted event yields and the efficiencies for $D^+ \to \rho^0 \ell^+ \nu_\ell$ and $D^+ \to \overline{K}^{*0} \ell^+ \nu_\ell$ decays, the following branching ratios are determined:

$$\frac{\mathcal{B}(D^+ \to \rho^0 e^+ \nu_e)}{\mathcal{B}(D^+ \to \overline{K}^{*0} e^+ \nu_e)} = (4.5 \pm 1.4 \pm 0.9)\%,$$

$$\frac{\mathcal{B}(D^+ \to \rho^0 \mu^+ \nu_\mu)}{\mathcal{B}(D^+ \to \overline{K}^{*0} \mu^+ \nu_\mu)} = (5.1 \pm 1.5 \pm 0.9)\%.$$

We combine the results from the electronic and muonic modes, taking correlated errors into account, to obtain a final result of

$$\frac{\mathcal{B}(D^+ \to \rho^0 \ell^+ \nu_\ell)}{\mathcal{B}(D^+ \to \overline{K}^{*0} \ell^+ \nu_\ell)} = (4.7 \pm 1.3)\%,$$

where the error includes both statistical and systematic uncertainties. In Table 3 our result is compared with other previously-published experimental results, and with various theoretical predictions.

TABLE 3. Comparison of our results with previously-published experimental results and with theoretical predictions. The second column indicates the method used to obtain the results, where QM stands for quark model, HQET for heavy quark effective theory, SR for QCD sum rule, and LQCD for lattice QCD.

Group	Method	ℓ	$\frac{\Gamma(D^+\to\rho^0\ell^+\nu_\ell)}{\Gamma(D^+\to K^{*0}\ell^+\nu_\ell)}$	$\Gamma(D^+\to\rho^0\ell^+\nu_\ell)$ $(10^{10}s^{-1})$
E791 [23]	Exp.	e	$0.045\pm0.014\pm0.009$	$0.20\pm0.07\pm0.05$
E791 [23]	Exp.	μ	$0.051\pm0.015\pm0.009$	$0.22\pm0.07\pm0.05$
E791 [23]	Exp.	ℓ	0.047 ± 0.013	0.21 ± 0.06
E687 [24]	Exp.	μ	$0.079\pm0.019\pm0.013$	
E653 [25]	Exp.	μ	$0.044^{+0.031}_{-0.025}\pm0.014$	$0.19^{+0.14}_{-0.11}\pm0.07$
ISGW2 [8]	QM	ℓ	0.022	0.12
Jaus [26]	QM	ℓ	0.030	0.33
Bajc [27]	HQET	ℓ	—	0.21 ± 0.02

			$\frac{1}{2}\frac{\Gamma(D^0\to\rho^-\ell^+\nu_\ell)}{\Gamma(D^0\to K^{*-}\ell^+\nu_\ell)}$	$\frac{1}{2}\Gamma(D^0\to\rho^-\ell^+\nu_\ell)$ $(10^{10}s^{-1})$
BSW [9]	QM	ℓ	0.037	0.35
ELC [14]	LQCD	ℓ	0.047 ± 0.032	$0.3\pm0.15\pm0.05$
APE [28]	LQCD	ℓ	0.043 ± 0.018	0.3 ± 0.1
UKQCD [15]	LQCD	ℓ	$0.036^{+0.010}_{-0.013}$	0.215 ± 0.055
LMMS [13]	LQCD	ℓ	0.040 ± 0.011	0.20 ± 0.045
Casalbuoni [29]	HQET	ℓ	0.06	0.225
Ball [30]	SR	ℓ	—	0.12 ± 0.035

SUMMARY

Fermilab experiment E791 has measured the form factor ratios in the decay channel $D^+\to\overline{K}^{*0}\ell^+\nu_\ell$ to be $r_V=1.87\pm0.08\pm0.07$ and $r_2=0.73\pm0.06\pm0.08$. The data sample has about 6000 events (for the sum of electron and muon modes) after subtracting about 1200 background events. As a first measurement of r_3 (performed only in the muon mode) we obtain $r_3=0.04\pm0.33\pm0.29$.

A similar measurement was performed for the decay channel $D_s^+\to\Phi\ell^+\nu_\ell$. In this case the data sample contains 215 events (for electron and muon modes) after subtracting 56 background events. The E791 results measured for this decay channel are $r_V=2.27\pm0.35\pm0.22$ and $r_2=1.57\pm0.25\pm0.19$.

We also have measured the branching ratio $\mathcal{B}(D^+\to\rho^0\ell^+\nu_\ell)/\mathcal{B}(D^+\to\overline{K}^{*0}\ell^+\nu_\ell)$. Combining the results from both electron and muon modes we obtain $\mathcal{B}(D^+\to\rho^0\ell^+\nu_\ell)/\mathcal{B}(D^+\to\overline{K}^{*0}\ell^+\nu_\ell)=0.047\pm0.013$ where the errors include statistical and systematic uncertainties.

Each of the measurements reported here represents a significant reduction in uncertainties of the results.

REFERENCES

1. Appel, J. A, *Ann. Rev. Nucl. Part. Sci.* **42**, 367 (1992); Summers, D. J. et al., XXVII Rencontre de Moriond, Les Arcs, France, 417 (15-22 March 1992).
2. Isgur, N. and Wise, M. B., *Phys. Rev. D* **42** 2388 (1990).
3. Ligeti, Z., Stewart, I. W. and Wise, M. B., *Phys. Lett. B* **420** 359 (1998).
4. Körner, J. G., and Schuler, G. A., *Phys. Lett. B* **226** 185 (1989).
5. Fermilab E687 Collaboration, Frabetti, P. L. et al., *Phys. Lett. B* **307** 262 (1993).
6. Fermilab E653 Collaboration, Kodama, K. et al., *Phys. Lett. B* **274** 246 (1992).
7. Fermilab E691 Collaboration, Anjos, J. C. et al., *Phys. Rev. Lett.* **65** 2630 (1990).
8. Scora, D., and Isgur, N., *Phys. Rev. D* **52** 2783 (1995).
9. Wirbel, M., Stech, B., and Bauer, M., *Z. Phys. C* **29** 637 (1985).
10. Altomari, T., and Wolfenstein, L., *Phys. Rev. D* **37** 681 (1988).
11. Gilman, F. G., and Singleton, R. L., Jr., *Phys. Rev. D* **41** 142 (1990).
12. Stech, B., *Z. Phys. C* **75** 245 (1997).
13. Lubicz, V., Martinelli, G., McCarthy, M. S., and Sachrajda, C. T, Phys. Lett. B **274** 415 (1992).
14. Abada, A. et al., *Nucl. Phys. B* **416** 675 (1994).
15. Bowler, K. C. et al., *Phys. Rev. D* **51** 4905 (1995).
16. Bhattacharya, T. and Gupta, R., *Nucl. Phys. B* bf 47 481 (1996).
17. Fermilab E791 Collaboration, Aitala, E. M. et al., *Phys. Rev. Lett.* **80** 1393 (1998).
18. Fermilab E791 Collaboration, Aitala, E. M. et al., xxx.lanl.gov e-print archive hep-ex/9809026, to appear in *Phys. Lett. B* (1998).
19. CLEO Collaboration, Avery P. et al., *Phys. Lett. B* **337** 405 (1994).
20. Bernard, C. W., El-Khadra, Z. X., and Soni, A., *Phys. Rev. D* **45** 869 (1991).
21. Particle Data Group, *Phys. Rev. D* **50** 1173 (1994).
22. Fermilab E687 Collaboration, Frabetti, P. L. et al., *Phys. Lett. B* **307** 262 (1993).
23. Fermilab E791 Collaboration, Aitala, E. M. et al., *Phys. Lett. B* **397** 325 (1997).
24. Fermilab E687 Collaboration, Frabetti, P. L. et al., *Phys. Lett. B* **391** 235 (1997).
25. Fermilab E653 Collaboration, Kodama, K et al., *Phys. Lett. B* **316** 455 (1993).
26. Jaus, W., Phys. Rev. D **53** 1349 (1996).
27. Bajc, B., Fajfer, S. and Oakes, R. J., *Phys. Rev. D* **53** 4957 (1996).
28. Allton, C. R. et al., *Phys. Lett. B* **345** 513 (1995).
29. Casalbuoni, R. et al., *Phys. Lett. B* **299** 139 (1993).
30. Ball, P., Braun, V. M., and Dosch, H. G., *Phys. Rev. D* **44** 3567 (1991).

Charm Semileptonic Decays from FOCUS/E687:
A Survey of Previous E687 Results, and Expectations from the FOCUS data

Will E. Johns
for the FOCUS Collaboration[1]

University of Puerto Rico
Department of Physics
College Station
Mayaguez, PR 00681

Abstract. The Fermilab Photoproduction Experiment E831 (FOCUS) will reconstruct over 1 million decays containing a charm quark. This yield is approximately 15 times the number of charm decays reconstructed in Experiment E687, the previous instance of the FOCUS spectrometer. From 23% of the total sample available for analyses, we demonstrate that for semileptonic analyses, the FOCUS collaboration can expect 30–40 times the events used in E687 semileptonic analyses. We review the E687 semileptonic results and make predictions for the physics from the FOCUS sample.

E687 SEMILEPTONICS

Since unitarity constraints [2] on the Cabibbo-Kobayashi-Maskawa (CKM) matrix elements were well below the statistical range possible for an E687 measurement, the analyses of semileptonic decays (see Figure 1) of mesons in E687 centered primarily on measuring form factors.

When a charm meson decays to a vector particle and a lepton–neutrino, there are 4 form factors to measure. Pole dominance is assumed for the q^2 dependence of the form factors. This choice is somewhat justified since the q^2 range of the decay, and hence the influence of the form factor at different values of q^2, is limited by the large mass of the vector particle.

For the analyses of the Cabibbo allowed vector decays, $D^+ \to \overline{K}^{*0} \mu^+ \nu_\mu$ [4] and $D_s^+ \to \phi \mu^+ \nu_\mu$ [5,6], in E687 (see Figure 2), the value of one form factor, $A_1(0)$, was set by the measured relative rate of the decay to a well measured hadronic mode and two other form factors were measured in ratio to $A_1(0)$, R_2 and R_V. The sensitivity to the forth form factor, $A_3(0)$, is diminished since $A_3(0)$ enters the

TABLE 1. E687 Charm vector semileptonic results compared to current best.

Decay Mode	Experiment	R_2	R_V
$D^+ \to \overline{K}^{*0} l^+ \nu_l$	E791 [13]	0.73 ± 0.10	1.87 ± 0.11
$D^+ \to \overline{K}^{*0} \mu^+ \nu_\mu$	E687 [4]	0.78 ± 0.20	1.74 ± 0.39
$D_s^+ \to \phi l^+ \nu_l$	E791 [13]	1.57 ± 0.30	2.28 ± 0.40
$D_s^+ \to \phi \mu^+ \nu_\mu$	E687 [6]	1.0 ± 0.5	1.8 ± 0.8

Decay Mode	Experiment	BR(rel. to $D^+ \to \overline{K}^{*0} \mu(l)^+ \nu_{\mu(l)}$)	
$D^+ \to \rho_0 l^+ \nu_l$	E791 [12]	0.047 ± 0.013	
$D^+ \to \rho_0 \mu^+ \nu_\mu$	E687 [8]	0.079 ± 0.023	

matrix element for the decay with a strong lepton mass dependence, and $A_3(0)$ was not measured by E687. For the analyses of the Cabibbo suppressed vector decay $D^+ \to \rho_0 \mu^+ \nu_\mu$, E687 measured the rate relative to $D^+ \to \overline{K}^{*0} \mu^+ \nu_\mu$. Most of these results (see Figure 2) have only recently been eclipsed by Fermilab Experiment E791 (see Table 1).

When the charm meson decays to a pseudoscalar particle and a lepton–neutrino, there are 2 form factors to measure. Again, the q^2 dependence of the form factors is assumed. For the pseudoscalar case though, there is enough q^2 range to measure hadronic structure. In the past, the measurement of hadronic structure has been limited to measuring a parameter, such as the pole mass, of the form factor choice (see Equation 1).

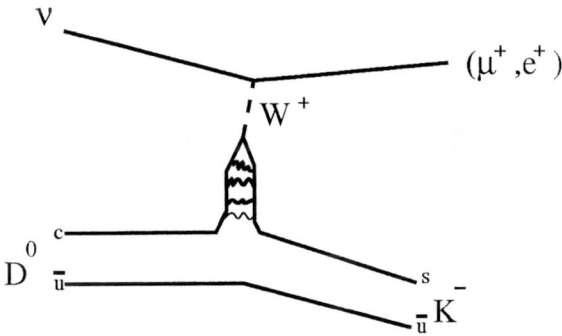

FIGURE 1. Anatomy of the pseudoscalar to pseudoscalar semileptonic decay $D^0 \to K^- l^+ \nu$. As the virtual W^+ particle is emitted, the D^0 meson undergoes a quark flavor change, the charm(c) quark becomes a strange(s) quark, and the D^0 becomes a kaon(K^-). During the flavor change, the quarks of the D^0 and the K^- interact(indicated by the wavy lines between the c and the s quark). The strength of this interaction is a function only of the invariant mass of the virtual W^+ particle and hence the lepton and the neutrino that the W^+ decays into. Since the decay of the virtual W^+ into a lepton and a neutrino is well described by $V - A$ theory, a kinematic reconstruction of the decay products gives us insight into the hadronic behavior of the c–s transition.

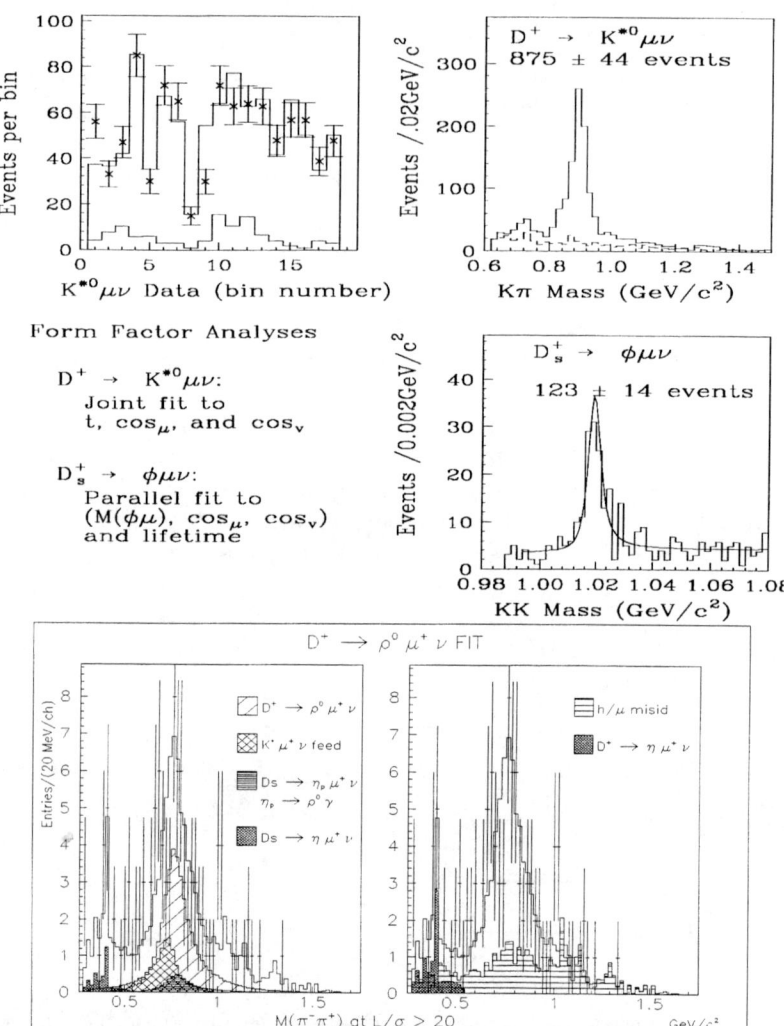

FIGURE 2. In the upper left hand plot we show the 18 bins used in the form factor analysis of $D^+ \to \overline{K}^{*0} \mu^+ \nu_\mu$. The variables used are $\cos\theta_\nu$, the angle between the π and the D direction in the $\overline{K}^{*0}$ rest frame, $\cos\theta_l$, the angle between the ν and the D direction in the $\mu\nu$ rest frame, and t, the square of the $\mu\nu$ mass. In the upper right, we show the fit to the $K^-\pi^+$ mass distribution (background is shown as a dashed line) from the same publication [4], and below the $K^-\pi^+$ mass we show the fit to the K^-K^+ mass used in the $D_s^+ \to \phi\mu^+\nu_\mu$ analysis [5]. In the lower two plots, the components of the fit used in the analysis of $D^+ \to \rho_0\mu^+\nu_\mu$ are shown [8]. Note the η contribution.

TABLE 2. E687 Charm pseudoscalar semileptonic results compared to CLEO.

Experiment	Decay	M_{pole} (GeV/c^2)	BR(rel. to $D^0 \to K^-\pi^+$)
E687 [3]	$D^0 \to K^-\mu^+\nu_\mu$	$1.87^{+0.11+0.07}_{-0.08-0.06}$	$0.852 \pm 0.034 \pm 0.028$
CLEO(93) [9]	$D^0 \to K^-l^+\nu_l$	$2.0 \pm 0.12 \pm 0.18$	$0.978 \pm 0.027 \pm 0.044$
Experiment	Decay	BR(rel. to $D^0 \to K^-l^+\nu_l$)	$\left|f_+^\pi(0)/f_+^K(0)\right|$
E687 [7]	$D^0 \to \pi^-l^+\nu_l$	$0.101 \pm 0.020(stat) \pm 0.003(syst)$	$1.00 \pm 0.11 \pm 0.02$
CLEO [10]	$D^0 \to \pi^-l^+\nu_l$	$0.103 \pm 0.039(stat) \pm 0.013(syst)$	$1.01 \pm 0.20 \pm 0.07$

$$f_\pm(q^2) = \frac{f_\pm(0)}{1 - \frac{q^2}{M_{pole}^2}} \quad (1)$$

For the E687 analysis [3] of the decay $D^0 \to K^-\mu^+\nu_\mu$, one form factor, $f_+(0)$, was set by the measurement of the rate relative to the decay $D^0 \to K^-\pi^+$. The pole mass and the other form factor, set in ratio to $f_+(0)$ ($f_-/f_+ = -1.30 \pm 3.5$), were determined using the *distribution* of the reconstructed decay variables E_μ and q^2 (see Figure 3). For the E687 analysis [7] of the Cabibbo suppressed decay $D^0 \to \pi^-l^+\nu_l$, the rate relative to $D^0 \to K^-l^+\nu_l$ was measured, and, assuming pole dominance, the ratio $f_+^\pi(0)/f_+^K(0)$ was estimated. The q^2 sensitivity of the decay was specifically addressed, and care was taken to account for the sensitivity of the result to the assumed q^2 dependence (see Figure 4). These E687 results remain competitive even now (see Table 2).

FOCUS SEMILEPTONICS

In Figure 5 we show the results of a preliminary, non-optimized semileptonic analysis of the decay modes $D^+ \to \overline{K}^{*0}l^+\nu_l$ and D^* tagged $D^0 \to K^-l^+\nu_l$ with 23% of the FOCUS data set. Based on the yields in Figure 5 and the yields using similar analysis cuts from E687 for the modes published by E687, we compare the expected total yield from the entire FOCUS data set to the yields from E687 in Table 3. Conservatively, for most modes, a factor of 40 improvement over the complimentary yields in the E687 analyses is expected.

With the large sample of data from FOCUS, we will be able to significantly improve the measurement of the Cabibbo allowed semileptonic decays, and the results for the Cabibbo suppressed decays should be of a similar to superior quality to the E687 Cabibbo allowed results. Additionally, there are new challenges to overcome to get the maximum benefit from the FOCUS data.

The differences in rate between the electron and muon modes must be determined to better than 1% accuracy if we wish to combine the two modes for measurements of form factors or CKM matrix elements. For instance, one can express the ratio of rates for the pseudoscalar decays (we have assumed pole dominance and a pole mass of 1.98 GeV/c^2) $D^0 \to K^-\mu^+\nu_\mu$ and $D^0 \to K^-e^+\nu_e$ in an expansion of form factors.

$$\frac{\Gamma_\mu}{\Gamma_e} = 0.984 + 0.0209 \left(\frac{f_-}{f_+}\right) + 0.003 \left|\frac{f_-}{f_+}\right|^2 \qquad (2)$$

In order to combine the electron and muon result, a measurement of $\frac{f_-}{f_+}$ to an accuracy of ± 0.5 is indicated. Since only some 400 events were used in the E687

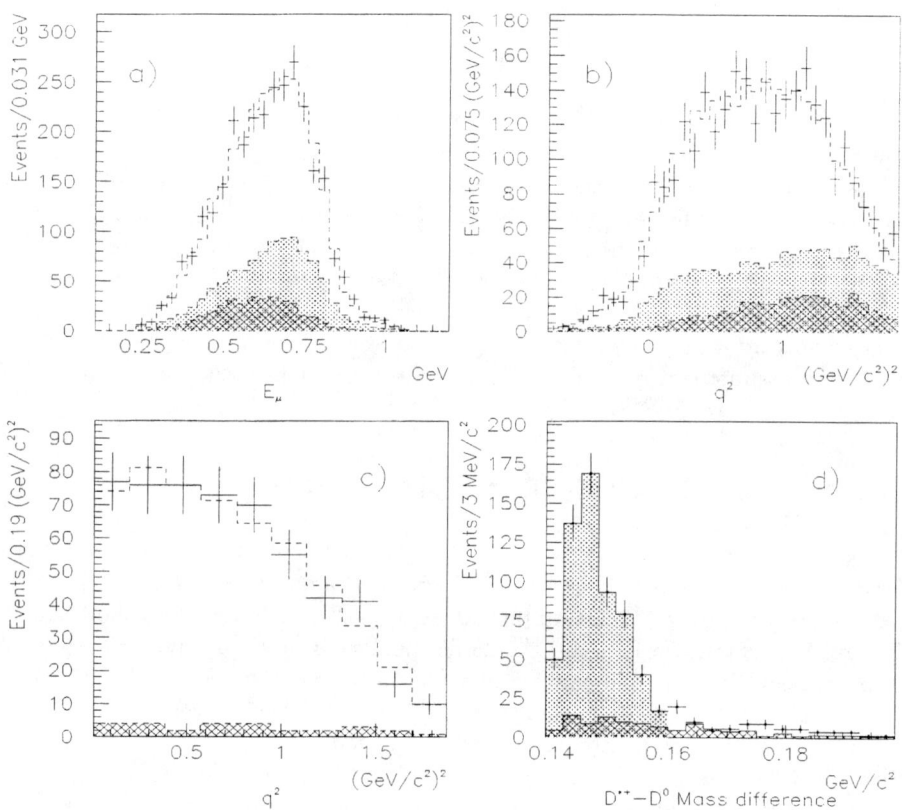

FIGURE 3. We show a fit (upper histogram), the total background (shaded histogram), and the background component due to Cabibbo allowed vector decays (hatched histogram) for the, a), μ momentum in the D center of mass and the, b), q^2 distribution of the E687 data (error bars) used for part of the non-D^*-tagged $D^0 \to K^-\mu^+\nu_\mu$ analysis. In the lower two plots we show: the a), q^2 and, b), D^* mass difference for a portion of the D^* tagged $D^0 \to K^-\mu^+\nu_\mu$ analysis (used to measure f_-/f_+ and as a check on the non-tagged result). The fit is the upper histogram in c), the wrong sign data is shown as a hatched histogram for both c) and d), and the shaded histogram in d) represents the limit of the data used in the f_-/f_+ analysis.

analysis of $\frac{f_-}{f_+}$ and we are expecting about 20000 events from FOCUS (before optimization of the analysis in FOCUS) for the same mode, a factor of 6–7 reduction in the error of the E687 result $f_-/f_+ = -1.30 \pm 0.35$ is reasonable. Further, instead of combining the electron and muon rates, one could compare the measurement of the electron rate to the muon rate and improve the measurement of $\frac{f_-}{f_+}$. This is especially appealing given the large statistical boost one gets from measuring the $D^0 \to K^- l^+ \nu_l$ decay without D^* tag as was done in E687 for the measurement of the rate $D^0 \to K^- \mu^+ \nu_\mu$ relative to $D^0 \to K^- \pi^+$. No other experiments have made these measurements (see Table 4).

FOCUS should be able to measure R_3, the $A_3(0)$ form factor in ratio to $A_1(0)$, for $D^+ \to \overline{K}^{*0} \mu^+ \nu_\mu$ and determine the difference in rate between the electron and muon vector decays. This is based on the E791 result [13], $R_3 = 0.04 \pm 0.44$, from about 3000 $D^+ \to \overline{K}^{*0} \mu^+ \nu_\mu$ events. For FOCUS, we expect to have over 40000 events in this mode which is a substantial statistical advantage.

Indeed, FOCUS is the only experiment that has data where the electron and muon modes have yields greater than 10000 events each. In Table 4 we compare current experiments to the preliminary FOCUS estimates using the D^* tagged analysis of $D^0 \to K^- l^+ \nu_l$ in both muon and electron modes. The FOCUS yields are larger, especially in the muon mode. We have also included a very preliminary estimate of the yield for a repeat of the non D^* tagged analysis we did in E687. Substantial work needs to be done before the non D^* tagged estimates are firm, regardless, the statistics should be staggering.

In general, it is reasonable to take E687 results and divide the total error by a factor of 6. This simple operation leads to some very interesting results. Namely, a simple repeat measurement of the decay rate for the D^* tagged analysis of $D^0 \to \pi^- l^+ \nu_l$ could produce a result on the ratio of form factors $\left|\frac{f_+^\pi(0)}{f_+^K(0)}\right|$ to 2% accuracy, likewise, if theoretical estimates of $\left|\frac{f_+^\pi(0)}{f_+^K(0)}\right|$ improve, perhaps with a measurement of the q^2 dependence of the form factors (see below), one could $measure$ $\left|\frac{V_{cd}}{V_{cs}}\right|$ to 2% accuracy (the current measured value from the Particle Data Group for V_{CS} has a 7% accuracy [14]).

Perhaps the most exciting semileptonic result from FOCUS will be the non-parametric measurement of the q^2 dependence of the $f_+(q^2)$ form factor for $D^0 \to K^- l^+ \nu_l$ decays. This is important! Theory makes reliable predictions at maximal values of q^2 (when the kaon is at rest in D^0 rest frame) while the data is best measured at low values of q^2 where data is plentiful (see Figure 3). In order to compare theory and experiment an extrapolation, such as Equation 1, is used. The correct extrapolation is not known, hence, the extrapolation is a significant source of systematic uncertainty when comparing theory and experiment. As presented by Jim Wiss at last years Varenna Summer School, FOCUS will measure the $D^0 \to K^- l^+ \nu_l$, q^2 behavior non-parametrically with reliable results on $|f_+(q^2)|^2$ to $\pm 3\%$ as close as 0.15 GeV^2 to q^2_{max} [15]. We are also studying the measurement of hadronic

TABLE 3. FOCUS projections Compared to Approximate E687 yields

Exp	$Kl\nu$	$\pi l\nu$	$K^{*0}l\nu$	$\phi l\nu$
E687	800	80	900	120
FOCUS	$\sim 40K$	$\sim 5K$	$\sim 50K$	$\sim 10K$

structure using the decays $D^0 \to \pi^- l^+ \nu_l$ and $D^+ \to K_s^0 l^+ \nu_l$. Perhaps the increased sensitivity (see Figure 4) of the decay $D^0 \to \pi^- l^+ \nu_l$ to higher values of q^2 will compensate the lower statistics. As well, the increase in resolution for long lived D^+ *tracks* measured with the FOCUS vertex detector may improve measurements near q_{max}^2.

We remind the reader that semileptonic analyses are a stepping-stone process. For instance, a measurement of the form factors for $D^+ \to \overline{K}^{*0} l^+ \nu_l$ will yield an improved result for $D^+ \to \overline{K}^{*0} n(\pi^0) l^+ \nu_l$, $D^+ \to (K^-\pi^+)_{n.r.} l^+ \nu_l$, $D^+ \to K^-\pi^+ n(\pi^0) l^+ \nu_l$, and $\frac{BR(D \to K^* l\nu)}{BR(D \to Kl\nu)}$. A measurement of $D^+ \to \rho l^+ \nu_l$ will yield results for decays containing an η (see Figure 2). We also point out that several unpublished results from E687 will receive the needed statistical bonus to warrant publication when the FOCUS data set is used. These analyses include $D^+ \to K_s^0 l^+ \nu_l$ and $D^0 \to K^{*-} l^+ \nu_l$.

Finally, one of the improvements to the FOCUS spectrometer [16] allows us to reconstruct charm particle *tracks* for long-lived D^+'s. This capability is being studied to try to increase the resolution of the D^+ direction for our semileptonic and fully leptonic analyses. Further, we can look for "one prong" decays such as $D^+ \to K_L^0 l^+ \nu_l$.

CONCLUSIONS

The FOCUS data set is large. We will produce precision semileptonic measurements of form factors and branching ratios. The relative rates for the Cabibbo allowed decays $D^0 \to K^- l^+ \nu_l$ and $D^+ \to \overline{K}^{*0} l^+ \nu_l$ will be measured to better than 1%. The q^2 behavior of $f_+(q^2)$ for the decay $D^0 \to K^- l^+ \nu_l$ will be measured non-parametrically, with high statistics, for the first time near q_{max}^2. Finally, a 2% result for the ratio of form factors, $\left|\frac{f_+^\pi(0)}{f_+^K(0)}\right|$, from the analysis of the Cabibbo suppressed decay $D^0 \to \pi^- l^+ \nu_l$ appears likely. Along with the other analyses mentioned, the suite of charm semileptonic results from FOCUS will be very interesting.

ACKNOWLEDGEMENTS

We wish to acknowledge the assistance of the staffs of the Fermi National Accelerator Laboratory, the INFN of Italy, and the physics departments of the collaborating institutions. This research was supported in part by the U.S. National

TABLE 4. Expected FOCUS yields compared to other current experiments.

Experiment	CLEO II [9]	CLEO II.5 [9]	E791 [11]	FOCUS ("tag")	FOCUS ("no tag")
Data Taking Done	yes	end 1998	yes	yes	yes
Reconstruction Done	yes	partial	yes	end 1998	end 1998
Est. $D^0 \to K^-\mu^+\nu_\mu$	1900	3100	1237	24000	88000?
Est. $D^0 \to K^-e^+\nu_e$	6100	9700	1267	18000	68000?
Tot. $D^0 \to K^-l^+\nu_l$	8100	12800	2504	42000	156000?

Science Foundation, the U.S. Department of Energy, the Italian Istituto Nazionale di Fisica Nucleare and Ministero dell'Università e della Ricerca Scientifica e Tecnologica, the Brazilian Conselho Nacional de Desenvolvimento Cieantífico e Tecnológico, CONACyT-México, the Korean Ministry of Education, and the Korean Science and Engineering Foundation.

REFERENCES

1. FOCUS (FNAL–E831) collaboration, in *Heavy Quarks at Fixed Target*, 1998.
2. Particle Data Group, K.Hikasa et al., Phys. Rev. D45, No. 11, Pt II, (1992)
3. E687, P. L. Frabetti et al., Phys. Lett. B364, 127 (1995).
4. E687, P. L. Frabetti et al., Phys. Lett. B307, 262 (1993).
5. E687, P. L. Frabetti et al., Phys. Lett. B313, 253 (1993).
6. E687, P. L. Frabetti et al., Phys. Lett. B328, 187 (1994).
7. E687, P. L. Frabetti et al., Phys. Lett. B382, 312 (1996).
8. E687, P. L. Frabetti et al., Phys. Lett. B391, 235 (1997).
9. CLEO, A. Bean et al., Phys. Lett. B317:647-654,(1993). The CLEO II and II.5 estimates are based on the yields in this paper and a total integrated CLEO II and II.5 luminosity of $13fb^{-1}$.
10. CLEO, F. Butler et al., Phys. Rev. D52:2656-2660, (1995).
11. E791, E. M. Aitala et al., Phys. Rev. Lett. 77:2384-2387,(1996).
12. E791,E. M. Aitala et al., Phys. Lett. B 397,325-332 (1997).
13. Mihalcea, D.,*Charm Semileptonic Decays from E791*, Talk presented in *Heavy Quarks at Fixed Target*, 1998.
14. C. Caso et al., Eur. Phys. Jour. C3:1-794,(1998).
15. Wiss J. E., *Proceedings of the International Summer School of Physics, Varenna 1997, Course CXXXVII*, IOS Press, Amsterdam:39-39, (1998).
16. Link, J., *The FOCUS spectrometer and hadronic signals in FOCUS*, Talk presented in *Heavy Quarks at Fixed Target*, 1998.

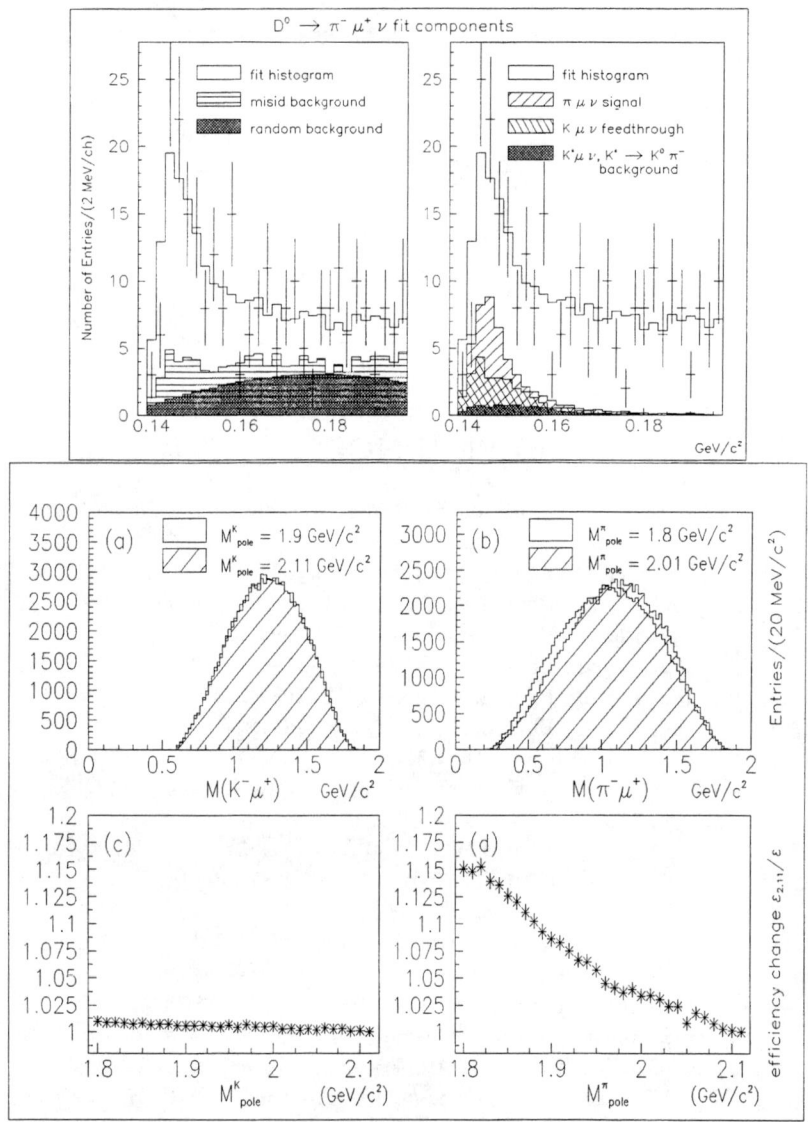

FIGURE 4. In the upper 2 plots we show the components of the fit, plotted as a mass difference, to the D^* tagged $D^0 \to \pi^- \mu^+ \nu_\mu$ signal. (The electron and combined plots can be seen in reference [7].) The components are shown as patterned histograms and the fit is the uppermost solid line. In the 4 bottom plots we demonstrate the heightened sensitivity of the $D^0 \to \pi^- \mu^+ \nu_\mu$ signal to hadronic structure in comparison to the decay $D^0 \to K^- \mu^+ \nu_\mu$ using Equation 1. Notice how the higher q^2 reach of the $D^0 \to \pi^- \mu^+ \nu_\mu$ brings us closer to the virtual pole of Equation 1 and alters the overall rate as we change the pole mass in d).

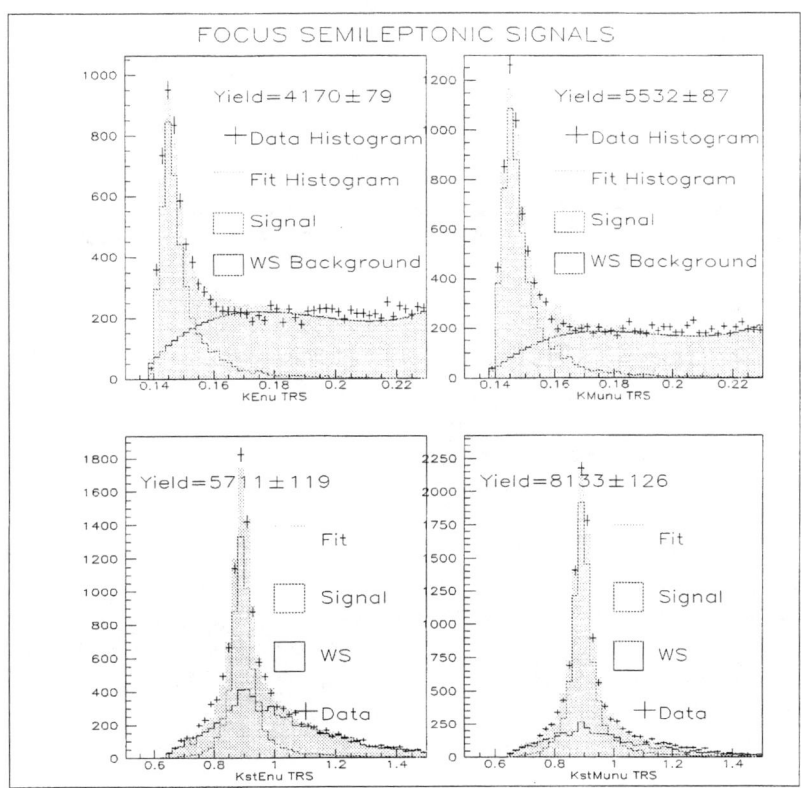

FIGURE 5. Yields of the decay modes $D^0 \to K^- e^+ \nu_e$, $D^0 \to K^- \mu^+ \nu_\mu$, $D^+ \to \overline{K}^{*0} e^+ \nu_e$ and $D^+ \to \overline{K}^{*0} \mu^+ \nu_\mu$ based on a very preliminary, non-optimized analysis from 23% of the FOCUS sample. For the D^0 analyses, we have utilized the decay chain $D^{*+} \to D^0 \pi^+$, where the charge of the pion "tags" the event "right sign" if $Q(\pi) = Q(l)$ and "wrong sign"(WS Background) if $Q(\pi) \neq Q(l)$. For the D^+ analyses, we use the charge correlation between the pion from the $K^{*0}(892)$ resonance and the lepton. The event is "right sign" if $Q(\pi) = Q(l)$ and "wrong sign"(WS Background) if $Q(\pi) \neq Q(l)$. In the plots, the solid histogram represents the fit to the data. The fit histogram has been constructed from a Monte Carlo signal histogram (resonance) and a "wrong sign" background histogram.

Semileptonic B Decays from CLEO

Karl M. Ecklund

Laboratory of Nuclear Studies
Cornell University, Ithaca, New York 14853
e-mail: kme@mail.lns.cornell.edu

Abstract. I report recent results in semileptonic decays of B mesons from the CLEO collaboration. Results for exclusive reconstruction of $B \to D\ell\nu$ and $B \to \rho\ell\nu$ are given including the q^2 dependence of the form factors. Two preliminary analyses using inclusive techniques measure the lepton momentum spectrum and hadronic recoil mass spectrum in $B \to X_c \ell \nu$ decays, showing promise for future precision measurements of V_{cb}.

INTRODUCTION

Study of semileptonic decays of B mesons allows measurement of the CKM matrix elements V_{ub} and V_{cb}. The determination of CKM matrix elements from the partial width for a semileptonic decay is free of theoretical uncertainty from final state interactions; comparable hadronic decays do suffer from larger theoretical uncertainties. Accurate measurement of CKM matrix elements becomes increasingly important as we approach the era of the B-Factories and studies of CP violation in B meson decays. Semileptonic decays also set constraints on the unitarity triangle of the CKM matrix: measurement of the ratio $|V_{ub}|/|V_{cb}|$ gives an annular constraint in the complex ρ-η plane. In recent years heavy quark effective theory (HQET) has provided detailed and rather robust predictions for the dynamics of heavy quark decay. These predictions are beginning to be tested in semileptonic decays of B mesons.

CLEO EXPERIMENT

CLEO is a 4π solenoidal detector located at the interaction region of the Cornell Electron Storage Ring (CESR). CESR is an e^+e^- collider operating on the $\Upsilon(4S)$ resonance at a center of mass energy of 10.58 GeV, just above the threshold for $B\bar{B}$ production. $\Upsilon(4S)$ decays are essentially 100% $B_d^0 \bar{B}_d^0$ and B^+B^- pairs. At threshold the B's are produced nearly at rest: $p_B \approx 300$ MeV/c. In addition to $\Upsilon(4S)$ production with a cross section of 1.0 nb, there is continuum production

(3.0 nb) of $e^+e^- \to$ hadrons. CLEO also collects data 60 MeV below the $\Upsilon(4S)$ for use in subtraction of this continuum from on-resonance data.

The central region of the CLEO detector consists of three concentric cylindrical drift chambers, a scintillator time-of-flight system and a CsI calorimeter all inside a superconducting coil and 1.5 T magnetic field. Endcap time-of-flight and CsI calorimeters provide forward and backward coverage for a total of 95% of the solid angle. The drift chambers provide excellent tracking and momentum resolution, and the calorimeter has excellent photon and π^0 identification. In the flux return for the superconducting solenoid, proportional tube counters provide muon identification at depths of 3, 5 and 7 interaction lengths. The CLEO detector is described in detail elsewhere [1].

$B \to D\ell\nu$ BRANCHING FRACTION AND FORM FACTORS

This analysis is complementary to the $B \to D^*\ell\nu$ decay which currently gives the best measurement of V_{cb}. It also provides a test of the HQET predictions for the dynamics of heavy quark decay.

In a preliminary analysis [2] reported at ICHEP 98, we identify events containing a D^+ or D^0 (and charge conjugates) and an electron or muon (ℓ). The D-ℓ combinations give a sample including $B \to D\ell\nu$, $B \to D^*\ell\nu$, $B \to D^{**}\ell\nu$ and $B \to D^{(*)}\pi\ell\nu$. We separate $B \to D\ell\nu$ from the other semileptonic modes using the energy and momentum of the particle(s) recoiling against the D-ℓ pair. The yield of $D\ell\nu$ events in bins of q^2, the invariant mass of the virtual W, gives information on the partial width and form factors in the decay $B \to D\ell\nu$.

We select events with at least five charged tracks and reconstruct D candidates in the modes $D^0 \to K^-\pi^+$ or $D^+ \to K^-\pi^+\pi^+$, separating K and π tracks by using time-of-flight and dE/dx measurements. The invariant mass of D^0 candidates must satisfy 1.835 GeV/$c^2 < M(K\pi) <$ 1.893 GeV/c^2; that of D^+ candidates must be between 1.846 GeV and 1.890 GeV/c^2. To suppress D's from the continuum, we require the D candidates to have momentum $p_D < 2.5$ GeV/c.

We select electron candidates of momentum 0.8 GeV/$c < p_\ell <$ 2.4 GeV/c using the CsI calorimeter. Muon candidates must have associated hits in the muon counters, penetrating at least 5 interaction lengths of material, which increases the lower momentum cut for muons to 1.4 GeV/c. For 90% of signal $D\ell\nu$ events, the lepton and D lie in opposite hemispheres; we require this of all D-ℓ pairs.

For each D-ℓ pair we compute $\cos\theta_{B-D\ell}$, the angle between the $D\ell$ momentum and the B momentum assuming that the only missing particle is a massless neutrino.

$$\cos\theta_{B-D\ell} = \frac{2E_B E_{D\ell} - M_B^2 - M_{D\ell}^2}{2|\vec{p}_B||\vec{p}_{D\ell}|} \quad (1)$$

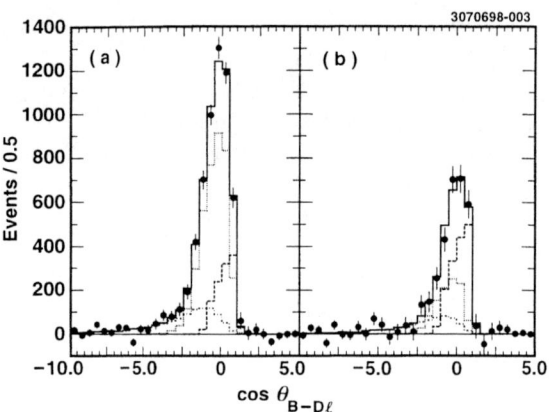

FIGURE 1. The $\cos\theta_{B-D\ell}$ distribution for (a) $D^0 X\ell\nu$ and (b) $D^+ X\ell\nu$ candidates. The data (solid circles) are overlaid with simulated $B \to D\ell\nu$ decays (dashed histogram), $B \to D^*\ell\nu$ decays (dotted histogram), $B \to D^{**}\ell\nu + D^{(*)}\ell\nu$ decays (dash-dotted histogram), and their total (solid histogram).

For $B \to D\ell\nu$ decays $\cos\theta_{B-D\ell}$ lies between -1 and 1. When final state particles other than the neutrino are missing, it is shifted towards negative values. Thus we may use this quantity to distinguish $D\ell\nu$ from $DX\ell\nu$. Before doing so other backgrounds must be subtracted.

Background sources yielding a D-ℓ pair may arise from (1) random $K\pi(\pi)$ combinations (fake D), (2) a D paired with a lepton from the other B decay (uncorrelated), (3) a D paired with a lepton that is a granddaughter of the same B (correlated), (4) misidentification of a hadron as a lepton, or (5) $e^+e^- \to q\bar{q}$ events. We remove backgrounds from fake D candidates by using events in the D mass sidebands. The uncorrelated background contribution is estimated from our data by flipping the direction of leptons in the same hemisphere as the D candidate. The small amount of correlated background (e.g. from $B \to D^{(*)}\tau\nu$, $\tau \to \ell\nu\bar{\nu}$) is removed using Monte Carlo (MC) simulation. Fake leptons and continuum events are subtracted using measured fake rates and off-resonance data respectively.

In Figure 1 the resulting $\cos\theta_{B-D\ell}$ distributions are shown along with a fit to the data. We model the distributions in the fit using MC simulation and various models for $b \to c$ semileptonic decay: for $B \to D\ell\nu$ we use ISGW2 [3,4]; for $B \to D^*\ell\nu$ we use CLEO form factors [6,7]; for $B \to D^{**}\ell\nu$ we use ISGW2; and for non-resonant $B \to D^{(*)}\pi\ell\nu$ we use the results of Goity and Roberts [8].

To extract form factor results we perform the fit to $\cos\theta_{B-D\ell}$ in ten bins of the HQET variable $w = (m_B^2 + m_D^2 - q^2)/(2m_B m_D)$, where q^2 is the invariant mass of the D-ℓ pair. The $D\ell\nu$ yield in each w bin is shown in Figure 2. We fit the differential decay rate

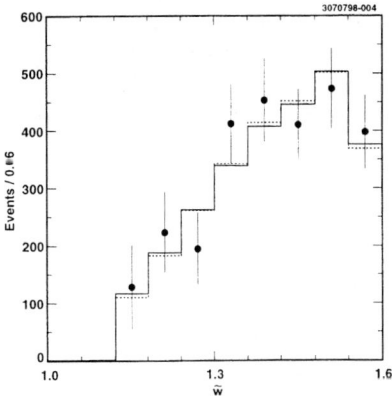

FIGURE 2. The sum of $B^- \to D^0 \ell \bar\nu$ and $\bar B^0 \to D^+ \ell \bar\nu$ yields as a function of $\tilde w$, for the data (solid circles) and using the best fit linear form factor (dashed histogram) or dispersion relation inspired form factor of Boyd et al. (solid histogram).

TABLE 1. Summary of $B \to D\ell\nu$ form factor fits.

Form Factor	ρ_D^2	c_D	$10^2 \lvert V_{cb} \rvert F_D(1)$
Linear	$0.76 \pm 0.16 \pm 0.09$	-	$4.05 \pm 0.45 \pm 0.32$
Free curvature	$0.77^{+1.18}_{-2.83} \pm 0.09$	$0.01^{+1.70}_{-3.96} \pm 0.001$	$4.05^{+1.51}_{-1.63} \pm 0.32$
Boyd	$1.30 \pm 0.27 \pm 0.16$	$1.21 \pm 0.31 \pm 0.15$	$4.48 \pm 0.61 \pm 0.37$
Caprini	$1.27 \pm 0.25 \pm 0.15$	$1.18 \pm 0.26 \pm 0.14$	$4.44 \pm 0.58 \pm 0.36$

$$\frac{d\Gamma}{dw} = \frac{G_F^2 \lvert V_{cb} \rvert^2}{48\pi}(m_B + m_D)^2 m_D^3 (w^2 - 1)^{3/2} F_D(w)^2 \qquad (2)$$

assuming different form factors $F_D(w)$. The fit accounts for detector acceptance and smearing in the reconstruction of w due to motion of the B and detector resolution.

The preliminary results of the fit are given in Table 1. We first parameterize the form factor as a Taylor expansion about $w = 1$: $F_D(w) = F_D(1)(1 - \rho_D^2(w-1) + c_D(w-1)^2)$. We first fit using only a linear term ($c_D = 0$) and then include the curvature term. We also parameterize the form factor using the dispersion relation inspired result of Boyd et al. [5], which contains terms of higher order in $(w-1)$. Similar results are obtained using the parameterization of Caprini et al. [9,10].

We obtain the total decay rate for $B \to D\ell\nu$ by integrating $d\Gamma/dw$ over w using best fit values to Boyd et al.'s parameterization of the the form factor. We find $\Gamma(B \to D\ell\nu) = (14.1 \pm 1.0 \pm 1.2)$ ns^{-1}, where we have combined $B^- \to D^0 \ell^- \bar\nu$ and $\bar B^0 \to D^+ \ell^- \bar\nu$) samples by assuming that $B^0 \bar B^0$ and $B^+ B^-$ saturate the decays of the $\Upsilon(4S)$. Using measured B lifetimes this implies the branching fractions, $\mathcal{B}(B^- \to D^0 \ell^- \bar{\nu}) = (2.32 \pm 0.17 \pm 0.20)\%$ and $\mathcal{B}(\bar{B}^0 \to D^+ \ell^- \bar{\nu}) = (2.20 \pm 0.16 \pm 0.19)\%$, where the errors are statistical and systematic respectively. Since we derive the branching fractions from the decay width, the errors are completely correlated. These results are combined with the previous CLEO measurement [11] in Equation 3, taking into account all correlations.

$$\begin{aligned}\Gamma(B \to D\ell\bar{\nu}) &= (13.4 \pm 0.8 \pm 1.2) \text{ ns}^{-1} \\ \mathcal{B}(B^- \to D^0 \ell^- \bar{\nu}) &= (2.21 \pm 0.13 \pm 0.19)\% \\ \mathcal{B}(\bar{B}^0 \to D^+ \ell^- \bar{\nu}) &= (2.09 \pm 0.13 \pm 0.18)\%\end{aligned} \quad (3)$$

From the best fit parameters to the dispersion relation form factor we also obtain

$$|V_{cb}| F_D(1) = (4.48 \pm 0.61 \pm 0.37) \times 10^{-2} \quad (4)$$

Theoretical expectations for $F_D(1)$ range from 0.98 ± 0.07 [9] to 1.03 ± 0.07 [3,4]. A recent lattice calculation finds the preliminary value 1.069 ± 0.029 [12]. Using $F_D(1) = 1$ we find $V_{cb} = 0.045 \pm 0.006 \pm 0.004 \pm 0.005$, where the errors are statistical, systematic and due to theoretical uncertainty in $F_D(1)$. This value of V_{cb} is consistent with that obtained in studies of the decay $B \to D^* \ell \nu$. If we use, instead, the best fit parameters to a linear form factor, the value of V_{cb} decreases by 10%. Linear models for the form factor have been used in extracting V_{cb} from $B \to D^* \ell \nu$ decays. Including higher order terms in the $B \to D^* \ell \nu$ form factor may also have an important effect in such analyses.

$B \to \rho \ell \nu$ BRANCHING FRACTION AND Q^2 DEPENDENCE

CLEO can also measure $b \to u \ell \nu$ decays which are sensitive to V_{ub}. Experimentally such measurements are more difficult due to large backgrounds from the Cabibbo favored $b \to c \ell \nu$ decays.

In a preliminary analysis [13] reported at ICHEP 98, we analyze the decay $B \to \rho \ell \nu$ using high momentum leptons paired with π, ρ and ω candidates. In the high momentum region we are able to measure the q^2 distribution of $B \to \rho \ell \nu$ events.

We select events with leptons of energy $E_\ell > 1.7$ GeV/c accompanied by a hadronic system consistent with a ρ ($\pi^+\pi^-$ or $\pi^\pm\pi^0$), ω ($\pi^+\pi^-\pi^0$) or π ($\pi^\pm$ or π^0). To reduce background from $b \to c \ell \nu$ decays we divide the sample into three lepton energy bins: HILEP (2.3–2.7 GeV/c), LOLEP (2.0–2.3 GeV/c) and LOLOLEP (1.7–2.0 GeV/c). Leptons in the HILEP bin have energy above the kinematic endpoint for $b \to c \ell \nu$ decays. The LOLEP bin contains mostly $b \to c \ell \nu$ events but still has some sensitivity to $b \to u \ell \nu$. The lowest energy bin provides a normalization of the $b \to c \ell \nu$ background.

The dominant source of background in the highest energy bin comes from continuum production of hadrons: $e^+ e^- \to q\bar{q}$, $q = u, d, s, c$. Since the decays of B

mesons at rest are more spherical than jet-like $q\bar{q}$ events, we suppress this background using event shape variables. We obtain additional suppression by requiring $\cos\theta_{B-\rho\ell}$ to be physical. (See Equation 1.)

For each $\rho\ell\nu$ candidate, we compute $\Delta E = E_\rho + E_\ell + |\vec{p}_{\text{miss}}| - E_{\text{beam}}$, where $\vec{p}_{\text{miss}}$ is the net missing momentum in the event. For signal events, ΔE should peak near zero since $\vec{p}_{\text{miss}}$ gives a measure of the neutrino energy and momentum. Because we rely on the hermeticity of the detector for this measurement, we require the missing momentum not to point down the beam pipe. We also require $\vec{p}_{\text{miss}}$ to be within 35° of the neutrino direction inferred from the $\rho + \ell$ candidate; the later is known up to an azimuthal ambiguity about the B momentum direction.

To measure the $\rho\ell\nu$ branching fraction, we perform a simultaneous maximum likelihood fit for all five modes in all three lepton energy bins. We fit in two variables, ΔE and $m(\pi\pi(\pi))$, for the ρ and ω modes; for the π modes, we fit only to ΔE. The fit contains contributions from the physics processes $B \to \rho(\omega)\ell\nu$, $B \to \pi\ell\nu$, $b \to u\ell\nu$ (modes other than ρ, ω and π) and $b \to c\ell\nu$. The fit also contains background contributions from continuum and fake leptons; we measure these contributions using off-resonance data and known fake rates. The signal shapes for the fit are taken from Monte Carlo simulation using various form factor models for $B \to \rho\ell\nu$ and $B \to \pi\ell\nu$, the ISGW2 [4] model for $b \to u\ell\nu$ and a combination of ISGW2 and CLEO form factor results [6] for $b \to c\ell\nu$. Isospin and quark model relations are used to constrain the relative normalizations of the three vector modes ($B^0 \to \rho^-\ell^+\nu$, $B^+ \to \rho^0\ell^+\nu$ and $B^+ \to \omega\ell^+\nu$) and, separately, the normalizations of the pseudoscalar modes ($B^0 \to \pi^-\ell^+\nu$ and $B^+ \to \pi^0\ell^+\nu$). Our fit also accounts for the large cross-feed between the various vector signal modes.

Figure 3 shows projections of the maximum likelihood fit for $\pi^+\pi^-$ and $\pi^\pm\pi^0$ modes in the high energy bin onto the variables ΔE and $M(\pi\pi)$ overlayed with the data. We average over the various form factor models for $\rho\ell\nu$ and $\pi\ell\nu$, finding

$$\mathcal{B}(B \to \rho\ell\nu) = (2.8 \pm 0.4 \pm 0.4 \pm 0.6) \times 10^{-4} \qquad (5)$$
$$|V_{ub}| = (3.2 \pm 0.3^{+0.2}_{-0.3} \pm 0.6) \times 10^{-3},$$

where the errors are statistical, systematic and due to model dependence. The results for $\mathcal{B}(B \to \rho\ell\nu)$ and V_{ub} are consistent with an earlier CLEO analysis [15]. The two results are statistically independent, but the systematic and model dependence uncertainties are largely correlated.

We are also able to measure the q^2 distribution for $B \to \rho\ell\nu$ events with $E_\ell > 2.3$ GeV/c. Figure 4 shows the data distribution of q^2 after requiring $|M(\pi\pi) - M_\rho| < \Gamma_\rho$ and $\Delta E < 500$ MeV. The results of the maximum likelihood fit are superimposed showing clearly that a measurement of the q^2 distribution with this data is possible.

MOMENT ANALYSIS OF $B \to X_C \ell \nu$

Inclusive measurements of $b \to c\ell\nu$ also give information on V_{cb}. CLEO has two preliminary results based on inclusive techniques for measuring semileptonic de-

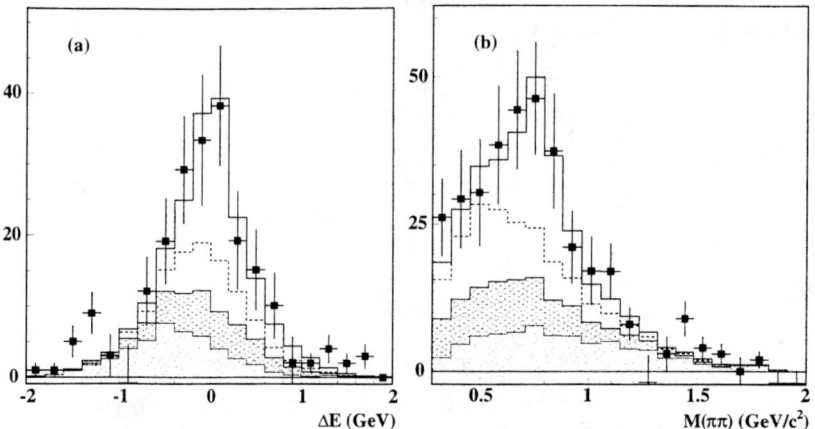

FIGURE 3. Projections of maximum likelihood fit for HILEP energy bin: (a) ΔE distribution with the cut $|M(\pi\pi) - M_\rho| < \Gamma_\rho$ and (b) $M(\pi\pi)$ distribution with the cut $\Delta E < 500$ MeV. The points are the continuum subtracted data. The solid histogram is the fit, represented as the sum of three components: signal and cross-feed (open regions), background from non-signal $b \to u\ell\nu$ (double-hatched region) and background from $b \to c\ell\nu$ (single-hatched region).

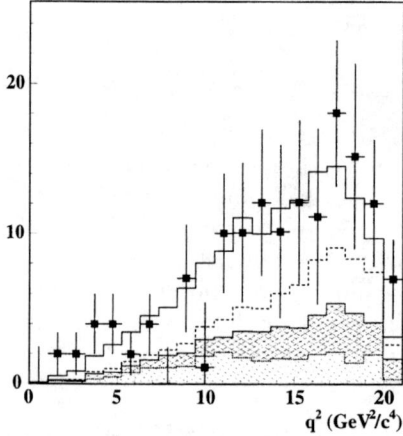

FIGURE 4. Projection of the fit onto q^2 for HILEP after the cuts $\Delta E < 500$ MeV and $|M(\pi\pi) - M_\rho| < \Gamma_\rho$. The points show the on-resonance data after continuum subtraction, while the histogram shows the projection of the fit. The fit components are shown as specified in Fig. 3.

cays [16,17]. Both rely on heavy quark effective theory (HQET) and the operator product expansion (OPE) to interpret the results. Within this framework an inclusive measurement summed over many final states is readily interpreted from quark level calculations. What is interesting in both of these analyses is that they each help constrain parameters of HQET and thus show promise to decrease the theoretical uncertainty in converting inclusive rate measurements into measurements of V_{cb}. Combining the two analyses over-constrains the theory parameters thus allowing a test of the theoretical framework and experimental understanding of b-quark decays.

Hadronic Mass Moments

For decays $B \to X_c \ell \nu$, the first method measures the first and second hadronic mass moments. Falk et al. [18] give an expansion for the moments of the hadronic mass (M_{X_c}) distribution in the variables $1/M_B$ and α_s. Equation 6 gives the expression for the first moment to order $1/M_B^2$.

$$\langle M_{X_c}^2 - \bar{M}_D^2 \rangle = M_B^2 [0.0272 \tfrac{\alpha_s}{\pi} + 0.207 \tfrac{\bar{\Lambda}}{M_B}(1 + 0.43 \tfrac{\alpha_s}{\pi}) + 0.193 \tfrac{\bar{\Lambda}^2}{M_B^2} \qquad (6)$$
$$+ 1.38 \tfrac{\lambda_1}{M_B^2} + 0.203 \tfrac{\lambda_2}{M_B^2}]$$

Here λ_1, λ_2 and $\bar{\Lambda}$ are the HQET parameters. The B^*-B mass splitting determines λ_2, so by measuring the first two moments and inverting the equations one may determine or constrain the remaining HQET parameters λ_1, which is proportional to the kinetic energy of the b-quark in the B meson, and $\bar{\Lambda}$, which relates the b-quark mass and the B meson mass.

To measure the hadronic mass moments in semileptonic B decays we select events with one lepton of momentum $p_\ell > 1.5$ GeV/c. We "reconstruct" the neutrino using the hermeticity of the detector, imposing strict event quality cuts to insure no particles are missed. The net charge of the event must be zero, and the missing mass must be consistent with a neutrino. The mass recoiling against the lepton and neutrino is:

$$M_{X_c}^2 = M_B^2 + M_{\ell\nu}^2 - 2 E_B E_{\ell\nu} + 2|\vec{P}_B||\vec{P}_{\ell\nu}|\cos\theta_{\ell\nu-B}. \qquad (7)$$

Since the B momentum is small but the direction is unknown, we approximate $M_{X_c}^2$ by dropping the last term.

$$\widetilde{M_{X_c}^2} = M_B^2 + M_{\ell\nu}^2 - 2 E_B E_{\ell\nu} \qquad (8)$$

The resulting distribution shown in Figure 5a has contributions from $b \to c\ell\nu$ (96%), $b \to c \to s\ell\nu$ (3%) and $b \to u\ell\nu$ (1%). We compute the moments after background subtraction using MC shapes. We further correct for a bias in the reconstructed hadronic mass due to asymmetric resolution of the neutrino reconstruction. We find $\langle M_{X_c}^2 - \bar{M}_D^2 \rangle = 0.286 \pm 0.023 \pm 0.080$ GeV^2 and

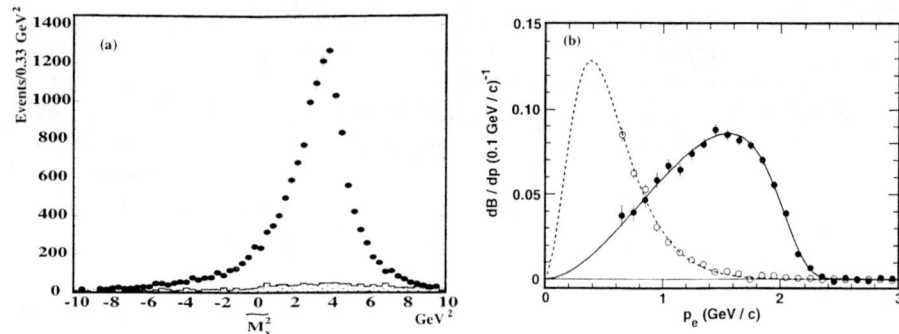

FIGURE 5. (a) Measured $\widetilde{M^2_{X_c}}$ distributions for on-resonance data (points) and scaled off-resonance data (hatched histogram). (b) Electron momentum spectrum from $B \to Xe\nu$ (solid circles) and $b \to c \to ye\nu$ (open circles). The curves show the best fit to the ISGW model with 23% $B \to D^{**}\ell\nu$.

$\langle (M^2_{X_c} - \bar{M}^2_D)^2 \rangle = 0.911 \pm 0.066 \pm 0.309$ GeV^4, where $\bar{M}_D$ is the spin-averaged charm meson mass (1.975 GeV). Inverting the theoretical expression for the moments, calculated to order $1/M^3_B$, and solving for HQET parameters gives the results in Equation 9.

$$\bar{\Lambda} = +0.33 \pm 0.02 \pm 0.08 \text{ GeV}$$
$$\lambda_1 = -0.13 \pm 0.01 \pm 0.06 \text{ (GeV}/c)^2 \tag{9}$$

Equivalently, each moment measurement provides a constraint in the λ_1-$\bar{\Lambda}$ plane. The allowed bands and overlap region are shown in Figure 6.

Lepton Energy Moments

The second method uses the inclusive electron spectrum from B decays measured by CLEO [19]. Theoretical expressions for the moments of the lepton spectrum are given by Voloshin et al. [20]. As in the case of the hadronic mass moments, these expressions may be inverted to place constraints on λ_1 and $\bar{\Lambda}$.

The $B \to Xe\nu$ electron spectrum measurement [19] shown in Figure 5b is an observed spectrum above 0.6 GeV. To measure the moments and compare to theory, we must apply corrections to the observed spectrum. We must extrapolate below 0.6 GeV and correct for detector smearing (including bremsstrahlung) and motion of the B in the lab frame. There are also electromagnetic radiative corrections which are not included in the theoretical expressions for the moments. After all corrections we find the preliminary results in Equation 10.

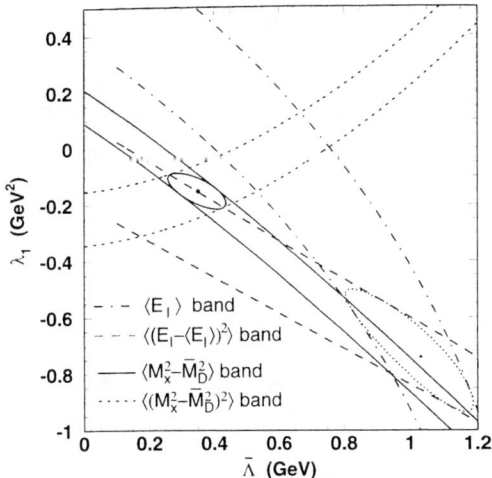

FIGURE 6. Combined constraints on HQET parameters λ_1 and $\bar{\Lambda}$ from hadronic recoil mass moments and lepton energy spectrum moments. The allowed bands and error ellipses give the 1 σ contours (including systematic uncertainties) for the constraints from the moment measurements.

$$\langle E_\ell \rangle = 1.36 \pm 0.01 \pm 0.02 \text{ GeV} \tag{10}$$
$$\langle (E_\ell - \langle E_\ell \rangle)^2 \rangle = 0.19 \pm 0.004 \pm 0.005 \text{ GeV}^2$$

The moment measurements can be converted to allowed bands in the λ_1-$\bar{\Lambda}$ plane. Figure 6 shows the two bands obtained from the lepton energy moment measurements along with the bands from the hadronic moment analysis. In these preliminary analyses, the agreement among the four allowed bands is poor. Barring experimental error, a resolution of the discrepancy may require higher order expansions for the lepton energy moments. These moments are presently only calculated to second order in $1/M_B$, while the hadronic mass moments are calculated to third order. If the preliminary results of the two methods continue to disagree, we may have to question the assumption of quark-hadron duality implicit in such inclusive analyses.

CONCLUSION

The new CLEO measurements of the $B \to D\ell\nu$ form factors and q^2 distribution in $B \to \rho\ell\nu$ show progress in the experimental understanding of the dynamics of heavy quark decay. This understanding, coupled with more theoretical work, should make possible more precise determinations V_{cb} and V_{ub}. Likewise the new moment based analyses of inclusive semileptonic B decays seek to use data to constrain theory parameters and thereby reduce the uncertainties in extracting V_{cb} from the inclusive rate for $b \to c\ell\nu$.

REFERENCES

1. Y. Kubota et al., Nucl. Instrum. Methods A **320**, 66 (1992).
2. M. Artuso et al., ICHEP98 856, CLEO CONF 98-12 (1998); Submitted to Phys. Rev. Lett.
3. N. Isgur et al., Phys. Rev. D **39**, 799 (1989).
4. D. Scora and N. Isgur, Phys. Rev. D **52**, 2783 (1995).
5. C. G. Boyd, B. Grinstein, and R. F. Lebed, Phys. Rev. D **56**, 6895 (1997).
6. B.Barish et al., Phys. Rev. D **51**, 1014 (1995).
7. J.E. Duboscq et al., Phys. Rev. Lett. **79**, 3898 (1996).
8. J. Goity and W. Roberts, Phys. Rev. D **51**, 3459 (1995).
9. I. Caprini and M. Neubert, Phys. Lett. B **380**, 376 (1996).
10. I. Caprini, L. Lellouch and M. Neubert, CERN-TH/97-21, hep-ph/9712417.
11. M. Athanas et al., Phys. Rev. Lett. **79**, 2208 (1997).
12. S. Hashimoto, Sixteenth International Symposium on Lattice Field Theory (Lattice'98), Boulder Colorado, July 13–18, 1998.
13. M. Artuso et al., ICHEP98 855, CLEO CONF 98-18 (1998).
14. Particle Data Group, Phys. Rev. D54, 477 (1996).
15. J. Alexander et al., Phys. Rev. Lett. **77**, 5000 (1996).
16. M. Artuso et al., ICHEP98 1013, CLEO CONF 98-21 (1998).
17. S.E.Roberts, Ph.D. thesis, Univ. of Rochester (1997).
18. A. Falk, M. Luke, M. Savage, Phys. Rev. D **53**, 2491; ibid. D **53**, 6316 (1996).
19. B. Barish et al., Phys. Rev. Lett. **76**, 1570 (1996).
20. M. Voloshin, Phys. Rev. D **51**, 4934 (1995).

Lattice QCD Calculations of Leptonic and Semileptonic Decays

Andreas S. Kronfeld

Fermi National Accelerator Laboratory,[1] Batavia, Illinois, U.S.A.

Abstract. In lattice QCD, obtaining properties of heavy-light mesons has been easier said than done. Focusing on the B meson's decay constant, it is argued that towards the end of 1997 the last obstacles were removed, at least in the quenched approximation. These developments, which resulted from a fuller understanding and implementation of ideas in effective field theory, bode well for current studies of neutral meson mixing and of semileptonic decays.

INTRODUCTION

Eleven years ago at *Lattice '87*, three talks gave birth to a new field of research, the study of heavy quarks in lattice QCD. Estia Eichten [1] emphasized that lattice QCD should provide reliable information on hadrons with heavy quarks, which would help determine the Cabibbo-Kobayashi-Maskawa (CKM) matrix. He suggested starting with the *static* approximation and adding corrections in $1/m_Q$. One of the developments to grow from this suggestion is the heavy-quark effective theory (HQET), an enormous subject in its own right. Peter Lepage [2] introduced non-relativistic QCD (NRQCD) as a tool for studying spectroscopy and matrix elements of systems with one or more heavy quarks. This effective theory also has enjoyed widespread application. Finally, Luciano Maiani [3] presented, among other things, the first well-known calculation from lattice QCD of the leptonic decay constant of the B meson. Maiani and his collaborators took Wilson's action for light quarks and bravely applied it to the b quark, even though their lattice's ultraviolet cutoff was below the mass m_b.

The idea that lattice QCD could play a role in interpreting future experiments was exciting, and it attracted much attention. In addition to the spectra of quarkonium and "heavy-light" hadrons, considerable effort has been devoted to the hadronic matrix elements for leptonic and semileptonic decays of heavy-light mesons (B, B_s, D, and D_s) and neutral meson mixing. The decay constant of the

---

[1] Fermilab is operated by Universities Research Association Inc., under contract with the United States' Department of Energy.

generic heavy-light pseudoscalar meson, denoted f_P, parameterizes the hadronic amplitude for leptonic decays. It has received the most attention, especially f_B. It was expected to be the most straightforward of matrix elements and, thus, a bellwether. Unfortunately, a reliable calculation could not be done quickly. Fortunately, the technical and conceptual difficulties have been largely overcome, and the calculation of f_B in lattice QCD has grown up at last.

Computationally f_B is indeed just as easy as f_π, but the interpretation of the results has not been obvious. Consequently, the literature contains a wide range of estimates, some of which should not be taken seriously. For example, several early calculations in the static limit contained too much contamination from radial excitations B [4], and the results are misleadingly high.

More recently, a fuller understanding of the interplay between the effective theories and the lattice has helped to reduce the effects of lattice artifacts. The effective theories NRQCD and HQET are derived from QCD by lowering the renormalization point, or cutoff, μ until $|\boldsymbol{p}| \ll \mu \approx m_Q$, where $|\boldsymbol{p}|$ is the heavy quark's typical momentum and m_Q its mass. Because $|\boldsymbol{p}|/m_Q \ll 1$ the interactions in the effective Lagrangian can be organized in powers of $|\boldsymbol{p}|/m_Q$. (In quarkonium $|\boldsymbol{p}| \sim \alpha_s m_Q$; in heavy-light systems $|\boldsymbol{p}| \sim \Lambda_{\rm QCD} \sim 200$ MeV.) The two effective theories share the same Lagrangian, although the power of $|\boldsymbol{p}|/m_Q$ assigned to operators of higher dimension can differ. One can take the effective theories' Lagrangian and introduce the lattice as an ultraviolet regulator, choosing the lattice spacing a so that $m_Q \approx \mu \sim \pi/a$. This lattice theory is often called lattice NRQCD [5]. There are discretization effects, which are the higher-dimension operators multiplied by calculable coefficients. Although some of these operators are new, many are the same as in the $1/m_Q$ expansion. Consequently, physical $1/m_Q$ effects and artificial a effects have become intertwined. Lattice practitioners must disentangle them and remove the lattice artifacts, at least to the desired accuracy.

Alternatively, one can start with an action derived for $m_q a \ll 1$, such as Wilson's, and ask what happens when one applies it for $m_Q a \approx 1$. This, essentially, is the way of Ref. [3]. Since then, many experts have said (and still do, out of habit), that a heavy quark cannot be put directly on the lattice because $m_Q a \approx 1$. On the other hand, numerous calculations have been published with $m_Q a \approx 1$, so there must be more to the story. Indeed, the lattice theory does not break down. Instead, the lattice artifacts are again intertwined with the $1/m_Q$ effects, as in lattice NRQCD. The same operators appear, but the coefficients are different, though still calculable. For a class of actions based on the Wilson action, it has been shown, to all orders in perturbation theory, that the coefficients remain small, for all $m_Q a$ [6].

In preparing a brief review of a subject one is faced with the choice between a catalog of all recent results or a synthesis of developments over a longer period of time. The proceedings of the Lattice conferences provide excellent examples of the former [7–9]. By contrast, this paper gives a view of the (theoretical and computational) progress, focusing on f_B. Owing to space limitations the material presented on the allied subject of neutral meson mixing and on phenomenologically promising form factors of semileptonic decays is brief.

NUMERICAL LATTICE CALCULATIONS

When one thinks of lattice QCD, one usually thinks of large-scale numerical calculations. This approach computes the functional integrals of quantum field theory by applying a Monte Carlo method with importance sampling. Statistical errors arise here, and with more and more computer time these errors can be made arbitrarily small. Over the years various clever techniques have been devised to enhance the "signal-to-noise" ratio, that is, to reduce the statistical error for fixed computing resources.

To use Monte Carlo methods, three modifications are introduced. First, spacetime becomes a finite box, usually with periodic boundary conditions. Second, the spacetime continuum becomes a discrete lattice. Last, the so-called quenched approximation is applied. The first two are common to many kinds of numerical analysis, but the last requires a short explanation. The quenched approximation treats a hadron's valence quarks and all the exchanged gluons fully, including retardation, but the back-reaction of closed quark loops on the gluons is omitted. For particle physics a more descriptive name (and one of the original names) would be the valence approximation, but the term "quenched," taken from statistical mechanics, is more commonly used. The back-reaction of closed quark loops is computationally burdensome. Because its omission saves computer time, the quenched approximation is a useful way to control the other errors and, thus, to teach theorists how to analyze uncertainties.

Let us return to the first two approximations. A volume larger than a few fm on a side should be good enough. After all, one does not expect the true size and boundary of the universe to effect the physics of hadrons. It is with the lattice itself that the subject becomes a craft. The continuum limit can be reached by taking the lattice spacing $a \to 0$ with brute force, or by improving the action to reduce discretization effects, or by a combination of the two. The crudest form of brute force, namely to take $m_Q a \ll 1$, would require, for the bottom quark, a fantastically small lattice spacing. This has never been done.

B, D, K & CKM

Table 1 contains a list of specific reactions with conventional parameterizations of hadronic matrix elements. Together each pair can determine the listed element of the CKM matrix. Every element of the CKM matrix appears except V_{tb} and the entire (3×3 unitary) matrix can be constrained.

In the Standard Model, the leptonic partial width of a generic pseudoscalar meson P, containing quarks of flavors p and q, is given by

$$\Gamma_{P \to l\nu} = \begin{pmatrix} \text{known} \\ \text{factors} \end{pmatrix} f_P^2 |V_{pq}|^2. \tag{1}$$

If one could compute f_P (i.e., f_K or f_D or f_B) with a reliably estimated uncertainty, a measurement of the partial width is tantamount to a measurement of $|V_{pq}|$.

TABLE 1. How to combine exclusive experimental measurements with calculations in (lattice) QCD to obtain the Cabibbo-Kobayashi-Maskawa matrix.

Leptonic decays			Semileptonic decays		
measure	compute	determine	measure	compute	determine
$\pi \to \mu\nu$	f_π	$|V_{ud}|$			
$K \to \mu\nu$	f_K	$|V_{us}|$	$K \to \pi e\nu$	$f_\pm^{K\pi}(q^2)$	$|V_{us}|$
$B \to \tau\nu$	f_B	$|V_{ub}|$	$B \to \pi e\nu$ $\to \rho e\nu$	$f_\pm^{B\pi}(q^2)$ $A_1^{B\rho}(q^2)$	$|V_{ub}|$
$D \to \mu\nu$	f_D	$|V_{cd}|$	$D \to \pi e\nu$ $\to \rho e\nu$	$f_\pm^{D\pi}(q^2)$ $A_1^{D\rho}(q^2)$	$|V_{cd}|$
$D_s \to \mu\nu$	f_{D_s}	$|V_{cs}|$	$D \to K e\nu$ $\to K^* e\nu$	$f_\pm^{DK}(q^2)$ $A_1^{DK^*}(q^2)$	$|V_{cs}|$
$B_c \to \mu\nu$	f_{B_c}	$|V_{cb}|$	$B \to D e\nu$ $\to D^* e\nu$	$h_\pm(w)$ $h_{A_1}(w)$	$|V_{cb}|$
Neutral meson mixing					
$K^0 \leftrightarrow \bar{K}^0$	$\frac{8}{3} m_K^2 f_K^2 B_K$	$\varepsilon_K(\rho, \eta)$			
$B_d^0 \leftrightarrow \bar{B}_d^0$	$\frac{8}{3} m_{B_d}^2 f_{B_d}^2 B_{B_d}$	$|V_{td}|$	$B_s^0 \leftrightarrow \bar{B}_s^0$	$\frac{8}{3} m_{B_s}^2 f_{B_s}^2 B_{B_s}$	$|V_{ts}|$

Similarly, the differential decay rate of a semileptonic decay is given by

$$\frac{d\Gamma_{P \to Hl\nu}}{dq^2} = \left(\begin{array}{c} \text{known} \\ \text{factors} \end{array} \right) \mathcal{F}^2(q^2)|V_{pq}|^2, \quad (2)$$

where q is momentum carried off by the leptons, and $\mathcal{F}$ is the appropriate form factor. These processes do not suffer the helicity suppression of the leptonic decays, so the statistical error of the experimental measurements is smaller. Furthermore, lattice QCD can provide the q^2 dependence, at least when the momentum of the daughter hadron is not too large. If theory and experiment exhibit the same shape as a function of q^2, one's confidence in the systematics increases qualitatively.

Neutral meson mixing reveals a glimpse of the third row of the CKM matrix. For example, the mass difference of neutral B mesons is given by

$$x_q = \frac{\Delta M_{B_q^0}}{\Gamma_{B_q^0}} = \left(\begin{array}{c} \text{known} \\ \text{factors} \end{array} \right) \frac{8}{3} m_{B_q}^2 f_{B_q}^2 B_{B_q} |V_{tq}^* V_{tb}|^2, \quad (3)$$

where the flavor q can be either down or strange. The notation employing $\frac{8}{3} m_{B_q}^2$, $f_{B_q}^2$, and the "bag parameter" B_{B_q} is historical and is taken from the kaon system. (In the so-called vacuum saturation approximation, $B_B = 1$.) Nevertheless, this formula, more so than leptonic B decay, motivates the interest in f_{B_d} and f_{B_s}.

Figure 1 shows a time-line of calculations of f_B with lattice QCD [10–30]. The results included in Fig. 1 have been selected with two criteria: Conference proceedings, which are almost always followed (eventually) by papers in refereed jour-

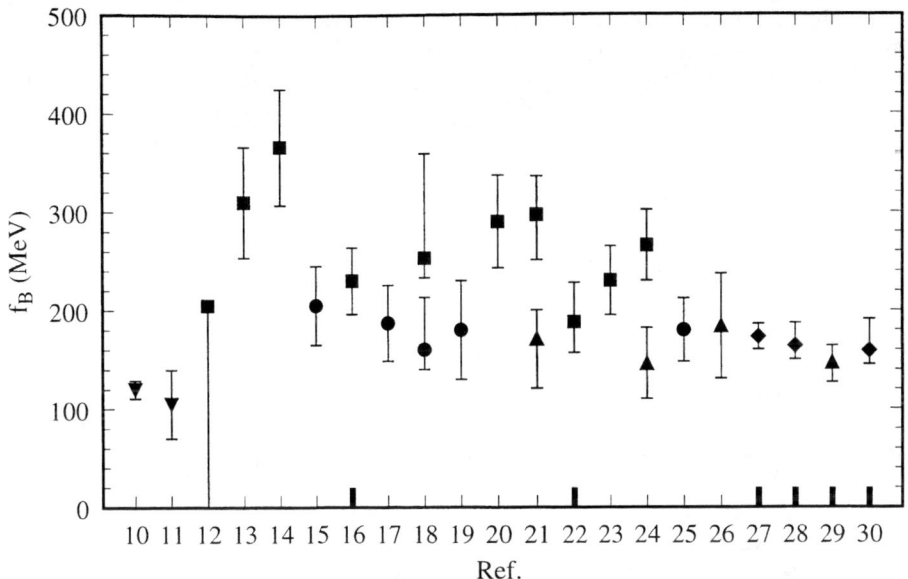

FIGURE 1. Time-line of calculations of f_B with lattice QCD. Methods shown are extrapolation from $m_Q \leq m_c$ to m_b (inverted triangles), the static limit $m_Q \to \infty$ (squares), interpolation between m_c and ∞ (circles), NRQCD (triangles), and that of Ref. [6]. Entries with a thick tick-mark on the horizontal axis control lattice spacing effects, as explained in the text.

nals, have been omitted. Otherwise, I have taken papers whose authors were self-confident enough to quote a result in the abstract. The second criterion is not necessarily fair to cautious innovators, but, on the other hand, it generates the picture that is seen by outsiders. Figure 1 does contain a few exceptions to the second criterion, to include calculations that offered new technical developments.

One can divide the time-line into infancy [10–14], childhood [15–22], youth [23–26], and adulthood [27–30]. (A similar, more discriminating classification has been made by Bernard [31].) In adulthood, with refereed publications starting in November 1997, the scatter that characterizes the field through its youth has settled down. The results with a thick tick-mark on the horizontal axis use several lattice spacings [16,22,27,28,30] or a fully consistent implementation of lattice NRQCD [29]. Thus, one could say that Refs. [16,22] are mature results of the static approximation but, because the contribution of order $1/m_b$ is not negligible, not of f_B.

HEAVY QUARKS AND LATTICE FIELD THEORY

In the introduction, I explained that physical $1/m_Q$ effects and artificial a effects are intertwined and that a better appreciation of the intertwining was required before calculations of f_B could mature. In particular, a theoretical analysis [6] of Wilson quarks away from the small-mass limit was needed to obtain the results in Refs. [27,28,30]. This section summarizes some of the main ideas by comparing and contrasting the effective theories in the continuum and on the lattice.

The effective Hamiltonian of QCD for heavy quarks can be written

$$H = M_1 \bar{\Psi}\Psi + \bar{\Psi}\gamma_0 A_0 \Psi + \bar{\Psi} h \Psi, \qquad (4)$$

where Ψ is the fermion field and h is given by an expansion in $1/m_Q$, namely

$$h = -\frac{\boldsymbol{D}^2}{2M_2} - z_B \frac{i\boldsymbol{\Sigma} \cdot \boldsymbol{B}}{2M_2} + z_{\text{s.o.}} \frac{\{\boldsymbol{\gamma} \cdot \boldsymbol{D}, \boldsymbol{\alpha} \cdot \boldsymbol{E}\}}{8M_2^2} - z_4 \frac{(\boldsymbol{D}^2)^2}{8M_2^3} + \cdots, \qquad (5)$$

corresponding to the kinetic energy, hyperfine splitting, spin-orbit splitting, relativistic corrections, etc.[2] This result follows from a series of Foldy-Wouthuysen-Tani transformations, or from noticing the heavy-quark symmetries of QCD, writing down allowed operators with arbitrary coefficients, and matching physical observables to standard QCD. With radiative corrections, one finds $z = 1 + O(g^2)$.

The rest mass of a quark is M_1 and the kinetic energy of a quark is $\boldsymbol{p}^2/2M_2$. It is convenient to call M_2 the kinetic mass, even though Lorentz invariance implies $M_2 = M_1$. Let us write $H = M_1 \bar{\Psi}\Psi + H_Q$. The rest-mass term and H_Q commute, even in the interacting theory, so eigenstates of H are simultaneously eigenstates of H_Q. Thus, one can drop the rest-mass term or readjust M_1 according to convenience, without changing the dynamics of heavy-quark QCD. The physically relevant parameter is the kinetic mass, which one adjusts so that $M_2 = m_Q$.

The lattice Hamiltonian takes the same structure as in Eqs. (4) and (5), but with changes to the operators' coefficients. One can express this by replacing h in Eq. (4) with $h_{\text{lat}} = h + \delta h$ and writing

$$\delta h = ab_B i\boldsymbol{\Sigma} \cdot \boldsymbol{B} + a^2 b_{\text{s.o.}} \{\boldsymbol{\gamma} \cdot \boldsymbol{D}, \boldsymbol{\alpha} \cdot \boldsymbol{E}\} + a^3 b_4 (\boldsymbol{D}^2)^2 + a^3 w_4 \sum_i D_i^4 + \cdots, \qquad (6)$$

where the coefficients $b = b(m_Q a, g^2)$ and $w = w(m_Q a, g^2)$ are functions of the (lattice) quark mass $m_Q a$ and the gauge coupling g^2. The same operators appear, as well as others, such as the last one, that break rotational symmetry. Part of the craft of the numerical work is to adjust the underlying lattice action so that these artifact coefficients b and w vanish, at least to some accuracy.

In lattice NRQCD and HQET this pattern arises by construction, in particular one sets $M_1 = 0$. It is fairly straightforward to adjust the bs and ws to vanish at the tree level of perturbation theory in g^2. Beyond the tree level it is still possible, but arduous perturbative calculations are needed. Power-law divergences appear in loop diagrams, so one cannot take $a \to 0$ by brute force [2]. That means that the lattice artifacts can be removed only by further refinements of the NRQCD

---

[2] Analogous expansions are introduced for the operators mediating electroweak transitions.

action. For heavy-light mesons, the action of the light quarks must be improved to a consistent level [32], as in Ref. [29].

With actions for Wilson quarks, such as the clover action[3] or the Wilson action itself, the pattern sketched here is less immediate. It is guaranteed, however, by the heavy-quark symmetry of the lattice action. Because the lattice violates Lorentz invariance, the rest and kinetic masses differ; in practice $M_1 < M_2$. The bs and ws are, however, *bounded* for all masses. As $m_Q a \to 0$, these coefficient functions go to a constant[4] or vanish like $(m_Q a)^{s_0}$, for some integer s_0, by Symanzik's standard analysis of cutoff effects. On the other hand, as $m_Q a \to \infty$, they vanish like $(1/m_Q a)^{s_\infty}$, for some s_∞, by heavy-quark symmetry [6]. The full functional dependence on $m_Q a$ is *calculable* in perturbation theory and, sometimes, nonperturbatively. The coefficient functions bs and ws depend on the details of the lattice action. For example, with the Wilson action $b_B \neq 0$, but with the clover action one can adjust an unphysical coupling until b_B vanishes.

LEPTONIC DECAYS

Let us now return to Fig. 1. The plotting symbols distinguish methods. Squares denote calculations in the static approximation, and triangles denote calculations with lattice NRQCD. Inverted triangles [10,11], circles, and diamonds denote treatments of numerical data from the Wilson or clover action.

Because the latter connects naturally with the effective theories, Ref. [6] suggested treating the bottom quark on the lattice by adjusting the bare mass until $M_2 = m_b$. In addition, a suitably normalized operator, essentially one built from the heavy-quark field Ψ, must be used for the current. Results with this approach are denoted with diamonds. The discretization errors are of order $\Lambda_{\text{QCD}} a$ to some power, from matrix elements of the operators in Eq. (6). (These effects are then multiplied by a coefficient b or w, which is a number of order 1 or, in limiting cases, smaller still.) With this interpretation of the numerical data from the clover action, the lattice results for f_B are nearly independent of the lattice spacing [27,28].

In earlier work with Wilson quarks, the tuning of the mass and the normalization of the current introduced discretization effects of order $m_Q a$ or $(m_Q a)^2$. The authors minimized these lattice artifacts by reducing the quark mass of the lattice calculations below that of the charmed quark. This is still large: $m_c a \sim 5\Lambda_{\text{QCD}} a$. Then the results were extrapolated up to m_b with fits to $1/m_Q$ expansions, either with (circles) or without (inverted triangles) the help of the static value. This intertwines lattice artifacts and $1/m_Q$ effects in ways that depend on details of the fits, not least because one cannot verify whether the $1/m_Q$ expansion converges for the low quark masses, on which the fits are based.

Table 2 tabulates the mature results [27–30] for f_B, f_{B_s}, f_D, and f_{D_s}, along with an average [34] of results from experimental measurements of $\Gamma_{D_s \to \mu\nu}$, combined

---

[3] The clover action adds a term to the Wilson action to eliminate the leading lattice artifact [33].
[4] If the limit is a constant, the lattice artifact still vanishes, owing to the explicit powers of a.

TABLE 2. Compendium of recent results for decay constants of heavy-light mesons. The experimental average comes, in fact, from measurements of $|V_{cs}|f_{D_s}$, which then take $|V_{cs}|$ from unitarity or from neutrino production of charm with $|V_{cs}|$.

method	Ref.	f_B				f_{B_s}			
		MeV	stat	syst	quench	MeV	stat	syst	quench
[6]	[27]	173	04	09	09	199	03	10	10
[6]	[28]	164	$^{+14}_{-11}$	08	$^{+10}_{-00}\%$	185	$^{+13}_{-08}$	09	$^{+10}_{-00}\%$
NRQCD	[29]	147	11	11	$^{+08}_{-12}$	175	08	13	10
[6]	[30]	159	11	$^{+22}_{-09}$	$^{+21}_{-00}$	175	10	$^{+28}_{-10}$	$^{+25}_{-01}$

method	Ref.	f_D				f_{D_s}			
		MeV	stat	syst	quench	MeV	stat	syst	quench
[6]	[27]	197	02	14	10	224	02	16	12
[6]	[28]	194	$^{+14}_{-10}$	10	$^{+10}_{-00}\%$	213	$^{+14}_{-11}$	11	$^{+10}_{-00}\%$
[6]	[30]	195	11	$^{+15}_{-08}$	$^{+15}_{-00}$	213	09	$^{+23}_{-09}$	$^{+17}_{-00}$
experiment	[34]					243		36 (total)	

While it is tempting to average the results, it is subtle to do so properly, because the systematic errors are largely, but not entirely, common. If you must have an average, consult Draper's review [9], or use your own eye.

The central values in Table 2 (except the experiment!) are in the quenched approximation. The *error bars*, however, reflect estimates of the associated uncertainty. The best estimation is that of the MILC Collaboration [30] who have some lattice results including up and down quark loops. (The strange and heavier quarks are still quenched.) The partially unquenched data sometimes lie higher than the quenched data, sometimes not, leading to the very asymmetric error estimate. These results are encouraging, not least because they suggest that the wait for a fully unquenched calculation will not be too much longer.

NEUTRAL MESON MIXING

Because the calculation is technically more demanding, the literature contains fewer calculations of the bag parameters of B_d^0 and B_s^0 meson mixing than of the decay constants. Recent publications report $B_{B_s}/B_{B_d} \approx 1$ and

$$\hat{B}_{B_d} = \begin{cases} 1.03 \pm 0.06 \pm 0.18 & [35] \\ 1.40 \pm 0.06 \, ^{+0.04}_{-0.26} & [36] \\ 1.23 \pm 0.05 \pm 0.15 & [36, 35] \\ 1.17 \pm 0.09 \pm 0.05 & [37] \end{cases}, \quad (7)$$

where $\hat{B}_{B_d}$ is the renormalization-scheme independent combination. The first three entries use the static approximation. The third entry comes from an analysis in Ref. [36] of the data in Ref. [35]. The last entry uses lattice NRQCD and finds, additionally, that the dependence on $1/m_Q$ is not large. See Ref. [9] for more results. There are also results for the ratio of matrix elements:

$$\frac{\frac{8}{3}m_{B_s}^2 f_{B_s}^2 B_{B_s}}{\frac{8}{3}m_{B_d}^2 f_{B_d}^2 B_{B_d}} = \begin{cases} 1.38 \pm 0.07 & [35] \\ 1.76 \pm 0.10 \,{}^{+57}_{-42} & [38] \end{cases}. \tag{8}$$

On the whole, the impression is that more work needs to be done to gain control over the systematic errors; calculations of B_B are not as mature as those of f_B.

The kaon's bag parameter B_K is needed to predict ε_K, a measure of indirect CP violation in $K \to \pi\pi$. Given B_K from (lattice) QCD, a measurement of ε_K traces a hyperbola in the complex V_{td} plane. The history of B_K has similarities to that of f_B: numerical work was a greater challenge than initially hoped, and theoretical insight is needed as a guide. For example, the lattice-spacing dependence (with staggered fermions) is surprisingly steep, but one now knows to extrapolate to the continuum limit in a^2 (rather than a) [39]. Two recent calculations find

$$B_K(\text{NDR}, 2\text{ GeV}) = \begin{cases} 0.62 \pm 0.02 \pm 0.02 & [40] \\ 0.628 \pm 0.042 & [41] \end{cases}, \tag{9}$$

where the main uncertainty comes from the continuum extrapolation. Quenching and degenerate quark-mass effects may each lead to underestimates of order 5% [42].

SEMILEPTONIC DECAYS

Semileptonic decays are wonderful for learning about the first two rows of the CKM matrix. Because a hadron is in the final state, lattice calculations of the form factors are more difficult than decay constants, but not much more difficult.

In these decays there is an additional kinematic variable, the momentum $|\bm{p}'|$ of the daughter hadron (in the parent's rest frame). The Lorentz invariant q^2 is linearly related to $|\bm{p}'|$. Until recently, calculations of these form factors were done with $m_Q \lesssim m_c$ and extrapolated up to m_c or m_b with $1/m_Q$ expansions. Since the kinematically allowed range of q^2 depends on the heavy-quark mass, another extrapolation is made. It is clear that the extrapolations once again intertwine artificial a effects with physical $1/m_Q$ and q^2 dependence. The details of the intertwining are not transparent (to me, anyway), so I reserve comment and direct the reader to reviews by Onogi [7] and by Draper [9].

More recently, calculations have been done with lattice NRQCD or with Wilson quarks interpreted [6] as suggested by Eqs. (4)–(6). These techniques are especially powerful here, because the ability to compute directly at the B mass decouples extrapolations in q^2 from the mass dependence. Calculations are underway for form factors of heavy-to-light transitions, such as $D \to \pi l\nu$ and $D \to K l\nu$, which together yield $|V_{cd}/V_{cs}|$ [43,44], and $B \to \pi l\nu$, which yields $|V_{ub}|$ [43,45,46].

I would like to conclude with a preliminary result [47] on the zero-recoil form factor for the decay $B \to D l\nu$, which will improve the determination of $|V_{cb}|$. A similar study of $B \to D^* l\nu$ is in progress. Until now there have been calculations of the shape of the form factors for $B \to D^{(*)} l\nu$, but the normalization has been computed only poorly, in perturbation theory. It is possible, however, to handle almost all of the normalization nonperturbatively. For example [48],

$$\frac{\langle D|\mathcal{V}_0|B\rangle\langle B|\mathcal{V}_0|D\rangle}{\langle D|\mathcal{V}_0|D\rangle\langle B|\mathcal{V}_0|B\rangle} = \frac{h_+^{B\to D}(1)h_+^{D\to B}(1)}{h_+^{D\to D}(1)h_+^{B\to B}(1)} = |h_+^{B\to D}(1)|^2, \tag{10}$$

and here at Fermilab we have found analogous ratios for $h_-^{B\to D}(1)$ and $h_{A_1}^{B\to D^*}(1)$. For the class of actions considered in Ref. [6], the remaining radiative corrections have been computed [49], and they are small.

In Eq. (2) for $B \to Dl\nu$ one requires $\mathcal{F} = h_+ - (m_B - m_D)h_-/(m_B + m_D)$. The advantage of the new method is that, in effect, it calculates not $\mathcal{F}(1)$ but the deviation of $\mathcal{F}(1)$ from 1. We find

$$\mathcal{F}(1) = 1.069 \pm 0.008 \pm 0.002 \pm 0.025, \tag{11}$$

where the uncertainties are Monte Carlo statistics, tuning of the quark masses, and a certain parametrically small contribution omitted from h_-. Uncertainties from lattice artifacts and from quenching have not yet been taken into account.

REFERENCES

1. Eichten, E., *Nucl. Phys. B Proc. Suppl.* **4**, 170 (1987).
2. Lepage, G. P., and Thacker, B. A., *Nucl. Phys. B Proc. Suppl.* **4**, 199 (1987). See also Caswell, W. E., and Lepage, G. P., *Phys. Lett.* **167B**, 437 (1986).
3. Gavela, M. B., Maiani, L., Martinelli, G., Péne, O., and Petrarca, S., *Nucl. Phys. B Proc. Suppl.* **4**, 466 (1987).
4. Hashimoto, S., and Saeki, Y., *Mod. Phys. Lett.* **A7**, 387 (1992).
5. Lepage, G. P., Magnea, L., Nakhleh, C., Magnea, U., and Hornbostel, K., *Phys. Rev.* **D46**, 4052 (1992).
6. El-Khadra, A. X., Kronfeld, A. S., and Mackenzie, P. B., *Phys. Rev.* **D55**, 3933 (1997).
7. Onogi, T., *Nucl. Phys. B Proc. Suppl.* **63**, 59 (1998).
8. Ali Khan, A., *Nucl. Phys. B Proc. Suppl.* **63**, 71 (1998).
9. Draper, T., `hep-lat/9810065`.
10. Gavela, M. B., Maiani, L., Petrarca, S., Martinelli, G., and Pène, O., *Phys. Lett.* **B206**, 113 (1988).
11. Bernard, C., Draper, T., Hockney, G., and Soni, A., *Phys. Rev.* **D38**, 3540 (1988).
12. Boucaud, Ph., Pène, O., Hill, V. J., Sachrajda, C. T., and Martinelli, G., *Phys. Lett.* **B220**, 219 (1989).
13. Allton, C. R., Sachrajda, C. T., Lubicz, V., Maiani., L., and Martinelli, G., *Nucl. Phys.* **B349**, 598 (1991).
14. Alexandrou, C., Jegerlehner, F., Güsken, S., Schilling, K., and Sommer, R. *Phys. Lett.* **B256**, 60 (1991).
15. Abada, A., *et al.*, *Nucl. Phys.* **B376**, 172 (1992).

16. Alexandrou, C., Güsken, S., Jegerlehner, F., Schilling, K., and Sommer, R., *Nucl. Phys.* **B414**, 815 (1994).
17. Bernard, C. W., Labrenz, J. N., and Soni, A., *Phys. Rev.* **D49**, 2536 (1994).
18. Baxter, R. M., *et al.* (UKQCD Collaboration), *Phys. Rev.* **D49**, 1594 (1994).
19. Alexandrou, C., Güsken, S., Jegerlehner, F., Schilling, K., Siegert, G., and Sommer, R., *Z. Phys.* **C62**, 659 (1994).
20. Allton, C. R., *et al.* (APE Collaboration), *Phys. Lett.* **B326**, 295 (1994).
21. Hashimoto, S., *Phys. Rev.* **D50**, 4639 (1994).
22. Duncan, A., Eichten, E., Flynn, J., Hill, B., Hockney, G., and Thacker, H., *Phys. Rev.* **D51**, 5101 (1995).
23. Allton, C. R., *Nucl. Phys.* **B437**, 641 (1995).
24. Ali Khan, A., Shigemitsu, J., Collins, S., Davies, C. T. H., Morningstar, C., and Sloan, J., *Phys. Rev.* **D56**, 7012 (1997).
25. Allton, C. R., Conti, L., Crisafulli, M., Giusti, L., Martinelli, G., and Rapuano, F., *Phys. Lett.* **B405**, 133 (1997).
26. Ishikawa, K.-I., Matsufuru, H., Onogi, T., Yamada, N., and Hashimoto, S., *Phys. Rev.* **D56**, 7028 (1997).
27. Aoki, S., *et al.* (JLQCD Collaboration), *Phys. Rev. Lett.* **80**, 5711 (1998).
28. El-Khadra, A. X., Kronfeld, A. S., Mackenzie, P. B., Ryan, S. M., and Simone, J. N., *Phys. Rev.* **D58**, 014506 (1998).
29. Ali Khan, A., Collins, S., Davies, C. T. H., Morningstar, C., Shigemitsu, J., and Sloan J., *Phys. Lett.* **B427**, 132 (1998).
30. Bernard, C., *et al.* (MILC Collaboration), *Phys. Rev. Lett.* **81**, 4812 (1998).
31. Bernard, C., hep-ph/9709460, to appear in the proceedings of the VIIth International Symposium on Heavy Flavor Physics.
32. Morningstar, C. J., and Shigemitsu, J., *Phys. Rev.* **D57**, 6741 (1998).
33. Sheikholeslami, B., and Wohlert, R., *Nucl. Phys.* **B259**, 572 (1985).
34. Parodi, F., Roudeau, P., and Stocchi, A., hep-ph/9802289.
35. Giménez, V., and Martinelli, G., *Phys. Lett.* **B398**, 135 (1997).
36. Christensen, J., Draper, T., and McNeile, C., *Phys. Rev.* **D56**, 6993 (1997).
37. Yamada, N., Hashimoto, S., Ishikawa, K.-I., Matsufuru, H., Onogi, T., and Tominaga, S., hep-lat/9809156.
38. Bernard, C., Blum, T., and Soni, A., *Phys. Rev.* **D58**, 014501 (1998).
39. Sharpe, S. R., *Nucl. Phys. B Proc. Suppl.* **34**, 403 (1994).
40. Kilcup, G., Gupta, R., and Sharpe S. R., *Phys. Rev.* **D57**, 1654 (1998).
41. Aoki, S., *et al.* (JLQCD Collaboration), *Phys. Rev. Lett.* **80**, 5271 (1998).
42. Sharpe, S R., *Nucl. Phys. B Proc. Suppl.* **53**, 181 (1997).
43. Aoki, S., *et al.* (JLQCD Collaboration), *Nucl. Phys. B Proc. Suppl.* **63**, 380 (1998).
44. Simone, J. N., *et al.*, hep-lat/9810040 [FERMILAB-CONF-98-333-T].
45. Matsufuru, H., Hashimoto, S., Ishikawa, K.-I., Onogi, T., and Yamada, N., *Nucl. Phys. B Proc. Suppl.* **63**, 368 (1998).
46. Ryan, S., *et al.*, hep-lat/9810041 [FERMILAB-CONF-98-269-T].
47. Hashimoto, S., *et al.*, hep-lat/9810056 [FERMILAB-CONF-98-340-T].
48. Mandula, J., and Ogilvie, M., *Nucl. Phys. B Proc. Suppl.* **34**, 480 (1994).
49. Kronfeld, A., and Hashimoto, S., hep-lat/9810042 [FERMILAB-CONF-98-332-T].

PART VI
SPECTROSCOPY

Charmonium Spectroscopy From Fermilab E835

Todd K. Pedlar [1]

Department of Physics and Astronomy
Northwestern University

Abstract. Fermilab E835 is dedicated to the study of charmonium spectroscopy via resonant formation in $\bar{p}p$ annihilations. This paper presents preliminary results on charmonium spectroscopy from E835 data taken in the 1996-7 Fixed Target running period. Specifically, results for the resonance parameters of η_c (1^1S_0), the $\gamma\gamma$ partial width of χ_2 (1^3P_2) and a search for η'_c (2^1S_0) are presented, as is the first evidence for χ_0 (1^3P_0) formation in $\bar{p}p$ annihilation.

INTRODUCTION

Precision spectroscopy of the charmonium system can yield quantitative results concerning the nature of QCD in the non-perturbative regime, where it is not well understood. The mass of the charm quark is small enough that $c\bar{c}$ cross sections are reasonably large, yet it is large enough that its motion within bound charmonium is nearly non-relativistic. The spectrum of charmonium states is shown in Fig. 1.

From the discovery of J/ψ in 1974 until the mid-1980's, nearly all charmonium spectroscopy was studied via e^+e^- annihilation at facilities like SLAC and DESY. Because e^+e^- annihilation proceeds via a virtual photon, only vector $J^{PC} = 1^{--}$ states may be formed directly in this manner. The study of non-vector states (such as the χ_J states) is possible only through the observation of the radiative decays ψ' (or J/ψ) $\rightarrow (c\bar{c}) + \gamma$. Thus the mass and width measurements for these states are limited by the photon energy resolution of the detector system. While most of the charmonium states were discovered in such experiments, their masses, and

---

[1] on behalf of the Fermilab E835 Collaboration: G. Garzoglio, K. E. Gollwitzer, A. Hahn, W. Marsh, J. Peoples Jr., S. Pordes, J. Streets, S. Werkema (*Fermi National Accelerator Laboratory*), M. Ambrogiani, M. Baldini, D. Bettoni, R. Calabrese, P. Dalpiaz, E. Luppi, R. Mussa, M. Savriè, G. Stancari (*INFN and University of Ferrara*), A. Buzzo, M. Lo Vetere, M. Macrì, M. Marinelli, M. Pallavicini, C. Patrignani, E. Robutti, A. Santroni (*INFN and University of Genova*), G. Lasio, M. Mandelkern, J. Schultz, M. Stancari, G. Zioulas (*University of California, Irvine*), X. Fan, S. Jin, J. Kasper, P. Maas, T. K. Pedlar, J. L. Rosen, K. K. Seth, A. Tomaradze(*Northwestern University*), S. Bagnasco, G. Borreani, R. Cester, F. Marchetto, E. Menichetti, N. Pastrone, P. Rumerio (*INFN and University of Torino*) and R. McTaggart[1].

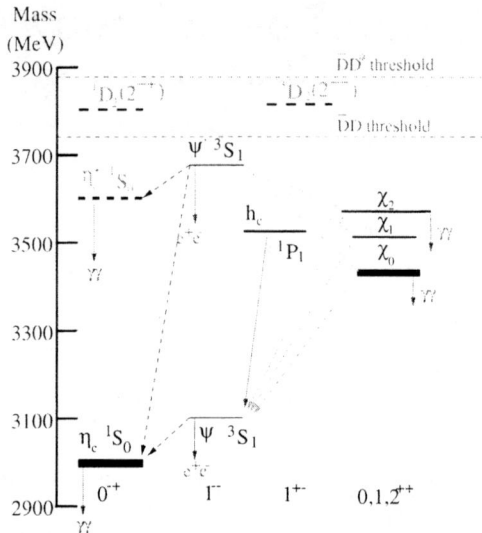
FIGURE 1. The spectrum of charmonium states.

especially widths, were not well measured. In order to do precision spectroscopy of charmonium, a new technique was required. Such a technique, charmonium via $\bar{p}p$ annihilation, was pioneered in 1984 by R704 at CERN [2]. Fermilab E760 and E835 have fully exploited this technique, and have obtained results which represent improvements in precision, in some cases by orders of magnitude.

The chief advantage of the technique of charmonium formation via $\bar{p}p$ annihilation is that all charmonium states may be formed directly - the $\bar{p}p$ annihilation process occurs through two or three gluons in the intermediate state, which may have any value of J^{PC}. Since every charmonium state can be formed directly, its resonance parameters may be determined with a precision dependent only on the knowledge of the beam energy and energy distribution, and event statistics.

FERMILAB EXPERIMENT E835

The E835 experiment is located in the Fermilab Antiproton Accumulator (See Fig. 2). The circulating beam of antiprotons intersects a hydrogen gas jet target to produce $\bar{p}p$ interactions. During Fermilab fixed target operations, the Accumulator stores antiprotons at 8.9 GeV, and decelerates them to the momentum required for the resonant formation of a $c\bar{c}$ state (antiproton momenta of 4 to 7 GeV, for charmonium masses from 3 to 4 GeV).

The stochastically-cooled $\bar{p}$ beam consists of up to $\sim 10^{12}$ $\bar{p}$ which orbit the 474.0454 m path length of the accumulator with a frequency of ~ 0.62 MHz. This frequency is measured using the Schottky noise spectrum of the beam to about 2 parts in 10^7. The beam intersects the gas jet target [3], whose density in the

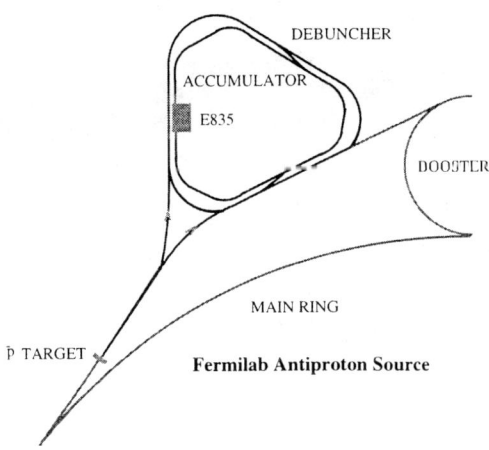

FIGURE 2.

interaction region can be varied between 10^{12} and 10^{14} atoms/cm^3, in order to keep the instantaneous luminosity nearly constant (typically, 2×10^{31} cm^{-2} sec^{-1}).

The center of mass energy $\sqrt{s}$ is determined by the revolution frequency and the orbit length, which is measured accurate to ± 0.7 mm by a calibration scan at the ψ', whose mass is 3686.00 ± 0.1 MeV. [4] Deviations of the beam from this reference orbit length are measured by beam position monitors to ± 1 mm, so that $\sqrt{s}$ may be known to $\sim \pm 50$ keV. The beam momentum spread, which is directly related to the width of the Schottky noise spectrum, is $\Delta p/p \simeq 1 - 2 \times 10^{-4}$. This Δp corresponds at the ψ' to a center of mass energy spread $\sigma(\sqrt{s}) \approx 200$ keV.

The E835 detector (See Fig. 3) is an upgrade of the E760 detector system, which has been described in several articles. [5]- [8] It is an azimuthally symmetric non-magnetic spectrometer with full azimuthal acceptance, and polar angle θ coverage from 2° to 70°. The main components of the detector are a 1280-element lead glass EM calorimeter, a large-acceptance segmented threshold Čerenkov counter, and several inner tracking detectors. Luminosity is measured to $\sim \pm 2.5\%$ by three solid-state detectors which observe elastically scattered protons at $\theta \approx 86.5°$. [8]

In E835, charmonium states are studied by decelerating the $\bar{p}$ beam so that $\sqrt{s}$ is varied in small steps across the resonance under study. At each step, the yield of the final state particles from the electromagnetic decays such as e^+e^-, J/ψ $\gamma \to e^+e^-\gamma$ or $\gamma\gamma$ is measured. The number of events divided by luminosity taken at each point, plotted versus the center of mass energy, gives the resonance excitation curve. This is just the convolution of a Breit-Wigner resonance cross section (which is proportional to the product of the branching ratios for the formation and the decay of the resonance, $B_{in} \times B_{out}$) and the measured beam energy distribution function. Fig. 4 shows an example of such an excitation curve from the scan of χ_2 in E760. [9]

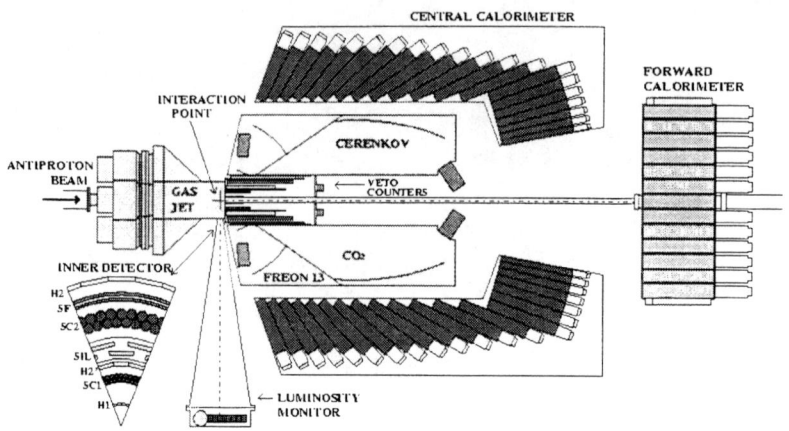

FIGURE 3. Schematic of the E835 detector system.

PRELIMINARY RESULTS

E835 collected an integrated luminosity of approximately 145 pb^{-1}, resulting in approximately 5 billion triggered events written to tape. From these data, we have obtained the following results (of which the first four are discussed here):

- Measurements of η_c resonance parameters M, Γ and $\Gamma(\eta_c \to \gamma\gamma)$

- Measurement of the partial width $\Gamma(\chi_2 \to \gamma\gamma)$

- Search for $\eta_c' \to \gamma\gamma$ in the region $\sqrt{s} = 3575$ to 3660 MeV

- First observation of χ_0 in $\bar{p}p$ annihilation

- Measurements of

    - $\psi' \to e^+e^-$ and $\psi' \to J/\psi\, X$ branching ratios
    - χ_2, χ_1 radiative decay angular distributions
    - Proton electromagnetic form factors
    - $J/\psi \to \bar{p}p$ and $\psi' \to \bar{p}p$ angular distributions

Detection of ($c\bar{c}$) via $\bar{p}p \to c\bar{c} \to \gamma\gamma$

Three of the four states which we address in this article are observed in E835 in their decay into two photons. These C-even states are difficult to study in $\bar{p}p$ collisions, because their $\gamma\gamma$ branching fractions are typically of the order of 10^{-4} (and cross sections for $\bar{p}p \to (c\bar{c}) \to \gamma\gamma$ of tens of picobarns). Furthermore, the limited acceptance of the E835 detector results in large backgrounds from prolific ($\sim 100\mu b$) all-neutral processes such as $\bar{p}p \to \pi^0\gamma$ and $\bar{p}p \to \pi^0\pi^0$. These backgrounds arise

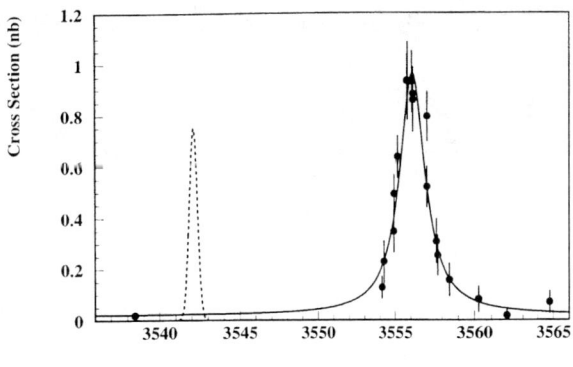

FIGURE 4. Example of a scan of a $c\bar{c}$ state. The solid curve is a fit to the acceptance and efficiency-corrected data, while the dotted curve shows the beam energy spread. (From Ref. [9])

when either one or both π^0 decay asymmetrically, and the low-energy γ is not detected, either falling below threshold or being outside the detector acceptance. This results in typical signal-to-background ratios of 1:1 to 1:3.

The $\gamma\gamma$ partial widths of charmonium states are of great theoretical interest. The $c\bar{c} \to \gamma\gamma$ process is simply calculable to zeroth order in QED. However, there are large (up to 75%) first-order QCD radiative corrections [10]. Unfortunately, most existing $c\bar{c} \to \gamma\gamma$ measurements have large errors; hence the present measurements.

In E835 $\gamma\gamma$ events are selected with a topological trigger designed for efficient detection of a high-mass pair of electromagnetic clusters in the CCAL, with no charged tracks identified in the tracking hodoscopes. In order to minimize the problem of pile-up, each channel of the CCAL is instrumented with TDCs, to reject out-of-time interactions which contaminate our event sample due to running at high luminosity. Of the neutral two-body trigger sample, events were accepted which contained only two on-time CCAL clusters, and which had no additional clusters which paired with one of the on-time clusters to form an invariant mass consistent with m_{π^0}. A 4C kinematical fit was made on these events, and a χ^2 probability $\geq 5\%$ was required. Further, a cut on the center of mass angle θ^* was made in order to reduce the background contribution due to misidentification of $\pi^0\pi^0$ and $\pi^0\gamma$ events, and to thus increase the signal-to-background ratio.

Measurement of η_c (1^1S_0) Resonance Parameters

The observed cross section for the scan in the η_c mass region for $|cos(\theta^*)| \leq 0.20$ is shown in Fig. 5 plotted as a function of the center of mass energy, $\sqrt{s}$. The data have been fit to a Breit-Wigner resonance and a power-law background of form $\sigma_{BKG} = A \times (2984.8/\sqrt{s})^B$. The results of this fit are shown in Table 1 compared to those in PDG98 [4]. We note that compared to the PDG98 values for

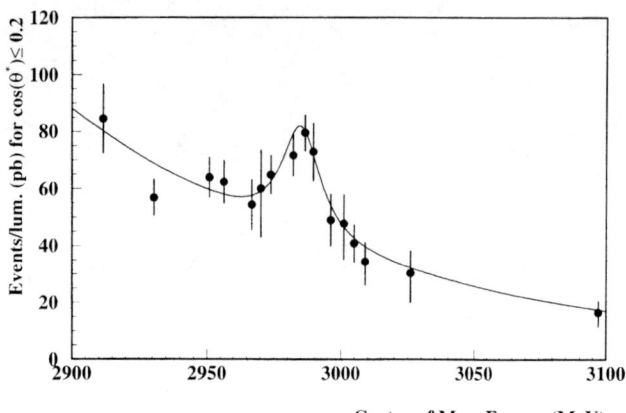

FIGURE 5. Observed cross section for $\bar{p}p \to \gamma\gamma$ for $|\cos\theta^*| \leq 0.2$ in the η_c region.

TABLE 1. Measurements of η_c resonance parameters.

Parameter	PDG98	E835
Mass	2979.8 ± 2.1 MeV	2984.8 ± 1.9 MeV
Γ	$13.2^{+3.8}_{-3.2}$ MeV	$17.8^{+7.2}_{-5.9}$ MeV
$B(\bar{p}p \to \eta_c) \times \Gamma(\eta_c \to \gamma\gamma)$	-	$4.4^{+1.8}_{-1.6}$ eV
$\Gamma(\eta_c \to \gamma\gamma)$	$7.5^{+1.6}_{-1.4}$ keV	$3.7^{+1.5}_{-1.3} \pm 1.2$ keV

η_c, our mass is ~ 5 MeV larger, the width is ~ 5 MeV larger, though with larger uncertainty, and our two photon width is nearly a factor two smaller.

Search for $\eta_c' \to \gamma\gamma$

η_c', the radial excitation of η_c, has been reported in only one experiment, Crystal Ball, in 1982, with a mass of $M(\eta_c') = 3594 \pm 4$ MeV. [12] E760 searched for η_c' at 6 data points from 3591 to 3621 MeV, and obtained no conclusive result. E835 has searched in a much wider mass region (3575 to 3660 MeV), investing ~ 30 pb^{-1}. We have also failed to find any evidence for η_c' excitation. From these data we can only derive an upper limit for the branching fraction product $B(\bar{p}p \to \eta_c') \times B(\eta_c' \to \gamma\gamma)$ of 9×10^{-8} at 95% confidence for an assumed η_c' width of 5 MeV. The observed $\gamma\gamma$ cross sections for $|\cos\theta^*| \leq 0.4$ are plotted from 3570 to 3700 MeV in Fig. 6.

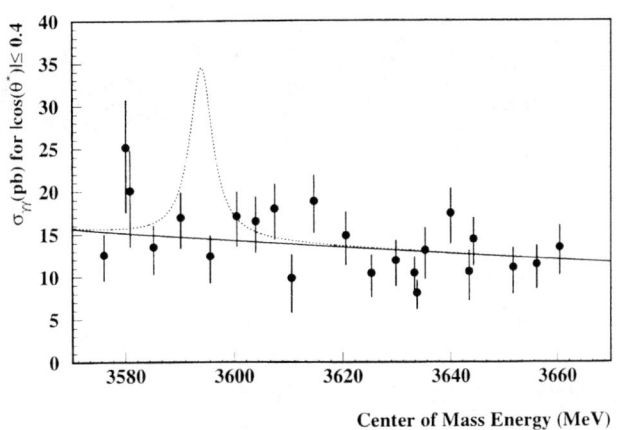

FIGURE 6. Observed cross sections for $\bar{p}p \to \gamma\gamma$ for $|\cos\theta^*| \leq 0.4$ in the η_c' search region. The dashed curve represents the expected signal at $M(\eta_c') = 3495$ MeV for $\Gamma(\eta_c') = 5$ MeV, and $B(\gamma\gamma) \times B(\bar{p}p)$ equal to those of η_c.

Measurement of $\Gamma(\chi_2 \to \gamma\gamma)$

E760 reported a measurement of $\Gamma(\chi_2 \to \gamma\gamma) = 0.321 \pm 0.078 \pm 0.054$ [13] which differed significantly from values obtained in other experiments, such as CLEO. [14] We have made a new measurement of the same quantity which confirms the E760 value, and has smaller errors. We obtain $\Gamma(\chi_2 \to \gamma\gamma) = 0.311 \pm 0.041 \pm 0.031$ keV. The data for this measurement are shown in Fig. 7.

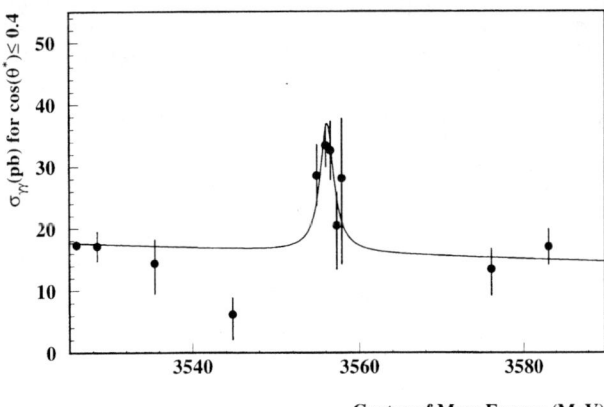

FIGURE 7. Observed cross sections for $\bar{p}p \to \gamma\gamma$ for $|\cos\theta^*| \leq 0.4$ in the χ_2 region.

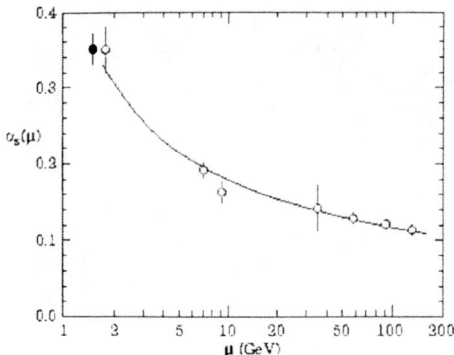

FIGURE 8. Measurements of the running coupling constant α_s at various values of Q^2.

α_s from charmonium decays

Frank Wilczek has emphasized the need for measurements of α_s at small values of Q^2. [15] Our data for $\eta_c \to \gamma\gamma$ and $\chi_2 \to \gamma\gamma$ give us two separate ways of calculating α_s at $Q^2 = (m_c)^2$, using the pQCD formulae with first order radiative corrections: [10]

$$\frac{\Gamma(\eta_c \to gg)}{\Gamma(\eta_c \to \gamma\gamma)} = \frac{\Gamma_{TOT}(\eta_c)}{\Gamma(\eta_c \to \gamma\gamma)} = \frac{9\alpha_s^{\ 2}}{8\alpha^2} \times \frac{(1 + 4.8\alpha_s/\pi)}{(1 - 3.4\alpha_s/\pi)} \qquad (1)$$

$$\frac{\Gamma(\chi_2 \to gg)}{\Gamma(\chi_2 \to \gamma\gamma)} = \frac{\Gamma(\chi_2) - \Gamma(\chi_2 \to \gamma J/\psi)}{\Gamma(\chi_2 \to \gamma\gamma)} = \frac{9\alpha_s^{\ 2}}{8\alpha^2} \times \frac{(1 - 2.2\alpha_s/\pi)}{(1 - 16\alpha_s/3\pi)}. \qquad (2)$$

Using our measured values for $\Gamma_{TOT}(\eta_c)$ and $\Gamma(\eta_c \to \gamma\gamma)$, we calculate $\alpha_s(m_c) = 0.33 \pm 0.05$. Using the PDG98 values of $\Gamma_{TOT}(\chi_2) = 2.00 \pm 0.18$ MeV and $B(\chi_2 \to \gamma J/\psi) = 0.135 \pm 0.011$, and our measured value of $\Gamma(\chi_2 \to gg)$, we find $\alpha_s(m_c) = 0.36 \pm 0.02$. The two results are consistent, giving an average of $\alpha_s(m_c) = 0.35 \pm 0.02$. This result is in excellent agreement with $\alpha_s(m_\tau) = 0.35 \pm 0.03$. [4] For comparison with other data on α_s, our value of $\alpha_s(m_c)$ is plotted along with the compilation from PDG98 [4] in Fig. 8.

Detection of χ_0 via $\bar{p}p \to c\bar{c} \to \gamma J/\psi$

The extremely background-free channels for the investigation of $c\bar{c}$ states in E835 are those which include a single high-mass e^+e^- pair, such as $c\bar{c} \to e^+e^-$ or $c\bar{c} \to \gamma J/\psi \to (e^+e^-)\gamma$. These events are triggered by a high mass pair of clusters in the CCAL, associated with two tracks in the inner charged tracking detectors and at least one associated signal in the Čerenkov. This trigger was successfully used in E760 to measure the resonance parameters of χ_1 and χ_2. The χ_0 state was not investigated.

A preliminary study of χ_0 was performed with limited luminosity at the end of the 1997 run of E835, a very clean $J/\psi\,\gamma$ signal in the region was observed.

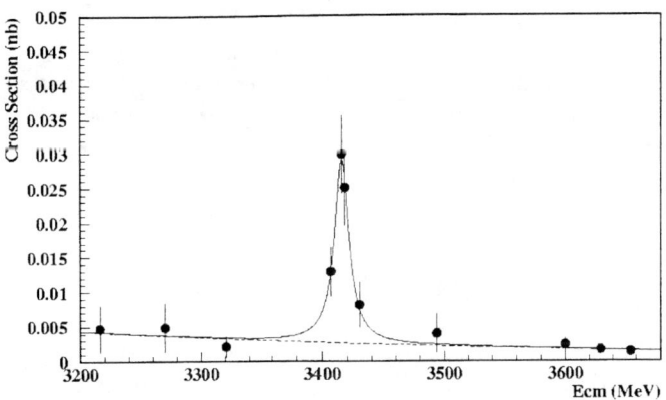

FIGURE 9. Cross section for $\bar{p}p \to \gamma J/\psi$ in the χ_0 region.

TABLE 2. Measurements of χ_0 resonance parameters.

Parameter	PDG98	E835
Mass	3417.3 ± 2.8 MeV	$3415^{+2.1}_{-1.7}$ MeV
Γ	13.5 ± 5.3 MeV	$13.9^{+5.3}_{-3.9}$ MeV
$B(\bar{p}p \to \chi_0) \times B(\chi_0 \to \gamma J/\psi)$	-	$2.80^{+0.63}_{-0.46} \times 10^{-6}$
$B(\chi_0 \to \gamma\gamma)$	$\leq 9 \times 10^{-4}$	$4.24^{+0.96}_{-0.70} \pm 1.16 \times 10^{-4}$

This represents the first measurement of the branching fraction $B(\chi_0 \to \bar{p}p)$. The cross section for this reaction appears in Fig. 9, and the resonance parameters derived from the maximum likelihood fit to the data are found in Table 2. PDG98 gives only an upper limit of $B(\chi_0 \to \bar{p}p) \leq 9 \times 10^{-4}$, while we have obtained $B(\chi_0 \to \bar{p}p) = (4.24^{+0.96}_{-0.70} \pm 1.16) \times 10^{-4}$.

FUTURE PLANS FOR E835

The Fermilab PAC has approved E835 for running again in 1999. Since several of the stated priorities of E835 were not achieved during the 1996-97 fixed target running period, these measurements are the first scheduled for the extension run. These are, in order of priority,

- Confirm the $^1P_1$ discovery by E760 and measure its resonance parameters.

- Measure χ_0 resonance parameters with better precision.

- Make a new search for η'_c.

- Scan above $D\bar{D}$ threshold for $^3D_2$, $^1D_2$ states.

CONCLUSIONS

E760 was the first experiment to fully exploit the technique of charmonium formation in $\bar{p}p$ annihilation. It made several contributions to the field of precision charmonium spectroscopy. E835 has advanced these efforts further, and has obtained new results. We look forward to running again in 1999, and completing our program of study of the charmonium spectrum.

ACKNOWLEDGEMENTS

The E835 Collaboration would especially like to thank our colleagues in Fermilab Beams Division, without whose efforts this experiment would not have been possible. We also acknowledge the support of the U.S. Department of Energy (DOE), the U.S. National Science Foundation (NSF), and the Italian Istituto Nazionale de Fisica Nucleare (INFN). The speaker wishes to thank the organizers of this workshop for their hard work in putting together an excellent conference.

REFERENCES

1. Pennsylvania State University.
2. C. Baglin, et al. , Nucl. Phys. **B286** (1987) 592
3. D. Allspach et al. , Nucl. Instr. Meth. **A410** (1998) 195.
4. C. Caso et al. (Particle Data Group) European Physical Journal **C3** (1998) 1.
5. T. A. Armstrong, et al. , Nucl. Phys. **B373** (1992) 35.
6. L. Bartosek et al. , Nucl. Instr. Meth. **A301** (1991) 47.
7. C. Biino et al. , Nucl. Instr. Meth. **A317** (1992) 135.
8. S. Trokenheim et al. , Nucl. Instr. Meth. **A355** (1995) 308.
9. T. A. Armstrong et al. , Phys. Rev. Lett. **68** (1992) 1468.
10. W. Kwong et al. , Phys. Rev. **D37** (1988) 3210.
11. T. A. Armstrong et al. ,Phys. Rev. **D52** (1995) 4839.
12. C. Edwards et al. , Phys. Rev. Lett. **48** (1982) 70.
13. T. A. Armstrong et al. , Phys. Rev. Lett. **70** (1993) 2988.
14. V. Savinov and R. Fulton, (CLEO Collaboration) Proc. 10th Int. Workshop on Photon-Photon Collisions, Sheffield, England, 1995, ed. by D. J. Miller et al. , (World Scientific, 1995) p. 203, J. Dominick et al. , (CLEO Collaboration), Phys. Rev. **D50** (1994) 4265.
15. F. Wilczek, Proc. Int. Symp. on Lepton and Photon Interactions, Ithaca, NY, 1993, ed. by P. Drell and D. Rubin, AIP Conf. Proc. 302 (1994) 593.

Charmed Baryon Spectroscopy from CLEO at CESR

M. Sajjad Alam

University at Albany and CLEO Collaboration
Physic Department, S.U.N.Y.A., Albany, NY 12222

Abstract. Charmed baryon spectroscopy has been unfolding since the discovery of the first charmed baryon in 1975. The Cornell Electron Storage Ring (CESR) has now established itself as a charmed particle factory. In this report, we present results on charmed baryon production at CESR using the CLEO detector.

INTRODUCTION

The discovery of J/ψ mesons at BNL[2] and SLAC[1] in 1974 heralded the era of particles containing a new quark carrying the charm quantum number and named the c quark. Immediately a rich, new spectroscopy of mesons and baryons containing the charm quark became possible. The first open charm mesons $D^0(c\bar{u})$ and $D^+(c\bar{d})$ were observed in 1976, and first evidence[3] for the ground state charm baryon $\Lambda_c^+(cud)$ appeared as a single neutrino interaction event in a bubble chamber at BNL in 1975. Soon clear signals for Λ_c^+ were observed in πp and in e^+e^- interactions at FNAL[6] and SLAC[7], respectively.

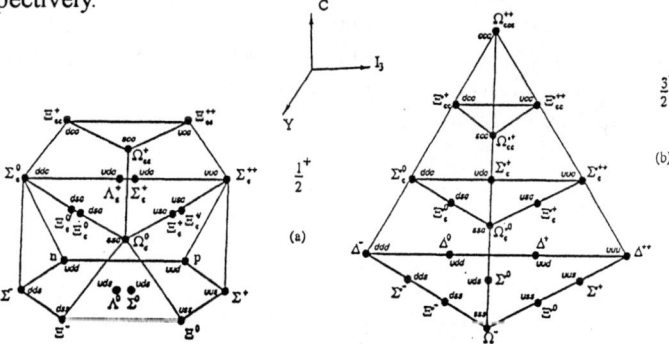

Figure 1. The ground state charmed baryon $J^P = \frac{1}{2}^+$ and $J^P = \frac{3}{2}^+$ multiplets.

Twenty five years after the observation of the first charmed baryon, charmed baryon spectroscopy is still unfolding. Just as the ground state meson nonets expanded to sixteen-plets, the ground state $J^P = \frac{1}{2}^+$ octet and $J^P = \frac{3}{2}^+$ decuplet of baryons expanded to a 20-plet in each case. They are shown in Figure 1.

At CESR, there are two ways that charmed baryons are produced. The virtual photon produced in e^+e^- annihilations from the continuum couple to $c\bar{c}$ pairs, which then hadronize into charmed mesons and baryons. Charmed baryons can also be produced from the secondary decays of $B/\bar{B}$ mesons produced at the $\Upsilon(4S)$. Let us define the production variable $x_p = p/p_{\max} = p/(\sqrt{E^2 - m^2})$, where p is the momentum of the particle and $p_{\max}$ its maximum value. E_b is the beam energy and m the mass of the charmed baryon under study. From measurements of continuum production of charmed mesons[12], it has been observed that 60% of charmed baryons produced from the continuum have $x_p > 0.5$. Charmed baryons produced from the secondary decays of $B/\bar{B}$ mesons are kinematically limited to x_p less than about 0.4 - 0.5 depending on the charmed baryon being considered. To avoid combinatorial background from low momentum combinations and only focus on continuum production, all CLEO analyses for charmed baryon studies have $x_p > 0.5$, typically.

THE LOWEST MASS $\Lambda_c^+(cud)$ CHARMED BARYON

Since the Λ_c^+ is the lowest mass charmed baryon, it can only decay weakly. A wide variety of final states are accessible through tree level spectator, exchange and annihilation diagrams corresponding to the $c \to sW^+$ coupling. In 1991, CLEO[16] published continuum production of the Λ_c^+ in the decay modes $pK^-\pi^+, p\overline{K^0}, p\overline{K^0}\pi^+\pi^-, \Lambda\pi^+, \Lambda\pi^+\pi^-\pi^+$, and $\Xi^-K^+\pi^+$ from about $430\,pb^{-1}$ of data around the $\Upsilon(4S)$ and the $\Upsilon(5S)$. As an example, Figure 2(a) shows the mass distributions corresponding to $pK^-\pi^+$ combinations with $x_p > 0.5$ with a fitted signal area of 512 ± 50 events. The weighted mass from fits to the mass distributions in all the above decay modes was reported to be $2284.7 \pm 0.6 \pm 0.7\,MeV/c^2$. Fitting to the Peterson[9] fragmentation function is a standard approach to parametrizing the shape of the x_p production spectrum in terms of the variable ϵ_Q. CLEO[16] has measured a value of $\epsilon_Q = 0.29 \pm 0.05$ from a fit to the x_p spectrum as shown in Figure 2(b).

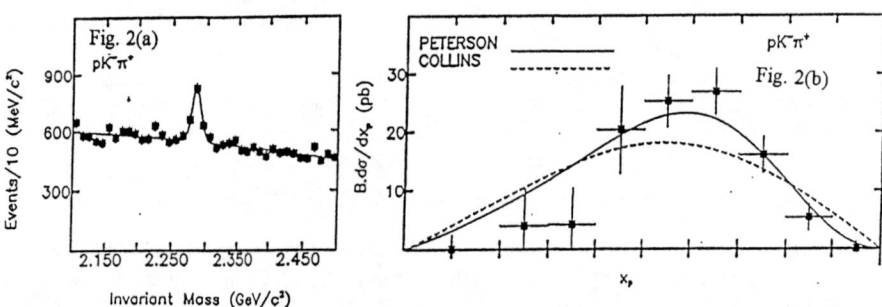

Figure 2. (a) Inv. mass distr. for $pK^-\pi^+$ combinations with $x_p > 0.5$. (b) x_p production spectrum with Peterson Function fit.

Using the inclusive decay $B \to \Lambda_c^+ X$ and assuming that all light baryons pro-

TABLE 1(a). Λ_c^+ Decay Modes		TABLE 1(b). Λ_c^+ Decay Modes Continued	
Decay Modes	Relative BR.	Decay Modes	Relative BR.
$pK^-\pi^+$ [16]	1.0	$pK^-\pi^+$ [16]	1.0
$p\overline{K}^0$	$0.44 \pm 0.07 \pm 0.05$	$p\phi$ [38]	$0.039 \pm 0.009 \pm 0.007$
$p\overline{K}^0\pi^+\pi^-$	$0.43 \pm 0.12 \pm 0.04$	$\Sigma^+\pi^0$ [22]	$0.20 \pm 0.03 \pm 0.03$
$\Lambda\pi^+$	$0.18 \pm 0.03 \pm 0.03$	$\Sigma^+\omega$	$0.54 \pm 0.13 \pm 0.06$
$\Lambda\pi^+\pi^-\pi^+$	$0.65 \pm 0.11 \pm 0.12$	$\Sigma^+\pi^+\pi^-$	$0.74 \pm 0.07 \pm 0.09$
$\Xi^-K^+\pi^+$	$015 \pm 0.04 \pm 0.03$	$\Sigma^+\rho^0$	< 0.27
$\Sigma^+K^+K^-$ [21]	$0.07 \pm 0.011 \pm 0.011$	$\Lambda\pi^+\pi^0$ [23]	$0.73 \pm 0.09 \pm 0.16$
$\Sigma^+\phi$	$0.07 \pm 0.020 \pm 0.016$	$\Sigma^0\pi^+\pi^0$	$0.36 \pm 0.09 \pm 0.10$
Ξ^0K^+	$0.08 \pm 0.013 \pm 0.013$	$\Sigma^0\pi^+\pi^-\pi^+$	$0.21 \pm 0.05 \pm 0.05$
$\Xi^-K^+\pi^+$	$0.08 \pm 0.014 \pm 0.014$	$\Sigma^0\pi^+$	$0.21 \pm 0.02 \pm 0.04$
$\Xi^{*0}\pi^+$	$0.05 \pm 0.016 \pm 0.010$	$p\overline{K}^0\eta$ [28]	$0.25 \pm 0.04 \pm 0.04$
$pK^-\pi^+\pi^0$ [40]	$0.67 \pm 0.04 \pm 0.11$	$\Lambda\pi^+\eta$	$0.35 \pm 0.05 \pm 0.06$
$p\overline{K}^0$	$0.46 \pm 0.02 \pm 0.04$	$\Sigma^+\eta$	$0.11 \pm 0.03 \pm 0.02$
$p\overline{K}^0\pi^+\pi^-$	$0.52 \pm 0.04 \pm 0.05$	$\Sigma^{*+}\eta$	$0.17 \pm 0.04 \pm 0.3$
$p\overline{K}^0\pi^0$	$0.66 \pm 0.05 \pm 0.07$	$\Lambda K^0 K^+$	$0.12 \pm 0.02 \pm 0.02$
		$\Lambda l^+ \nu_l$ [26]	$0.52 \pm 0.03 \pm 0.09$
Absolute Branching Fraction: $Br(\Lambda_c^+ \to pK^-\pi^+) = (4.3 \pm 1.0 \pm 0.8)\%$			

duced in a B meson decays are from the secondary decay of predominantly the charmed baryon Λ_c^+ and that Ξ_c's decays are negligible, CLEO[18] has estimated the absolute branching fraction $Br(\Lambda_c^+ \to pK^-\pi^+)$ to be $(4.3 \pm 1.0 \pm 0.8)\%$. A similar result has also been obtained by ARGUS using similar model assumptions. Since then, CLEO has observed several decay modes of the Λ_c^+; these measurements are summarized in Table 1(a) and (b).

OBSERVATION OF Σ_c^{++}, Σ_c^+, AND Σ_c^0

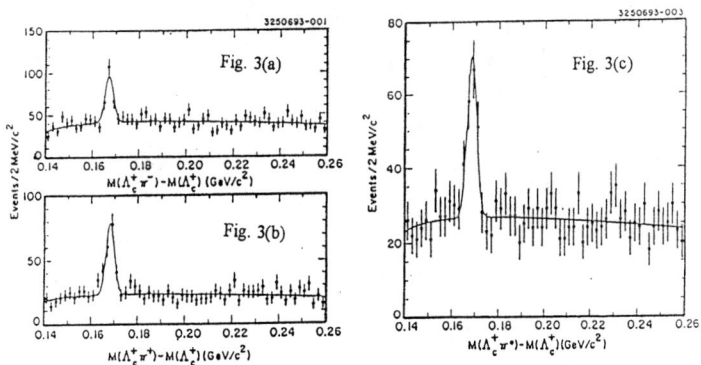

Figure 3. Inv. Mass difference distributions for $\Lambda_c\pi$ combinations relative to Λ_c.

The first evidence of the Σ_c^{++} (cuu) was reported as a single bubble chamber event[3] in 1975 observed as $\nu p \to \mu^- \Sigma_c^{++} \to \mu^- \Lambda_c^+ \pi^+ \to \mu^- (\Lambda\pi^+\pi^-\pi^+) \pi^+$. The neutral

$\Sigma_c^0(cdd)$ was first produced[6] in γ Be collisions in 1976. For a long time, the only evidence for the Σ_c^+ (cud) was a single νp interaction observed in the Big European Bubble Chamber at CERN in 1980[8]. The first observations of Σ_c^{++} and Σ_c^0 at CESR were presented by CLEO[13] in 1989 using about 418 pb^{-1} of data in the region of the $\Upsilon(4S)$. A more precise measurement[20] was presented in 1993 using 1.48 fb^{-1}, which also included the first convincing evidence for the Σ_c^+ observed in the decay mode $\Lambda_c^+\pi^0$. In Figures 3 (a), (b), and (c) are shown the mass difference distributions $M(\Lambda_c\pi) - M(\Lambda_c^+)$ for the $\Sigma_c^{++}, \Sigma_c^+$, and Σ_c^0, respectively. From a fit to these mass difference distributions, the corresponding mass differences are measured to be $168.2 \pm 0.3 \pm 0.2$, $168.5 \pm 0.4 \pm 0.2$, and $167.1 \pm 0.3 \pm 0.2$ MeV/c^2, respectively. It may be noted that the $\Lambda_c^+(c[u,d])$ and the $\Sigma_c^+(c\{u,d\})$ have the same quark structure, but the wave functions are antisymmetric (denoted by $[u,d]$) and symmetric (denoted by $\{u,d\}$) with respect to the interchange of the light quarks, respectively. This sets the scale of mass splitting for the two light quarks to be in the antisymmetric or symmetric configurations.

OBSERVATION OF $\Xi_c^+(csu)$ AND $\Xi_c^0(csd)$

The charmed strange baryon Ξ_c^+ was first (1983) observed[10] in $\Sigma^- +$ Be collisions at CERN at a mass of 2460 ± 25 MeV/c^2 in the decay mode $\Lambda K^-\pi^+\pi^+ + X$. Its isospin partner, the Ξ_c^0 was first (1989) observed[14] at CESR in e^+e^- collisions in the decay mode $\Xi^-\pi^+$ at a mass of $2471 \pm 3 \pm 4$ MeV/c^2.

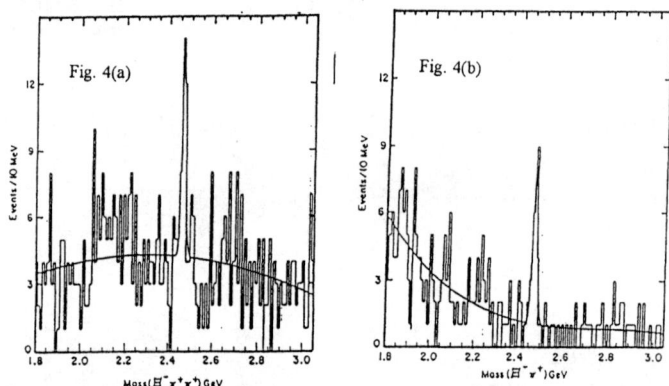

Figure 4. Inv. mass distrib. for $\Xi^-\pi^+\pi^+$ and $\Xi^-\pi^+$ combinations with $x_p > 0.5$.

Using a larger data sample of 430 pb^{-1} of data in the region of the $\Upsilon(4S)$, CLEO reported in 1989 the first observation[15] of Ξ_c^+ in e^+e^- collisions in the decay mode $\Xi^-\pi^+\pi^+$ at a mass of $2467 \pm 3 \pm 4$ MeV/c^2. In the same experiment, the isospin mass-splitting of the Ξ_c^+ relative to Ξ_c^0 was measured to be $\Delta M = -5 \pm 4 \pm 1$ MeV/c^2 and the production cross-sections were measured to be $\sigma.Br(\Xi_c^0 \to \Xi^-\pi^+) = 0.39 \pm 0.1$ pb for the Ξ_c^0 and $\sigma.Br(\Xi_c^+ \to \Xi^-\pi^+\pi^+) = 0.57 \pm 0.16$ pb for the Ξ_c^+ with $x_p > 0.5$ for both cases. In Figures 4 (a) and (b), we show the mass distributions for $\Xi^-\pi^+\pi^+$

and $\Xi^-\pi^+$ combinations with $x_p > 0.5$, where fitting yields signal sizes of 23.0 ± 6.3 and 18.8 ± 4.9 events corresponding to Ξ_c^+ and Ξ_c^0, respectively.

Since then, CLEO has observed the Ξ_c^+ in several decay modes[37]·[41] and relative to the $\Xi^-\pi^+\pi^+$ decay mode, branching fractions for the decay modes $\Xi_c^+ \to \Sigma^+ K^-\pi^+$, $\Sigma^+ K^{*0}$, $\Lambda K^-\pi^+\pi^+$, $\Xi^0\pi^+$, $\Xi^0\pi^+\pi^0$, and $\Xi^0\pi^+\pi^-\pi^+$ have been measured to be $1.18 \pm 0.26 \pm 0.17, 0.92 \pm 0.27 \pm 0.14, 0.58 \pm 0.16 \pm 0.07, 0.55 \pm 0.13 \pm 0.09, 2.34 \pm 0.57 \pm 0.37$, and $1.74 \pm 0.42 \pm 0.27$, respectively. A fit to x_p production spectrum of the Ξ_c^+ with the Peterson function yields $\epsilon_Q = 0.23^{+0.06}_{-0.05} \pm 0.3$.

CLEO has also observed[17] the Ξ_c^0 in the decay modes $\Omega^- K^+$, $\Xi^-\pi^+\pi^0$, $\Xi^0\pi^+\pi^-$ and $\Xi^-\pi^+\pi^-\pi^+$. The branching fractions for these modes relative to that into $\Xi^-\pi^+$ have been measured to be $0.51 \pm 0.21 \pm 0.05, 3.0 \pm 0.6 \pm 0.5, 2.2 \pm 0.6 \pm 0.4, 1.8 \pm 0.6 \pm 0.5$, respectively. The last three measurements are preliminary. CLEO has also observed[29] the decay modes $\Xi_c^+ \to \Xi^0 e^+\nu_e$ and $\Xi_c^0 \to \Xi^- e^+\nu_e$ using $\Xi - e^+$ correlations. They have measured $Br(\Xi_c^+ \to \Xi^-\pi^+\pi^+)/Br(\Xi_c^+ \to \Xi^0 e^+\nu_e)$ and $Br(\Xi_c^0 \to \Xi^-\pi^+)/Br(\Xi_c^0 \to \Xi^- e^+\nu_e)$ to be $0.44 \pm 0.11^{+0.11}_{-0.06}$ and $0.32 \pm 0.10^{+0.05}_{-0.03}$, respectively. Further, assuming Ξ_c^+ and Ξ_c^0 production to be equal in e^+e^- collisions, the lifetime ratio of Ξ_c^+ to Ξ_c^0 is estimated to be $2.46 \pm 0.70^{+0.33}_{-0.23}$. Using the world measurement of Ξ_c lifetimes, and the semileptonic branching fraction measurements, the absolute branching fractions $Br(\Xi_c^+ \to \Xi^-\pi^+\pi^+)$ and $Br(\Xi_c^0 \to \Xi^-\pi^+)$ are estimated to be $f_c(2.1 \pm 0.8 \pm 0.4)\%$ and $f_c(0.43 \pm 0.15 \pm 0.10)\%$, respectively. The factor $f_c = Br(\Xi_c \to \Xi l^+\nu_l)/Br(\Xi_c \to X l^+\nu_l)$ and is expected from theoretical models to be between 0.6 to 1.0.

OBSERVATION OF $\Xi_c^{+'}$ AND $\Xi_c^{0'}$

The charmed strange baryons $\Xi_c^{+'}(c\{s,u\})$ and $\Xi_c^{0'}(c\{s,d\})$ have the same quark content as the $\Xi_c^+(c[s,u])$ and $\Xi_c^0(c[s,u])$, but their wave-functions are symmetric under interchange of the light quarks. The mass difference of Ξ_c' relative to Ξ_c baryons is predicted[32][35] to be between $100 - 110$ MeV/c^2; consequently, only photonic transitions between them are possible. Based on 4.96 fb^{-1} of data in the region of the $\Upsilon(4S)$, CLEO observes a yield of (225 ± 21) Ξ_c^+ events in the decay modes $\Xi^-\pi^+\pi^+$, $\Xi^0\pi^+\pi^0$ and (289 ± 44) Ξ_c^0 events in the decay modes $\Xi^-\pi^+$, $\Xi^-\pi^+\pi^0$, $\Omega^- K^+$, and $\Xi^0\pi^+\pi^-$. Combinations are formed of above Ξ_c^+ and Ξ_c^0 candidates with clean and isolated photons with energy greater than 100 MeV. In Figures 5 (a) and (b), we show the mass difference distributions $\Delta M^+ = M(\Xi_c^+\gamma) - M(\Xi_c^+)$ and $\Delta M = M(\Xi_c^0\gamma) - M(\Xi_c^0)$, where the contributions from the different decay modes have been added. Fitting to the observed mass enhancements in these figures yields signal areas of (25.5 ± 6.5) and (28.0 ± 7.1) events, respectively. The mass difference peaks are measured to be $\Delta M^+ = (107.8 \pm 1.7 \pm 2.5)$ MeV/c^2 and $\Delta M^- = (107.0 \pm 1.4 \pm 2.5)$ MeV/c^2. Since the Ξ_c^{*+} and Ξ_c^{*0} have already been observed, the most likely interpretation of the above resonances would be as the $J^P = \frac{1}{2}^+$ charmed strange baryons $\Xi_c^{+'}$ and $\Xi_c^{0'}$, respectively. A fit to the x_p production spectrum averaged over the two charged states with the Peterson function yields $\epsilon_Q = 0.20^{+0.23}_{-0.09} \pm 0.07$. We also measure that the fraction of Ξ_c from Ξ'_c baryons, averaged over both charged states, to be $(35 \pm 9 \pm 7)\%$.

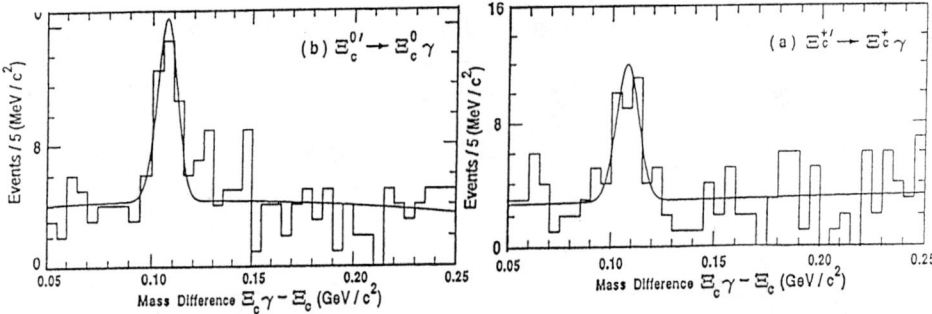

Figure 5. Inv. mass distrib. for $\Xi^0_c \gamma$ and $\Xi^+_c \gamma$ combinations with $x_p > 0.5$.

SEARCH FOR THE DOUBLY STRANGE CHARMED BARYON $\Omega^0_c(css)$

The Ω^0_c was first reported as three events in the decay mode $\Xi^- K^- \pi^+ \pi^+$ at a mass of 2746 ± 20 MeV/c^2 from the WA62 hyperon beam experiment at CERN[11] in 1985. In e^+e^- collisions, the ARGUS collaboration was first to present evidence[19] with 12.2 ± 4.5 events in the decay mode $\Xi^- K^- \pi^+ \pi^+$ at a mass of $2719.0 \pm 7.0 \pm 2.5$ MeV/c^2. Their data sample consisted of 389 pb^{-1}. But using 1.8 fb^{-1} of data, CLEO failed to observe any events in this decay mode and placed a 90% C.L. upper limit of $\sigma.Br(\Omega^0_c \to \Xi^- K^- \pi^+ \pi^+) < 0.4$ pb to be compared to ARGUS's reported measurement of $2.41 \pm 0.90 \pm 0.3$ pb for this value. While the observation of Ω^0_c in e^+e^- collisions is not clear, FNAL experiment E687 has now reported observations[25] [?] of 10.3 ± 3.9 and 42 ± 9 events in the decay modes $\Omega^- \pi^+$ and $\Xi^- K^- \pi^+ \pi^+$, respectively. The masses measured from these decay modes are reported to be $2705.9 \pm 3.3 \pm 2.2$ and 2699 ± 2.9 MeV/c^2, respectively.

OBSERVATION OF THE $\Xi^{*0}_c(csd)$ AND $\Xi^{*+}_c(csu)$

In 1995, CLEO[33] reported the observation of the Ξ^{*0}_c, the $J^P = \frac{3}{2}^+$ partner of the Ξ^0_c, where the two light quarks are in spin $S = 1$ state. Following this, CLEO[34] reported the observation of its isospin partner Ξ^{*+}_c in 1996. The data sample consisted of 3.7 fb^{-1} and 4.1 fb^{-1} of data in the region of the $\Upsilon(4S)$, respectively. They were detected by forming $\Xi^+_c \pi^-$ and $\Xi^0_c \pi^+$ combinations, respectively. The Ξ^+_c was reconstructed in the decay modes $\Xi^- \pi^+ \pi^+$, $\Xi^0 \pi^+ \pi^0$ and $\Sigma^+ K^{*0}$, while the Ξ^0_c was reconstructed in the decay modes $\Xi^- \pi^+$, $\Omega^- K^+$, $\Xi^- \pi^+ \pi^0$ and $\Xi^0 \pi^- \pi^+$. To obtain improved mass resolution, instead of the invariant mass distributions, the mass difference $\Xi_c\pi - \Xi_c$ distributions were plotted for combinations with $x_p > 0.4 - 0.6$ depending on the decay mode of the Ξ_c used. Figures 6 (a) and (b) show the corresponding mass difference distributions with clear peaks.

For the first figure, a fit to the resonance peak yields a signal size of 54.6 ± 12.1 events at $\Delta M^+ = 178.2 \pm 0.5 \pm 1.0~MeV/c^2$. The natural width is estimated to be $\Gamma^+ = 2.6^{+1.7}_{-1.4}~MeV/c^2$. A fit to the measured x_p production spectrum with the Peterson function yields $\epsilon_Q = 0.22^{+.15}_{-.08}$ and extrapolating to $x_p < 0.5$, CLEO calculates that $(27 \pm 6 \pm 6)\%$ of all observed Ξ_c^+'s are produced from the secondary decays of the higher Ξ_c^{*0} states. Fitting to the second figure yields a signal size of $34.2^{+8.9}_{-7.9}$ events at $\Delta M^0 = 174.3 \pm 0.5 \pm 1.0~MeV/c^2$ with natural width $\Gamma^0 < 3.1~MeV/c^2$. The Peterson function fit to the x_p production spectrum gives a value of $\epsilon_Q = 0.23^{+0.06}_{-0.05} \pm 0.03$ and using it to extrapolate the measured spectrum below $x_p < 0.5$, the fraction of Ξ_c^0's produced from the secondary decays of Ξ_c^{*+} baryons is calculated to be $(17 \pm 5^{+4}_{-3})\%$.

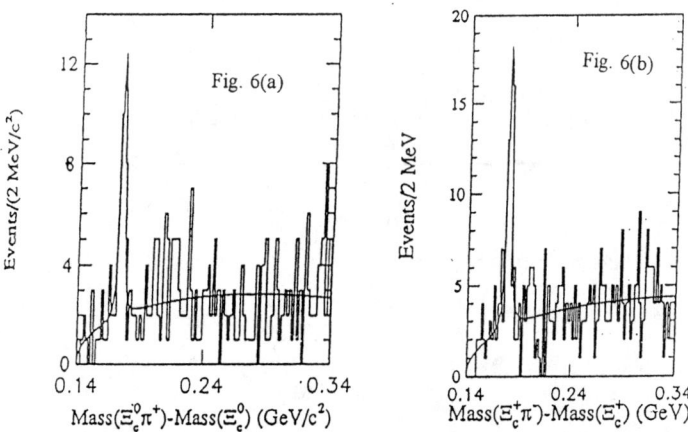

Figure 6. Inv. mass difference distrib. for (a) $\Xi_c^0\pi^+$ and (b) $\Xi_c^+ \pi^-$ combinations relative to Ξ_c^0 and Ξ_c^+ combinations.

OBSERVATION OF Σ_c^{*++} AND Σ_c^{*0}

The first observations of the Σ_c^{*++} and Σ_c^{*0} were reported[39] by the CLEO collaboration in 1995 in the decay modes $\Lambda_c^+\pi^+$ and $\Lambda_c^+\pi^-$. Based on 4.8 fb^{-1} of data around the $\Upsilon(4S)$ region, Λ_c^+ candidates were reconstructed in thirteen different decay modes. In Figures 7 (a) and (b), the mass difference distributions for the above combinations with respect to the Λ_c^+ are plotted. In each case, a narrow peak corresponding to the $J^P = \frac{1}{2}^+$ charmed baryon Σ_c and the broader $J^P = \frac{3}{2}^+$ partner Σ_c^{**} can be seen. Fitting these peaks to a Gaussian resolution and a Breit-Wigner function for both the resonance yields signal sizes of 677^{+101}_{-93} and 504^{+93}_{-83} events for the Σ_c^{*++} and Σ_c^{*0}, respectively. CLEO reports $\Delta M^{++} = 234.5 \pm 1.1 \pm 0.8~MeV/c^2$ and

$\Delta M^0 = 232.6 \pm 1.0 \pm 0.8 \; MeV/c^2$ with natural widths $\Gamma^{++} = 18^{+4}_{-3} \; MeV/c^2$ and $\Gamma^0 = 13^{+4}_{-3} \; MeV/c^2$. A fit to the x_p charge-averaged production spectrum with the Peterson function yields $\epsilon_Q = 0.30^{+0.10}_{-0.07}$ and extrapolating the measured spectrum to $x_p < 0.5$, it is estimated that $(12.8^{+1.5}_{-1.3} \pm 3.2)\%$ of Λ_c^+ baryons are produced from the decay of the two Σ_c^{**} baryons.

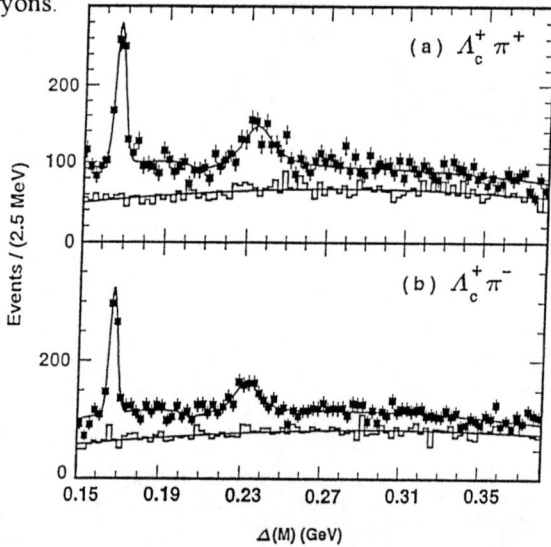

Figure 7. Inv. mass difference distrib. of (a) $\Lambda_c^+ \pi^+$ and (b) $\Lambda_c^+ \pi^-$ combinations with $x_p > 0.5$.

OBSERVATION OF $\Lambda_{c1}^{**+}(1/2^-)$ AND $\Lambda_{c1}^{**+}(3/2^-)$

In the case of the ground state $\Lambda_c^+ (cud)$ charmed baryon, the orbital angular momentum quantum number between the light quarks and that between the heavy charm quark and the light diquark, denoted as l and l', are both equal to zero. If the orbital angular momentum between the charm quark and the diquark is excited to $l = 1$, two orbitally excited states $\Lambda_c^{**+}(1/2^-)$ and $\Lambda_c^{**+}(3/2^-)$ are expected, with the $(3/2^-)$ state expected to be more massive than the $(1/2^-)$ one. Both states can decay via $\Sigma_c \pi$; but the $(1/2^-)$ state has an S-wave decay while the $(3/2^-)$ state has a D-wave decay. In 1993, ARGUS[24] reported the observation of a resonant state decaying into $\Lambda_c^+ \pi^+ \pi^-$ at a mass difference $\Delta M^+ = M(\Lambda_c^+ \pi^+ \pi^-) - M(\Lambda_c^+) = 341.5 \pm 0.6 \pm 1.6 \; MeV/c^2$. Not long after, E687[27] at FNAL reported a similar observation with a mass difference of $340.4 \pm 0.6 \pm 0.3 \; MeV/c^2$.

The CLEO[31] analysis is based on 3 fb^{-1} of data in the region of the $\Upsilon(4S)$. A Λ_c^+ sample is obtained using six different decay modes: $pK^-\pi^+$, $p\overline{K^0}$, $\Lambda\pi^+$, $\Lambda\pi^+\pi^0$, $\Lambda\pi^+\pi^+\pi^-$ and $\Sigma^+\pi^+\pi^-$. In Figure 8, we show the mass difference distribution $\Delta M^+ = M(\Lambda_c^+\pi^+\pi^-) - M(\Lambda_c^+)$ for all $\Lambda_c^+\pi^+\pi^-$ combinations with $x_p > 0.7$. Two clear peaks can be seen.

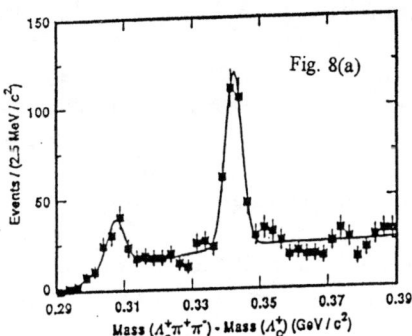
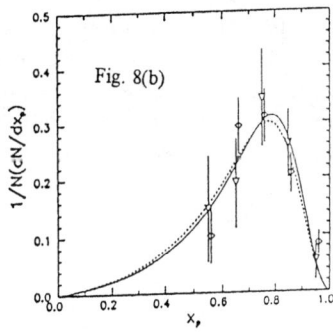

Figure 8. (a) Inv. mass distrib. for $\Lambda_c^+\pi^+\pi^-$ combinations with $x_p>0.5$. (b) x_p production spectrum with Peterson function fit.

A fit using Briet-Wigner line shape convoluted with a Gaussian detector mass resolution yields signal sizes of 112.5 ± 16.5 and 244.6 ± 19.0 events at mass differences of $307.5\pm0.4\pm1.0$ and $342.2\pm0.2\pm0.5$ MeV/c^2 and natural widths $\Gamma_l = 3.9^{+1.4+2.0}_{-1.2-1.0}$ and $\Gamma_h < 1.9$ MeV/c^2, respectively. The two resonances are commonly referred to as the $\Lambda_c^{**+}(2593)$ and $\Lambda_c^{**+}(2625)$ excited states. The fraction of Λ_c^+'s from the secondary decays of these states has been measured to be $(1.44 \pm 0.24 \pm 0.30)\%$ and $(3.51 \pm 0.34 \pm 0.28)\%$, respectively. From the study of $\Lambda_c\pi$ mass distributions, CLEO measures the branching fraction of $\Lambda_c^{**+}(2593)$ to $\Sigma_c^{++}\pi^-$ and $\Sigma^0\pi^+$ to be $(36 \pm 09 \pm 09)\%$ and $(42 \pm 09 \pm 09)\%$, respectively. There is no evidence for the $\Lambda_c^{**+}(2625)$ to be decaying through the $\Sigma_c\pi$ channel. Thus the lower mass state is identified as the $(1/2^-)$ and the higher mass state as the $(3/2^-)$ state.

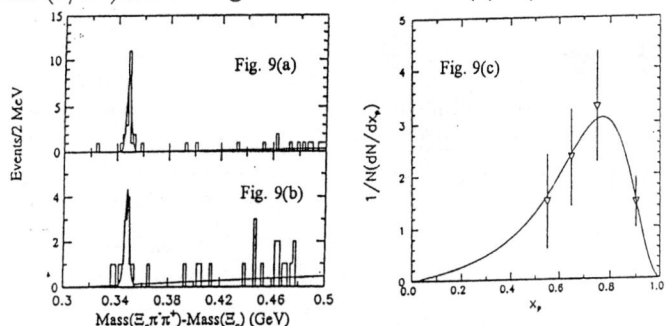

Figure 9. Inv. mass distrib. for $\Xi_c^{*0}\pi^+$ and $\Xi_c^{*+}\pi^-$ combinations with $x_p>0.5$. x_p production spectrum averaged over the two states and fit to Peterson function.

OBSERVATION OF $\Xi_c^{**+}(3/2^-)$ AND $\Xi_c^{**0}(3/2^-)$

The lowest orbitally excited states of the $\Xi_c(csq)$ are obtained by putting the heavy quark with an angular momentum quantum number $l = 1$ with respect to the light diquark. Two sets of states with $J = (1/2^-)$ and $J = (3/2^-)$ are expected. CLEO[42] reported the first observation of two resonances decaying to $\Xi_c^{*0}\pi^+$ and $\Xi_c^{*+}\pi^-$, which may be interpreted as the most likely candidates for the $\Xi_c^{**}(3/2^-)$ isospin doublet. Using 4.8 fb^{-1} of data in the region of the $\Upsilon(4S)$, the analysis starts by defining Ξ_c^{*+} and Ξ_c^{*0} candidates using the decay sequence $\Xi_c^0\pi^+$ and $\Xi_c^+\pi^-$, respectively. The Ξ_c^0 is detected in the decay modes: $\Xi^-\pi^+$, $\Xi^-\pi^+\pi^0$, $\Lambda\overline{K^0}$, $\Xi^0\pi^+\pi^-$, $\Omega^-\pi^+$ and $\Lambda K^-\pi^+$. The decay modes $\Xi^-\pi^+\pi^+$, $\Xi^0\pi^+\pi^0$ and $\Lambda\overline{K^0}\pi^+$ are used for reconstructing Ξ_c^+. Combining Ξ_c^* candidates with the charged pions in the event, the mass differences $\Delta M^+ = M(\Xi_c^{*0}\pi^+) - M(\Xi_c^{*0})$ and $\Delta M^0 = M(\Xi_c^{*+}\pi^-) - M(\Xi_c^{*+})$ are calculated. In Figure 9, the mass difference distributions ΔM^+ and ΔM^0 for combinations with $x_p > 0.6$ are plotted. Two very clean peaks with little background can be seen. Fitting these peaks with a Breit-Wigner convoluted with a Gaussian detector resolution yields signal areas of 18.8±4.4 and 9.5±3.2 events at mass differences of 348.5±0.5±1.0 and 348.1 ± 0.8 ± 1.0 MeV/c^2, respectively. The corresponding natural widths are measured to be $\Gamma^+ < 3.5$ and $\Gamma^0 < 8.1$ MeV/c^2, respectively. Assuming the production spectrum of the charged and neutral member to be the same, a fit to the averaged x_p spectrum with a Peterson function returns the value $\epsilon_Q = 0.07^{+0.03}_{-0.02}$, which is similar to the corresponding value for the orbitally excited Λ_c^{**+}'s reported earlier. These states are associated with the $(3/2^-)$ states rather than the $(1/2^-)$ states, as in this case, the decays proceed through an S-wave rather than a D-wave that would be required in the second case, which would be suppressed. The measured mass differences are consistent with recently published theoretical calculations[43] [44].

SUMMARY AND CONCLUSION

We may summarize our results as follows. Most of the singly-charmed ground-state baryons have been found. Only the Σ_c^{**0} and the Ω_c^{*0} have not been seen. Using the newly reprocessed data along with data from CLEO II.5 may yield these signals soon. Masses and isospin mass splittings have been measured. Using the lowest masses as input, various models are successful at predicting the higher mass members. The isospin mass splitting measurements are limited in statistics. The theoretical calculations are also limited in their predictive power, even the sign of the splitting varies from model to model. Orbitally excited charmed baryons are beginning to pop up. The Peterson function is a good representation of the x_p production spectrum. The x_p spectra of all the ground-state baryons can be fit to Peterson functions with ϵ_Q parameter between 0.2 – 0.3, while those of the orbitally-excited members require values of ϵ_Q between 0.05 – 0.07. A substantial fraction of ground-state baryons are produced from the secondary decays of higher mass states and so they have softer spectra. Doubly charmed baryons may remain out of reach for the present CLEO data set. Fermi National Accelerator Lab and the LHC at CERN may have easier access to these higher mass states. Although the spectroscopy of ground-state baryons appears to be nearing completion, measurement of the branching fractions has just begun. **In conclusion, it would appear that CESR has already become a charmed baryon factory.**

REFERENCE

[1] J. E. Augustin et al., Phy. Rev. Lett., **33**, 1406 (1974).
[2] J. J. Aubert et al., Phy. Rev. Lett., **33**, 1404 (1974).
[3] E. G. Cazzoli et al., Phy. Rev. Lett., **17**, 1125 (1975).
[4] G. Goldhaber et al., Phy. Rev. Lett., **37**, 255 (1976).
[5] I. Peruzzi et al., Phy. Rev. Lett., **37**, 569 (1976).
[6] B. Knapp et al., Phy. Rev. Lett., **37**, 882 (1976).
[7] G. S. Abrams et al., Phy. Rev. Lett., **44**, 10 (1980).
[8] M. Calicchio et al., Phy. Ltrs. **93B**, 521 (1980).
[9] C. Peterson, Phy. Rev **D27**, 105 (1983).
[10] S. F. Biagi et al., Phy. Letts. **122**, 455 (1983).
[11] S. Biagi et al., Zeitschr. f. Phys. **C28**, 175 (1985).
[12] CLEO Collaboration: D. Bortolleto et al., Phy. Rev **D37**, 1719 (1988).
[13] CLEO Collaboration: T. Bowcock et al., Phy. Rev. Lett., **62**, 1240 (1989).
[14] CLEO Collaboration: P. Avery et al., Phy. Rev. Ltrs., **62**, 863 (1989).
[15] CLEO Collaboration: M. S. Alam et al., Phy. Ltrs. **B226**, 401, (1989).
[16] CLEO Collaboration, P. Avery et al., Phy. Rev. **D43**, 3599 (1991).
[17] CLEO Collaboration: S. Henderson et al., Phy. Lett. **B283**, 161 (1992).
[18] CLEO Collaboration: G. Crawford et al., Phy. Rev **D45**, 752 (1992)
[19] H. Albrecht et al., Phys. Letts. **B288**, 367 (1992).
[20] CLEO Collaboration: G. Crawford et al., Phy. Rev. Lett., **71**, 3259 (1993)
[21] CLEO Collaboration: P. Avery et al., Phy. Rev. Lett., **71**, 2391 (1993).
[22] CLEO Collaboration: Y. Kubota et al., Phy. Rev. Lett., **71**, 3255 (1993).
[23] CLEO Collaboration: P. Avery et al., Phy. Letts. **B325**, 267, (1993).
[24] ARGUS Collaboration: H. Albrecht et al., Phy. Letts. **B317**, 227 (1993).
[25] E-687 Collaboration: P. L. Frabetti et al., Phy. Letts. **B300**,190 (1993)
[26] CLEO Collaboration: T. Bergfeld, et al., Phy. Letts. **B323,** 219 (1994).
[27] E-687 Collaboration: P. L. Frabetti et al., Phy. Letts. **B338**,106 (1994).
[28] CLEO Collaboration: R. Ammar et al., Phy. Rev. Letts., **74**, 3534 (1995).
[29] CLEO Collaboration: J. P. Alexander et al., Phy. Rev. Letts. **74**, 3113 (1995).
[30] CLEO Collaboration: J. P. Alexander et al., Phy. Rev. Letts. **74**, 3113 (1995) Erratum.
[31] CLEO Collaboration: K. W. Edwards et al., Phy. Rev. Letts. **74**, 3331 (1995).
[32] M. J. Savage, Phy. Letts., **B359**, 189 (1995).
[33] CLEO Collaboration: P. Avery et al., Phy. Rev. Let. **75**, 4365 (1995)
[34] CLEO Collaboration: L. Gibbons et al., Phy. Rev. Let. **77**, 810 (1996).
[35] A. Falk, Phy. Rev. Lett., **77**, 223 (1996).
[36] CLEO Collaboration: T. Bergfeld et al., Phy. Letts., **B365**, 431 (1996).
[37] CLEO Collaboration: K. Edwards et al., Phy. Letts. **B373**, 261 (1996).
[38] CLEO Collaboration: J. P. Alexander, et al., Phy. Rev **D53**. Rapid Comm., R1013, Feb., 1996.
[39] CLEO Collaboration: T. Bergfeld et al., Phy. Rev Lett , **77**, 4503 (1996).
[40] CLEO Collaboration: M. S. Alam et al., Phy. Rev **D57**, 4467, 1998.
[41] CLEO Collaboration: T. Bergfeld et al., European Physical Society Conference, 1997.
[42] CLEO Collaboration: M. Athanas et al., Int. Conference on High Energy Physics 98/866, 1998.
[43] D. Pirjol and A. FalkFalk, Phy. Rev. **D56**, 5483 (1997).
[44] G. Chiladze and T-M. Yan, Phy. Rev. **D56**, 6738 (1997).

The COMPASS Experiment at CERN
Prospects for Charm and Structure Function Measurements

Alessandro Bravar, for the COMPASS collaboration

Institut für Kernphysik, Universität Mainz, D-55099 Mainz, Germany

Abstract. The COMPASS experiment at CERN is aimed at the study of the nucleon spin structure and hadron spectroscopy. It attempts a measurement of the gluon polarization around $x_g \simeq 0.1$ with a precision better than $\delta(\Delta g/g) < 0.1$. The experiment uses muo-production of open charm and correlated high-p_T hadron pairs to tag the photon-gluon fusion process. In parallel COMPASS will perform also a rich spin-physics program in polarized DIS (*muon* program). The *hadron* physics program covers studies of pion and kaon polarizabilities in Primakoff scattering, glueballs and hybrid searches in the gluon rich central production, studies of charmed hadron production and (semi-leptonic and leptonic) decays. COMPASS uses a double forward spectrometer for best acceptance and momentum resolution. Both spectrometers are equipped with RICH detectors, electromagnetic and hadronic calorimeters, and muon walls for particle identification.

COMPASS is a fixed-target experiment using a "Common Muon and Proton Apparatus for Structure and Spectroscopy". It has been recently approved by the CERN Research Board as NA58 and will be commissioned in year 2000. First results are expected soon after that year.

The COMPASS experiment covers a rich and broad physics program. Important and intriguing questions of perturbative and non-perturbative QCD will be addressed from the polarizabilities of hadrons in Primakoff scattering to centrally produced glueballs and exotic mesons, from the hadro-production of charm and the studies of semi-leptonic and leptonic decays of charm hadrons (*hadron* program), to open charm production in polarized deep inelastic scattering (DIS) and photo-production limit, from longitudinal to transverse polarized quark distributions in DIS (*muon* program).

One of the most intriguing issues in understanding the nucleon's spin structure is the polarization of gluons, $\Delta g/g$. A sizable value of Δg could explain the smallness of the contribution of the quark spins to the nucleon spin, $\Delta\Sigma$ [1], as well as the correlated appearance of a negatively polarized strange sea. Indications of a large value of Δg of about two units in $\hbar$ at $Q^2 = 4$ GeV^2 come from recent QCD analyses [2] of existing g_1 data. However, it is generally agreed that a direct measurement of Δg is urgently needed. A particularly clean process involving the gluon distribution in leading order is the open charm production via the photon-gluon fusion process, $\gamma g \to c\bar{c}$, shown in Fig. 1. The hard scale is set by the charm quark mass, $4m_c^2 \simeq 10$ GeV^2.

One fundamental prediction of QCD is the existence of states containing valence gluons: glueballs and hybrids [3]. For many years experiments tried to establish their existence, with the best candidate, a scalar glueball $f_0(1500)$ [4]. For a positive identification of exotic states the detection of as many decay modes and production mechanisms as possible is required, extending well into the 2 GeV mass region, where lattice QCD predicts the tensor glueball 2^{++} around 2.3 GeV/c. Central collisions of protons on protons provide a gluon rich environment in the double Pomeron scattering. This is believed to efficiently produce glueballs and exotic non-$q\bar{q}$ states predicted by QCD.

Apart from the measurement of $\Delta g/g$ COMPASS offers a rich spin-physics program at high Q^2 with a high luminosity including the measurement of g_1 and g_2, the transversity structure function h_1, spin-flavor decomposition of the structure functions, and $\Lambda/\bar{\Lambda}$ polarization in both the current and target fragmentation regions. It exploits the existing polarized muon beam at CERN and the proved DNP polarized target technology. Details can be found in the full COMPASS proposal [5].

In COMPASS [5] we will access the gluon polarization $\Delta g/g$ by measuring the open charm cross section spin-asymmetry with a polarized muon beam of 100 – 200 GeV scattering off a fixed polarized target using the full virtual photon spectrum down to quasi-real photons. At these muon energies the contributions of resolved photons, intrinsic charm, and diffractive components to the total charm photoproduction cross section are expected to be very small.

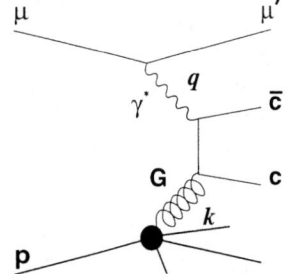

FIGURE 1. The PGF process.

The photon-nucleon cross section spin-asymmetry for charm production via PGF by real (or quasi-real) photons, $A_{\gamma N}^{c\bar{c}}$, is obtained by integrating the cross sections over the kinematically allowed range $x_g^{\min} = 4m_c^2/2M\nu \le x_g \le 1$

$$A_{\gamma N}^{c\bar{c}}(\nu) = \frac{\Delta\sigma^{\gamma N \to c\bar{c}X}}{\sigma^{\gamma N \to c\bar{c}X}} = \frac{\int_{4m_c^2}^{2M\nu} d\hat{s}\, \Delta\hat{\sigma}(\hat{s})\, \Delta g(x_g,\hat{s})}{\int_{4m_c^2}^{2M\nu} d\hat{s}\, \hat{\sigma}(\hat{s})\, g(x_g,\hat{s})} \sim \langle a_{LL}\rangle \cdot \langle \frac{\Delta g}{g}\rangle. \quad (1)$$

$x_g = \hat{s}/2M\nu$ denotes the nucleon momentum fraction carried by the gluon (note that $x_g \ge x_{Bj}$) and $\hat{s}$ is the invariant mass of the two emerging quarks.

The PGF spin-averaged cross sections, $\hat{\sigma}(\hat{s})$, is known to next-to-leading order since some time [6], while the spin-dependent cross section, $\Delta\hat{\sigma}(\hat{s})$, only recently has become available to next-to-leading order [7]. Both terms, $\hat{\sigma}$ and $\Delta\hat{\sigma}$ (Fig. 2a), rise sharply at the threshold, $\hat{s}_{th} = 4m_c^2$. The PGF scattering spin-asymmetry $a_{LL}(\hat{s}) = \hat{\sigma}/\Delta\hat{\sigma}$ measures the analysis power of this process. At threshold $a_{LL} = 1$,

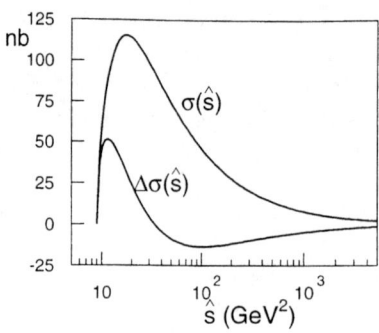 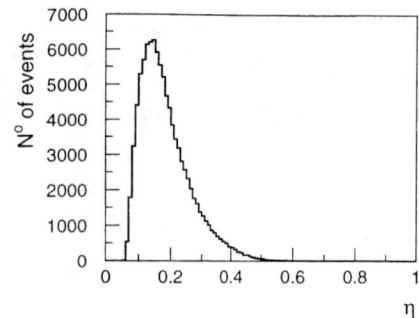

FIGURE 2. a) The photon-gluon cross sections $\hat{\sigma}$ and $\Delta\hat{\sigma}$ as a function of $\hat{s}$. b) The covered x_g (= η in the figure) range for $35 \leq \nu \leq 85$ GeV.

it crosses zero at about $2 \times \hat{s}_{th}$ and in the asymptotic limit of very high $\hat{s}$ reaches the value -1 of the massless case. Knowing $\hat{\sigma}(\hat{s})$, $\Delta\hat{\sigma}(\hat{s})$ (or $a_{LL}(\hat{s})$), and $g(x_g)$ allows to evaluate $\Delta g(x_g)$ from the measured asymmetry $A_{\gamma N}^{c\bar{c}}$ (Eq. 1).

Open charm photo-production leads to D mesons and/or charmed baryons in the final state. For the kinematics of the COMPASS experiment we find in average $1.2 \cdot (D^0 + \overline{D}^0)$ mesons per initial $c\bar{c}$ pair. The most important problem in the measurement of the asymmetry of $c\bar{c}$ production is the unambigous identification of charmed hadrons. We concentrate on the two-body decays $D^0 \to K^-\pi^+ + c.c.$ with branching ratios of about 4 %. A major concern is the combinatorial background of πK pairs within the D^0 mass window. In the COMPASS apparatus the mass of the D^0 meson will be reconstructed with a resolution of $\sigma_{M_D} \simeq 10$ MeV/c^2. In the range $35 < \nu < 85$ GeV the open charm production cross section amounts to 2 nb compared to 500 nb for the non-diffractive photoproduction. Methods employing secondary vertex detection are not applicable in the COMPASS experiment, because the distance between the primary and decay vertex of a few mm's cannot be resolved due to multiple scattering in the target. Kaons emitted at large angles, θ_K^*, in the D's rest frame with respect to the D's direction of flight in the laboratory frame have large transverse momenta. On the other hand, kaons from ordinary fragmentation have small transverse momenta and thus dominantly mimic collinear decays. Three strategies for background reduction are therefore employed: (i) particle identification especially for $K^\pm$-mesons, (ii) high momentum resolution for charged particles leading to $\sigma_{M_D} \simeq 10$ MeV/c^2, and (iii) kinematical cuts on the cm-decay angle θ_K^* of the $K\pi$ pair, relative to the D^0 direction. The background rejection was studied in Monte Carlo simulations. The best result is obtained with the requirements $|\cos\theta_K^*| \leq 0.5 \pm 0.1$ and $z_D = E_D/\nu \geq 0.25 \pm 0.05$ (see Fig. 3), which improve the background-to-signal ratio by a factor 1750 to about $N^B/N^{c\bar{c}} \simeq 3.7$ at the expense of losing about 2/3 of the D^0 mesons. The accuracy of this measurement will be limited by statistics even with the five times higher muon intensity compared to the SMC experiment.

The COMPASS apparatus is based partially on the existing SMC setup. Figure 4

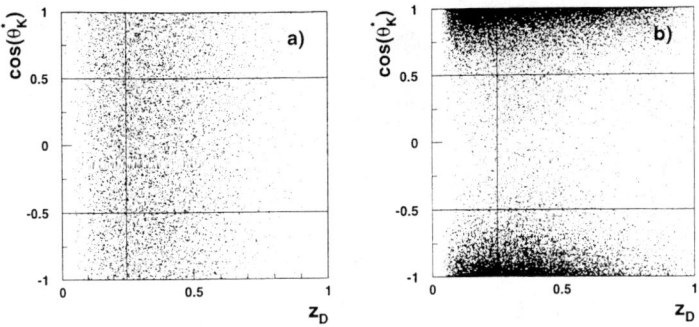

FIGURE 3. Simulated distribution of apparent D^0-momentum fraction z_D versus c.m.-decay angle $\cos\theta_K^*$ for true $D \to K\pi$ (a) and for the πK combinatorial background in the D^0 mass window (b). The lines indicates the cuts used in the rate estimates.

shows the initial setup of the COMPASS experiment. It is somewhat reduced compared to the full setup, but adequate to start with the muon program in year 2000. The wide angular range of hadrons produced in DIS event, in particular from D meson decays, requires a two stage magnetic spectrometer with particle identification in both spectrometer stages, and a new super-conducting solenoid for the polarized target with an opening matching that of the first spectrometer stage. Downstream of the polarized target the new hadron stage of the spectrometer covers hadron angles up to ± 200 mrad. Its large-aperture dipole magnet provides a bending power of 1 Tm. In the second stage of the spectrometer the scattered muon and fast hadrons are measured. It uses the present SMC spectrometer magnet with a $\int Bdl$ of 2 Tm.

The very high beam intensity of 2×10^8 muons per SPS spill (2×10^8/s) poses stringent requirements on the tracking detectors from the rate resistance, to readout speed and radiation hardness. Tracking in the inner beam region will be performed with scintillating fibre detectors. In the inner high rate part of the acceptance close to the beam newly developed gaseous detectors, the *micromegas* [10] and GEMs [11], will be used. The micromega chamber implements a fine copper mesh with a 25 μm grid about 50 – 100 μm above an anode plane with strip structure to achieve an additional gas amplification in a high electric field. This detector promises high rate capacity at good resolution. The gas electron multiplier or GEM detector employs one or two copper coated foils with 70 μm diameter holes spaced at 140 μm as amplification stage about 1 mm above an anode plane with segmented structure. Both detectors provide better rate resistance than ordinary wire chambers. Tracking in the outer region will be performed mainly with straw tubes of 6 mm and 10 mm diameter, and more conventional MWPC and drift chambers.

Charged particle identification will be performed by two ring-imaging Cherenkov detectors, one for each spectrometer stage. The RICH's will be filled with different gases and will provide π/K separation in the momentum range 3–65 GeV/c and

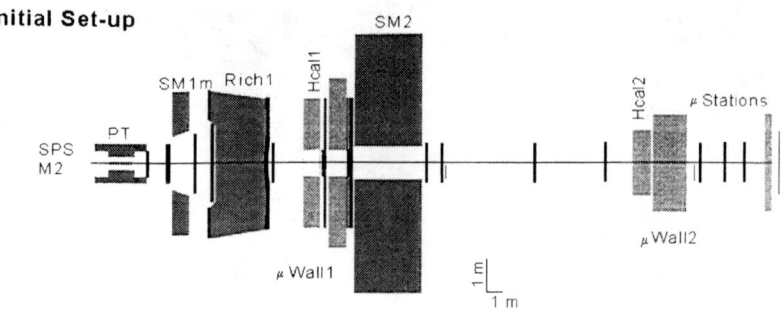

FIGURE 4. COMPASS initial setup. A second RICH detector downstream of SM2, electromagnetic calorimeters, and additional trackers will complete the apparatus.

30–120 GeV/c, respectively. Lead-glass arrays will be used for the electromagnetic calorimetry. Both spectrometer stages end with a muon wall consisting of a hadron absorber followed by tracking chambers and trigger hodoscopes. The trigger condition of the experiment will ask for a muon with a minimum energy loss of 30 GeV and a minimum hadron energy of 5 GeV deposited in the hadronic calorimeters.

The polarized target consists of a dilution refrigerator containing the target material with a microwave cavity included in a 2.5 T, 60 cm diameter superconducting solenoid magnet. In order to achieve high luminosity and good invariant mass resolution we will use a long target made out of two oppositely polarized 60 cm long cells and 3 cm diameter. Two different target materials, lithium deuteride (^{6}LiD) for the *deuteron* and ammonia (NH$_3$) for the *proton*, are foreseen. ^{6}LiD can be polarized up to 50 % with a dilution factor f of 0.50 compared to NH$_3$ which can reach a much higher polarization of 90 % but with a much less favorable f of 0.17. The materials are polarized by dynamic nuclear polarization (DNP) in magnetic field with microwaves at a temperature of a few mK only.

The planned muon beam intensity is 2×10^8 muons per SPS spill (2×10^8/s), with a polarization of ~ -80 %. The nominal luminosity amounts to $\int \mathcal{L} dt \simeq$ 2 fb^{-1}/year assuming a combined efficiency of 0.25 for the muon beam and the experimental apparatus with 150 days of data taking per year.

For a running time of 2 years shared between ^{6}LiD and NH$_3$ targets polarized longitudinally, the projected statistical error of the measured asymmetry, based only on the D$^0 \to$ K$^-\pi^+$ + c.c. detection, is $\delta A^{c\bar{c}}_{\gamma N} = 0.076$. Due to the higher figure of merit of the ^{6}LiD target a similar precision will be already reached after the first year. The result can be further improved by tagging the D^0's from D$^{*+} \to$ D$^0 \pi_s^+ \to$ (K$^-\pi^+)\pi_s^+$ decays. The D^{*+} reconstruction leads to a charm signal with very little background allowing thus to release most of the strong kinematical cuts discussed above (Fig. 3). Taking into account the acceptances and possible re-interactions in the target the statistical error of the asymmetry reduces to

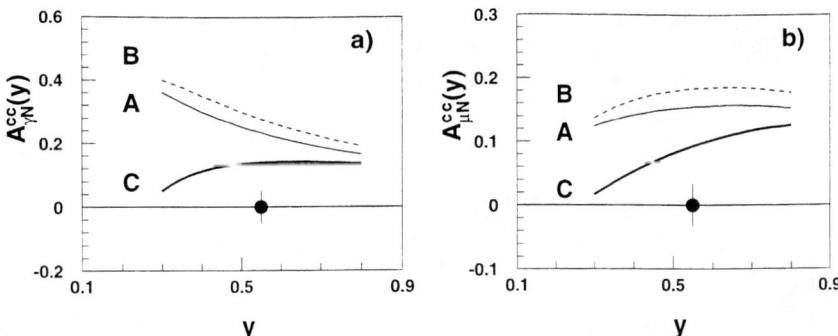

FIGURE 5. Photon-nucleon asymmetry (a), and muon-nucleon asymmetry (b), as a function of $y = \nu/E_\mu$ ($E_\mu = 100$ GeV) for the three gluon distributions of Ref. [8]. The data points indicate the projected precision of the COMPASS measurement.

$$\delta A^{c\bar{c}}_{\gamma N} = 0.05 \quad \text{corresponding to} \quad \delta\left(\frac{\Delta g}{g}\right) \simeq 0.14. \qquad (2)$$

As in the inclusive case the muon-nucleon asymmetry is reduced from the virtual-photon asymmetry by the depolarization factor $A^{c\bar{c}}_{\mu N} = D A^{c\bar{c}}_{\gamma N}$ ($\overline{D} \simeq 0.66$ for this measurement). Figure 5 shows both asymmetries, $A^{c\bar{c}}_{\gamma N}$ and $A^{c\bar{c}}_{\mu N}$, as a function of $y = \nu/E_\mu$ ($E_\mu = 100$ GeV). The correlation between the D meson observables in the final state and the $c\bar{c}$ pair in the c.m.s., which survives the hadronization, can be exploited to enhance the analysis power of the process. A larger sensitivity on the gluon polarization can be obtained by analyzing the dependence of $A^{c\bar{c}}_{\gamma N}$ on the D mesons' p_T, and rejecting D mesons with $p_T > 1$ GeV/c will yield

$$\delta\left(\frac{\Delta g}{g}\right) \simeq 0.10. \qquad (3)$$

The three and four-body decay channels of the D mesons, like $D^0 \to K^-\pi^+\pi^0$ (B.R. ~ 13.8 %), $D^0 \to K^-\pi^+\pi^+\pi^-$ (B.R. ~ 8.1 %), $D^+ \to K^-\pi^+\pi^+$ (B.R. ~ 9.1 %), etc., may further improve the precision of the measurement, in particular if the D* tagging can be applied.

Another, very interesting and promising channel to access ΔG is the cross section spin-asymmetry for correlated high-p_T hadron pair (or *pseudo* jets) production in the forward hemisphere [9]. This channel is not statistic limited, but the precision of the measurement is mostly affected by the *background* coming from gluon radiation. Selecting oppositely charged kaon pairs will allow to substantially suppress this background at the expense of a lower statistics. In one year of running at 200 GeV/c, COMPASS can achieve a precision of

$$\delta\left(\frac{\Delta g}{g}\right) < 0.05 \qquad (4)$$

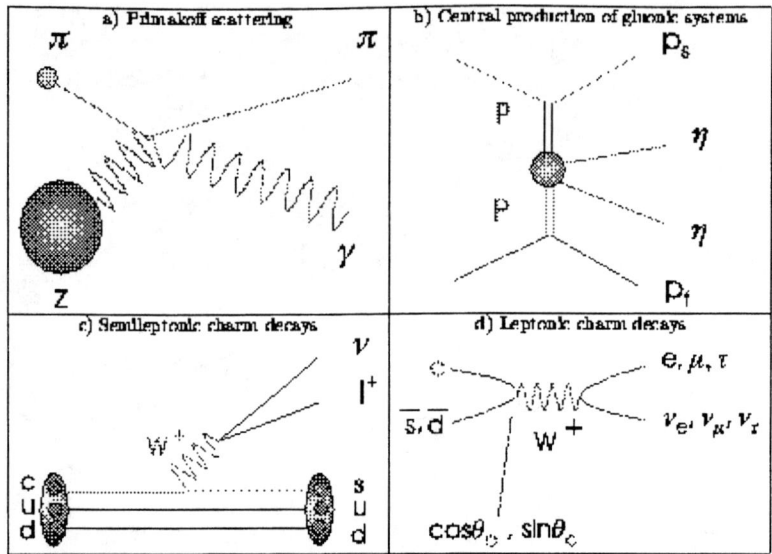

FIGURE 6. Diagrams of processes that will be addressed in COMPASS using hadron beams: Primakoff scattering, central production of gluonic states via double Pomeron exchange, semi-leptonic decays of charmed baryons, and leptonic decays of charmed mesons.

for several x_g bins in the region $0.04 < x_g < 0.2$

COMPASS will also address several physics issues of hadronic structure and dynamics, and hadron spectroscopy in both the perturbative and non-perturbative QCD domains, by studying different reactions with a variety of hadron beams. The *hadron* physics program is summarized below, and the diagrams for the relevant process are shown in Fig. 6.

- HADRON STRUCTURE
    - Polarizabilities in Primakoff reactions (π, K, p)
- LIGHT QUARK SPECTROSCOPY AND GLUONIC STATES
    - Search for glueballs in Pomeron - Pomeron scattering
    - Search for hybrid / exotic states
- CHARMED HADRONS
    - production phenomena (π, K, p beams)
    - semi-leptonic decays ($\Lambda_c^+ \to \Lambda^0 + \mu^+ + \nu$)

- leptonic decays ($D^+ \to \mu^+ + \nu$, $D_s^+ \to \mu^+ + \nu$)

- precision measurements of c - baryon lifetimes

- production and spectroscopy of cc - baryons (Ξ_{cc}, Ω_{cc})

For these measurements we plan to use a variety of hadron beams up to 300 GeV/c at a very high intensity up to 5×10^7 particles/s. A slightly modified set-up compared to the *muon* one of Fig. 4 will be used. The major differences concern the target region and the first spectrometer stage, where different targets and silicon microstrip detectors will be used.

The studies of charm hadrons cover a wide spectrum from the investigation of charm production itself, spectroscopy, decay studies up to searches for rare processes. Assuming 100 days of effective data taking at maximal beam intensity, COMPASS will collect and fully reconstruct about 5×10^6 charmed hadrons

Among these, a particularly clean sample of about 3,000 charmed baryon semi-leptonic decays ($\Lambda_c^+ \to \Lambda^0 + \mu^+ + \nu + c.c.$), accompanied by the tagged decay of the second charmed hadrons produced in the event, are expected. This sample will allow to probe predictions of HQET, although the mass of the charmed quark ($m_c \simeq 1.5$ GeV/c) is still small, and its effects are estimated to be around 20 %. Also, about 500 $D_s^\pm$ and about 100 $D^\pm$ leptonic decays will be reconstructed, yielding to a precision of about 10 % and 20 % for the $D_s^\pm$ and the $D^\pm$ decay constants f_{D_s} and f_D, respectively, allowing to confront predictions of lattice QCD calculations.

A further goal of the charm program is to look for doubly charmed baryons never observed so far.

For the study of semi-leptonic and leptonic charmed hadron decays it is very important to *see* the decay vertex of the charmed hadron inside the vertex detector, because the neutrino carries away momentum, and only the additional information from the charm track allows the kinematical reconstruction of the decay. This poses strong requirements on the target design. The vertex detector (Fig. 7) will consist of a 2.5 % thick copper target followed by multiplicity jump trigger counters and an array of 150 μm thick silicon microstrip detectors with double side readout. These detectors will have a very fine pitch of about 12 μm and will be narrowly spaced at a distance of 2 mm along the beam axis.

In conclusion, COMPASS offers a broad and rich physics program addressing the nucleon structure and hadron spectroscopy. COMPASS is a challenging experiment under construction at CERN, with a planned commissioning in year 2000. First results are expected soon after. At present, is the only new major experiment approved to run at CERN in the so called pre-LHC era. The richness of the physics program suggests, however, a further prolongation well into the LHC era.

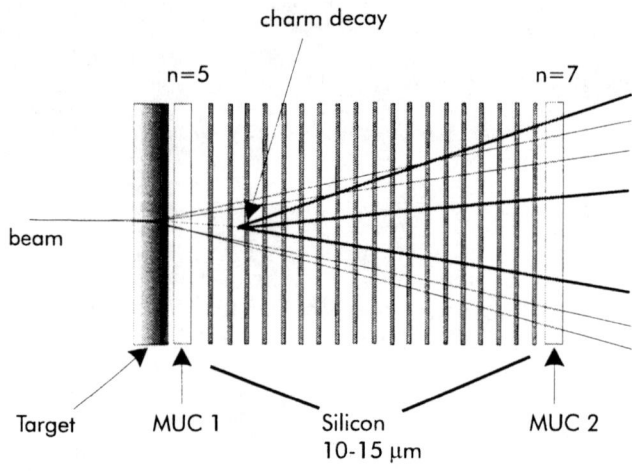

FIGURE 7. The target for charm production with fine pitch silicon microstrip detectors.

REFERENCES

1. B. Adeva *et al.* (SM Collaboration), Phys. Rev. D **58**, 112001 (1998);
   K. Abe *et al.* (E143 Collaboration), Phys. Rev. D **58**, 112003 (1998).
2. G. Altarelli *et al.*, Nucl. Phys. B **496**, 337 (1997);
   B. Adeva *et al.* (SM Collaboration), Phys. Rev. D **58**, 112002 (1998).
3. H. Fritzsch and U. Gell-Mann, in *Proccedings of the XVI Int. Conf. on High Energy Physics*, Chicago, 1972.
4. C. Amsler *et al.*, Phys. Lett. B **342**, 433 (1995).
5. The COMPASS Collaboration, COMPASS proposal, CERN/SPSLC 96-14, SPSC/P297 (Geneva, March 1996); CERN/SPSLC 96-30 (Geneva, May 1996).
6. S. Frixione *et al.*, Nucl. Phys. B **431**, 453 (1994); and references therein.
7. I. Bojak and M. Stratmann, Phys. Lett. B **433**, 411 (1998).
8. T. Gehrmann and W.J. Stirling, Z. Phys. C **65**, 461 (1995).
9. A. Bravar, D. von Harrach, and A. Kotzinian, Phys. Lett. B **421**, 349 (1998).
10. Y. Giomataris *et al.*, Nucl. Instrum. Meth. A **376**, 29 (1996).
11. F. Sauli *et al.*, Nucl. Instrum. Meth. A **386**, 531 (1997).

B and D Spectroscopy at LEP

Franz Muheim*

*Université de Genève
Département de physique nucléaire et corpusculaire
24, quai Ernest-Ansermet, 1211 Genève 4, Switzerland
Email: Franz.Muheim@cern.ch

Abstract. Results from the four LEP experiments ALEPH, DELPHI, L3, and OPAL on the spectroscopy of B and charmed mesons are presented. The predictions of Heavy Quark Effective Theory (HQET) for the masses and the widths of excited $L = 1$ B mesons are supported by a new measurement from L3. A few B_c^+ candidate events have masses consistent with the recent CDF observation and the predictions. New results on D^{**} production and $B \rightarrow D^{**} \ell \nu$ are also presented. The evidence for a $D^{*'}$ meson reported recently by DELPHI is not supported by OPAL and CLEO.

INTRODUCTION

Detailed understanding of the spectroscopy of orbitally excited heavy mesons containing a b or a c quark provides important information regarding the underlying theory. A flavor-spin symmetry arises from the fact that the mass of a heavy quark Q is large relative to Λ_{QCD}. In this approximation, the spin $\vec{s}_Q$ of the heavy quark Q is conserved in the interactions, independently of the total angular momentum $\vec{j}_q = \vec{s}_q + \vec{L}$ of the light quark q. Corrections to this symmetry are a series expansion in $1/m_Q, 1/m_Q^2$, calculable in Heavy Quark Effective Theory (HQET) [1].

TABLE 1. $L = 1$ B mesons containing a u or a d light quark with corresponding spin states, relative production rates, prediction for masses and widths, and two-body decay modes.

Name	j_q	J^P	Production	Mass [MeV]	Width [MeV]	Decay Mode	
B_0^*	1/2	0^+	1/12	$M_{B_1^*} - 12$	~ 150	$B\pi$	S-wave
B_1^*	1/2	1^+	3/12	$M_{B_2^*} \pm 100$	~ 150	$B^*\pi$	S-wave
B_1	3/2	1^+	3/12	5759	21	$B^*\pi$	D-wave
B_2^*	3/2	2^+	5/12	5771	25	$B^*\pi, B\pi$	D-wave

The $L = 0$ mesons, for which $j_q = 1/2$, have two possible spin states: a pseudoscalar P ($J^P = 0^-$) and a vector V ($J^P = 1^-$). If the spin of the heavy quark is conserved independently, the relative production rate of these states is expected to be $V/(V+P) = 0.75$. Corrections due to the decay of higher excited states are predicted to be small. Recent measurements of this rate for the B system [2,3] agree well with this ratio.

In the case of $L = 1$ orbitally excited B mesons two sets of doublets are expected: the B_0^* and B_1^* ($j_q = 1/2$) and the B_1 and B_2^* ($j_q = 3/2$) mesons (see Table 1). Their relative production rate follows from spin state counting (2J+1 states) [4]. For the dominant two-body decays, the $j_q = 1/2$ states can decay via an S-wave transition and their decay widths are expected to be broad in comparison to those of the $j_q = 3/2$ states which must decay via a D-wave transition. Many measurements exist for $L = 1$ orbitally excited charm mesons. All six narrow states, a doublet (D_2^* and D_1) for each quark content ($c\bar{u}, c\bar{d}$ and $c\bar{s}$) are well established [17]. The wide $L = 1$ states are hard to measure and have not been clearly identified.

Several models based on HQET and on the charmed $L = 1$ meson data, have made predictions for the masses and widths of orbitally excited B^{**} mesons [5–9] (see Table 1). Some of these models place the average mass of the $j_q = 3/2$ states above that of the $j_q = 1/2$ states, while others predict the opposite ("spin-orbit inversion"). The mass splitting within each doublet is predicted to be 12 MeV.

B^{**} SPECTROSCOPY

At LEP excited states of B mesons are produced. Each of the four experiments has collected about 4×10^6 hadronic events out of which 0.9×10^6 events contain $B\bar{B}$ pairs. In all B^{**} analyses, first, the b-quark purity of the data sample is increased by applying a lifetime based event tag. B mesons are reconstructed inclusively. Typically the two most energetic jets of the event are considered B candidates. The decay products of the B meson are separated from the background due to fragmentation particles using secondary vertex tagging (for charged decay particles) or the rapidity of the decay products with respect to the B-jet axis. An alternative method is to fully reconstruct the B-meson decay chain which improves the resolution B-mass resolution but suffers from low statistics.

The decay of a B^{**} meson ($B^{**} \to B^{(*)}\pi$) is carried out via a strong interaction and thus the transition pion originates at the primary event vertex. In addition, the predicted masses for the $L = 1$ states correspond to relatively small Q values, so that the pion direction is forward with respect to the B-meson direction. The track with the largest component of momentum in the direction of the B jet is selected.

A first measurement of $L = 1$ orbitally excited B mesons has been presented by OPAL [10]. Using secondary vertex charge tagging to inclusively reconstruct B mesons, the invariant mass distributions of $B^{(*)+}\pi^-$ and $B^{(*)+}K^-$ combinations show enhancements consistent with the decay of B^{**} resonances as shown in Fig. 1. An excess of 1738 ± 121 $B^{(*)+}\pi^-$ and 149 ± 30 $B^{(*)+}K^-$ candidates is observed in the mass ranges 5.60 - 5.85 GeV and 5.80 - 6.00 GeV, respectively. The background

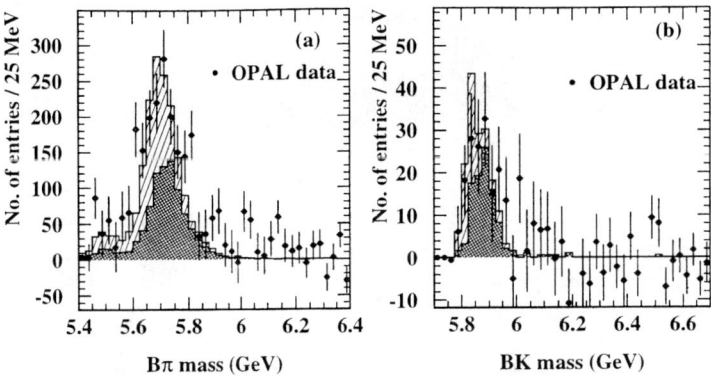

FIGURE 1. The invariant mass distribution for (a) $B^{(*)+}\pi^-$ and (b) $B^{(*)+}K^-$ combinations after subtraction of the background, respectively. The solid histograms shows the contribution from the B_2^* and the hatched histogram shows the contribution from the B_1 state.

is estimated with the wrong (like)-sign combinations. Fitting the excess with a single Breit-Wigner function yields an average mass $M(B_{u,d}^{**})$ and a production rate $f_{B^{**}} = \mathcal{B}(b \to B_{u,d}^{**})/\mathcal{B}(b \to B_{u,d})$. Throughout this paper, isospin symmetry is always employed to account for B^{**} decays via neutral pions. DELPHI and ALEPH have made similar measurements using rapidity to inclusively reconstruct B mesons [11,12]. The results are summarized in Table 2.

TABLE 2. Inclusive B^{**} measurements.

	$M(B_{u,d}^{**})$ [MeV]	$f_{B^{**}}$
OPAL	5681 ± 11	0.270 ± 0.056
DELPHI	$5734 \pm 5 \pm 17$	$0.32 \pm 0.018 \pm 0.06$
ALEPH	$5703 \pm 4 \pm 10$	$0.279 \pm 0.016 \pm 0.059\,^{+0.039}_{-0.056}$

A new measurement using an exclusive method is presented by ALEPH [13]. Using many decay modes ($B \to D^{(*)}X$, where $X \in [\pi, \rho, a_1]$ and $B \to J/\psi(\psi')K^{(*)}$) 238 charged and 166 neutral B meson candidates have been fully reconstructed. The sample has a B meson purity of 85 %. Each B candidate is then combined with a charged pion from the primary vertex. An excess of 45 ± 13 events is seen in the right-sign sample compared to the wrong-sign sample. Fig. 2 shows a fit to the right-sign mass spectrum where the signal shape consists of five Breit-Wigner peaks. The relative masses, the widths, and the relative production rates of the individual B^{**} mesons have been fixed to the predictions from HQET. The mass of the B_2^* meson and the overall production rate are measured to be:

$$M_{B_2^*} = 5739 \,^{+8}_{-11} \,^{+6}_{-4} \text{ MeV}$$

$$f_{B^{**}} = 0.31 \pm 0.09 \,^{+0.06}_{-0.05} \quad .$$

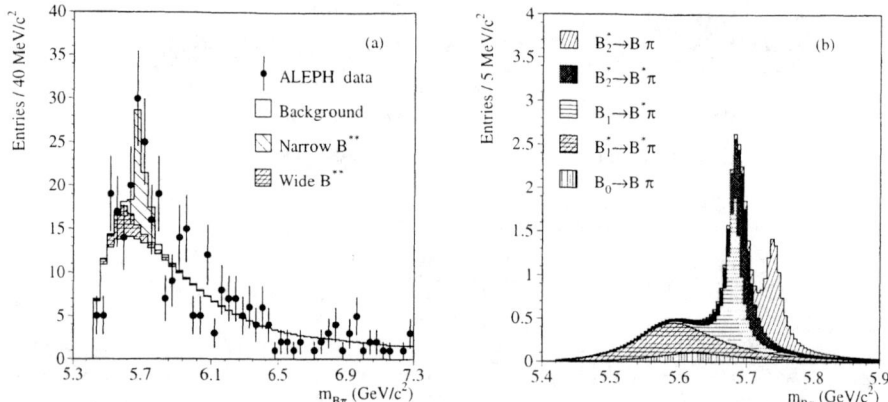

FIGURE 2. (a) $B\pi$ mass spectrum from data. The fit (histogram) includes the expected background plus contributions from the narrow and wide B^{**} states. (b) An expanded view of the signal region.

A new analysis using an inclusive method is presented here by the L3 experiment [14]. Several techniques are used to both improve on the resolution of the $B\pi$ mass spectrum and to unfold this resolution from the signal components. As a result, L3 is able to extract measurements for the masses and widths of both the D-wave B_2^* decays and the S-wave B_1^* decays.

B meson candidates are reconstructed inclusively from all charged and neutral particles with rapidity $y > 1.6$ relative to the original jet axis. The measurement of the direction of the B meson is determined by an error-weighted average of the direction of the measured secondary decay vertex and of the direction of the B candidate. The angular resolution obtained is $\sigma_1 = 12$ mrad for ϕ, and $\sigma_1 = 18$ mrad for θ, respectively. The energy of the B meson candidate, E_B, is estimated by taking advantage of the known center-of-mass energy at LEP, E_{cm}, to be

$$E_B = \frac{E_{cm}^2 - M_B^2 + M_{recoil}^2}{2E_{cm}} \quad , \qquad (1)$$

where M_{recoil} is the mass of all particles in the event recoiling against the B candidate. The difference between reconstructed and generated values for the B-meson energy can be described by an asymmetric Gaussian with widths of 1.9 GeV and 2.8 GeV.

Fig. 3a) shows the resulting $B\pi$ invariant mass spectrum together with the expected background from Monte Carlo. A clear signal due $B^{**} \to B^{(*)}\pi$ decays is seen above the background which is well described by the simulation. Thus the background is parameterized by a threshold function, the shape of which is determined from the Monte Carlo.

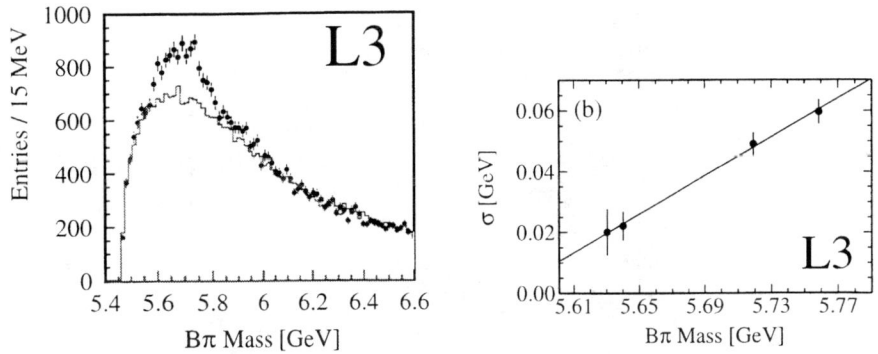

FIGURE 3. (a) Mass spectrum for selected $B\pi$ pairs. The dots are data and the shaded histogram represents the expected background from Monte Carlo normalized to the sideband region 6.0 – 6.6 GeV. (b) Linear fit to the extracted $B\pi$ mass resolution for Monte Carlo signal components generated at four different B^{**} mass values.

To resolve the underlying structure of the signal, it is necessary to unfold effects due to detector resolution. The dominant sources of uncertainty for the mass measurement are, with about equal magnitude, the angular and energy resolution of the B meson. The dependence of the $B\pi$ mass resolution on the B^{**} mass is studied by simulating signal events at four different values of B^{**} mass and Breit-Wigner width. Each signal $B\pi$ mass distribution is then fit to a Breit-Wigner function convoluted with a Gaussian resolution. The Breit-Wigner width is fixed to its generated value and the Gaussian resolution is extracted from the fit and shown in Fig. 3b) as a function of the $B\pi$ mass together with a linear parameterization. The mass resolution is increasing with increasing $B\pi$ mass.

The fit function for the signal consists of five Breit Wigner mass peaks one for the each of the five decay modes allowed by spin-parity rules: $B_2^* \to B\pi, B^*\pi$, $B_1 \to B^*\pi$, $B_1^* \to B^*\pi$, and $B_0^* \to B\pi$. Each Breit-Wigner width is convoluted with the mass dependent Gaussian resolution. No attempt is made to tag subsequent $B^* \to B\gamma$ decays, as the efficiency for selecting the soft photon is low. The relative production rate, and the mass splittings and the relative width within each doublet are constrained to the predictions from HQET (see Table 1).

The $B\pi$ invariant mass distribution fit with the signal and background function described above is shown in Fig. 4. The results of the fit provide the first mea-

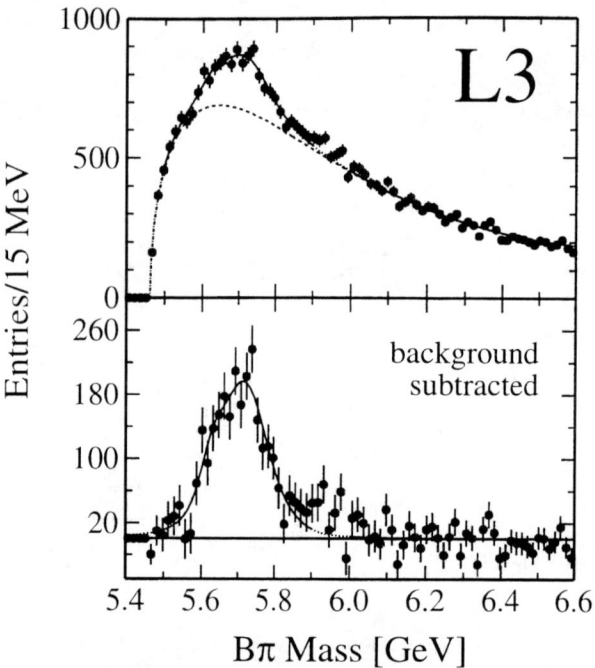

FIGURE 4. Fit to the data Bπ mass distribution with the five-peak signal function and the background function described in the text.

surements of the masses and decay widths of the B_2^* ($j_q = 3/2$) and B_1^* ($j_q = 1/2$) mesons:

$$M_{B_2^*} = (5770 \pm 6 \text{ (stat.)} \pm 4\text{(syst)}) \text{ MeV}$$
$$\Gamma_{B_2^*} = (21 \pm 24 \text{ (stat.)} \pm 15\text{(syst)}) \text{ MeV}$$
$$M_{B_1^*} = (5675 \pm 12 \text{ (stat.)} \pm 4\text{(syst)}) \text{ MeV}$$
$$\Gamma_{B_1^*} = (75 \pm 28 \text{ (stat.)} \pm 15\text{(syst)}) \text{ MeV} \quad .$$

A total of 2652 ± 232 events that occupy the signal region correspond to a relative production rate $f_{B^{**}}$ for all $L = 1$ spin states of

$$f_{B^{**}} = 0.39 \pm 0.06 \text{ (stat.)} \pm 0.06\text{(syst)} \quad .$$

Systematic errors are mainly due to the modelling of the background, the limited knowledge of the signal function and the mass constraint within the doublets.

These results disfavor recent theoretical models proposing spin-orbit inversion [8,9], but do agree well with several earlier models [5–7] and provide strong support for HQET.

B_c^+ STUDIES

DELPHI, ALEPH, and OPAL have published searches for B_c^+ mesons in Z decays [15]. No signals have been found. Table 3 shows the number of candidate events and the obtained upper limits on the production rates $\mathcal{B}(Z \to B_c^+ X) \times \mathcal{B}(B_c^+ \to J/\psi\pi^+, J/\psi\ell^+\nu, J/\psi\pi^+\pi^-\pi^+)$. The 3 $B_c^+ \to J/\psi\pi^+$ candidates are consistent with

TABLE 3. B_c^+ studies at LEP.

Decay mode	Candidates			Prod. Limit [10^{-5}] at 90% CL		
	DELPHI	ALEPH	OPAL	DELPHI	ALEPH	OPAL
$B_c^+ \to J/\psi\pi^+$	1	0	2	10.5 to 8.4	3.6	10.6
$B_c^+ \to J/\psi\ell^+\nu$	0	2	1	5.8 to 5.0	5.2	6.96
$B_c^+ \to J/\psi\pi^+\pi^-\pi^+$	1	-	0	17.5	-	5.53

a background estimate of 2.3 expected events. A fit to the mass yields the following values: $M_{J/\psi\pi^+} = 6.342 \pm 0.027$ GeV (DELPHI) and $M_{J/\psi\pi^+} = 6.32 \pm 0.06$ GeV (OPAL average), respectively. Predictions for the B_c^+ mass are in the range 6.24 to 6.31 GeV. The CDF experiment at the Tevatron has recently reported the observation of the B_c^+ meson in the decay channel $B_c^+ \to J/\psi\ell^+\nu$ [16]. They find $20.4^{+6.2}_{-5.8}$ events and obtain a mass value of $M(B_c^+) = 6.40 \pm 0.39 \pm 0.13$ GeV.

D^{**} SPECTROSCOPY

During the last year several new results on D^{**} production have been presented by the LEP collaborations. D^{**0} mesons are fully reconstructed in the decay chain

TABLE 4. D^{**} production fractions $\mathcal{B}$ [%].

Production mode	OPAL	DELPHI	ALEPH
$b \to D_1^0$	$5.0 \pm 1.4 \pm 0.6$	2.0 ± 0.6	2.3 ± 0.7
$b \to D_2^{*0}$	$4.7 \pm 2.4 \pm 1.3$	4.8 ± 2.0	< 2.0
$c \to D_1^0$	$2.1 \pm 0.7 \pm 0.3$	1.9 ± 0.4	1.6 ± 0.5
$c \to D_2^{*0}$	$5.2 \pm 2.2 \pm 1.3$	4.7 ± 1.3	4.7 ± 1.0
$c \to D_{s1}^+$	$1.6 \pm 0.4 \pm 0.3$	-	$0.77 \pm 0.20 \pm 0.08$
$c \to D_{s2}^{*+}$	-	-	$1.3 \pm 0.5 \pm 0.2$
$b \to D_{s1}^+$	-	-	$1.1 \pm 0.3 \pm 0.2$
$b \to D_{s2}^{*+}$	-	-	$2.2 \pm 0.8 \pm 0.5$

$D^{**0} \to D^{*+}\pi^-$, $D^{*+} \to D^0\pi^+$, $D^0 \to K^-\pi^+$. High momentum D^{**0} together with short decay lengths are selected to obtain $c\bar{c}$ enriched samples whereas B and D meson vertexing is used to select $b\bar{b}$ enriched samples. Table 4 shows the results for the D^{**} production fractions measured by OPAL, DELPHI, and ALEPH [18–20].

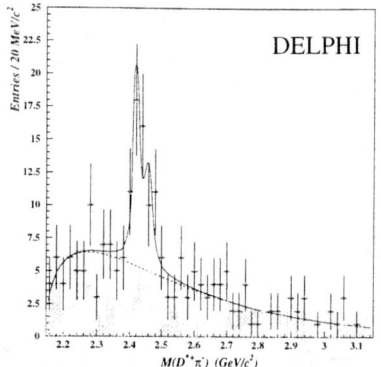
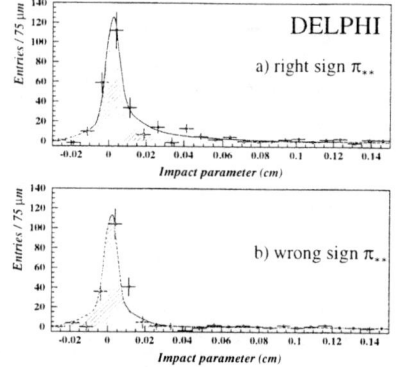

FIGURE 5. (a) Experimental $D^{*+}\pi^-$ invariant mass distribution for $D^{*+}\pi^-\ell^-$ events. The solid line is a fit to the narrow D^{**} states plus background. The wrong sign $D^{*+}\pi^-\ell^+$ candidates are shown in the hatched histogram. (b) Impact parameter relative to the primary event vertex for right charge and wrong charge π_{**} candidates. The fit is described in the text.

DELPHI has also presented a new $B \to D^{**}\ell\nu$ analysis [21]. In semileptonic events, the decay chain $D^{*+} \to D^0\pi^+$, $D^0 \to K^-\pi^+, K^-\pi^+\pi^-\pi^+, K^-\pi^+(\pi^0)$ is fully reconstructed. The D^{*+} candidates are then combined with opposite sign π^- and the $D^{*+}\pi^-$ mass, shown in Fig. 5a), is fit to the narrow D^{**} states, resulting in the following branching fraction: $\mathcal{B}(B^- \to D_1^0 \ell^- \bar{\nu}) = 0.72 \pm 0.22 \pm 0.13\%$.

A fit to the impact parameter distribution of the bachelor pion π_{**} stemming from the $D^{**} \to D^*\pi$ transition for right (unlike)-sign and wrong (like)-sign combinations, as shown in Fig. 5b), allows to extract the following branching fraction $\mathcal{B}(B^- \to D^{*+}\pi^-\ell^-\bar{\nu}) = 1.15 \pm 0.17 \pm 0.14\%$ where the signal comprises narrow and wide D^{**} resonances plus non-resonant $D^{*+}\pi^-$ combinations. These results are in agreement with previous LEP and CLEO measurements.

$D^{*'}$ STUDIES

DELPHI has recently reported an excess of events in the $D^{*+}\pi^-\pi^+$ mass spectrum as shown in Fig. 6a) [22]. The fit yields $N = 66 \pm 14$ events, corresponding to a production rate $f_{D^{*'}}/f_{D^{**}} = 0.49 \pm 0.18 \pm 0.10$, a mass $M = 2637 \pm 2 \pm 6$ MeV and a width consistent with the experimental resolution. This mass value is consistent with predictions of a radial excited $D^{*'}$ meson.

OPAL has performed a similar analysis [23]. The resulting $D^{*+}\pi^-\pi^+$ mass spectrum is shown in Fig. 6b) for data and Monte Carlo events, where a DELPHI-like signal has been added in the simulation. No excess is seen in the data ($N < 32.8$ at 95 % CL) corresponding to a limit on the production rate of $f_{D^{*\prime}}/f_{D^{**}} < 0.21$ at 95 % CL, thus not confirming the DELPHI result. CLEO also has examined their $D^{*+}\pi^-\pi^+$ mass spectrum and does not confirm the DELPHI evidence [24].

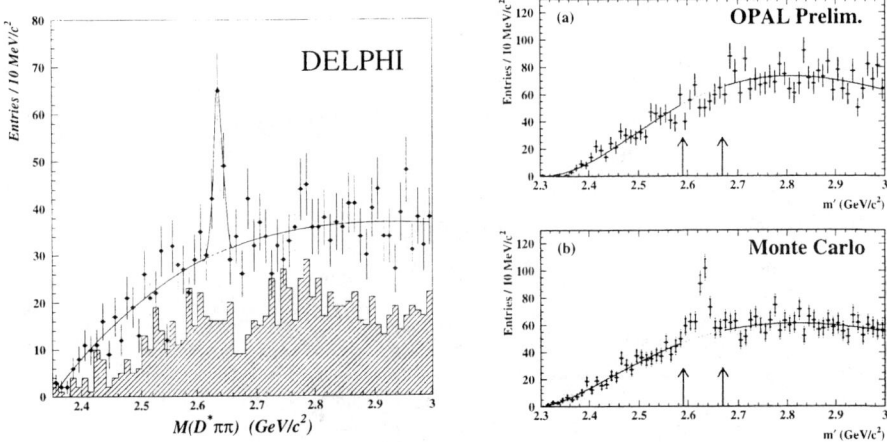

FIGURE 6. (a) DELPHI invariant mass distributions $D^{*+}\pi^+\pi^-$ (dots) and $D^{*+}\pi^-\pi^-$ (hatched histogram). (b) OPAL $D^{*+}\pi^+\pi^-$ mass distribution for data and Monte Carlo. A DELPHI-like signal has been added in the simulation.

ACKNOWLEDGEMENTS

I wish to thank my colleagues from the other LEP collaborations for providing me with the results and the figure files. I also thank S. Goldfarb for his help when preparing this presentation.

REFERENCES

1. N. Isgur and M.B. Wise, Phys. Lett. **B 232** (1989) 113; Phys. Lett. **B 237** (1990) 527; Their overview "Heavy Quark Symmetry" is included in *B Decays*, revised 2nd edition, edited by S. Stone (World Scientific, 1994) 231 and in *Heavy Flavors*, edited by A.J. Buras and M. Lindner (World Scientific, 1992) 234.
2. L3 Collaboration, Phys. Lett. **B 345** (1995) 589; Contribution to EPS Conference, Brussels EPS95 (1995).

3. DELPHI Collaboration, Z. Phys. **C 68** (1995) 353;
   ALEPH Collaboration, Z. Phys. **C 69** (1996) 393;
   OPAL Collaboration, Z. Phys. **C 74** (1997) 413.
4. J.L. Rosner, Comments Nucl. Part. Phys. **16** (1986) 109;
   N. Isgur and M.B. Wise, Phys. Rev. Lett. **66** (1991) 1130.
5. M. Gronau, A. Nippe and J.L. Rosner, Phys. Rev. **D 47** (1993) 1988;
   M. Gronau and J.L. Rosner, Phys. Rev. **D 49** (1994) 254.
6. E.J. Eichten, C.T. Hill and C. Quigg, Phys. Rev. Lett. **71** (1993) 4116; Fermilab-Conf-94/118-T (1994).
7. A.F. Falk and T. Mehen, Phys. Rev. **D 53** (1996) 231.
8. N. Isgur, Phys. Rev. **D 57** (1998) 4041.
9. D. Ebert, V.O. Galkin and R.N. Faustov, Phys. Rev. **D 57** (1998) 5663.
10. OPAL Collaboration, Z. Phys. **C 66** (1995) 19.
11. DELPHI Collaboration, Phys. Lett. **B 345** (1995) 598; Contribution to EPS conference, Brussels EPS95 (1995).
12. ALEPH Collaboration, Z. Phys. **C 69** (1996) 393.
13. ALEPH Collaboration, Phys. Lett. **B 425** (1998) 215.
14. L3 Collaboration, to be submitted to Phys. Lett. **B**.
15. DELPHI Collaboration, Phys. Lett.**B 398** (1997) 207;
    ALEPH Collaboration, Phys. Lett.**B 402** (1997) 213;
    OPAL Collaboration, Phys. Lett.**B 420** (1998) 157.
16. CDF Collaboration, Phys. Rev. Lett. **81**(1998) 2432; Phys. Rev. **D 58**(1998) 112004.
17. ARGUS Collaboration, Phys. Rev. Lett. **56** (1986) 549; Phys. Lett. **B 221** (1989) 422; Phys. Lett. **B 231** (1989) 208; Z. Phys. **C 69** (1996) 405;
    CLEO Collaboration, Phys. Rev. **D 41** (1990) 774; Phys. Lett. **B 303** (1993) 377; Phys. Lett. **B 331** (1994) 236; Phys. Lett. **B 340** (1994) 194; Phys. Rev. Lett. **72** (1994) 1972;
    E691 Collaboration, Phys. Rev. Lett. **62** (1989) 1717;
    E687 Collaboration, Phys. Rev. Lett. **72** (1994) 324;
    BEBC Collaboration, Z. Phys. **C 61** (1994) 563;
    ALEPH Collaboration, Phys. Lett. **B 345** (1994) 103; Z. Phys. **C 62** (1994) 1; Z. Phys. **C 73** (1997) 601.
18. OPAL Collaboration, Z. Phys. **C 76** (1997) 425.
19. DELPHI Collaboration, Contribution to ICHEP98 conference, Vancouver, (1998), paper # 240.
20. ALEPH Collaboration, Contribution to ICHEP98 conference, Vancouver, (1998), papers # 943 and # 944.
21. DELPHI Collaboration, Contribution to ICHEP98 conference, Vancouver, (1998), paper #239.
22. DELPHI Collaboration, Phys. Lett. **B 426** (1998) 231.
23. OPAL Collaboration, Contribution to ICHEP98 conference, Vancouver, (1998), paper # 1037
24. CLEO Collaboration, Contribution to ICHEP98 conference, Vancouver, (1998).

PART VII
RARE AND FORBIDDEN DECAYS

Rare Kaon Decays: Brookhaven E871

David A. Ambrose

Department of Physics, University of Texas, Austin, Texas 78712

Abstract. This paper describes recent results from the rare kaon decay Experiment 871 at the Brookhaven AGS, which searched for the lepton flavor violating decay $K_L^0 \to \mu^\pm e^\mp$, and measured branching ratios for the GIM and helicity suppressed modes $K_L^0 \to \mu^+\mu^-$ and $K_L^0 \to e^+e^-$. Observing no candidate μe events, E871 established a new upper limit (90% CL) on the branching fraction of $B(K_L^0 \to \mu^\pm e^\mp) < 4.8 \times 10^{-12}$, representing the most sensitive search to date for this decay mode. E871 measured over 6200 $\mu\mu$ candidates, a factor of six improvement on all previous measurements combined, giving a preliminary branching fraction of $B(K_L^0 \to \mu^+\mu^-) = (7.23 \pm 0.22) \times 10^{-9}$, which should reduce the uncertainty in this decay mode by a factor of three. Lastly, E871 recorded four candidate ee events, the first observation of this rare decay, leading to a branching fraction of $B(K_L^0 \to e^+e^-) = 8.7^{+5.7}_{-4.1} \times 10^{-12}$, which equals the smallest branching fraction ever measured in particle physics.

THEORETICAL MOTIVATION

The study of kaon decays has had a profound impact on our understanding of elementary particle physics. Early observations of K^+ decays into different states of parity (the "$\tau - \theta$ puzzle"), led to the postulate and subsequent experimental verification that weak interactions violate the conservation of parity. In 1964, an experiment at the Brookhaven Alternating Gradient Synchrotron measured the decay of the long-lived neutral kaon K_L^0 into two pions, proving that weak interactions also violate the conservation of CP. Today, the Brookhaven AGS produces the most intense kaon beam in the world, allowing experiments to study extremely rare or forbidden decays with high precision, and search for physics beyond the Standard Model.

$K_L^0 \to \mu^\pm e^\mp$

Lepton flavor violation searches are motivated by the fact that these additive conservation rules do not derive from any dynamical laws associated with a gauge principal, as with the conservation of charge from local gauge invariance of the

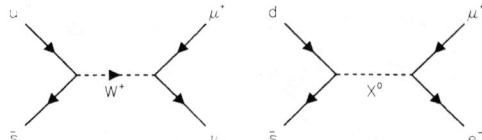

FIGURE 1. First-order diagrams for $K^+ \to \mu^+ \nu$ and $K_L^0 \to \mu^\pm e^\mp$.

electromagnetic interaction, so that lepton flavor violation is not strictly forbidden. If neutrinos masses are not zero, then lepton flavor mixing in the weak eigenstates, in analogy with quark mixing, implies a branching ratio for $K_L^0 \to \mu^\pm e^\mp$ on the order of 10^{-25}, which is not experimentally observable. Therefore, any observation of this decay is evidence for new physics beyond the Standard Model. In fact, many extensions to the Standard Model include the possibility for lepton flavor violation. While models such as right-left symmetry and supersymmetry predict levels of lepton flavor violation which are still unobservable, extensions such as technicolor, compositeness, and other horizontal symmetry models call for rates for $K_L^0 \to \mu^\pm e^\mp$ closer to or above current limits [1], and can therefore be tested experimentally.

In general, rare decay searches can probe very high mass regions. We can compare the rate for $K_L^0 \to \mu^\pm e^\mp$ mediated by some horizontal gauge boson X to the analogous decay $K^+ \to \mu^+ \nu$ (see Figure 1). Representing the average of the quark and lepton couplings by g_X, the ratio of decay rates becomes:

$$\frac{\Gamma(K_L^0 \to \mu e)}{\Gamma(K^+ \to \mu^+ \nu)} = \left[\frac{g_X^2/M_X^2}{g^2 \sin\theta_c/M_W^2}\right]^2 \implies M_X \approx (220 TeV) \left[\frac{10^{-12}}{B(K_L^0 \to \mu e)}\right]^{1/4} \quad (1)$$

Thus, inserting the known branching fraction for the K^+ decay, and assuming the coupling constants are similar, implies a lower limit on M_X in hundreds of TeV for a measured μe branching fraction of 10^{-12}, which is well beyond existing accelerators.

$K_L^0 \to \mu^+\mu^-$ and $K_L^0 \to e^+e^-$

The decay $K_L^0 \to \mu^+\mu^-$ is important historically because its low rate of occurrence led in part to the Glashow-Illiopoulos-Maini (GIM) mechanism [2], which invoked a fourth quark charm whose presence cancels flavor changing neutral currents at tree level, and strongly suppresses the second order loop diagrams involving the up quark, shown in Figure 2. The remaining decay rate is known to be dominated by an absorptive process with a two-photon intermediate state, where the photons are real, and can be calculated through QED [3] and the measured $K_L^0 \to \gamma\gamma$ branching fraction [4]:

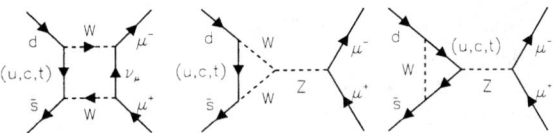

FIGURE 2. Second-order weak interaction diagrams for $K_L^0 \to \mu^+\mu^-$.

$$B(K_L^0 \to \mu\mu)_{\gamma\gamma} = \frac{1}{2}\alpha^2 \left(\frac{m_\mu}{m_K}\right)^2 \left(\ln\frac{1+\beta}{1-\beta}\right)^2 \times B(K_L^0 \to \gamma\gamma) \quad (2)$$

equaling $(7.07 \pm 0.18) \times 10^{-9}$. This value establishes the minimum rate we should expect for $K_L^0 \to \mu^+\mu^-$, and is called the *unitarity bound*.

The total $\mu\mu$ decay rate is then the sum of this absorptive (or imaginary) part of the amplitude with a dispersive (real) part, which contains the short-distance electroweak contributions from Figure 2 and a long-distance electromagnetic term:

$$B(K_L^0 \to \mu^+\mu^-) = |ReA|^2 + |ImA|^2 \implies ReA = A_{SD} + A_{LD}. \quad (3)$$

The short-distance diagrams are dominated by the heavy top quark, so that the amplitude is roughly proportional to the V_{ts} and V_{td} elements of the CKM matrix:

$$|A_{SD}|^2 \sim |ReV_{ts}^*V_{td}|^2 \implies Re(V_{ts}^*V_{td}) = -A^2\lambda^5(1-\rho) \quad (4)$$

and therefore to $(1-\rho)$ in the Wolfenstein parameterization [5]. Success in extracting information on ρ from the $\mu\mu$ branching fraction depends on the ability to estimate the long-distance contribution, which is dominated by an intermediate state of two virtual photons, and has theoretical uncertainty [6].

The decay $K_L^0 \to e^+e^-$ shares the physics of GIM suppression of the short-distance electroweak interaction, as with the $\mu\mu$ mode, except that the decay is further suppressed by helicity constraints on the spin-0 kaon decay. This short-distance helicity suppression can be compared with that of the K-plus decay:

$$\frac{B(K_L^0 \to ee)_{SD}}{B(K_L^0 \to \mu\mu)_{SD}} \approx \frac{B(K^+ \to e\nu)}{B(K^+ \to \mu\nu)} \approx \left(\frac{m_e}{m_\mu}\right)^2 \approx 2.4 \times 10^{-5} \quad (5)$$

and is approximately equal to the ratio of lepton masses squared. The absorptive contribution to the decay rate from the two-photon intermediate state is also helicity suppressed, but is enhanced by an order of magnitude due to the logarithmic singularity in the limit of zero lepton mass [1]:

$$\frac{B(K_L^0 \to \gamma\gamma \to ee)}{B(K_L^0 \to \gamma\gamma \to \mu\mu)} = \left(\frac{m_e}{m_\mu}\right)^2 \frac{\beta_e \left(\ln\frac{1+\beta_e}{1-\beta_e}\right)^2}{\beta_\mu \left(\ln\frac{1+\beta_\mu}{1-\beta_\mu}\right)^2} \implies \beta_\ell = \sqrt{1 - 4m_\ell^2/m_K^2} \quad (6)$$

giving a *unitarity bound* for $K_L^0 \to e^+e^-$ of about 3×10^{-12}, so that the short-distance contributions are negligible. The long-distance dispersive term is also enhanced over the short-distance effects, and is in fact larger than the absorptive part, as opposed to the $\mu\mu$ decay. Recent calculations of the long-distance amplitude involving Chiral Perturbation Theory [7,8] predict a total decay rate for $K_L^0 \to e^+e^-$ equal to $(9.0 \pm 0.5) \times 10^{-12}$.

EXPERIMENTAL APPARATUS

Brookhaven E871 represented an upgrade to the previous experiment BNL E791 [9–11], which measured roughly 700 $\mu\mu$ events giving a branching fraction $B(K_L^0 \to \mu^+\mu^-) = (6.86 \pm 0.37) \times 10^{-9}$, while observing no candidate μe or ee events and setting upper limits of 3.3×10^{-11} and 4.1×10^{-11}, respectively. E871 took advantage of a factor of four increase in primary beam due to the addition of the AGS booster, giving a nominal 15×10^{12} protons per beam spill for roughly 10^8 kaons per pulse. To accommodate the increase in beam intensity, several modifications to the apparatus were performed. The upstream spectrometer drift chambers were replaced with straw chambers with $5\,mm$ diameter using fast ionizing gas. A neutral beam stop was added within the spectrometer to reduce rates in the downstream drift chamber and particle identification detectors, which could then be combined from a two-arm system to single modules spanning the beamline. Also, the trigger scintillator hodoscope and electronics were configured to allow for a parallel trigger selection favoring two-body decays over the three-body semileptonic background. E871 was proposed in 1990, and after two brief engineering runs took data in 1995 and 1996.

The neutral kaon beam was produced with $24\,GeV/c$ momentum protons incident on a platinum target. A series of sweeping magnets and collimators defined the neutral beam, consisting of mostly neutrons and kaons in a ratio of about 60:1. The kaons then entered a vacuum decay tank about 10 meters long (see

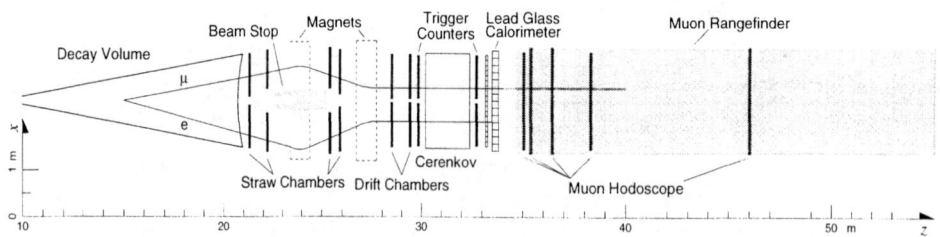

FIGURE 3. BNL E871 apparatus plan view.

Figure 3). Secondary decay particles exiting the decay tank proceeded through a double-magnet spectrometer containing 6 left-right modules of drift chambers, the first four of which were straws. The neutral beam terminated in the beam plug within the first magnet. The magnets had opposite polarity giving an overall p_T kick inwards of about $200\,MeV/c$, tuned to the dilepton decay momenta to allow the particles to emerge from the spectrometer with roughly parallel trajectories. Two banks of narrow scintillator hodoscope counters then provided a first-level two-body trigger. A segmented threshold Čerenkov counter using hydrogen gas triggered electrons, while a muon hodoscope placed downstream of an $18\,in$ iron filter triggered muons. The segmentation of the Čerenkov and muon hodoscopes were aligned with roads within the trigger scintillators to provide the trigger electronics with signals to form parallel dilepton events. Prescaled minimum bias triggers using only the trigger scintillators then allowed for the simultaneous measurement of $K_L^0 \to \pi^+\pi^-$ events for normalization. Offline, E871 used redundant electron and muon identification from a lead glass calorimeter and muon rangefinder, consisting of proportional counters interspersed between slabs of iron, marble, and aluminum to give a 5% range measurement.

DATA ANALYSIS

Backgrounds for the dilepton kaon decays typically involve some form of the semileptonic decays $K_L^0 \to \pi\mu\nu$ and $K_L^0 \to \pi e\nu$ (referred to as $K_{\mu 3}$ and K_{e3}, respectively), where the pion subsequently decays to mimic $\mu\mu$ or μe events, or a K_{e3} pion is mis-identified as an electron to produce an ee signal. These three-body backgrounds are suppressed by cutting on the net transverse momentum (p_T) of the decay with respect to the kaon direction, which should equal zero for two-body decays and non-zero for semileptonics, with missing momentum carried off by the neutrino. A cut on the reconstructed invariant dilepton mass $m_{\ell\ell'}$ also removes the majority of semileptonic backgrounds:

$$m_{\ell\ell'}^2 = m_\ell^2 + m_{\ell'}^2 + 2(E_\ell E_{\ell'} - \mathbf{p}_\ell \cdot \mathbf{p}_{\ell'}) \implies m_{max}^2 = m_K^2 + m_\ell^2 - m_\pi^2 \qquad (7)$$

where the kinematic threshold m_{max} for K_{e3} (in the limit of zero neutrino momentum), with mis-identification of a pion as a lepton, lies almost $10\,MeV/c^2$ below the kaon mass for μe and $\mu\mu$, and over $20\,MeV/c^2$ for ee reconstruction. Therefore, E871 required good momentum and mass resolution to separate these K_{e3} background events from the signal region.

Figure 4 shows the reconstructed $\pi\pi$ invariant mass and p_T^2 from the normalization sample (solid line). Since the kinematic threshold for semileptonic events reconstructed as $\pi\pi$ falls above the kaon mass, an electron veto from the lead glass array is used to remove K_{e3} events, and the remaining $K_{\mu 3}$ events are simulated with Monte Carlo (dashed line in Figure 4) and normalized to side bands of the data to perform the background subtraction. From the single-wire straw and drift chamber resolutions of 120-160 μm, an event fitting algorithm using a three-dimensional

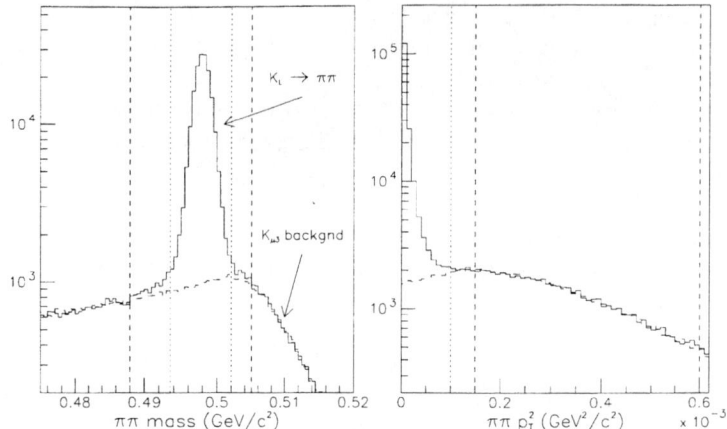

FIGURE 4. Reconstructed $\pi\pi$ invariant mass and p_T^2 from normalization sample (solid line), with Monte Carlo $K_{\mu 3}$ distributions (dashed line). Vertical lines denote signal (dotted) and sideband (dashed) regions.

field map of the analyzing magnets and accounting for multiple-scattering within the spectrometer provides a mass resolution for $\pi\pi$ events of roughly $1.1\,MeV/c^2$. Monte Carlo simulation of dilepton decays reveal mass resolutions for $\mu\mu$, μe, and ee modes ranging from 1.3-$1.4\,MeV/c^2$, so that the reconstructed kaon mass lies at least 6 standard deviations above the semileptonic background thresholds.

Additional dilepton backgrounds remained after application of the mass and p_T^2 cuts, involving either large, non-Gaussian scatters within the spectrometer or "pile-up" events with multiple kaon decays, confusing the event reconstruction or particle identification. These anomalous backgrounds were reduced by placing cuts on the returned χ^2 values from the event fitter, ensuring that the reconstructed decay vertices fell within the angular region defined by the neutral beam collimators, and searching for spurious tracks within the spectrometer or trigger scintillator hodoscope. Tight timing cuts on the Čerenkov and muon hodoscope signals, along with an electron veto for the μe and $\mu\mu$ muon tracks, further suppressed these accidental backgrounds.

$K_L^0 \to \mu^+\mu^-$ Results

Since several thousand $K_L^0 \to \mu^+\mu^-$ events were expected for the E871 experimental sensitivity, the appearance of a small fraction of background events within the signal region was not deleterious to the measurement. Therefore, a relatively loose set of kinematic and particle identification cuts were applied to reduce the background level to roughly 1% (where errors on the subtraction method became negligible), while minimizing potential systematic errors introduced through tighter cuts. The relative branching ratio for $K_L^0 \to \mu^+\mu^-$ with respect to $K_L^0 \to \pi^+\pi^-$ was then calculated from:

$$\frac{B(K_L^0 \to \mu\mu)}{B(K_L^0 \to \pi\pi)} = \frac{1}{P} \left(\frac{N_{\mu\mu}}{N_{\pi\pi}}\right) \left(\frac{A_{\pi\pi}}{A_{\mu\mu}}\right) \left(\frac{1}{\varepsilon_{\mu\mu}^{trig}}\right) \left(\frac{1}{\varepsilon_{\mu\mu}^{PID}}\right) \ldots \quad (8)$$

where $N_{\mu\mu}$ and $N_{\pi\pi}$ equal the numbers of $\mu\mu$ and $\pi\pi$ candidate events observed, and P denotes the total prescale applied to the normalization sample. The acceptances A correct for the mode dependence of the detector geometry, including the kinematic cuts and the trigger parallelism, while various efficiency terms ε account for inefficiencies in the hardware and software triggering, particle identification and electron vetoes, and the loss of $\pi\pi$ events from hadronic interactions within the apparatus.

The acceptance and efficiency terms within the branching ratio were calculated with a high degree of precision ($\sim$ 0.1-0.2%) so that systematic errors would not limit the result. To prevent potential bias in these calculations, E871 performed a "blind" analysis by applying a random prescale factor to the normalization sample, which was revealed only after all cuts were fixed and branching ratio correction terms finalized.

Figure 5 plots the resulting reconstructed $\mu\mu$ invariant mass and p_T^2 after all cuts were applied. The $\mu\mu$ mass peak appears well separated from the tail of the $K_{\mu 3}$ background distribution, with a small amount of background extending under the signal peak. The majority of background events were due to "pile-up" of two or more kaon decays, and was reduced by checking for extra tracks within the trigger hodoscope. To subtract this background, a linear fit was performed on the p_T^2 axis within a high-p_T sideband, and extrapolated into the signal region, resulting in

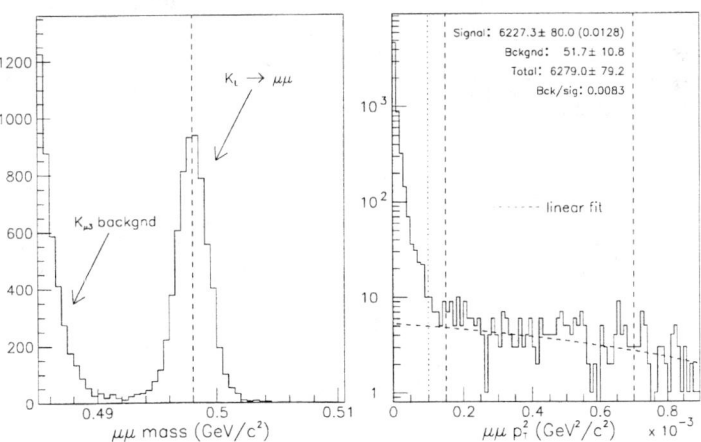

FIGURE 5. Reconstructed $\mu\mu$ invariant mass and p_T^2 after cuts, with background subtraction results from a linear fit in p_T^2.

over 6200 candidate $\mu\mu$ events with a relative error of roughly 1.3%. Revealing the blind prescale factor and applying the acceptance and efficiency corrections gives a preliminary relative branching ratio:

$$\frac{B(K^0_L \to \mu\mu)}{B(K^0_L \to \pi\pi)} = (3.50 \pm 0.11) \times 10^{-6} \text{ (preliminary)} \qquad (9)$$

corresponding to a branching fraction (using the measured value for $K^0_L \to \pi^+\pi^-$ [4]) of:

$$B(K^0_L \to \mu^+\mu^-) = (7.23 \pm 0.22) \times 10^{-9} \text{ (preliminary)} \qquad (10)$$

A final result for the E871 $\mu\mu$ measurement awaits further checks on systematic errors.

$K^0_L \to \mu^\pm e^\mp$ Results

The E871 search for $K^0_L \to \mu^\pm e^\mp$ involved a much tighter set of cuts than those used in the $\mu\mu$ analysis, to reduce the possibility of a background event appearing in the signal region. To avoid biasing these cuts through knowledge of potential signal events, an extended area surrounding the signal box was hidden from view until all cuts were finalized, and an estimate of the background performed. High mass and p_T events outside this "blind" exclusion region were then studied to optimize the kinematic and particle identification cuts.

These studies, along with Monte Carlo simulations, showed that the majority of μe background resulted from large scatters of K_{e3} events in the downstream vacuum decay tank window. A scatter occurring within the decay plane of the event could artificially increase the decay opening angle, boosting the reconstructed μe invariant mass into the signal region. To suppress these large scatter events, an additional cut was placed on the event p_T component parallel to the decay plane. A more aggressive cut on accidental tracks from "pile-up" using the drift chamber hits also inhibited potential μe backgrounds. The predicted background level using the full set of analysis cuts was roughly 0.1 events.

Revealing the exclusion region, Figure 6 shows the reconstructed invariant mass versus p_T^2 for the μe data sample. The actual signal region was chosen to maximize acceptance, while minimizing the potential background from large scatters using an elliptical boundary at low mass. Observing no candidate μe events, E871 set a 90% confidence level upper limit on the $K^0_L \to \mu^\pm e^\mp$ branching fraction of [12]:

$$B(K^0_L \to \mu^\pm e^\mp) < 4.8 \times 10^{-12} \text{ (90\%CL)} \qquad (11)$$

which improves on the previous limit [10] by a factor of almost seven.

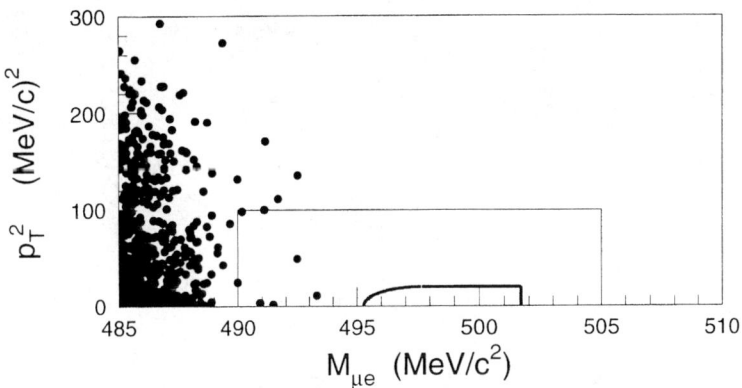

FIGURE 6. Reconstructed μe invariant mass versus p_T^2 after cuts, showing exclusion region (rectangle) and elliptical signal box.

$K_L^0 \to e^+e^-$ Results

The E871 search for $K_L^0 \to e^+e^-$ followed a strategy similar to the μe analysis, with a set of tight kinematic and particle identification cuts optimized from high mass and p_T events surrounding an exclusion region about the signal. While the ee data sample was much cleaner than μe or $\mu\mu$, due to the lower threshold for misidentified K_{e3} events, there were additional physics background sources from $K_L^0 \to e^+e^-\gamma$ (in the limit of zero photon energy) and $K_L^0 \to e^+e^-e^+e^-$, where [4]:

$$B(K_L^0 \to e^+e^-\gamma) = (9.1 \pm 0.5) \times 10^{-6} \tag{12}$$
$$B(K_L^0 \to e^+e^-e^+e^-) = (4.1 \pm 0.8) \times 10^{-8} \tag{13}$$

Since the E871 apparatus was not optimized for photon detection, no attempt was made to suppress $K_L^0 \to e^+e^-\gamma$ background. However, $4e$ events were actively removed by searching for partial tracks in the upstream straw drift chambers which projected to the decay vertex.

After applying the full set of analysis cuts, including the $K_L^0 \to e^+e^-e^+e^-$ "stub" cut, Figure 7 shows the reconstructed ee invariant mass versus p_T^2 for the remaining data sample. The inset plots the predicted physics backgrounds as a function of mass, which combine to produce a level of approximately 0.2 events. An elliptical signal region (within the hatched exclusion zone) reveals four candidate ee events. The actual number of signal events was estimated using a maximum likelihood fit to the data, and found to be $4.20^{+2.69}_{-1.94}$, resulting in a measured branching fraction for $K_L^0 \to e^+e^-$ of [13]:

$$B(K_L^0 \to e^+e^-) = 8.7^{+5.7}_{-4.1} \times 10^{-12} \tag{14}$$

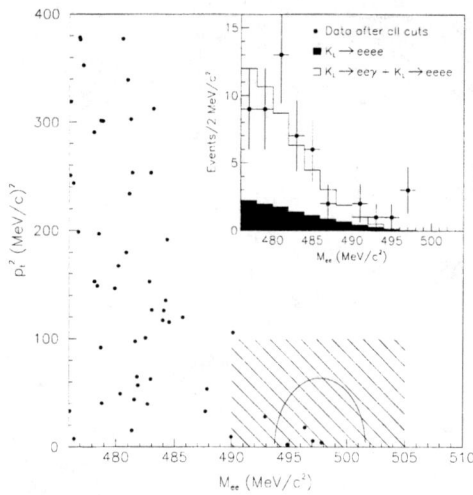

FIGURE 7. Reconstructed ee invariant mass versus p_T^2 after cuts, with predicted physics background levels (inset) by mass.

consistent with theoretical predictions [7,8]. E871 represents the first observation of this rare kaon decay, which is the smallest branching fraction ever measured in particle physics.

REFERENCES

1. Ritchie, J.L., Wojcicki, S.G., Rev. Mod. Phys. **65**, 1149 (1993).
2. Glashow, S.L., Iliopoulus, J., Maiani, L., Phys. Rev. **D2**, 1285 (1970).
3. Sehgal, L.M., Phys. Rev. **183**, 1511 (1969).
4. Particle Data Group, C. Caso *et al.*, Eur. Phys. J.**C3**, 1-794 (1998).
5. Wolfenstein, L., Phys. Rev. Lett. **51**, 1945 (1983).
6. D'Ambrosio, G.D., Isidori, G., Portolés, J., hep-ph/9708326, (1997).
7. Valencia, G., Nucl. Phys. **B517**, 339 (1998).
8. Dumm, D.G., Pich, A., Phys. Rev. Lett. **80**, 4633 (1998).
9. Heinson, A.P., *et al.* (BNL E791 Collaboration), Phys. Rev. **D51**, 985 (1995).
10. Arisaka, K., *et al.* (BNL E791 Collaboration), Phys. Rev. Lett. **70**, 1049 (1993).
11. Arisaka, K., *et al.* (BNL E791 Collaboration), Phys. Rev. Lett. **71**, 3910 (1993).
12. Ambrose, D., *et al.* (BNL E871 Collaboration), "New Limit on Muon and Electron Lepton Number Violation from $K_L^0 \to \mu^\pm e^\mp$", (submitted to Phys. Rev. Lett., September, 1998).
13. Ambrose, D., *et al.* (BNL E871 Collaboration), "First Observation of the Rare Decay Mode $K_L^0 \to e^+e^-$", (accepted for publication in Phys. Rev. Lett., October, 1998).

Evidence for $K^+ \to \pi^+ \nu \bar{\nu}$

Steve Kettell
(for the E787 and E949 collaborations[1])

*Brookhaven National Laboratory
Upton, NY 11973*

Abstract.
The decay $K^+ \to \pi^+ \nu \bar{\nu}$ has been observed for the first time. The E787 experiment has presented evidence for the $K^+ \to \pi^+ \nu \bar{\nu}$ decay, based on the observation of a single clean event from data collected during the 1995 run of the AGS (Alternating Gradient Synchrotron at Brookhaven National Laboratory). The branching ratio indicated by this observation, $B(K^+ \to \pi^+ \nu \bar{\nu}) = 4.2^{+9.7}_{-3.5} \times 10^{-10}$, is consistent with the Standard Model expectation although the central experimental value is four times larger. The final E787 data sample, from the 1995–99 runs, should reach a sensitivity of about eight times that of the 1995 run alone. A new experiment, E949, has been approved to run, starting in the year 2000, and is expected to achieve a sensitivity of more than an order of magnitude below the prediction of the Standard Model.

INTRODUCTION

The rare decay $K^+ \to \pi^+ \nu \bar{\nu}$ is a flavor changing neutral current, mediated in the Standard Model (SM) by heavy quark loop diagrams [1,2], and in particular, is sensitive to the Cabibbo-Kobayashi-Maskawa matrix element $|V_{td}|$ [3]. This sensitivity arises from the heavy top quark mass, the hard GIM suppression, and the very small long distance contribution [4–8]. The hadronic matrix element can be determined to the percent level from the $K^+ \to \pi^0 e^+ \nu_e$ rate, with the inclusion of isospin violating and electroweak radiative effects [9,10]. The relatively small QCD corrections have been calculated to next to leading log [11]. QCD corrections to the charm quark contribution are the major source of the intrinsic theoretical uncertainty, leading to a 7% uncertainty in the calculation of the branching ratio [3]. If $B(K^+ \to \pi^+ \nu \bar{\nu})$, m_t, $|V_{ub}/V_{cb}|$, and $|V_{cb}|$ were perfectly known, $|V_{td}|$ could be determined to $\sim 5\%$. Given the current uncertainties in m_t, m_c, V_{cb}, $|V_{ub}/V_{cb}|$, ϵ_K and $\bar{B} - B$ mixing, the standard model prediction for the branching ratio, is 0.6—1.5×10^{-10}.

---

[1] see Acknowledgements

A measurement of B($K^+ \to \pi^+ \nu \bar{\nu}$) is one of the theoretically cleanest ways of determining $|V_{td}|$. Combined with the other 'Golden Mode', the neutral analog — $K_L^\circ \to \pi^\circ \nu \bar{\nu}$, the CKM triangle can be completely determined from the K system. With new measurements of the CKM parameters in the B system expected from the B-factories, additional tests of the Standard Model will be possible. In many extensions to the Standard Model the effects on the K and B system turn out to be discernibly different.

THE E787 EXPERIMENT

The experimental signature for $K^+ \to \pi^+ \nu \bar{\nu}$ is a single incident charged kaon track and a single outgoing charged pion track, with two missing neutrinos. The separation of this signal from the background requires that the particle identification and kinematics of the π must be very well measured and any additional particles must be vetoed with high efficiency. This is most readily solved in the center of mass frame; therefore, E787 [12] runs in a stopped kaon beam at the AGS.

A drawing of the E787 detector is shown in Fig. 1. The detector is located in the C4 beam line at the AGS. The C4 beam line, or Low Energy Separated Beam (LESB3) [13], transports kaons of up to 830 MeV/c. At 690 MeV/c it can transmit more than $3 \times 10^6 K^+$/spill with a K/π ratio of >3:1 for 10^{13} protons on the production target. The K^+ are stopped in a scintillating fiber target in the center of the detector. The decay π^+ are down the beamline and tracked through the target and drift chamber [14] into the plastic scintillator range stack (RS). The detector is located in a 1 T, solenoidal magnetic field.

The two most significant backgrounds are the two dominant K^+ decay modes, $K^+ \to \pi^+ \pi^\circ$ ($K_{\pi 2}$) and $K^+ \to \mu^+ \nu_\mu$ ($K_{\mu 2}$), which produce mono-energetic charged particles. The search region for $K^+ \to \pi^+ \nu \bar{\nu}$ is away from these two kinematic peaks. The E787 detector uses redundant measures of the kinematics: momentum (P), energy (E) and range (R). The $K_{\pi 2}$ can also be suppressed by vetoing on the π^0 photons, so the detector is surrounded by a nearly 4π photon veto (PV). The primary components of the PV are the barrel veto (BV) and endcap (EC) [15–18], but, in fact, almost the entire detector not traversed by the π^+ is used as a veto. The $K_{\mu 2}$ can also be suppressed by dE/dx and by requiring that the π decay to a μ in the RS. The entire $\pi^+ \to \mu^+ \to e^+$ decay chain is observed with 500 MHz, 8-bit transient digitizers(TD) [19] sampling the output of the RS scintillators. The only other significant background comes from scattered beam pions or K^+ charge exchange (CEX). The primary tools for rejecting these events are good particle identification in the beam counters, identification of both the K^+ and π^+ in the fiber target, and the requirement that the π^+ track occur later than the K^+ track.

The search for $K^+ \to \pi^+ \nu \bar{\nu}$ requires an identified K^+ to stop in the target followed, after a delay of at least 2 ns, by a single charged-particle track that is unaccompanied by any other decay product or beam particle. This particle must be identified as a π^+ with P, R and E between the $K_{\pi 2}$ and $K_{\mu 2}$ peaks. To elude

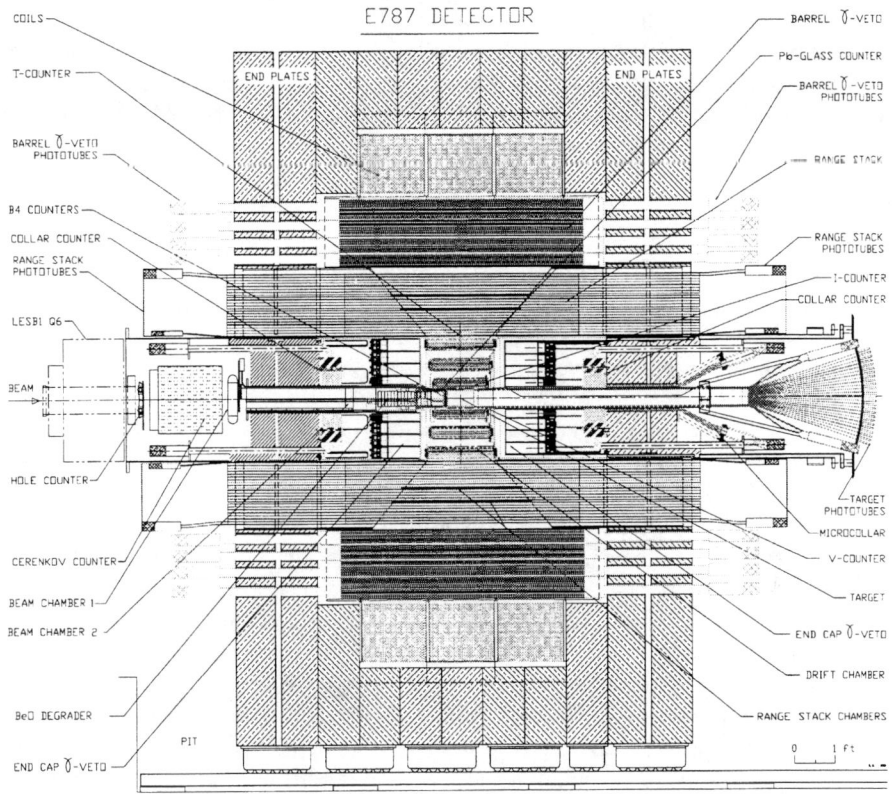

FIGURE 1. The E787 detector.

rejection, $K_{\mu 2}$ and $K_{\pi 2}$ events have to be reconstructed incorrectly in P, R and E. In addition, any event with a muon has to have its track misidentified as a pion — the TD's provide a suppression factor 10^{-5}. Events with photons, such as $K_{\pi 2}$ decays, are efficiently eliminated — the suppression of π^0's is 10^{-6} (photon energy threshold of ~ 1 MeV). A scattered beam pion can survive the analysis only by misidentification as a K^+ and if the track is mismeasured as delayed, or if the track is missed entirely by the beam counters after a valid K^+ stopped in the target. CEX background events can survive only if the K_L^0 is produced at low enough energy to remain in the target for at least 2 ns, if there is no visible gap between the beam track and the observed π^+ track, and if the additional charged lepton evades detection.

Reliable estimation of backgrounds is one of the most important aspects of the measurement of $K^+ \to \pi^+ \nu \bar{\nu}$ at the 10^{-10} level. For each source of background two independent sets of cuts are established, by taking advantage of the redundancy of detector measurements. One set of cuts is relaxed or inverted to enhance the background (by up to three orders of magnitude) so that the other group can be evaluated to determine its power for rejection. In this fashion backgrounds can be studied at sensitivities up to 1000 times greater than the experimental sensitivity. For example, $K_{\mu 2}$ (including $K^+ \to \mu^+ \nu_\mu \gamma$) are to be studied by separately measuring the rejections of the TD particle identification and kinematic cuts. The background from $K_{\pi 2}$ is evaluated by separately measuring the rejections of the photon detection system and kinematic cuts. The background from beam pion scattering is evaluated by separately measuring the rejections of the beam counter and timing cuts. Measurements of K^+ charge exchange in the target were performed in E787. This data combined with Monte Carlo simulation of semi-leptonic K_L decays, allows the CEX background to be determined. Small correlations in the separate groups of cuts are investigated for each background source and corrected for if they existed.

Further confidence in the background estimates and in the measurements of the background distributions near the signal region is provided by extending the method described above to estimate the number of events expected to appear when the cuts were relaxed in predetermined ways so as to allow higher levels of all background types. Confronting these estimates with measurements from the full $K^+ \to \pi^+ \nu \bar\nu$ data, where the two sets of cuts for each background type were relaxed simultaneously, tested the independence of the two sets of cuts. The background level for the 1995 data set, b, was measured to be 0.08 events. At the level of $20 \times b$, two events were observed where 1.6 ± 0.6 were expected, and at $150 \times b$, 15 events were found where 12 ± 5 were expected. Under detailed examination, the events admitted by the relaxed cuts were consistent with being due to the known background sources. Within the final signal region, additional background rejection capability is available. Therefore, prior to looking in the signal region, several sets of increasingly tighter criteria were established, to be used only to interpret any events that fell into the signal region.

THE EVENT

E787 has recently published the first observation of the $K^+ \to \pi^+ \nu \bar\nu$ decay [20], based on data from the 1995 run of the AGS. The range and energy of event candidates passing all other cuts is shown in Fig. 2. The box (which encloses the upper 16.2% of the $K^+ \to \pi^+ \nu \bar\nu$ phase space) indicates the signal region. One event consistent with the decay $K^+ \to \pi^+ \nu \bar\nu$ was observed. The expected number of background events from all sources was 0.08 ± 0.03 events (branching ratio equivalent of 3×10^{-11}). The separate levels of background were — $K_{\mu 2}$: 0.02 ± 0.02, $K_{\pi 2}$: 0.03 ± 0.02, beam π^+: 0.02 ± 0.01 and CEX: 0.01 ± 0.01. This event was in a particularly clean region where the expected background was 0.008 ± 0.005 and which contained 55% of the acceptance of the full signal region. A reconstruction of the event is shown in Fig. 3. The kaon decayed to a pion at 23.9 ns, followed by a clean $\pi^+ \to \mu^+$ decay 27.0 ns later, as can be seen in the upper insert in Fig. 3;

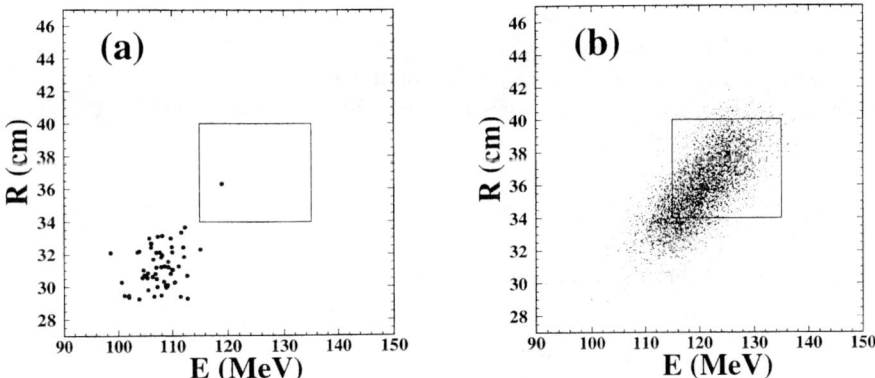

FIGURE 2. Final event candidate for $K^+ \to \pi^+ \nu \bar{\nu}$: a) Data b) Monte-Carlo

there was also a clean $\mu^+ \to e^+$ decay at 3201.1 ns.

There was no significant activity anywhere else in the detector at the time of the K^+ decay. The lower insert in Fig. 3 shows one of the target fibers that contains the incident kaon track, yet does not lie on the outgoing π^+ track; there is no activity at the K^+ decay time. The branching ratio for $K^+ \to \pi^+ \nu \bar{\nu}$ implied by the

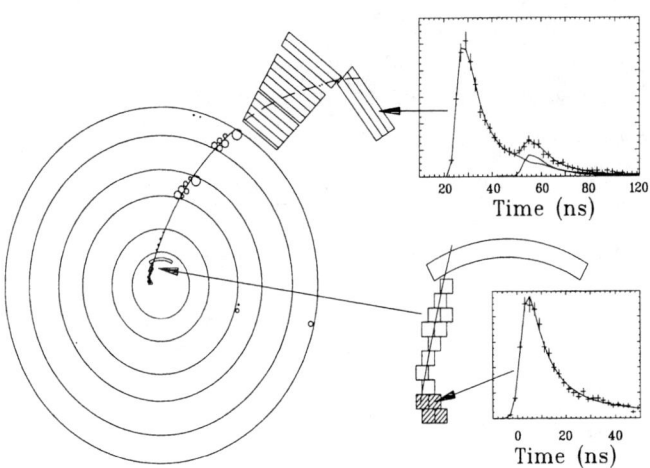

FIGURE 3. $K^+ \to \pi^+ \nu \bar{\nu}$ event.

observation of this event is $B(K^+ \to \pi^+ \nu \bar{\nu}) = 4.2^{+9.7}_{-3.5} \times 10^{-10}$.

TABLE 1. Running conditions for E787. The number of stopped kaons with the detector live (KB_L) is the online measure of the experimental sensitivity. As the kaon momentum is lowered the fraction of useful kaons that stop (sf) increases. The duty factor (DF) of the AGS steadily increased. The single event sensitivity (S.E.S.) is measured from 1995 and estimated for the later years based on KB_L and the acceptance calculated from the detector rates. The background levels (bck) for 1996–99 are estimated from the reanalysis of the 1995 data.

year	KB_L (10^{12})	$|\vec{p_K}|$	DF (%)	sf (%)	(S.E.S.)$^{-1}$ (10^{10})	bck events
1995	1.49	790	41	18.7	0.24	0.08
1995–97	3.33	670–790	43	22	0.6	0.09
1998–99	3.5	710	52	27	0.8	0.1
1995–99	7	670–790	48	25	1.5	0.15

EXPECTATIONS FOR THE E787 FINAL RESULT

The E787 experiment has had three runs: 1995, 1996 and 1997. The fourth and final running period, 1998–99, is ongoing and is expected to terminate at the end of December 1998.

The typical conditions for the 1995 run were 13 Tp/spill, 5.3 MHz of incident K^+, a stopped kaon rate of 1.2 M/spill, a deadtime of 25%, and an acceptance of 0.16%. The rate dependence of the acceptance has been measured; the acceptance is 60% of what the acceptance would have been with zero rate. The rates in most detector elements have been measured to be proportional to the incident flux and not to the stopped kaon flux. This implies that the sensitivity can be increased by increasing the fraction of stopped kaons/incident kaons. So E787 has lowered the momentum of the incident kaons in subsequent years, increasing the sensitivity without increasing the rates in most detector elements. In addition, the duty factor of the AGS has been increased from 37% at the start of 1995 to 55% at the end of 1998–99.

A summary of the E787 sensitivity, including the published 1995 result, is shown in Table 1, along with some of the running conditions.

With improvements in the online trigger efficiency and in acceptance from running at lower momentum, the 1996 data set should have comparable sensitivity to the 1995 data set, even though the useful period of data collection was only $2/3$ as long. Further improvements in live time, through trigger and DAQ upgrades [21-24], and in acceptance from lower momentum, give an expected sensitivity for the 1997 data set of ∼65% of the 1995 data set, even though the useful period of running was less than $1/2$ that of 1995.

The expected sensitivity from the 1995–97 runs ∼2.5 times that of the 1995 data alone. A preliminary re-analysis of the E787 1995 data with improvements in the analysis software have demonstrated a background rejection that is ~3 times larger with no loss in acceptance. This background level (roughly equivalent to a branching ratio of 1×10^{-11}) is sufficient for all future measurements of the $K^+ \to \pi^+ \nu \bar{\nu}$ branching ratio. Results of the analysis of the larger data set are expected this year.

With improved running conditions, including an increased duty factor, improved DAQ [25] and a long running period of almost five months in 1998–99, the E787 sensitivity for $K^+ \to \pi^+ \nu \bar{\nu}$ should extend well below the most probable SM level ($\sim 1 \times 10^{-10}$), to less than 7×10^{-11}, sufficient to observe 1–2 SM events.

FUTURE PLANS — E949

A new experiment to measure the branching ratio B($K^+ \to \pi^+ \nu \bar{\nu}$), E949 [26], was recently approved to run at the AGS starting in the year 2000. This experiment is designed to reach a sensitivity of $(8–14) \times 10^{-12}$, an order of magnitude below the Standard Model prediction and to determine $|V_{td}|$ to better than 27%. It is built around the existing E787 detector to take advantage of the extensive analysis of E787 data and allowing a reliable projection of the new experiment to the required sensitivity with a high level of confidence.

The E949 detector will have significantly upgraded photon veto systems, DAQ and trigger compared to the E787 experiment. The PV upgrade includes a barrel veto liner that will replace the outer layers of the RS and fill a gap between the RS and BV. It is 2.3 X_o thick and will add substantially to the thin region at 45°. Additional PV upgrades will be installed along the beam direction. The most important DAQ upgrade will be to instrument the RS with TDC's to extend the search time for the Michel electron ($\mu^+ \to e^+$) and to allow the TD range to be shortened. The shortening of the TD range should allow a reduction of deadtime by 30–40%. Trigger upgrades should reduce the deadtime further and reduce the acceptance loss due to the online PV, by improving the timing on the RS and BV. Compared to the 1995 data set of E787, an improvement of 54% has already been realized. Additional improvements in these areas and in offline software are expected to gain another 90%. Additional sensitivity gains can be realized by including the region of phase space below the $K_{\pi 2}$ peak and by reoptimizing the analysis algorithms to run at higher rates. Each of these should provide a factor of 2 more sensitivity. The proposal assumes that only one of these factors will be realized.

The operating conditions will be significantly upgraded. E949 will run with a 700 MeV/c K^+ beam, with 100 Tp (10^{14} protons per spill) and with an AGS duty factor of close to 70%. These conditions are all within the expected AGS operating parameters for the year 2000 [27]. They will require a new target station and replacements for the upstream magnets in the LESB3 beam line. The gain in sensitivity from these conditions will be a factor of 2.2. A summary of the

TABLE 2. Sensitivity improvement factors for E949, compared to the published E787 result.

Upgrade	Improvement factor
Lower momentum	1.44
Higher duty factor	1.53
Established improvements	1.54
Additional efficiency improvements	1.9
Phase space below $K^+ \to \pi^+\pi^\circ$	2
Total	13

improvement factors is given in Table 2 The total gain in sensitivity per hour will be 6–13 times over the E787 published result on the 1995 data set. All of the measurements of deadtime and acceptance as a function of rate; and of the stopping fraction and kaon flux have been included in a calculation of the optimum running conditions. The sensitivity of the 1995 conditions is 1.46×10^6/hr and for E949 is 9.6×10^6/hr, not including the plans for reoptimizing for higher rates or the phase space below the $K_{\pi 2}$ peak.

CONCLUSIONS

The prospects for further improvement in the determination of $B(K^+ \to \pi^+\nu\bar{\nu})$ is bright. The first observation of this rare and interesting decay has recently been published. The data on hand, or soon to be available, from the E787 experiment, should provide almost an order of magnitude more sensitivity. The recently approved experiment E949 should reach at least a factor of five further than E787 and make a very interesting measurement of $|V_{td}|$. There is also a proposal, CKM, at the FNAL Main Injector, to push even further, to 10^{-12} by looking for the decay in flight. A plot showing the progress from past, current and approved experiments for $B(K^+ \to \pi^+\nu\bar{\nu})$ is shown in Figure 4. The search for this decay, with its very clean and well understood theory, has had a long and now very fruitful history.

ACKNOWLEDGEMENTS

The members of the E787 or E949 collaborations:

- **University of Alberta**: P. Kitching, H.-S. Ng, R. Soluk

- **Brookhaven National Laboratory**: S. Adler, M.S. Atiya, I-H. Chiang, M.V. Diwan, J.S. Frank, J.S. Haggerty, V. Jain, S.H. Kettell, T.F. Kycia, K.K. Li, L.S. Littenberg, C. Ng, A. Sambamurti, A.J. Stevens, R.C. Strand, C. Witzig

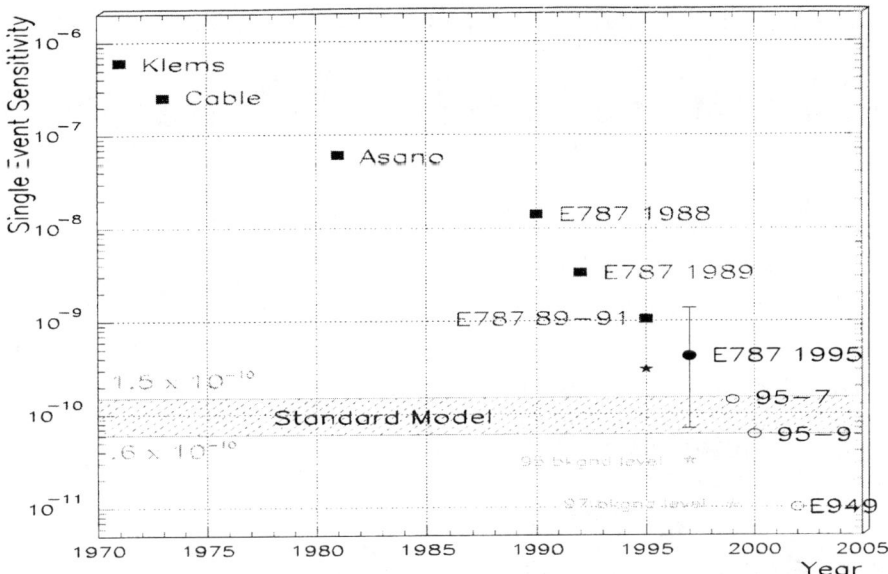

FIGURE 4. History of progress in the search for $K^+ \to \pi^+ \nu \bar{\nu}$. The sensitivity of experiments setting limits is shown in solid squares. The first actual measurement is shown as a solid circle and the projected future measurements are shown as open circles. The background levels are shown as stars for the recent data.

- **Fukui University**: M. Miyajima, J. Nishide, K. Shimada, T. Shimoyama, Y. Tamagawa

- **KEK—Tanashi**: T.K. Komatsubara, M. Kuriki, N. Muramatsu, K. Omata, S. Sugimoto

- **KEK—Tsukuba**: M. Aoki, T. Inagaki, S. Kabe, M. Kobayashi, Y. Kuno, T. Sato, T. Shinkawa, Y. Yoshimura

- **Osaka University**: Y. Kishi, T. Nakano, M. Nomachi

- **Princeton University**: M. Ardebili, A.O. Bazarko, M.R. Convery, M.M. Ito, D.R. Marlow, R.A. McPherson, P.D. Meyers, F.C. Shoemaker, A.J.S. Smith, J.R. Stone

- **TRIUMF**: P.C. Bergbusch, E.W. Blackmore, D.A. Bryman, S. Chen, A. Konaka, J.A. Macdonald, J. Mildenberger, T. Numao, P. Padley, J.-M. Poutissou, R. Poutissou, G. Redlinger, J. Roy, A.S. Turcot

REFERENCES

1. M.K. Gaillard and B.W. Lee, *Phys. Rev.* **D10**, 897 (1974).
2. T. Inami and C. S. Lim, *Prog. Theor. Phys.* **65**, 297 (1981).
3. A.J. Buras and R. Fleischer, *Heavy Flavours II*, World Scientific, 1998, ed. Buras and Linder, p65.
4. J.S. Hagelin and L.S. Littenberg, *Prog. Part. Nucl. Phys.* **23**, 1 (1989).
5. D. Rein and L.M. Sehgal, *Phys. Rev.* **D39**, 3325 (1989).
6. M. Lu and M.B. Wise, *Phys. Lett.* **B324**, 461 (1994).
7. C.Q. Geng et al., *Phys. Lett.* **B355**, 569 (1995).
8. S. Faijfer, *Nuovo. Cim.* **110A**, 397 (1997).
9. W.J. Marciano and Z. Parsa, *Phys. Rev.* **D53**, R1 (1996).
10. G. Buchalla and A.J. Buras, *Phys. Rev.* **D57**, 216 (1998).
11. G. Buchalla and A.J. Buras, *Nucl. Phys.* **B412**, 106 (1994).
12. M.S. Atiya et al., NIM **A321**, 129 (1992).
13. I.-H. Chiang et al., to be submitted to NIM **A**.
14. E.W. Blackmore et al., *NIM* **A404**, 295 (1998).
15. I.-H. Chiang et al., *IEEE Trans. Nucl. Sci.* **NS–42**, 394 (1995).
16. T.K. Komatsubara et al., *NIM* **A404**, 315 (1998).
17. M. Kobayashi et al., *NIM* **A337**, 355 (1994).
18. D.A. Bryman et al., *NIM* **A396**, 394 (1997).
19. M.S. Atiya et al., *NIM* **A279**, 180 (1989).
20. S. Adler et al., *Phys. Rev. Lett.* **79**, 2204 (1997).
21. M. Burke et al., *IEEE Trans. Nucl. Sci.* **NS–41**, 131 (1994).
22. C. Witzig and S. Adler, *Real-Time Comput. Appl.*, 123 (1993).
23. S.S. Adler, *Inter. Conf. Electr. Part. Phys.*, 133 (1997).
24. C. Zein et al., *Real-Time Comput. Appl.*, 103 (1993).
25. S.S. Adler et al., to be submitted to IEEE.
26. M. Aoki et al., *AGS Proposal 949*, August 1998.
27. J.M. Brennan and T. Roser, 'High intensity performance of the Brookhaven AGS', 5th Europ.Part.Acc.Conf.(EPAC96), 530 (1996).

Recent Results from BNL E865

A. Sher[6], R. Appel[6], G.S. Atoyan[2], B. Bassaleck[5], D.N. Brown[6],
D.R. Bergman[8], N. Cheung[6], S. Dhawan[8], H. Do[8], J. Egger[3], S. Eilerts[5],
C. Felder[6], H. Fischer[5], M. Gach[6], W.D. Herold[3], V.V. Isakov[2],
H. Kaspar[3], D. Kraus[6], D. M. Lazarus[1], L. Leipuner[1], J. Lowe[5], J.
Lozano[8], H. Ma[1], W. Majid[8], W. Menzel[4], S. Pislak[8], A.A. Poblaguev[2],
A.L. Proskurjakov[2], P. Rehak[1], P. Robmann[7], R. Stotzer[5],
J.A. Thompson[6], P. Truöl[7], H. Weyer[4], M.E. Zeller[8]

[6]*University of Pittsburgh, Department of Physics and Astronomy, Pittsburgh, PA 15260*

[1]*Brookhaven National Laboratory,* [2]*Institute for Nuclear Research, Moscow,* [3]*Paul Scherrer Institute,*
[4]*University of Basel,* [5]*University of New Mexico,* [7]*University of Zürich,* [8]*Yale University.*

Abstract. Experiment E865 at the BNL AGS is a search for the Lepton Family Violating decay $K^+ \to \pi^- \mu^+ e^-$. We aim to reach a single event sensitivity below 10^{-11} for this decay. The E865 apparatus allow us to collect and study other rare and semi rare K^+ decays. Preliminary results are shown on $K^+ \to \pi^+ \mu^+ e^-$, $K^+ \to \pi^+ e^+ e^-$, $K^+ \to \pi^+ \mu^+ \mu^-$, $K^+ \to \pi^+ \pi^- e^+ \nu$, $K^+ \to e^+ \nu e^+ e^-$, $K^+ \to \mu^+ \nu e^+ e^-$, $K^+ \to \pi^0 e^+ \nu$ ($\pi^0 \to e^+ e^- \gamma$).

INTRODUCTION

Interactions between elementary particles are well described by the Standard Model of particle physics, and so far there is no proven experimental data contradicting it. Nevertheless, some qualities of the Standard Model justify ongoing experimental efforts to find facts not explained by or directly contradicting the Standard Model as we know it today. First, the Standard Model may be a low energy extension of a symmetry unbroken at the higher energies. Second, even though the Standard Model explains the fundamental particles masses through symmetry breaking (Higgs mechanism), it simultaneously introduces the problem of the fine-tuning. Also, it requires Lepton Flavor Conservation, a rule, which does not have a satisfactory explanation but has never been observed to be broken. Several theoretical extensions of the Standard Model, such as Technicolor, Leptoquarks, Horizontal Generation Changing Gauge Boson as well as Supersymmetry allow the violation of lepton flavor (LF). An experimental observation of such violation would give a valuable insight into the physics beyond the Standard Model. At the same time, improvement of the limits on the LF violating decays can rule out some of the theories or help to revise them.

E865 at Brookhaven National Laboratory's Alternating Gradient Synchrotron was designed primarily for the purpose of searching for the LF violating decay $K^+ \to \pi^+ \mu^+ e^-$. By the end of 1998 we expect to have enough data to reach single event sensitivity to this process below 10^{-11}. Aside from $K^+ \to \pi^+ \mu^+ e^-$, the E865 detector and trigger are designed to detect most K^+ decay modes containing at least three charged particles in the final state, thus enabling us to collect and study data on many rare and semi-rare kaon decays (Table 1). A study of these decays provides us with the means

CP459, *Heavy Quarks at Fixed Target*
edited by Harry W. K. Cheung and Joel N. Butler
© 1999 The American Institute of Physics 1-56396-864-9/99/$15.00

TABLE 1. Processes observed by E865

Process	BR (PDG)	# of previously observed events	# of events observed by E865	Why Study?
$K^+ \to \pi^+ \mu^+ e^-$	$< 2.1 \cdot 10^{-10}$ (90%CL)	0	0	LF violation
$K^+ \to \pi^+ e^+ e^-$	$(2.74 \pm 0.23) \cdot 10^{-7}$	1341	~10000	ChPT[c]
$K^+ \to \pi^+ \mu^+ \mu^-$	$(5.0 \pm 1.0) \cdot 10^{-8}$	203	~400	ChPT[c]
$K^+ \to e^+ \nu e^+ e^-$	$(3+3-1.5) \cdot 10^{-8}$	4	~380	ChPT[c]
$K^+ \to \mu^+ \nu e^+ e^-$	$(1.3 \pm 0.4) \cdot 10^{-7}$	14	~1500	ChPT[c]
$K^+ \to \pi^+ \pi^- e^+ \nu$	$(3.91 \pm 0.17) \cdot 10^{-5}$	~30000	~300000	$\pi\pi$ scattering
$K^+ \to \pi^0 e^+ \nu$[a]	$(4.82 \pm 0.06) \cdot 10^{-2}$	~4000[b]	~60000	V_{us}
$K^+ \to \pi^0 \mu^+ \nu$[a]	$(3.18 \pm 0.08) \cdot 10^{-2}$	~4000[b]	~60000	Form factors
$K^+ \to \pi^+ \pi^+ \pi^-$	$(5.59 \pm 0.05) \cdot 10^{-2}$			Calibration
$K^+ \to \pi^+ \pi^{0a}$	$(21.16 \pm 0.14) \cdot 10^{-2}$			Calibration

[a] In E865 the π^0 in these processes is detected through $\pi^0 \to e^+ e^- \gamma$ (BR=$(1.198 \pm 0.032) \cdot 10^{-2}$).
[b] Maximum number of events in the single experiment where branching ratio was obtained.
[c] ChPT stands for the Chiral Perturbation Theory

to probe further the existing theoretical models. Decays mentioned in Table 1 will be discussed in more details after a short description of the E865 detector and trigger.

E865 DETECTOR AND TRIGGER

The plan view of the E865 detector with a simulated $K^+ \to \pi^+ \mu^+ e^-$ event is depicted on Figure 1. Kaons are delivered in an unseparated beam with a momentum of 6GeV/c. The detector is designed to detect the products of the kaon decays occurring in the vacuum decay tank upstream from the apparatus. The tracking system of the detector consists of four Proportional Wire Chambers (PWCs). A dipole magnet D6, which delivers a transverse momentum kick of 250MeV/c, together with four PWCs, form a spectrometer. A dipole magnet D5 delivers a momentum kick equal in magnitude but opposite in direction, forcing particles of different charges to be on different sides of the detector and bending them out of the dead region left for the passage of the beam through the apparatus.

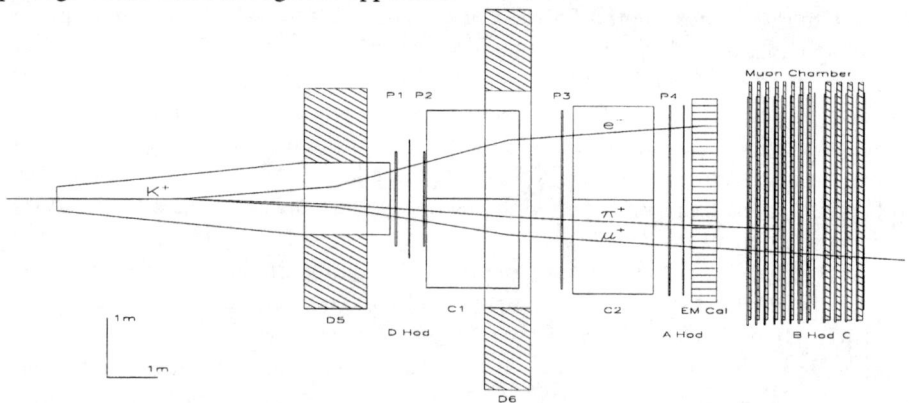

FIGURE 1. Plan view of the E865 detector with a simulated $K^+ \to \pi^+ \mu^+ e^-$ event. View is from above, thus left side of the apparatus is on the upper, while right is on the lower half of the diagram.

Two large atmospheric pressure Cerenkov counters (C1 and C2) are divided forming four independent cells, two on each side of the apparatus, and are filled with gaseous hydrogen on the left and methane on the right. The electromagnetic shower calorimeter, of the lead-scintillator shashlik type (1), plays an important role in trigger and particle identification. The muon system is located downstream of the electromagnetic calorimeter and consists of twenty four layers of proportional tubes, interlaid with two-inch thick iron plates and two trigger hodoscopes (B and C). Two other hodoscopes are located between P1 and P2 (D-hodoscope) and just upstream of the calorimeter (A-hodoscope).

Electrons (positrons) are identified by the signal in the Cerenkov counters and by the match of the energy deposited in the calorimeter with the particle's momentum. The energy deposition in the calorimeter and the signal in the muon system separate pions from muons.

The first level trigger uses a coincidence between the A-hodoscope, the shower calorimeter and the D-hodoscope to select events with three charged tracks. The second level trigger utilizes information from the Cerenkov counters and the muon system to select particular types of events. The third level software trigger is designed to do a more detailed selection.

STATUS OF INDIVIDUAL PROCESSES

$K^+ \to \pi^+ \mu^+ e^-$

The search for this decay is central to E865 experiment. The data for this process have been collected during 1995, 1996 and are being collected during the 1998 running periods. The data from 1995 have been analyzed and yielded a limit on the $K^+ \to \pi^+ \mu^+ e^-$ branching ratio of 2.1×10^{-10} (90% CL) (2, 3). This result is summarized in Figure 2.

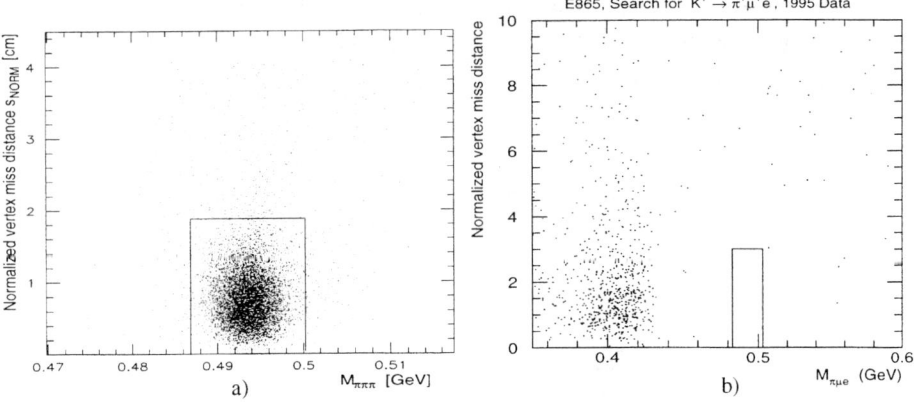

FIGURE 2. a) Quality of the vertex parameter versus the invariant $\pi\pi\pi$ mass of $K^+ \to \pi^+\pi^+\pi^-$ events. b) Quality of the vertex parameter versus invariant $\pi\mu e$ mass for the $K^+ \to \pi^+\mu^+e^-$ candidates from 1995 data. Box outlines $K^+ \to \pi^+\mu^+e^-$ signal region. Most of the background is from the $K^+ \to \pi^0\pi^+$ ($\pi^0 \to e^+e^-\gamma$) decays.

The data from the year 1996 are being analyzed now. Combining 1995, 1996 and 1998 data we expect to reach single event sensitivity below 10^{-11} for this decay.

$$K^+ \to \pi^+ \ell^+ \ell^-$$

Chiral Perturbation Theory (ChPT) contains parameters to be determined from the experiment. Measurement of $K^+ \to \pi^+ e^+ e^-$ and $K^+ \to \pi^+ \mu^+ \mu^-$ form factors and branching ratios are a strong test of this theory.

A parameterization of the $K^+ \to \pi^+ \ell^+ \ell^-$ decay mode, based on an assumption of a vector interaction and an energy dependent vector form factor is presented in Equation 1.

$$\frac{d\Gamma}{dM_{\ell\ell}} \propto M_{\ell\ell} \cdot P_\pi^3 \cdot |f(M_{\ell\ell})|^2, \quad f(M_{\ell\ell}) = f_0 \cdot \left(1 + \lambda \cdot \left(M_{\ell\ell}/M_\pi\right)^2\right) \qquad (1)$$

In Equation 1 $M_{\ell\ell}$ is the invariant mass of two leptons, P_π is the momentum of the pion in the rest frame of the kaon, and M_π is the pion mass.

ChPT of $O(P^4)$ gives more definitive parameterization of this decay (4) expressed in Equation 2.

$$\frac{d\Gamma}{dM_{\ell\ell}} \propto M_{\ell\ell} \cdot \overline{\Gamma} \cdot P_\pi^3 \cdot \frac{|\Phi_+|^2}{M_k^5}, \quad \Phi_+ = (\varphi_\pi + \varphi_k + \omega_+) \qquad (2)$$

In Equation 2, φ_π and φ_k are analytic functions of $M_{\ell\ell}$, and ω_+ is a parameter of ChPT of $O(P^4)$ to be determined from the experiment.

$$K^+ \to \pi^+ e^+ e^-$$

The analysis of 1995 and 1996 data yielded about 10000 $K^+ \to \pi^+ e^+ e^-$ events. Experimental distributions of these events are shown in Figure 3.

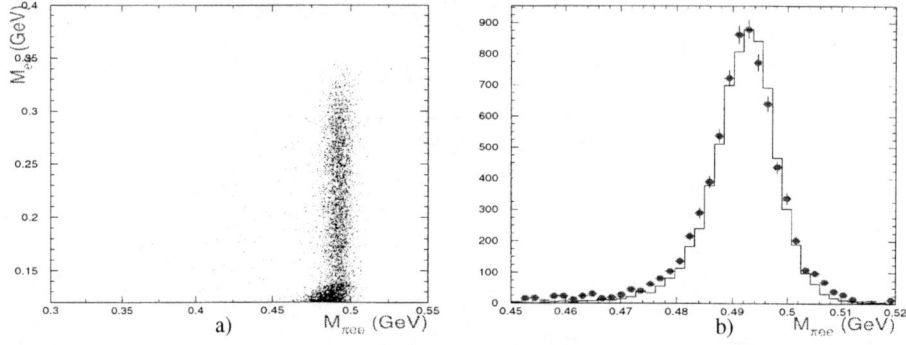

FIGURE 3. a) Invariant e^+e^- mass of $K^+ \to \pi^+ e^+ e^-$ candidates versus πee mass. Events with low e^+e^- mass are background from $K^+ \to \pi^+ \pi^0 (\pi^0 \to e^+ e^- \gamma)$. b) Comparison of πee mass distribution for events with e^+e^- mass larger 150MeV with Monte-Carlo. Histogram is Monte-Carlo. Points with errors represent data.

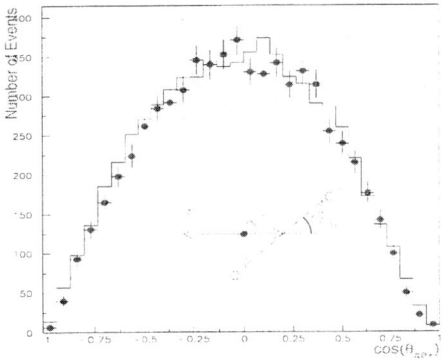

FIGURE 4. Comparison of the distribution of the cosine of the angle between the electron and pion in the rest frame of the e^+e^- pair with Monte-Carlo. Points with errors represent data; histogram represents Monte-Carlo.

Figure 4 presents the experimental distribution of the cosine of the angle between electron and pion in the rest frame of the e^+e^- pair compared to the Monte-Carlo distribution produced with the assumption of pure vector current. Data confirm that this process is a vector interaction.

A fit of the $K^+ \to \pi^+ e^+ e^-$ data normalized to $K^+ \to \pi^+ \pi^0 (\pi^0 \to e^+ e^- \gamma)$ with the Monte-Carlo simulation according to Equation 1 (Figure 5) yields a preliminary result of $\lambda = 0.20 \pm 0.02$ and $BR(K^+ \to \pi^+ e^+ e^-) = (2.69 \pm 0.2) \cdot 10^{-7}$ (5). A fit to the normalized M_{ee} spectrum according to Equation 2 can yield the ChPT parameter ω_+. Even though the ChPT parameterization fits the M_{ee} spectrum shape well, it does not simultaneously fit the branching ratio. A comparison of our results for the branching ratio and ω_+ with theoretical predictions and results of the previous experiments is shown in Figure 5b. From this comparison we conclude that Chiral Perturbation Theory of $O(P^4)$ order is insufficient to describe our data.

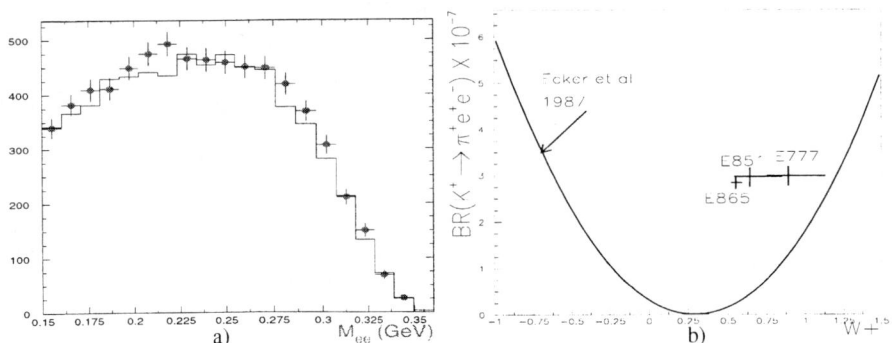

FIGURE 5. a) Fit to the M_{ee} spectrum by parameterization (1). b) Comparison of E865 results in the branching ratio/ω_+ plane with ChPT of $O(P^4)$ prediction and previous experiments (6), (7). The solid line represents the Chiral Perturbation Theory of $O(P^4)$ prediction according to (4).

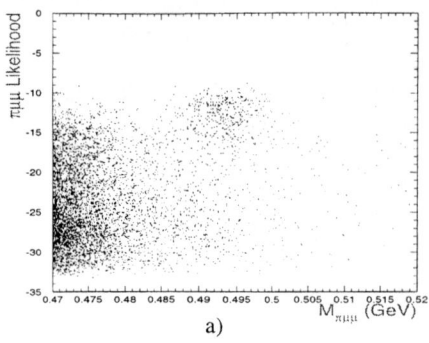 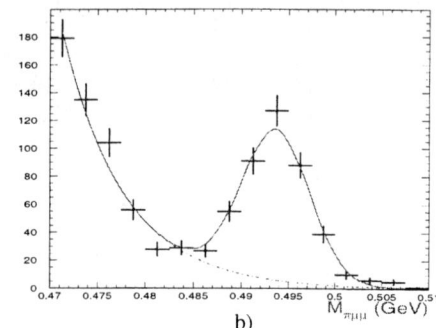

FIGURE 6. a) The Combined Likelihood function of the event being $K^+\to\pi^+\mu^+\mu^-$ versus the invariant $\pi\mu\mu$ mass of the event. b) $\pi\mu\mu$ invariant mass spectrum after Likelihood cut. Points represent data; the line is a gaussian plus exponent fit. Background events on both plots are $K^+\to\pi^+\pi^+\pi^-$ events with $\pi^\pm\to\mu^\pm\nu$.

$K^+\to\pi^+\mu^+\mu^-$

Our sample of $K^+\to\pi^+\mu^+\mu^-$ events was collected during a dedicated period of the 1997 run. The main background for this decay is $K^+\to\pi^+\pi^+\pi^-$, where $\pi^\pm\to\mu^\pm\nu$. Even though these events tend to have lower invariant $\pi\mu\mu$ mass, misreconstruction of the particle momentum can place the invariant mass of such events in the signal region for $\pi\mu\mu$. The Combined Likelihood function using information from the different parts of the detector was constructed to maximize the signal to noise ratio. The likelihood value versus the invariant mass of $\pi\mu\mu$ events and the distribution of the invariant mass after a likelihood cut are presented in Figure 6.

Comparisons of Monte-Carlo with the data were made for the cosine of the angle between the muon and pion in the rest frame of the kaon and the $M_{\mu\mu}$ spectrum (Figure 7).

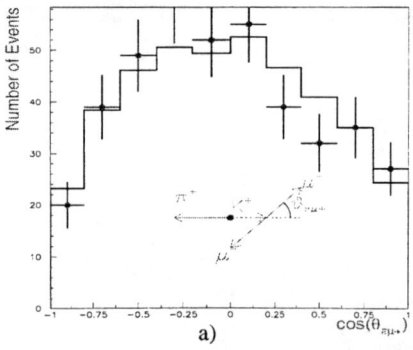 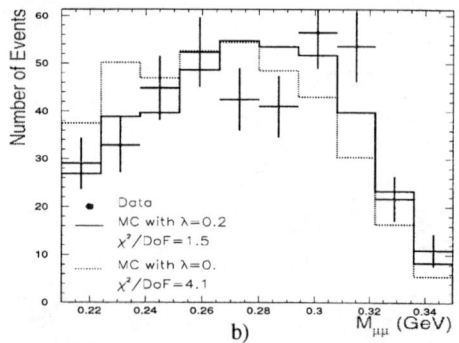

FIGURE 7. a) Distribution of the cosine of the angle between muon and pion in the rest frame of the kaon compared to Monte-Carlo assuming a vector current b) Comparison of the data with Monte-Carlo for the $M_{\mu\mu}$ spectrum. On both plots points with errors represent the data and histograms represent Monte-Carlo.

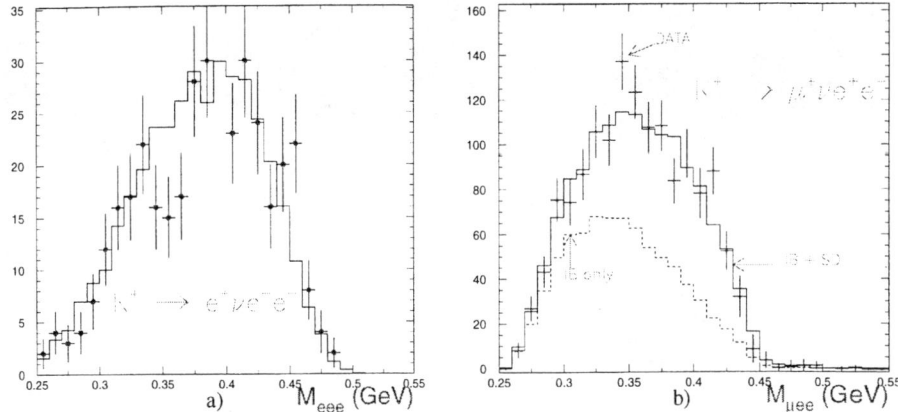

FIGURE 8. a) M_{eee} spectrum compared to Monte-Carlo dominated by the structure dependent part of the decay. b) $M_{\mu ee}$ spectrum compared to Monte-Carlo with both structure dependent and inner bremsstrahlung parts. Points with errors represent data, histograms represent Monte-Carlo.

The data are consistent with the slope parameter λ obtained from $K^+\to\pi^+e^+e^-$. Work is in progress on obtaining the $K^+\to\pi^+\mu^+\mu^-$ branching ratio.

$K^+\to\ell^+\nu e^+e^-$

$K^+\to\ell^+\nu e^+e^-$ consists of two parts. One is the inner bremsstrahlung (IB), determined by the kaon decay constant and quantum electrodynamics. The other is the structure dependent (SD) part. The contribution of the SD part was calculated in the framework of Chiral Perturbation Theory. The SD part is parameterized by the form factors F_V, F_A and R (8).

We observe about 380 $K^+\to e^+\nu e^+e^-$ events and about 1500 $K^+\to\mu^+\nu e^+e^-$ events. Figure 8 shows distributions of the invariant $\ell^+e^+e^-$ mass for both decay modes compared to Monte-Carlo done for both inner bremsstrahlung and structure dependent parts. Work is in progress on obtaining the SD parameters and branching ratios for both decays.

$K^+\to\pi^+\pi^-e^+\nu$

Studies of low energy $\pi\pi$ scattering are important for understanding hadronic interactions. The $K^+\to\pi^+\pi^-e^+\nu$ decay (Ke_4) has no strongly interacting particles in the final state other than $\pi^+\pi^-$, thus allowing to study $\pi\pi$ interaction without introducing theoretical uncertainties present in pion-nucleon scattering experiments.

To determine the kaon's position before its decay containing a neutrino ($K^+\to\pi^+\pi^-e^+\nu$, $K^+\to\pi^0 e^+\nu$, $K^+\to\pi^0\mu^+\nu$), a pixel beam hodoscope was installed upstream of the decay tank prior to the 1997 run. It improved our ability to study $\pi\pi$ scattering parameters in the collected Ke_4 data. E865 acquired about 300,000 Ke_4 events during 1997.

The experimental distribution of the $\pi^+\pi^-$ mass compared to the Monte-Carlo

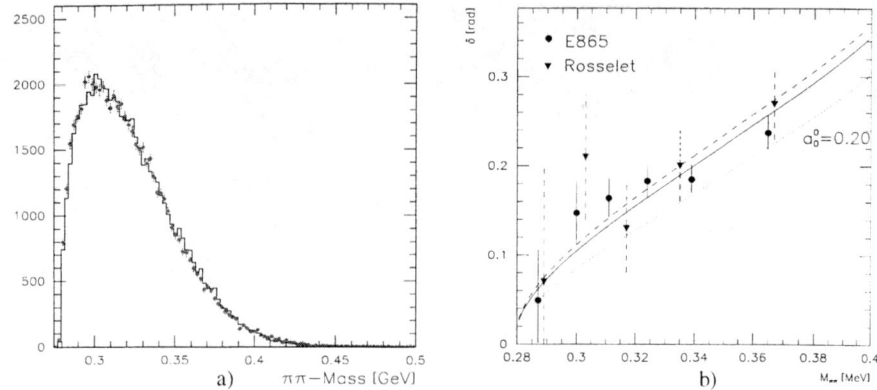

FIGURE 9. a) Distribution of the $\pi\pi$ invariant mass compared to Monte-Carlo. Points represent data, histogram represents Monte-Carlo. b) $\pi\pi$ scattering phase δ as a function of the $\pi\pi$ invariant mass. Dash line represents result obtained by Rosselet et al (9). Solid line represents preliminary E865 result. Dotted line represents Chiral Perturbation Theory of $O(P^4)$ order prediction (10).

simulation is presented in Figure 9.

One of the parameters obtained from the $\pi^+\pi^-$ invariant mass spectrum is the $\pi^+\pi^-$ scattering phase δ. The preliminary result for δ as a function of $\pi^+\pi^-$ mass compared to the previous experimental result from (9) is shown in Figure 9. Only half of the E865 data was used in the fit, and only statistical errors are shown for the E865 results. Work is in progress on understanding systematic effects.

$$K^+ \to \pi^0 e^+ \nu$$

Unitarity of the Cabibbo-Kobayashi-Maskawa (CKM) matrix implies certain relations between its elements, providing an additional test of the Standard Model. One such relation is the sum of the magnitudes of the elements in the first row of the CKM matrix, which should be equal to one for the matrix to be unitary.

As pointed out in reference (11), the most precise value of $|V_{ud}|=0.9740\pm0.0005$ is obtained from nuclear beta decays. The unitarity sum calculated using this value for V_{ud} is more than two standard deviations less than one (11) (Equation 3).

$$|V_{ud}|^2 + |V_{us}|^2 + |V_{ub}|^2 = 0.9968 \pm 0.0014 \quad (3)$$
$$|V_{ud}|=0.9740\pm0.0005$$
$$|V_{us}|=0.2196\pm0.0023$$
$$|V_{ub}|=0.0032\pm0.008$$

The values of V_{us} and V_{ub} are taken from the latest Review of Particle Physics (12). The V_{us} value used is based solely on the $K^+ \to \pi^0 e^+ \nu$ (K_{e3}) data, which introduces less theoretical uncertainty in the derivation of V_{us} than hyperon decays (12). The contribution of the error of V_{us} to the error of the sum is as large as the contribution of the V_{ud} error. Thus, an improvement of the V_{us} error through a new high precision

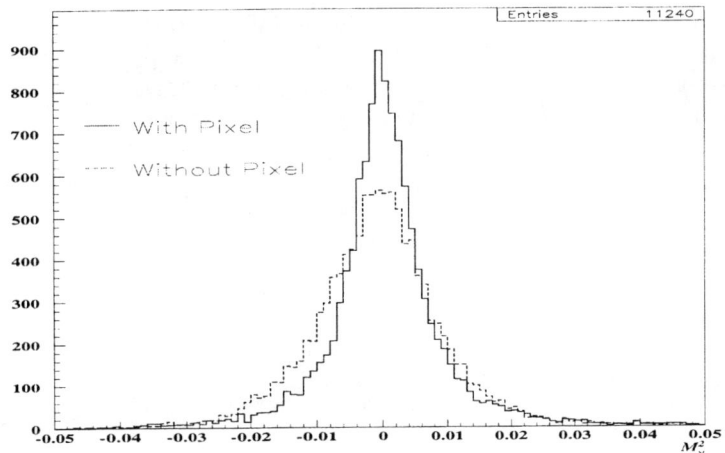

FIGURE 10. Distribution of neutrino mass squared in $K^+ \rightarrow \pi^0 e^+ \nu$ events obtained without (dotted line) and with (solid line) use of the pixel hodoscope. The plot represents about 15% of E865 Ke_3 data.

measurement of Ke_3 branching ratio and form factors is desirable.

Experiment E865 conducted a dedicated Ke_3 run during the 1997 running period and collected about 60,000 Ke_3 events, which were detected through $K^+ \rightarrow \pi^0 e^+ \nu, \pi^0 \rightarrow e^+ e^- \gamma$. The data taking was done with the pixel beam hodoscope in place which allows the study of the form factors. Figure 10 shows the distribution of the neutrino mass squared in the sample of Ke_3 events that represents about 15% of E865 data.

Work is in progress on obtaining high precision (up to 1%) relative branching ratio $BR(K^+ \rightarrow \pi^0 e^+ \nu, \pi^0 \rightarrow e^+ e^- \gamma)/BR(K^+ \rightarrow \pi^0 \pi^+, \pi^0 \rightarrow e^+ e^- \gamma)$ and Ke_3 form factors.

CONCLUSION

Since 1995, experiment E865 collected data on many rare and semi-rare charged kaon decays.

Data taking and analysis are in progress for the decay $K^+ \rightarrow \pi^+ \mu^+ e^-$. Analysis of 1995 data yielded $2.1 \cdot 10^{-10}$ (90%CL) limit on this Lepton Flavor violating decay mode. We expect to reach a single event sensitivity to this process in the range of 10^{-11}.

Large samples of fully reconstructed $K^+ \rightarrow \pi^+ e^+ e^-$ (about 10,000 events) and $K^+ \rightarrow \pi^+ \mu^+ \mu^-$ (about 400 events) decays were accumulated. Analysis of $K^+ \rightarrow \pi^+ e^+ e^-$ data yielded a preliminary result for the vector form factor slope parameter $\lambda = 0.20 \pm 0.02$ and $BR(K^+ \rightarrow \pi^+ e^+ e^-) = (2.69 \pm 0.2) \cdot 10^{-7}$. We find that our $K^+ \rightarrow \pi^+ e^+ e^-$ data can not be described by the Chiral Perturbation Theory of $O(P^4)$. Analysis of the $K^+ \rightarrow \pi^+ \mu^+ \mu^-$ decay mode is underway.

The E865 sample of $K^+ \rightarrow e^+ \nu e^+ e^-$ (about 380 events) and $K^+ \rightarrow \mu^+ \nu e^+ e^-$ (about 1,500 events) decay modes provides an opportunity for the detailed study of the structure dependent part of these decays.

During the 1997 running period, about 300,000 events of $K^+ \to \pi^+\pi^- e^+\nu$ were collected. Improved resolution of the kaon position due to the pixel beam hodoscope installed for this running period enabled us to study low energy $\pi\pi$ scattering in this process. Work is in progress on extracting the $\pi\pi$ scattering parameters for this decay.

A high precision measurement of the $K^+ \to \pi^0 e^+ \nu$ (K_{e3}) branching ratio and form factors is desirable for a precision Cabibbo-Kobayashi-Maskawa matrix unitarity test. E865 accumulated about 60,000 events of K_{e3} decays. Analysis of these data is in progress with the goal of obtaining the relative branching ratio $BR(K^+ \to \pi^0 e^+\nu, \pi^0 \to e^+e^-\gamma)/BR(K^+ \to \pi^0\pi^+, \pi^0 \to e^+e^-\gamma)$ with a precision of up to 1%.

REFERENCES

1. Atoyan, G.S. et al., *Nucl. Instr. and Meth.* A **320**, 144 (1992).
2. Pislak, S., *Experiment E865 at BNL, A search for the decay* $K^+ \to \pi^+ \mu^+ e^-$, Ph. D. thesis, University of Zürich, 1997.
3. Bergman, D., *A search for the decay* $K^+ \to \pi^+ \mu^+ e^-$, Ph. D. thesis, Yale University, 1997.
4. Ecker, G., Pich, A., and de Rafael, E., *Nucl. Phys.* B **291**, 692 (1987).
5. Ma, H. et al., "Study of $K^+ \to \pi^+ e^+ e^-$ and $K^+ \to \pi^+ \mu^+ \mu^-$ decays in E865 at the AGS," presented at the International Conference on High Energy Physics: ICHEP98, Vancouver, Canada, July 23-29, 1998.
6. Alliegro, C. et al., *Phys. Rev. Lett.* **68**, 278 (1992).
7. Despande, A.L., *A study of the decay of positively charged kaon into a positively charged pion, a positron and an electron, and a measurement of the decay of a neutral pion into a positron and an electron*, Ph. D. thesis, Yale University, 1995.
8. Bijnens, J., Ecker, G., and Gasser, J., *Nucl. Phys.* B **396**, 81 (1993).
9. Rosselet, L. et al., *Phys. Rev.* D **15**, 574 (1977).
10. Gasser, J. and Leutwyler, H., *Phys. Lett.* B**125**, 325 (1983).
11. Towner, I.S. and Hardy J. C., "The current status of V_{ud}," presented at the . 5th International Symposium on Weak and Electromagnetic Interactions in Nuclei: WEIN '98, Santa Fe, NM, USA; June 14-21, 1998.
12. Caso, C. et al., *The European Physical Journal* C **3**, 1 (1998).

Review of Rare Decays and CP violation in Charm

Simon W.L. Kwan

Fermi National Accelerator Laboratory, Batavia, IL 60510, USA

Abstract. A review of current experimental results on rare and forbidden decays, and searches for CP violation in charm decay is presented. Future prospects will be discussed.

I INTRODUCTION

Despite the many successes of the Standard Model, it is generally agreed that it is an incomplete model and that it will break down at some higher mass scales that is not presently directly accessible. In recent years, a lot of experimental effort has been devoted to the search for physics beyond the Standard Model. This provides the physics motivation and interest in the study of rare and forbidden decays and the search for CP violation in charm [1]. One can probe very high mass scales through the contribution of virtual states to such rare processes [2].

By rare decay, we refer to a process which cannot proceed via simple charged-current W-exchange and hence can only proceed via an internal quark loop. Such flavor changing neutral current (FCNC) decays are expected to be very rare in the Standard Model. Forbidden decay, such as lepton number violation (LNV) or lepton family number violation (LFNV), refers to a process which is strictly not allowed by the Standard Model. CP violation in the charm sector is also expected to be very small in the Standard Model. Hence, the physics interest in searching for rare and forbidden decays and CP violation in charm comes from the window of opportunity for observing new physics - physics beyond the Standard Model. An anomalously large rate for any of the above would be a strong indication for new physics.

The current limits on these rare processes in the charm sector are orders of magnitude away from the level where Standard Model effects are expected. Table 1 summarizes the results and the predictions. Besides the large window for discovery evident in the table, more importantly, the charm quark is the only quark with charge $+2/3$ that is both unstable and yet lives long enough to form bound state particles that can be observed. Thus the charm sector provides a unique window to discover new physics effects that may couple only to up-type quarks.

CP459, *Heavy Quarks at Fixed Target*
edited by Harry W. K. Cheung and Joel N. Butler
© 1999 The American Institute of Physics 1-56396-864-9/99/$15.00

The latest round of results from fixed target experiments on searches for FCNC, LNV, and LFNV have approached branching ratio sensitivities in the range of $10^{-6} - 10^{-5}$. New results on searches for direct CP violation in charm decays have also been reported by fixed target experiments. These typically set limits of a few percent on the CP decay-rate asymmetry for some Cabibbo suppressed D^0 and D^+ decay modes.

II EXPERIMENTAL ISSUES

At current fixed target energies, the cross-section for charm production is rather low. In photoproduction, charm is produced roughly only once in every 200 interactions. The rate is even less favorable in hadroproduction where charm is produced only about once in every thousand interactions. Furthermore, the branching fraction of any particular decay mode is small (only a few % for the Cabibbo favored modes) and charm has a very short lifetime. Typically, the observed hadrons have decay lengths of only a few mm in the laboratory. On top of that, the combinatorial background is large and hence, stringent selection criteria have to be adopted to pick out the final event sample. Of the charm particles which are produced, only about a half percent of them are reconstructed in the fixed target experiments. Given such odds against the experiments, it is remarkable that current Fermilab fixed target experiments have reported results based on of the order of 100K fully reconstructed charm events.

The success of these experiments came about due to technological innovation, in particular:

- semiconductor vertex detectors

- powerful parallel processing computing systems and peripherials leading to fast DAQ, collection of massive volumes of data which can then be analysed by large computing farms

Nevertheless, it still takes experimenters many years of hard work to fully exploit such advances in technology. Silicon microstrip detectors were first introduced by a CERN fixed target experiment in 1982 [16]. The same collaboration also introduced the use of pixel detectors (in the form of CCD's) in high energy physics [17]. While the performance of the vertex detector of that experiment was really superb, it did not have the fast DAQ or the offline computing resources that are also needed for this kind of experiment. The use of fast DAQ and massive parallel computing processors were pioneered by Fermilab fixed target charm experiments [18], [19]. This opened up the possibility of collecting massive datasets. To analyse such a huge volume of data is of course not trivial. Even with powerful offline computing farms, the sheer number of tapes that need to be handled at each step of the analysis chain and the book-keeping involved could still be mind-boggling.

Because of all these challenges, only in the recent 3-4 years have fixed target experiments started to play a major role in rare decay searches, with published

TABLE 1. Window to new physics in charm decay.

Topic	90% CL Limit	ref	Std. Model prediction	ref	Typical NS Models
Direct CP Violation					
$D^0 \to K_s^0 \phi$	-0.028 ± 0.094	[3]			
$D^0 \to K_s^0 \pi^0$	$-.018 \pm 0.030$	[3]			
$D^0 \to K^- K^+$	0.026 ± 0.035	[4]			
$D^0 \to \pi^+ \pi^-$	-0.05 ± 0.08	[5]			
$D^+ \to K^- K^+ \pi^+$	-0.017 ± 0.027	[4]			
$D^+ \to \overline{K}^{*0} K^+$	-0.02 ± 0.05	[4]	$(2.8 \pm 0.8) \times 10^{-3}$	[6]	
$D^+ \to \phi \pi^+$	-0.014 ± 0.033	[4]			
$D^+ \to \pi^+ \pi^+ \pi^-$	-0.017 ± 0.042	[7]			
$D^+ \to \eta \pi^+$			$(-1.5 \pm 0.4) \times 10^{-3}$	[6]	
$D^+ \to K_s^0 \pi^+$			3.3×10^{-3}	[8]	
FCNC					
$D^0 \to \mu^+ \mu^-$	4.1×10^{-6}	[4]	$< 3 \times 10^{-15}$	[9]	4th Gen.,
$D^0 \to \pi^0 \mu^+ \mu^-$	1.8×10^{-4}	[10]			Tree-level
$D^0 \to \overline{K}^0 e^+ e^-$	1.1×10^{-4}	[4]	$< 2 \times 10^{-15}$	[9]	FCNC
$D^0 \to \overline{K}^0 \mu^+ \mu^-$	2.6×10^{-4}	[10]	$< 2 \times 10^{-15}$	[9]	
$D^+ \to \pi^+ e^+ e^-$	6.6×10^{-5}	[11]	$< 10^{-8}$	[9]	
$D^+ \to \pi^+ \mu^+ \mu^-$	1.8×10^{-5}	[11]	$< 10^{-8}$	[9]	
$D^+ \to K^+ e^+ e^-$	2.0×10^{-4}	[12]	$< 10^{-15}$	[9]	
$D^+ \to K^+ \mu^+ \mu^-$	9.7×10^{-5}	[12]	$< 10^{-15}$	[9]	
$D \to X_u + \gamma$			$\sim 10^{-5}$	[9]	
$D^0 \to \rho^0 \gamma$			$(1-5) \times 10^{-6}$	[9]	
$D^0 \to \phi \gamma$			$(0.1 - 3.4) \times 10^{-5}$	[9]	
LF or LN Violation					
$D^0 \to \mu^\pm e^\mp$	1.9×10^{-5}	[4]	0		LQ
$D^+ \to \pi^+ \mu^\pm e^\mp$	1.1×10^{-4}	[12]	0		
$D^+ \to K^+ \mu^\pm e^\mp$	1.2×10^{-4}	[12]	0		
$D^+ \to \pi^- \mu^+ \mu^+$	8.7×10^{-5}	[12]	0		
$D^+ \to K^- \mu^+ \mu^+$	1.2×10^{-4}	[12]	0		
$D^+ \to \rho^- \mu^+ \mu^+$	5.6×10^{-4}	[10]	0		
Mixing					
$\overset{(-)}{D^0} \to K^\mp \pi^\pm$	$r < 0.0037$	[13]			LQ, SUSY,
	$\Delta M_D <$				4th Gen.,
	1.3×10^{-4} eV	[13]	10^{-7} eV	[14]	Higgs
$\overset{(-)}{D^0} \to K\ell\nu$	$r < 0.005$	[15]			

limits typically 1-2 orders of magnitude better than previous results from e^+e^- collider experiments.

The current best results came predominantly from Fermilab fixed target experiments, in particular E653(1987/88), E687(1990-1992) and E791(1991/1992) run. Other experiments such as CLEO, E771, and WA92 also have contributed significantly to the measurements.

III RARE DECAY

Flavor changing neutral curents are forbidden in the Standard Model at the tree level. Furthermore, higher order charged-current processes are also highly suppressed. Examples of flavor-changing neutral currents in the charm systems are decays such as $D^0 \to l^+l^-$, $D \to \pi l^+l^-, \rho l^+l^-$, and $D_s^+ \to K^+l^+l^-$. The current best limit on these decays came from $D^0 \to \mu^+\mu^-$ decays in which the 90% CL limits have been measured to be 4.0×10^{-6} [20,21]. However, because of helicity suppression, $D^0 \to l^+l^-$ decay is expected to have a very small rate. Potentially, the three-body decays $D \to \pi l^+l^-$ are more sensitive to new physics. The predicted rate for such decays is of the order of 10^{-8} or below. Because of this small rate, there is a large discovery window: seeing this decay at a branching fraction above 10^{-7} would imply non-Standard-Model tree-level FCNC process or contributions to higher order loop diagrams and provide strong evidence for new physics.

LFNV and LNV decays such as $D^+ \to \pi^+e^+\mu^-$ or $D^+ \to \pi^-e^+e^+$ are forbbiden in the Standard Model. However, unlike charge conservation, which is required by gauge invariance, there is no fundamental principle which requires conservation of lepton number. Therefore, it is not unreasonable to assume that lepton number conservation will be violated at some energy scale which can be probed by searching for such decays.

Fig. 1a shows the Feynman diagram contributing to the forbidden decay $D^+ \to \pi^+\mu^+e^-$. The mediating particle, X^0, is referred to as the 'horizontal-gauge' boson since it couples quark to quark and lepton to lepton but horizontally across generations. The analogous SU(2)xU(1) process $D^+ \to \overline{K}^0\mu^+\nu$ is shown in Fig. 1b.

The ratio of partial widths for these two decays is given by:

$$\frac{\Gamma(D^+ \to \pi^+\mu^+e^-)}{\Gamma(D^+ \to \overline{K}^0\mu^+\nu)} = \frac{\frac{G_F^2|M_{X^0}|^2}{m_{X^0}^4}}{\frac{G_F^2|M_W|^2}{m_W^4}} = \left[\frac{m_W}{m_{X^0}}\right]^4 \times \frac{|M_{X^0}|^2}{|M_W|^2} \qquad (1)$$

where h represents a neutral or charged hadron, M represents matrix elements for the processes of interest, and m represents masses of the exchanged particles in those processes. ¿From this equation, one can then deduce from the measured branching ratio of the LFNV decay the mass scale being probed. Note that the

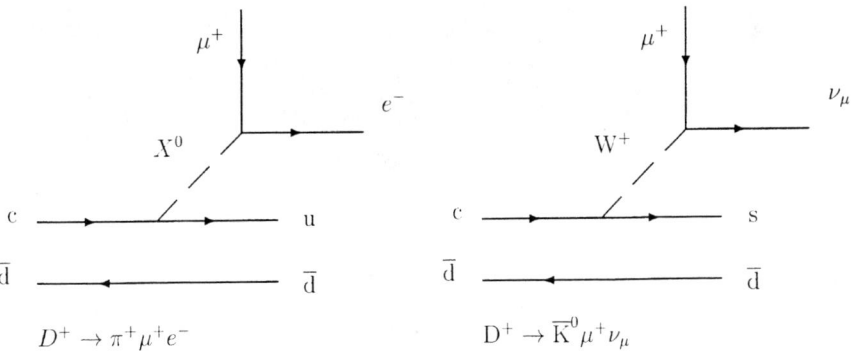

FIGURE 1. a) Tree level diagram contributing to the lepton number violating $D^+ \to \pi^+\mu^+e^-$ decay. b) The diagram for the allowed $D^+ \to \overline{K}^0\mu^+\nu$ decay.

sensitivity to higher-mass states increases only as the inverse fourth root of the measured branching ratio.

Considering the similarity of the tree-level diagrams, we can make comparisons for strange, charm, and beauty decays via exchanged neutral vector particles and the results are summarized in Table 2.

Table 3 and 4 summarize the current best limits on rare decays of D^0 and D^+ mesons.

TABLE 2. Mass scales probed in searches for lepton number violation.

Decay Mode	90% CL Limit	Implied Mass Limit
$K^+ \to \pi^+\mu^+e^-$	2×10^{-10}	$M_{X^0} > 9\ TeV/c^2$
$K^+ \to \pi^+\mu^-e^+$	7×10^{-9}	$M_{X^0} > 4\ TeV/c^2$
$D^+ \to \pi^+\mu^\pm e^\mp$	1×10^{-4}	$M_{X^0} > 410\ GeV/c^2$
$D^+ \to K^+\mu^\pm e^\mp$	1×10^{-4}	$M_{X^0} > 220\ GeV/c^2$
$B^+ \to \pi^+\mu^\pm e^\mp$	6×10^{-3}	Need $\Gamma(B \to h\mu\nu)$
$B^+ \to K^+\mu^\pm e^\mp$	6×10^{-3}	Need $\Gamma(B \to h\mu\nu)$

TABLE 3. 90% CL upper limits on the branching fractions for FCNC, lepton family number violating, and lepton number violating decays of the D^0 meson.

Decay Mode	E653	E771	WA92	CLEO
e^+e^-				1.3×10^{-5}
$\mu^+\mu^-$		0.42×10^{-5}	0.41×10^{-5}	
$\rho^0 e^+e^-$				10×10^{-5}
$\rho^0 \mu^+\mu^-$	23×10^{-5}			
ϕe^+e^-				5.2×10^{-5}
$\phi \mu^+\mu^-$				41×10^{-5}
$\pi^0 \mu^+\mu^-$	18×10^{-5}			
$K^0 e^+e^-$				11×10^{-5}
$K^0 \mu^+\mu^-$	26×10^{-5}			
$\mu^\pm e^\mp$				1.9×10^{-5}
$\pi^0 e^\pm \mu^\mp$				8.6×10^{-5}
$\phi e^\pm \mu^\mp$				3.4×10^{-5}
$\rho^0 e^\pm \mu^\mp$				4.9×10^{-5}

TABLE 4. 90% CL upper limits on the branching fractions for sixteen FCNC, lepton family number violating, and lepton number violating decays of the D^+ meson.

Decay Mode	E653	E687	E791
$\pi^+ e^+ e^-$		11×10^{-5}	6.6×10^{-5}
$K^+ e^+ e^-$		20×10^{-5}	
$\pi^+ \mu^+ \mu^-$	22×10^{-5}	8.9×10^{-5}	1.8×10^{-5}
$K^+ \mu^+ \mu^-$	33×10^{-5}	9.7×10^{-5}	
$\rho^+ \mu^+ \mu^-$	58×10^{-5}		
$\pi^+ \mu^+ e^-$		13×10^{-5}	
$K^+ \mu^+ e^-$		12×10^{-5}	
$\pi^+ \mu^- e^+$		11×10^{-5}	
$K^+ \mu^- e^+$		13×10^{-5}	
$\pi^- e^+ e^+$		11×10^{-5}	
$K^- e^+ e^+$		12×10^{-5}	
$\pi^- \mu^+ \mu^+$	20×10^{-5}	8.7×10^{-5}	
$K^- \mu^+ \mu^+$	33×10^{-5}	12×10^{-5}	
$\rho^- \mu^+ \mu^+$	60×10^{-5}		
$\pi^- \mu^+ e^+$	11×10^{-5}		
$K^- \mu^+ e^+$	13×10^{-5}		

IV CP VIOLATION

CP violation has only been observed in neutral kaon decay. In the Standard Model, CP violation requires the interference of two comparable (in magnitude) amplitudes and that the two have a large phase difference between them. In this model, strange quarks couple to the top quark in diagrams with internal loops, leading to observable CP violation. The comparable diagrams in the charm sector have bottom quarks in the internal loops, resulting in very small Standard Model contributions to CP violation. There is no significant CP violation in Cabibbo favored modes from the Standard Model. Moreover, prediction for CP violation in Cabibbo suppressed modes is $< 10^{-3}$ in the Standard Model. Thus the charm sector may be uniquely sensitive to physics outside the Standard Model [8]. New physics contributions can make CP violation in charm larger than the Standard Model prediction. Thus for example, one of the amplitudes could be for a Standard Model process. However, the second may be due to a process from new physics. The appearance of SUSY particles or extra Higgs particles in virtual loops leads to beyond-the-Standard-Model amplitudes.

The signal for CP violation is an absolute rate difference between decays of particle and antiparticle to charge-conjugate final states f and $\overline{f}$.

$$A_{CP} = \frac{\Gamma(D \to f) - \Gamma(\overline{D} \to \overline{f})}{\Gamma(D \to f) + \Gamma(\overline{D} \to \overline{f})}. \qquad (2)$$

At fixed target machines, there is a production asymmetry between D and $\overline{D}$ mesons and that must be corrected for. This is done by normalizing the observed rates in the single-Cabibbo-suppressed modes to those in the dominant Cabibbo-favored modes. This has the additional benefit of cancelling most of the corrections due to acceptances and efficiencies, thereby reducing systematic uncertainties. Experimentally, the measured quantity is given by:

$$\frac{\frac{N(D^0 \to f)}{N(D^0 \to K^-\pi^+)} - \frac{N(\overline{D}^0 \to \overline{f})}{N(\overline{D}^0 \to K^+\pi^-)}}{\frac{N(D^0 \to f)}{N(D^0 \to K^-\pi^+)} + \frac{N(\overline{D}^0 \to \overline{f})}{N(\overline{D}^0 \to K^+\pi^-)}} \qquad (3)$$

where, for each channel, $N = \epsilon n$ is the observed number of events, n is the produced number of events, and ϵ is the efficiency of the detector and analysis procedure. In this example, we use the Cabibbo favored $D^0 \to K^-\pi^+$ as the normalizing mode. Note that any CP asymmetry from interference between mixing and tree-level diagrams will not cancel through the $D^0 \to K^-\pi^+$ normalization. Thus the normalized A_{CP} is not a direct CP asymmetry parameter, but rather a measure of combined direct and indirect CP asymmetries. An implicit assumption in the analysis is that there is no measurable CP violation in the Cabibbo-favored decays of the D^0.

Exponentially-increasing charm event samples have led to substantially improved CP violation limits over time. The most sensitive limits come from Fermilab fixed-target experiments E791 [5,7], and E687 [22], and from CLEO [3]. No significant CP violation has been observed, and most limits are in the range of several percent. Table 5 summarizes the current CP violation limits in D meson decays. There is still a large window open for the discovery of non-Standard-Model effects.

TABLE 5. Summary of current CP asymmetry measurements for D mesons.

Decay mode	E687	E791	CLEO
$D^0 \to K^+K^-$	0.024 ± 0.084	$-0.010 \pm 0.049 \pm 0.012$	0.069 ± 0.059
$D^0 \to \pi^+\pi^-$		$-0.049 \pm 0.078 \pm 0.030$	
$D^0 \to K_s^0 \phi$			-0.007 ± 0.090
$D^0 \to K_s^0 \pi^0$			-0.013 ± 0.030
$D^+ \to K^+K^-\pi^+$	-0.031 ± 0.068	-0.014 ± 0.029	
$D^+ \to \phi\pi^+$	0.066 ± 0.086	-0.028 ± 0.036	
$D^+ \to \overline{K}^{*0}K^+$	-0.12 ± 0.13	-0.010 ± 0.050	
$D^+ \to \pi^+\pi^+\pi^-$		-0.017 ± 0.042	

V FUTURE PROSPECTS

E791 will have new results on rare and forbidden decays soon. These will include results on FCNC, LNV, and LFNV decays of D^0, D^+ and D_s, and will improve the currently published limits by factor of 2-3. Looking farther ahead, there are experiments such as FOCUS(E831) and CLEO II which are in the process of data reconstruction and analysis. Preliminary results from FOCUS [23] show that they will reach their stated goal of 1 million fully reconstructed charm events. Because of the improvements in the muon system, the silicon tracker and target setup, the improvement in rare and forbidden decay searches are expected to be substantial over the previous E687 results. They have also observed a large sample of decays with K_s^0 in the final states, which can be used to search for direct CP violation [24]. CLEO II [25] is still taking data. While their projected yield of charm events will not be as high as FOCUS, the cleanliness of the e^+e^- environment will certainly help. New results on CP and mixing from $D^0 \to K^+K^-, \pi^+\pi^-, \rho^0\pi^0, K_s^0X$ are expected.

There are new experiments such as CLEO III, B-factories, and HERA-B which will be commissioned next year. While these experiments are not proposed to do charm physics, the charm yield should be substantial. There is a new charm experiment (COMPASS) [26] now being prepared to run at the CERN SPS around the year 2000. In their EOI, they claimed that their expected yield is 10K $D^0 \to K^-\pi^+$ per day of data-taking. These experiments are expected to surpass current

TABLE 6. Current and approved future experiments with charm CP violation sensitivity.

Experiment	All charm decays		$\overline{D}^0 \to K^\mp \pi^\pm$		$\sigma(A_{CP})$
	prod.	rec.	prod.	rec.	(SCS)
E687		0.8×10^5			≈ 0.1
E791	10^8	2.5×10^5	1.2×10^6	3.7×10^4	≈ 0.05
CLEO II	2.7×10^6		1.0×10^5	1.8×10^4	≈ 0.05
FOCUS (E831)		10^6			$\approx 0.03?$
COMPASS				7×10^4	$\approx 0.03?$
HERA-B	few $\times 10^{10}$/yr				?
B Factories, CLEO III	3×10^7/yr				$\approx 0.01?$

sensitivities. As an example, the sensitivities for CP searches in D meson decays from current and future experiments is summarized in Table 6 [27].

The B Factories and CLEO III are expected to have the best charm CP reach among approved future experiments. (While HERA-B has the highest charm production rate, it is likely to have poor charm trigger efficiency due to p_t requirements imposed at the trigger level.) In multi-year runs, the combined reach of these experiments could be approximately an order of magnitude better than current limits, reaching the few $\times 10^{-3}$ level.

We have seen from Table 6 that the projected sensitivity for the approved experiments is unlikely to be sufficient to observe Standard Model CP violation in charm, though it may suffice for the discovery of non-SM effects if they are large. The proposed BTeV experiment at the Tevatron C0 interaction area [28] should be able to achieve $\sim 10^{-4}$ sensitivity, bringing even SM effects within reach.

VI CONCLUSION

Rare and forbidden charm decays and CP in charm offer a large "discovery" window for new physics. No CP violation in charm has been observed as yet, with the current sensitivity at a few %. Similarly, current limits on rare and forbidden decays of the D meson are at the 10^{-5} level. New experimental results are expected within the next year or two. There is still room for new physics.

REFERENCES

1. J. A. Appel, "From Symmetry Violation to Dynamics: the Charm Window", Fermilab-Conf-97/406, to appear in Proc. of Int. Conf. on Physics since Parity Symmetry Breaking - Conference in Memory of Prof. C.S. Wu, Nanjing, China, Aug. 16-18, 1997.
2. A. J. Schwartz, Mod. Phys. Lett. A, Vol. 8, No. 11 (1993) 967.
3. CLEO Collaboration, J. Bartelt *et al.*, Phys. Rev. D **52**, (1995) 4860.

4. Particle Data Group, Eur. Phys. J. **C3** (1998) 497.
5. E791 Collaboration, E.M. Aitala et al., Phys. Lett. **B403** (1997) 377.
6. F. Buccella et al., Phys. Lett. **B302**, 319 (1993);
   A. Pugliese and P. Santorelli, "Two Body Decays of D Mesons and CP Violating Asymmetries in Charged D Meson Decays," Proc. Third Workshop on the Tau/Charm Factory, Marbella, Spain, 1–6 June 1993, Edition Frontieres (1994), p. 387.
7. E791 Collaboration, E.M. Aitala et al., Phys. Rev. Lett. **76**, 364 (1996).
8. Z. Xing, Phys. Lett. B **353**, (1995) 313.
9. J. L. Hewett, "Searching for New Physics with Charm," SLAC-PUB-95-6821, hep-ph/9505246, to appear in Proc. LISHEP95 Workshop, Rio de Janeiro, Brazil, Feb. 20–22, 1995.
10. E653 Collaboration, K. Kodama et al., Phys. Lett. B **345**, 85 (1995).
11. E791 Collaboration, E.M. Aitala et al., Phys. Lett. **B421** (1998) 405.
12. E687 Collaboration, P.L. Frabetti et al., Phys. Rev. D **50**, R2953 (1994).
13. E691 Collaboration, J. C. Anjos et al., Phys. Rev. Lett. **60**, 1239 (1988);
14. G. Burdman, "Charm Mixing and CP Violation in the Standard Model," in **The Future of High-Sensitivity Charm Experiments: Proceedings of the CHARM2000 Workshop**, D. M. Kaplan and S. Kwan, ed.s, Fermilab, Batavia, Illinois, June 7-9, 1994, FERMILAB-Conf-94/190, p. 75; and "Potential for Discoveries in Charm Meson Physics," in **Workshop on the Tau/Charm Factory**, J. Repond, ed., Argonne National Laboratory, June 21–23, 1995, AIP Conference Proceedings No. 349 (1996), p. 409.
15. E791 Collaboration, E.M. Aitala et al., Phys. Rev. Lett. **77**, 2384 (1996).
16. B. Hyams et al., Nucl. Inst. and Meth., **205**, (1983) 99, J. Kemmer, Nucl. Inst. and Meth., **169**, (1980) 499.
17. S. Barlag et al., Phy. Lett. B 184 (1987), 283.
18. J. Biel et al., Proc. Int. Conf. Computing in High Energy Physics, Asilomar, Feb. 2-6, 1987, Computer Physics Communications, **45**, 331 (1987).
19. C. Stoughton and D.J. Summers, Computers in Physics, **6**, 371 (1992); F. Rinaldo and S. Wolbers, Computers in Physics, **7**, 184 (1993); S. Bracker et al., IEEE Trans. on Nucl. Sci., **43**, 2457 (1996).
20. E771 Collaboration, T. Alexopoulos et al., Phys. Rev. Lett. **77**, 2380 (1996).
21. WA92 Collaboration, M. Adamovich et al., Phys. Lett. B **408**, 469 (1997).
22. E687 Collaboration, P.L. Frabetti et al., Phys. Lett. B **398**, 239 (1997).
23. J. Link and E. Vaandering, these proceedings.
24. H. Lipkin, "Search for New Physics in $D^\pm \to K_s^0 X^\pm$ and $D^\pm \to K_s^0 K_s^0 K^{\pm}$", presented at the 4th Workshop on Heavy Quarks at Fixed Target, Fermilab, Batavia, Illinois, Oct. 9-12, 1998.
25. K. Ecklund, these proceedings.
26. A. Bravar, these proceedings.
27. D. Kaplan and V. Papavassiliou, "Experimental Prospects for CP violation in charm", hep-ph/9809399, to appear in Proc. of Workshop on CP Violation, Adelaide, Australia, July 3-8, 1998.
28. M. Artuso, these proceedings.

Rare B decays at CLEO

Peter B. Gaidarev,
CLEO Collaboration

Cornell University, Ithaca, NY 14853

Abstract. Studies of rare B decays provide an important tool for testing the Standard Model and searching for new physics. We present recent CLEO results including the improved measurement of the inclusive $b \to s\gamma$ process, as well as new measurements of charmless hadronic B decays into the $\eta' K$, $K\pi$ and $\pi\pi$ final states.

INTRODUCTION

Studies of rare B meson decays play an important role in improving our understanding of the Standard Model. Being strongly suppressed or strictly forbidden at the tree level, they provide an opportunity for new physics contributions to compete with Standard Model processes. Since these decays are usually dominated by loop diagrams, heavy particles from various extensions of the Standard Model can contribute at the significant level. They can either affect the overall decay rate leading to deviations from the Standard Model predictions, or introduce new phases and create a rich pattern of CP violating asymmetries.

The inclusive electromagnetic penguin decay $b \to s\gamma$, first reported by CLEO in 1995 [1], serves as a prime example of the former case, stimulating a lot of theoretical activity and imposing strong constraints on physics scenarios beyond the Standard Model [2]. The recent CLEO observation of $B^+ \to \eta' K^+$ [3] with unexpectedly large branching fraction generated a lot of theoretical discussions, which are still not settled. At the same time, additional gluonic penguin decays $B^0 \to K^+\pi^-$ and $B^+ \to K^0\pi^+$ were observed [4], providing further evidence for the dominance of gluonic penguin $b \to sg$ amplitude in charmless hadronic B decays.

In this paper we present results of improved measurements for these decays.

CLEO DETECTOR

The data were taken with the CLEO II and CLEO II.V detectors at the Cornell Electron Storage Ring (CESR). CLEO II and CLEO II.V are general purpose solenoidal magnet detectors, described in detail elsewhere [5]. In CLEO II, the momenta of charged particles are measured in a tracking system consisting of a 6-layer straw tube chamber, a 10-layer precision drift chamber, and a 51-layer main drift chamber, all operating inside a 1.5 T superconducting solenoid. The main drift chamber also provides a measurement of the specific ionization loss, dE/dx, used for particle identification. For CLEO II.V the 6-layer straw tube chamber was replaced by a 3-layer double-sided silicon vertex detector, and the gas in the main drift chamber was changed from an argon-ethane to a helium-propane mixture. The momentum resolution and $K - \pi$ separation from dE/dx at 2.6 GeV/c were both improved by $\sim$ 15%. Photons are detected using 7800-crystal CsI(Tl) electromagnetic calorimeter covering 98% of the solid angle. The energy resolution for photons near 2.5 GeV in the central angular region ($|\cos\theta_\gamma| < 0.7$) is 2%. Muons are identified using proportional counters placed at various depths in the steel return yoke of the magnet.

CLEO II data consist of 3.13 fb^{-1} taken on the $\Upsilon(4S)$ resonance and 1.60 fb^{-1} taken slightly below the resonance. The addition of CLEO II.V data brings the total to 5.6 and 2.9 fb^{-1}. The on-resonance sample contains 3.3 million $B\bar{B}$ events for CLEO II and 5.8 million for the combined CLEO II and CLEO II.V data. CLEO II data was recently reprocessed with improved calibration and new track-fitting algorithm.

INCLUSIVE $b \to s\gamma$

CLEO has previously observed inclusive radiative penguin decay $b \to s\gamma$, with the branching fraction of $(2.32 \pm 0.57 \pm 0.35) \times 10^{-4}$ [1]. Since our 1995 measurement, the theoretical understanding of the Standard Model rate greatly improved with full next-to-leading-log calculations [6] predicting a branching fraction of $(3.28 \pm 0.33) \times 10^{-4}$. Here we report an improved measurement of the inclusive $b \to s\gamma$ branching fraction [7], using 60% additional data and improved analysis techniques.

Our signature for $b \to s\gamma$ is a photon from B-meson decay with energy between 2.1 and 2.7 GeV (as compared with 2.2 – 2.7 GeV in the earlier analysis). Spectator model calculations [8] indicate that 85-94% of the signal lies in this range. Dominant backgrounds come from the continuum, both from the initial-state-radiation (ISR) process $e^+e^- \to q\bar{q}\gamma$, and from the reaction $e^+e^- \to q\bar{q}$. We suppress the continuum with the same two approaches used in the previous analyses, but combine them in a more effective way. Remaining continuum background is subtracted using off-resonance data.

We select general hadronic events containing a high energy calorimeter cluster with $|\cos\theta_\gamma| < 0.7$ and unmatched to a charged particle. We veto those high energy clusters which can form a π^0 or η when paired with another γ in the event. We also require that the lateral energy distribution of the cluster be consistent with that of a single isolated photon, suppressing random overlaps and merged π^0's.

The first approach for suppressing continuum background uses eight event-shape variables [9] which serve as inputs to the neural net. The output of the neural net is a single variable r which tends towards $+1$ for $b \to s\gamma$ and tends towards -1 for ISR and $q\bar{q}$.

In the second approach for suppressing the continuum, we search each event for combinations of particles that reconstruct to a $B \to X_s\gamma$ decay. We calculate the momentum P, the energy E, and the beam-constrained mass $M = \sqrt{E_{\text{beam}}^2 - P^2}$ of the combination, and also $\cos\theta_T$, where θ_T is the angle between the thrust axis of the candidate B and the thrust axis of the rest of the event. We discriminate between signal and background using $|\cos\theta_T|$ and χ_B^2, where

$$\chi_B^2 = \left(\frac{M - 5.28}{\sigma_M}\right)^2 + \left(\frac{E - E_{\text{beam}}}{\sigma_E}\right)^2. \tag{1}$$

For events containing a plausible reconstructed $B \to X_s\gamma$ (taken as $\chi_B^2 < 20$), we have three quantities which help discriminate between $b \to s\gamma$ signal and continuum background, namely r, $|\cos\theta_T|$ and χ_B^2. We combine those 3 variables into a single variable r_c, again using a neural network. For events containing no plausible reconstructed $B \to X_s\gamma$ ($\chi_B^2 > 20$, or no candidate at all), we use r alone to discriminate between signal and background.

We weight each event according to its value of r_c (or r), with a weight given by

$$w_i = s_i/[s_i + (1+\alpha)b_i],$$

where s_i is the expected signal yield in the r_c (or r) bin i, b_i is the expected continuum background yield in that bin, and α is the luminosity scale factor between on-resonance and off-resonance data samples. For the case where backgrounds other than from the continuum are negligible (a reasonable approximation in this case), this weighting gives the smallest statistical error on the background-subtracted yield.

Backgrounds from $b \to cW^-$ and $b \to uW^-$ in first approximation are taken from MC. We then correct for any difference between the π^0 momentum spectra from data and MC, and similarly for the η momentum spectra. The weighted yields are plotted against photon energy in Figure 1. The spectrum expected for $b \to s\gamma$, as given by a spectator model calculation, is also shown. The measured spectrum is perhaps somewhat softer than the spectator model expectations.

Because we are weighting our candidate events, we need a *weighted efficiency*. That is, for a Monte Carlo signal sample, we sum the weights of the events passing all cuts, and divide by the number of events in the sample. We denote this weighted efficiency ϵ_w. To model the decay $b \to s\gamma$, i.e., $B \to X_s\gamma$, we use the spectator model of Ali and Greub [8], which includes gluon bremsstrahlung and higher-order radiative effects.

We find

$$\epsilon_w = (4.43 \pm 0.29 \pm 0.22 \pm 0.03 \pm 0.31) \times 10^{-2},$$

where the first uncertainty is due to spectator model inputs, the second is due to recoil system hadronization, the third is due to the uncertainty in the production ratio of B^+B^- to $B^0\bar{B}^0$, and the last to detector modeling. Combining the

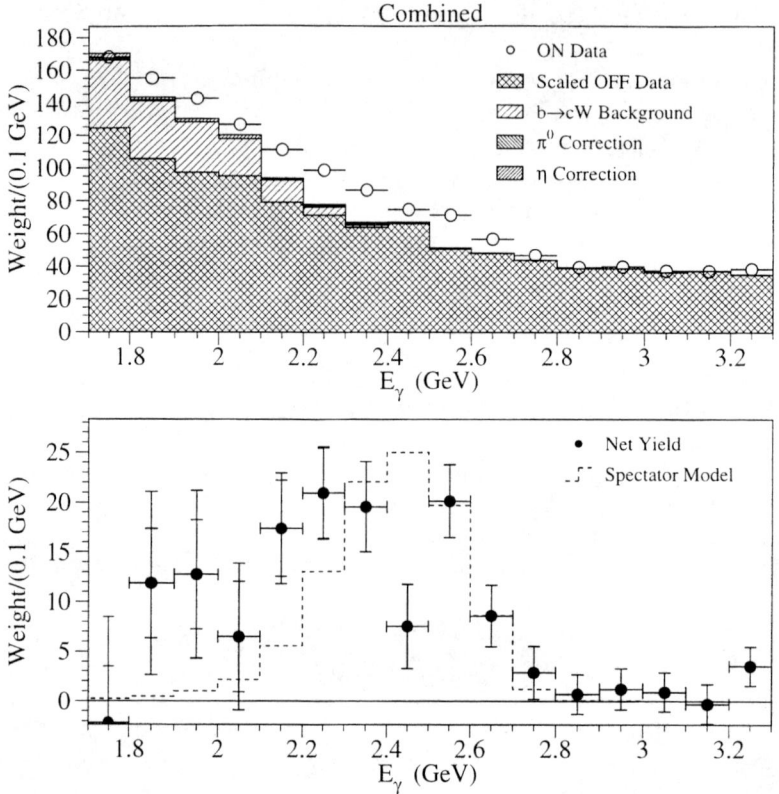

FIGURE 1. Weighted yields in $b \to s\gamma$ analysis. Top plot shows all backgrounds. Bottom plot shows weighted yields with backgrounds subtracted.

background-subtracted weighted yield with the weighted efficiency, we obtain a branching fraction

$$\mathcal{B}(b \to s\gamma) = (3.15 \pm 0.35 \pm 0.32 \pm 0.26) \times 10^{-4},$$

where the errors are statistical, systematic, and for model dependence, respectively.

This branching fraction is in good agreement with theoretical calculations based on the Standard Model and our previous measurement. The large yield in the 2.1–2.2 GeV region is the main reason for the increased central value of this measurement relative to the previous one. The statistical error is about 60% of that quoted previously.

CHARMLESS HADRONIC B DECAYS

The first observation of the decay $B^+ \to \eta' K^+$ (charge conjugate modes are always implied) with surprisingly large branching fraction of 6.5×10^{-5} was recently published by CLEO [3]. This was more than a factor of three larger than previous theoretical expectations [10] and four times larger than the branching fraction of the other measured gluonic penguin decay, $B^0 \to K^+ \pi^-$. The latter, with a branching fraction of 1.5×10^{-5}, when combined with an observation of $B^+ \to K^0 \pi^+$ at $\sim 2.3 \times 10^{-5}$ led theorists to suggest that nontrivial limit on angle γ of the unitarity triangle can be obtained from the study of rates of various $B \to K\pi$ decays [11]. The decay $B^0 \to \pi^+\pi^-$ can be used to measure angle α of the unitarity triangle, although measurements of other $B \to \pi\pi$ decays are required to extract α in a clean way [12]. No convincing evidence for any kind of $B \to \pi\pi$ decays has been seen yet.

In this paper we describe updated results for these modes, with 80% more data added.

Candidate selection

Charged tracks are required to pass track quality cuts based on the average hit residual and the impact parameters in both the $r - \phi$ and $r - z$ planes. Candidate K_S^0 are selected from pairs of tracks forming well measured displaced vertices. Isolated showers with energies greater than 40 MeV in the central region of the CsI calorimeter and greater than 50 MeV elsewhere, are defined to be photons. Pairs of photons with an invariant mass within 25 MeV ($\sim 2.5\sigma$) of the nominal π^0 mass are kinematically fitted with the mass constrained to the π^0 mass. To reduce combinatoric backgrounds we require the lateral shapes of the showers to be consistent with those from photons. To suppress further low energy showers from charged particle interactions in the calorimeter we apply a shower energy dependent isolation cut.

Charged particles are identified as kaons or pions using dE/dx. Electrons are rejected based on dE/dx and the ratio of the track momentum to the associated shower energy in the CsI calorimeter. We reject muons by requiring that the tracks do not penetrate the steel absorber to a depth greater than seven nuclear interaction lengths. We have studied the dE/dx separation between kaons and pions for momenta $p \sim 2.6$ GeV/c in data using D^{*+}-tagged $D^0 \to K^-\pi^+$ decays; we find a separation of $(1.7 \pm 0.1)\,\sigma$ for CLEO II and $(2.0 \pm 0.1)\,\sigma$ for CLEO II.V.

For each candidate we calculate a beam-constrained B mass M and define $\Delta E = E_1 + E_2 - E_{\text{beam}}$, where E_1 and E_2 are the energies of the daughters of the B meson candidate.

Background suppression

The dominant background for these modes comes from continuum $q\bar{q}$ production. Contributions from generic B decays and other rare decays are negligible.

The cut on $|\cos\theta_T|$ is used to suppress background from continuum $q\bar{q}$ events. Because the B mesons produced in $\Upsilon(4S)$ decays are almost at rest, the decay products of the $B\bar{B}$ pair tend to be spherically distributed. Particles from $q\bar{q}$ production tend to have a more jet-like distribution. In $|\cos\theta_T|$ variable, continuum background is strongly peaked near 1.0, while signal is approximately flat.

Additional discrimination between signal and $q\bar{q}$ background is provided by a Fisher discriminant technique as described in detail in Ref. [13]. The Fisher discriminant is a linear combination of N input variables y_i

$$\mathcal{F} \equiv \sum_{i=1}^{N} \alpha_i y_i,$$

where the coefficients α_i are chosen to maximize the separation between the signal and background Monte-Carlo samples. The 11 inputs used in $B \to K\pi, \pi\pi$ analysis are $|\cos\theta_{cand}|$ (the cosine of the angle between the candidate sphericity axis and beam axis), the ratio of Fox-Wolfram moments H_2/H_0 [14], and nine variables that measure the scalar sum of the momenta of tracks and showers from the rest of the event in nine angular bins, each of 10°, centered about the candidate's sphericity axis.

Maximum likelihood fit

We perform unbinned maximum-likelihood (ML) fits using variables like ΔE, M, $\mathcal{F}$, dE/dx, and resonance masses (where applicable) as input information for each candidate event to determine the signal yields. The likelihood of the event is parameterized by the sum of probabilities for all relevant signal and background hypotheses, with relative weights determined by maximizing the likelihood function ($\mathcal{L}$). The probability of a particular hypothesis is calculated as a product of the probability density functions (PDFs) for each of the input variables.

The parameters for the PDFs are determined from independent data and high-statistics Monte-Carlo samples. We estimate a systematic error on the fitted yield by varying the PDFs used in the fit within their uncertainties. These uncertainties are dominated by the limited statistics in the independent data samples we used to determine the PDFs.

The statistical significance of the yield is defined as

$$\mathcal{S} = \sqrt{-2ln(\mathcal{L}_0/\mathcal{L}_{\max})}, \qquad (2)$$

where $\mathcal{L}_{\max}$ denotes the value of the likelihood function at the point of maximum likelihood, and $\mathcal{L}_0$ denotes the value of the likelihood function at zero yield. The

90% confidence level upper limit on the yield is determined by finding the point at which the integral of the likelihood curve is at 90% of its total.

To illustrate results of the fits we make M (ΔE) projections for events in a signal region typically defined by $|\Delta E| < 2\sigma_{\Delta E}$ ($|M - 5.28| < 2\sigma_M$). We also make additional cut on $\mathcal{F}$ to further reduce background. Overlaid on these plots are the projections of the PDFs used in the fit, normalized according to the fit results multiplied by the efficiency of the additional cuts.

$$B \to \eta' K$$

We reconstruct η' candidates in two decay channels, $\eta' \to \pi^+ \pi^- \eta$ with $\eta \to \gamma\gamma$, and $\eta' \to \rho^0 \gamma$ with $\rho^0 \to \pi^+ \pi^-$.

Unbinned maximum likelihood fits are performed to obtain signal yields for the decay modes $B^+ \to \eta' K^+$, $B^+ \to \eta' \pi^+$, and $B^0 \to \eta' K^0$. For the modes $B^+ \to \eta' K^+$ and $B^+ \to \eta' \pi^+$ we perform a simultaneous fit. In addition to dE/dx information, we improve our discrimination between $\eta' K^+$ and $\eta' \pi^+$ signal by always assuming that the charged track is a kaon. In this case ΔE distribution for $\eta' \pi^+$ signal is shifted by +44 MeV. Before performing the likelihood fit, a number of preliminary cuts is applied. We require $|\Delta E| < 0.2$ GeV, $5.2 < M < 5.3$ GeV/c^2, $0.93 < M_{\eta'} < 0.99$ GeV/c^2, and, as required, $0.5 < M_\eta < 0.6$ GeV/c^2, $0.47 < M_{K_S} < 0.53$ GeV/c^2, and $0.4 < M_{\rho^0} < 0.95$ GeV/c^2. In addition, $|\cos\theta_T| < 0.9$ cut is imposed.

Table 1 gives the results of the maximum likelihood analyses [15]. Shown in the table are the the total reconstruction efficiency(ϵ) (including relevant branching fractions) with its systematic error, signal yields(N_{sig}), and the calculated branching fraction($\mathcal{B}$). Results for two η' decay modes are combined by adding the $-2ln(\mathcal{L}/\mathcal{L}_{\max})$ distributions. The first error on the quoted branching fractions is statistical and the second systematic.

TABLE 1. Maximum likelihood fit results for $B^+ \to \eta' K^+$ and $B^0 \to \eta' K^0$.

Mode	Efficiency ϵ(%)	Signal yield N_{sig}	$\mathcal{B}(\times 10^{-5})$
$\eta' K^+$	-	-	$7.4^{+0.8}_{-1.3} \pm 1.0$
($\eta' \to \pi^+\pi^-\eta$)	5.0 ± 0.6	18.4	$6.3^{+1.7}_{-1.5} \pm 0.8$
($\eta' \to \rho^0\gamma$)	10.8 ± 1.4	50.2	$7.8^{+1.5}_{-1.4} \pm 1.0$
$\eta' K^0$	-	-	$5.9^{+1.8}_{-1.6} \pm 0.9$
($\eta' \to \pi^+\pi^-\eta$)	1.6 ± 0.2	5.4	$5.4^{+2.8}_{-2.1} \pm 0.9$
($\eta' \to \rho^0\gamma$)	3.4 ± 0.5	12.7	$6.3^{+2.5}_{-2.1} \pm 0.8$
$\eta' \pi^+$	-	-	< 1.2
($\eta' \to \pi^+\pi^-\eta$)	5.0 ± 0.6	1.0	< 1.8
($\eta' \to \rho^0\gamma$)	10.7 ± 1.2	0.0	< 1.5

The systematic error includes a contribution from the efficiency and a contribution from changes in fit yield that depend on variations in the parameters of the probability distribution functions. To take systematic error on the efficiency into account, we reduce the efficiency by 12% (one standard deviation) before calculating the upper limit.

$$B \to K\pi, \pi\pi, KK$$

We accept events with $|\cos\theta_T| < 0.8$, M within $5.2 - 5.3$ GeV/c^2 and $|\Delta E| < 200(300)$ MeV for decay modes without (with) a π^0 in the final state. When calculating ΔE we always assume a pion mass hypothesis for charged tracks, thus shifting the ΔE peak position by -42 MeV for $B \to K\pi$ signal events. Three different ML fits are performed, one for each topology (h^+h^-, $h^+\pi^0$, and $h^+K_S^0$, h^+ referring to a charged kaon or pion). Results of the ML fits are summarized in Table 2 [16]. We find statistically significant signals for the decays $B^0 \to K^+\pi^-$, $B^+ \to K^+\pi^0$, and $B^+ \to K^0\pi^+$. For the $B^+ \to K^+\pi^0$ decay, which is the first observation, we also show M and ΔE projection plots in Figure 2. Only upper limits could be established for the $B \to \pi\pi, KK$ decays.

TABLE 2. Experimental results and theoretical model predictions [17] for $B \to K\pi, \pi\pi, KK$ decays.

Mode	Efficiency $\mathcal{E}(\%)$	Branching $\mathcal{B}(10^{-5})$	Theory $\mathcal{B}$
$\pi^+\pi^-$	53 ± 5	< 0.84	0.8–2.6
$\pi^+\pi^0$	42 ± 4	< 1.6	0.4–2.0
$K^+\pi^-$	53 ± 5	$1.4 \pm 0.3 \pm 0.2$	0.7–2.4
$K^+\pi^0$	42 ± 4	$1.5 \pm 0.4 \pm 0.3$	0.3–1.3
$K^0\pi^+$	15 ± 2	$1.4 \pm 0.5 \pm 0.2$	0.8–1.5
K^+K^-	53 ± 5	< 0.24	–
$K^+\bar{K}^0$	15 ± 2	< 0.93	0.07–0.13

CONCLUSIONS

In summary, the inclusive branching ratio for $b \to s\gamma$ has been determined using 60% additional data and improved analysis techniques relative to our previous measurement. Our measurement is in good agreement with Standard Model predictions, and places constraints on other models.

We confirmed previously observed decays $B^+ \to \eta'K^+$ and $B^0 \to \eta'K^0$ with branching fractions larger than most theoretical predictions. With more data added, we presented improved measurements of the $B^0 \to K^+\pi^-$ and $B^+ \to K^0\pi^+$ decays. We also made the first observation of the decay $B^+ \to K^+\pi^0$ with the branching fraction of $(1.5 \pm 0.4 \pm 0.3) \times 10^{-5}$.

So far, CLEO measured branching fractions for three of the four exclusive $B \to K\pi$ decays, while only upper limits could be established for the processes $B \to$

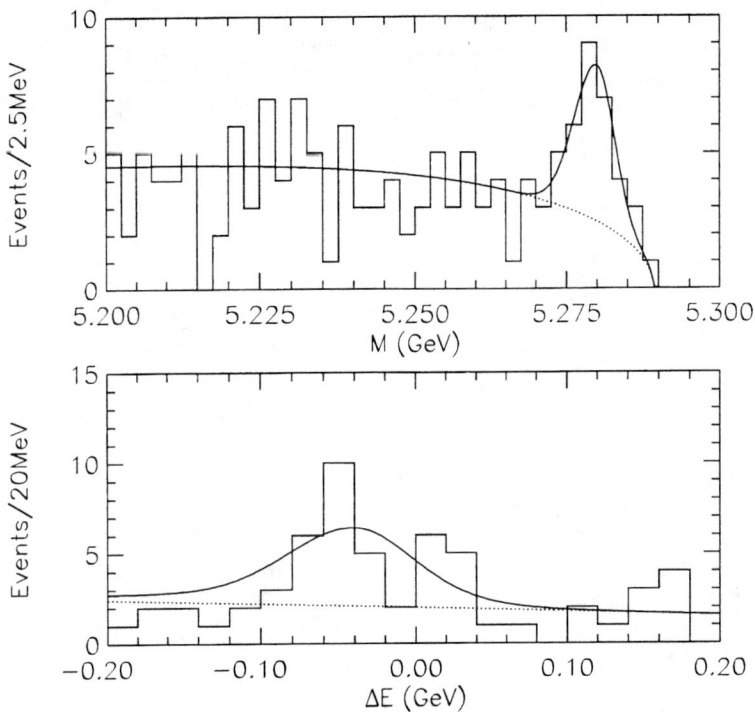

FIGURE 2. Projection plots for $B^+ \to K^+\pi^0$ decay.

$\pi\pi, KK$. Our results provide further evidence that the $b \to sg$ penguin amplitude dominates charmless hadronic B decays. Furthermore, the 90% confidence level upper limit $\mathcal{B}(B \to \pi^+\pi^-) < 8.4 \times 10^{-6}$ now rules out almost the entire range of theoretical predictions.

REFERENCES

1. M.S. Alam *et al.* (CLEO), Phys. Rev. Lett. **74**, 2885 (1995).
2. F. Borzumati and C. Greub, Phys. Rev. D **58** 074104 (1998); W. de Boer *et al.*, hep-ph/9805378; H. Baer *et al.*, Phys.Rev. D **58** 15007 (1998); J. Hewett, SLAC-PUB-6521 (1994).
3. B.H. Behrens *et al.* (CLEO Collaboration), Phys. Rev. Lett. **80**, 3710 (1998).
4. R. Godang *et al.* (CLEO Collaboration), Phys. Rev. Lett. **80**, 3456 (1998).

5. Y. Kubota et al. (CLEO Collaboration), Nucl. Instrum. Methods Phys. Res., Sec. A**320**, 66 (1992); T. Hill, 6th International Workshop on Vertex Detectors, VERTEX 97, Rio de Janeiro, Brazil.
6. K. Chetyrkin et al., Phys. Lett. B **400**, 206 (1997); Phys. Lett. B **425**, 414 (1998); A. Kagan and M. Neubert hep-ph/9805303.
7. S. Glenn et al. (CLEO Collaboration), "Improved measurement of $\mathcal{B}(b \to s\gamma)$," presented at the ICHEP 98, Vancouver, B.C., July 23-29, 1998.
8. A. Ali and C. Greub, Phys. Lett. B **259**, 182 (1991).
9. R. Ammar et al. (CLEO Collaboration), Phys. Rev. Lett. **71**, 674 (1993).
10. L-L. Chau, et al. Phys. Rev. D **43**, 2176 (1991), erratum Phys. Rev. D **58**, 019902 (1998); G. Kramer, W.F. Palmer, H. Simma, Z. Phys. C **66**, 429 (1995); D-S. Du, L. Guo, Z. Phys. C **75**, 9 (1997)
11. R. Fleischer and T. Mannel, Phys. Rev. D **57**, 2752 (1998).
12. M. Gronau and D. London, Phys. Rev. Lett. **65**, 3381 (1990).
13. D.M. Asner et al. (CLEO Collaboration), Phys. Rev. D **53**, 1039 (1996).
14. G. Fox and S. Wolfram, Phys. Rev. Lett. **41**, 1581 (1978).
15. B.H. Behrens et al. (CLEO Collaboration), "Update on the branching fraction of $B \to \eta' K$," presented at the ICHEP 98, Vancouver, B.C., July 23-29, 1998.
16. M. Artuso et al. (CLEO Collaboration), "First observation of the decay $B^{\pm} \to K^{\pm}\pi^0$," presented at the ICHEP 98, Vancouver, B.C., July 23-29, 1998.
17. N. G. Deshpande and J. Trampetic, Phys. Rev. D **41**, 895 (1990); L.-L. Chau et al., Phys. Rev. D **43**, 2176 (1991); A. Deandrea et al., Phys. Lett. B **318**, 549 (1993); A. Deandrea et al., Phys. Lett. B **320**, 170 (1994); G. Kramer and W. F. Palmer, Phys. Rev. D **52**, 6411 (1995); D. Ebert, R. N. Faustov, and V. O. Galkin, Phys. Rev. D **56**, 312 (1997); D. Du and L. Guo, Zeit. Phys. C **75**, 9 (1997); N.G. Deshpande, B. Dutta, and S. Oh, hep-ph/9712445; H. Cheng and B. Tseng, Phys. Rev. D **58** 094005 (1998); A. Ali, G. Kramer, and C. Lü, Phys. Rev. D **58** 094009 (1998) .

Theoretical Issues in Rare K, D and B Decays

Gustavo Burdman[†]

University of Wisconsin, Madison, WI 53706, USA.

Abstract. I discuss some theoretical aspects of rare K, D and B decays focusing mainly on their potential as tests of the one loop structure of the standard model. I concentrate on flavor changing neutral current processes and compare our ability to extract short distance physics in the three cases. Finally, I give some examples of the sensitivity of these decays to extensions of the standard model.

INTRODUCTION

The continuing success of the standard model (SM) has turned into a challenge both for theorists and experimentalists alike. On the one hand, theorists believe that the SM picture of the Higgs mechanism based on one elementary scalar doublet is unnatural, trivial and has come to be viewed as an effective description of a more complicated Higgs sector, one involving perhaps additional scalars and/or fermions or even new gauge interactions. In addition, the idea that fermion masses arise as a result of the interactions with this elementary scalar, requires for instance that its dimensionless couplings to two otherwise identical fermions such as the up and the top quarks, differ by more than four orders of magnitude. Understanding the mechanism for electroweak symmetry breaking (EWSB) and the origin of fermion masses calls for physics beyond the SM. At least in the case of EWSB, it is understood that this new physics must reside at an energy scale not far beyond 1 TeV. The experimental challenge of finding new physics in direct searches may still take some time if the new states or their effects only set in at several hundred GeV. A complement of these direct signals at the highest available energies is the measurement of the effects of the new particles in loops, either through precision measurements such as the ones performed at LEP, or through the detection of processes only occurring at one loop in the SM. Among these are the transitions induced by flavor changing neutral currents (FCNC), such as $K^0 - \bar{K}^0$ or $b \to s\gamma$. These are forbidden at tree level in the SM due to the presence of the Glashow-Illiopoulos-Maiani (GIM) mechanism. As a consequence, these processes

---

[†] burdman@pheno.physics.wisc.edu

are highly suppressed in the SM. Thus the one loop effect of a new heavy state may translate into a large effect in a branching ratio if this loop induces a FCNC transition. Here I discuss the FCNC decays of K, D and B and their potential as tools in searching for new physics at high energy scales $\gtrsim M_W$. One critical aspect in evaluating this potential is the extent to which a given process is determined by high energy scales, i.e. short distance physics, or it is contaminated by the more mundane effects of long distance dynamics such as the ones induced by propagating intermediate hadrons. Since the latter are not calculable in perturbation theory, the decay modes affected by long distance physics are less understood. Next, I will give examples of short and long distance contributions in the FCNC decays of K, D and B mesons. Once established which modes are the most likely to be testing grounds of the SM, I will turn to the discussion of two typical examples of its extensions: supersymmetry and the effects of anomalous triple gauge boson couplings.

RARE K DECAYS

Let us start the discussion with the decay $K^+ \to \pi^+ e^+ e^-$. In the SM this process receives one loop contributions from electroweak penguin and box diagrams involving up-type quarks. These are truly short distance diagrams and although some long distance physics enters in the hadronization from $s \to K$ and from $d \to \pi$ in the way of form-factors, the theoretical uncertainties this introduces are not by themselves large enough to obscure the interesting physics. However, the process $K^+ \to \pi^+ \gamma^*$ followed by $\gamma^* \to e^+ e^-$ gives another contribution, reflecting the long distance dynamics of the $K^+ \pi^+ \gamma$ vertex, and dominates the rate [1]. We conclude that this decay mode is not well suited for a test of the short distance physics entering the loops.

We now turn to consider $K_L \to \pi^0 e^+ e^-$. The short distance contributions to this process involve direct CP violation. This is extremely interesting since the measurement of this decay rate could be in principle a direct determination of the CP violating parameter η, the complex phase in the Cabibbo-Kobayashi-Maskawa (CKM) matrix. Furthermore, the one photon intermediate state cannot contribute here since the long distance vertex to one photon is CP conserving. Still, there is contamination from long distance physics. This comes from two sources. First, there is indirect CP violation given essentially by

$$BR(K_L \to \pi^0 e^+ e^-)_{\text{ICP}} = |\epsilon_K|^2 \frac{\tau(K_L)}{\tau(K_S)} BR(K_S \to \pi^0 e^+ e^-), \qquad (1)$$

where the K_S decay, being CP conserving, is dominated by the one photon intermediate state. In addition, there is a CP conserving contribution to the decay rate from $K_L \to \pi \gamma^* \gamma^*$ which gives a non interfering term resulting in $BR(K_L \to \pi^0 e^+ e^-)_{\text{CPC}} \simeq (1-2) \times 10^{-12}$, although this could be larger [2]. The indirect CP violating piece will be well known once the K_S decay mode is measured.

Progress to a better understanding of the CP conserving piece may be achieved in the near future. Thus, there is reasonable hope that in this process we may be able to disentangle the short and long distance physics, at least to some extent. For the moment, it remains in the list of long distance "polluted" modes.

Finally, we turn to the decay modes with neutrinos replacing the charged leptons in the final state. The processes $K^+ \to \pi^+ \nu \bar{\nu}$ and $K_L \to \pi^0 \nu \bar{\nu}$ are almost completely determined by short distance physics. The effective Hamiltonian for the charged mode is

$$\mathcal{H}_{\text{eff}} = \frac{4 G_F}{\sqrt{2}} \frac{\alpha}{2\pi s^2 \theta_W} \sum_\ell \left(\lambda_c X_c^\ell + \lambda_t X(x_t) \right) (\bar{s}_L \gamma_\mu d_L)(\bar{\nu}_L \gamma^\mu \nu_L)_\ell, \tag{2}$$

where $\lambda_i = V_{is}^* V_{id}$, ($\ell = e, \mu, \tau$), $X(x_t)$ is the result of the top quark loop contribution, and X_c^ℓ is the charm quark contribution and carries a dependence on the lepton flavor coming from the box diagram with charged leptons. The contribution from the charm loop, in addition to the charm quark mass dependence, introduces a sizeable scale dependence which is only reduced when including next-to-leading-order corrections. The remnant dependence on the scale m_c entering in the charm running mass results in an uncertainty $< 10\%$ in the $BR(K^+ \to \pi^+ \nu \bar{\nu})$. The hadronic matrix element needed to compute the matrix element of the exclusive mode can be obtained by isospin rotation from $K^+ \to \pi^0 \ell^+ \nu$ and therefore does not introduce additional hadronic uncertainties. The SM prediction for the branching ratio is obtained for $m_c = 1.3$ GeV, $m_t = 170$ GeV, $V_{cb} = 0.04$ and $V_{ub} = 0.032$, leading to [3] $Br(K^+ \to \pi^+ \nu \bar{\nu}) = (9.1 \pm 3.8) \times 10^{-11}$, where the uncertainty is mainly from the CKM parameters. The recent observation of one event in this channel by BNL E787, translates into [4] $Br(K^+ \to \pi^+ \nu \bar{\nu})_{\text{exp}} = (4.2 + 9.7 - 3.5) \times 10^{-10}$.

Finally, the neutral mode $K_L \to \pi^0 \nu \bar{\nu}$ is even cleaner than the charged one due to the fact that it is largely dominated by direct CP violation. As a result, only the top quark loop in Eq. (2) contributes so the uncertainties associated with the charm scale are not present. Furthermore, this mode constitutes a very unique and direct way to measure the CP violating phase of the CKM matrix. The amplitude depends on $Im[\lambda_t]$, which in the Wolfenstein parametrization can be written in term of the CP violating term η and V_{cb}, giving [3]

$$Br(K_L \to \pi^0 \nu \bar{\nu}) = 3.29 \times 10^{-5} \eta^2 |V_{cb}|^4 X^2(x_t), \tag{3}$$

with $\eta \equiv Im[V_{ub}^*/V_{cb}]/\lambda$. The branching ratio of the neutral mode is still too small in the SM when compared with current experimental limits. Future experiments expect to be sensitive to a few $\times 10^{-11}$ branching fraction.

There seems to be a compromise between experimental accessibility and the theoretical uncertainties in any of the modes discussed above: the most accessible modes tend to be affected by larger theoretical uncertainties. This tension is always present in rare FCNC decays. In the case of K decays, the neutrino modes are hard but still accessible experimentally. The charged kaon mode seems to be a good compromise, since is a short distance dominated process and the uncertainties seem to be surmountable. The additional interest of the neutral mode is the observation of direct CP violation and the direct measurement of CKM parameters.

RARE D VS. B DECAYS

We now turn to a comparative discussion of FCNC in D and B decays. The essential aspects can be framed as external-up-quark vs. external-down-quark processes. We will make the comparison using the radiative processes $c \to u\gamma$ vs. $b \to s\gamma$, for the sake of simplicity. Most of the conclusions can be extended to other modes. The short distance contributions to the radiative FCNC process $Q \to q\gamma$ result in the decay width

$$\Gamma^{(0)} = \frac{\alpha G_F^2}{128\pi^4} m_Q^5 \left|\sum_i \lambda_i F(x_i)\right|^2, \qquad (4)$$

where the superscript "0" denotes the absence of QCD corrections, $Q = (c, b)$, $x_i = (m_{q_i}/M_W)^2$, and the function $F(x)$ is the result of integrating the loop contribution of the internal quark i. This loop function is the same in the c and b cases. The main difference in Eq. (4) comes from the masses of the internal quarks and the CKM factors λ_i. In order to see how this affects the widths we turn to Table I, where we show separately the contribution of each quark flavor.

	i	$F(x_i)$	$\lambda_i F(x_i)$
$c \to u\gamma$	d	1.6×10^{-9}	3.4×10^{-10}
	s	2.9×10^{-7}	6.3×10^{-8}
	b	3.3×10^{-4}	3.2×10^{-8}
$b \to s\gamma$	u	2.3×10^{-9}	1.3×10^{-12}
	c	2.0×10^{-4}	7.3×10^{-6}
	t	0.4	1.6×10^{-2}

Table I: Contributions to $Q \to q\gamma$. From Ref. [5].

The CKM factors are $\lambda_i = V_{ci}^* V_{ui}$ for $c \to u\gamma$, and $\lambda_i = V_{ib}^* V_{is}$ for $b \to s\gamma$. As we can see, all three contributions are small in the $c \to u\gamma$ case, whereas for $b \to s\gamma$ the top quark loop gives the overwhelmingly dominant piece. The central point is that heavier quarks give the dominant contributions as long as their mixing with the external quarks is not highly suppressed. This is a consequence of the non-decoupling aspect of the SM, the fact that fermions that acquired masses from the Higgs mechanism do not decouple in loops involving the massive electroweak gauge bosons. In $c \to u\gamma$ the internal b quark contribution would dominate if it was not for the fact that V_{cb} and V_{ub} are extremely small. In any event, the QCD uncorrected $c \to u\gamma$ rate is very small due to the CKM dominance of the lighter intermediate states d and s. The $b \to s\gamma$ width is large due to the presence of a heavy top!

Although the QCD corrections to Eq. (4) are generally important, their impact also varies depending on the intermediate mass in the loop. They enhance the $c \to u\gamma$ rate by five orders of magnitude on the one hand, but the $b \to s\gamma$ rate goes up by less than a factor of three or so. The main source of these large corrections is the mixing of the short distance operators such as

$$\mathcal{O}_7 = \frac{e}{16\pi^2} m_Q (\bar{q}_L \sigma_{\mu\nu} Q_R) F^{\mu\nu}, \tag{5}$$

generated by the interesting short distance physics, with the more mundane four-fermion operators such as

$$(\bar{q}_L \gamma_\mu q'_L)(\bar{q}'_L \gamma^\mu Q_L), \tag{6}$$

that are generated at tree level by the SM charged currents. The mixing comes about when loop generated by gluons are taken into account. New physics, if present, will almost certainly appear in (5), not in (6), which is then a background for precision tests of the SM. Thus, the lesson from Table II is that in $b \to s\gamma$ there is still sensitivity to new physics affecting the operator (5), whereas even if it were possible to measure a branching ratio as low as 10^{-12} for $c \to u\gamma$, this would reflect the SM physics of operators such as the one in Eq. (6).

	No QCD	QCD Corrected
$Br(c \to u\gamma)$ [5]	1.5×10^{-17}	6.0×10^{-12}
$Br(b \to s\gamma)$	1.3×10^{-4}	3.3×10^{-4}

Table II: Leading order and QCD-corrected branching ratios for $Q \to q\gamma$. The QCD corrected rates involve important QCD uncertainties.

The important point is that the overwhelming dominance of the QCD corrections in $c \to u\gamma$ not only tells us that the short distance physics is not sensitive to the one loop FCNC operators of interest, but also signals that there will be even larger long distance contributions to the rate. After all, the QCD corrections were computed perturbatively. In general, the dominance of the perturbative one loop amplitude by light quark contributions hints the existence of large long distance effects. Although these cannot be computed from first principles, it is possible to estimate them phenomenologically. For instance, the operator (6) with $q' = s$ gives rise to the dominant short distance piece in $c \to u\gamma$, through a $\bar{s}s$ loop. But one could imagine the $\bar{s}s$ pair propagating a long distance, forming a "ϕ", which turns into a photon via vector meson dominance. These and other similar long distance mechanisms [5] give rise, for instance, to $Br(D^0 \to \rho^0 \gamma) \simeq 10^{-6}$, far above the level of the QCD-corrected short distance rates expected for $c \to u\gamma$ processes.

Many other long-distance dominated radiative D decays are at this level. Similar effects are expected in the leptonic modes. There, however, the gap between short and long distance physics is less dramatic. The inclusive short and long distance branching fractions are [6], respectively,

$$Br(c \to u\ell^+\ell^-)_{\text{SD}} \simeq 10^{-8}, \tag{7}$$
$$Br(c \to u\ell^+\ell^-)_{\text{LD}} \simeq 10^{-6}. \tag{8}$$

Then, although the long distance contributions still dominate in the SM, it is still conceivable that new physics contributions could overcome them. For instance, the SM prediction for the exclusive mode $Br(D^0 \to \pi^0 e^+ e^-) \simeq 7 \times 10^{-7}$, is still well below the current experimental bound, $Br(D^0 \to \pi^0 e^+ e^-)_{\text{exp}} < 4.5 \times 10^{-5}$. Some extensions of the SM may give large enhancements in charm processes. Although, these effects would be more noticeable in $D^0 - \bar{D}^0$ mixing, they could also result in $c \to u\ell^+\ell^-$ rates well above 10^{-6}. In general, however, one loop effects from new physics in D decays are likely to be small compared to long distance effects.

On the other hand, the analogous b decays are believed to be dominated by short distance physics. For instance, the next-to-leading order SM prediction for the short distance rate gives [7] $Br(b \to s\gamma) = (3.38 \pm 0.33) \times 10^{-4}$. Long distance contributions similar to those discussed in radiative charm decays, can proceed via the propagation of intermediate $\bar{c}c$ states, the off-shell "J/ψ". Estimates of the pollution due to these states in the inclusive rate [8] cannot be made reliably within controlled approximations. They tend to vary from one calculation to the next and can be as large as 20%. The current experimental measurements give [9]

$$Br(B \to X_s\gamma)_{\text{CLEO}} = (3.15 \pm 0.35 \pm 0.32 \pm 0.26) \times 10^{-4}, \tag{9}$$
$$Br(B \to X_s\gamma)_{\text{ALEPH}} = (3.11 \pm 0.80 \pm 0.72) \times 10^{-4}. \tag{10}$$

Thus, for the moment the potential long distance pollution is not problematic, but it should be taken into account in the future when precise enough measurements become available.

Similar considerations apply to the dilepton modes $b \to s\ell^+\ell^-$. In this case the long distance pollution comes in the form of "spill over" of the J/ψ and ψ' resonant peaks into the continuum [10]. But in principle these modes are short distance dominated and together with $b \to s\gamma$ constitute a stringent test of the SM. In addition to the dipole moment operator in (5), these modes receive contributions from the operators

$$\mathcal{O}_9 = \frac{e^2}{16\pi^2}(\bar{s}_L\gamma_\mu b_L)(\bar{\ell}\gamma^\mu\ell), \tag{11}$$

$$\mathcal{O}_{10} = \frac{e^2}{16\pi^2}(\bar{s}_L\gamma_\mu b_L)(\bar{\ell}\gamma^\mu\gamma_5\ell). \tag{12}$$

These receive contributions from Z penguin and box diagrams. As a result, there are three quantities to be measured: $C_7(m_b)$, $C_9(m_b)$ and $C_{10}(m_b)$, the Wilson coefficients evaluated at the relevant experimental scale. New physics effects enter as additions to the values of the coefficients at the high energy scale $E > M_W$. The extraction of these quantities is a research program involving the inclusive rates $B \to X_s\gamma$, $B \to X_s\ell^+\ell^-$, as well as exclusive modes such as $B \to K^{(*)}\ell^+\ell^-$ among others. A lot of theoretical effort has gone into understanding the inclusive decay rates [11], and the theoretical predictions are under control. On the other hand, the exclusive modes are, in principle, affected by large theoretical uncertainties due to our poor knowledge of the non-perturbative dynamics determining form-factors. Some sound theoretical predictions can be made based on symmetries [12]. In some cases [13], this is enough to extract the short distance physics. In any event, in the future all these form-factors will be obtained from first principle calculations on the lattice [14], where a lot of progress has been made recently in computing weak matrix elements [15].

SENSITIVITY TO NEW PHYSICS

Here we discuss two typical examples of extensions of the SM of very different kind: supersymmetry and anomalous triple gauge boson couplings (TGC).

Supersymmetry

Supersymmetry is perhaps one of the most popular extensions of the SM. However, as many other extensions, it has a FCNC problem: in its most general form it does not come with an automatic GIM mechanism, and therefore it may generate large FCNC effects [16]. Most of the trouble comes from the fact that the diagonalization of fermion mass matrices does not, in general, diagonalize the squark mass matrices. Thus flavor mixing in the sfermion sector "misaligned" with the fermions, are a potential disaster in general SUSY scenarios. Even if the sfermion sector is assumed to be diagonal (or aligned), there is an additional source of FCNC effects, coming from the charged Higgs and chargino-squark contributions, arising from the standard CKM matrix. These effects, then are expected to be present at most at the SM level, since they depend only on the masses of the charged Higgs, the charginos and the third generation squarks. In any case, some assumption about the sfermion mass matrices is necessary in order to accommodate the FCNC constraints. The two possibilities are: (i) sfermion mass matrices are diagonal at some high energy scale (e.g. $M_{\rm GUT}$) and small off diagonal elements are generated by the running down to the electroweak scale; (ii) they are (partially) aligned with the SM fermion mass matrices.

The vast parameter space of SUSY models includes these off diagonal elements, the superpartner and Higgs sector masses and mixings, and allows to accommodate the lack of deviations in FCNC processes such as $K^0 - \bar{K}^0$ mixing, $b \to s\gamma$, etc. However, one can argue that in most cases the SUSY effects should be "naturally" of the order of $(10 - 20)\%$ or larger. Such effects, for instance, in K, D and B

mixing, or $b \to s$ and $s \to d$ transitions, are hard to see at the moment. But larger effects are also possible. For example, even satisfying the current $b \to s\gamma$ and mixing bounds, we could still see enhancements of $\mathcal{O}(1)$ in $K \to \pi\nu\bar{\nu}$ [17] and $b \to s\ell^+\ell^-$ [18].

Moreover, it has been recently argued [19] that if large off diagonal sfermion mixings are allowed, the next to leading order expansion in theses mixings reveals the possibility of even larger effects in the $s \to dZ$ vertex. This would have a large impact in decay modes such as $K \to \pi\nu\bar{\nu}$, where the Z penguin plays a dominant role, resulting in enhancements of the branching ratios of one order of magnitude or more, depending on the modes. This is an interesting possibility and deserves further study, particularly the correlation with possible enhancements in D mixing and rare D decays that would result from very large mass insertions in the up-squark sector.

Anomalous Triple Gauge Boson Couplings

We now turn to examine the potential of rare FCNC decays to constrain anomalous triple gauge boson couplings (TGC). In general, we can assume in a model independent way, that extensions of the SM might modify some of the couplings of fermions and/or gauge bosons. In particular, the TGC are of interest since they have not been measured with such precision as some of the fermion couplings. We have in mind deviations from the SM values for the couplings of a pair of W to a photon or a Z. We would expect that the anomalous TGC encode the physics of some higher energy scale. Imposing CP conservation, the most general form of the WWN ($N = \gamma, Z$) couplings can be written as [20]

$$\mathcal{L}_{WWN} = g_{WWN} \left\{ i\kappa_N W^\dagger_\mu W_\nu N^{\mu\nu} + ig_1^N \left(W^\dagger_{\mu\nu} W^\mu N^\nu - W_{\mu\nu} W^{\dagger\mu} N^\nu \right) \right.$$
$$\left. + g_5^N \epsilon^{\mu\nu\rho\sigma} (W^\dagger_\mu \partial_\rho W_\nu - W_\mu \partial_\rho W^\dagger_\nu) N_\sigma + i\frac{\lambda_N}{M_W^2} W^\dagger_{\mu\nu} W^\nu_\lambda N^{\nu\lambda} \right\}, \quad (13)$$

with the conventional choices being $g_{WW\gamma} = -e$ and $g_{WWZ} = -g\cos\theta$. Additionally, the are three CP violating Lorentz invariant terms, resulting in other 6 parameters: $\tilde{\kappa}_N$ and $\tilde{\lambda}_N$, obtained from (13) by replacing $N_{\mu\nu}$ by the dual field strength; and g_4^N from a term similar to the second one in Eq. (13).

Gauge invariance implies $g_1^\gamma = 1$, $g_5^\gamma = g_4^\gamma = 0$. Then, in principle, there are 11 new free parameters. Two CP conserving ($\Delta\kappa_\gamma$, λ_γ) and two CP violating ($\Delta\tilde{\kappa}_\gamma$, $\tilde{\lambda}_\gamma$) affecting the $WW\gamma$ couplings; four CP conserving (Δg_1^Z, g_5^Z, λ_Z and $\Delta\kappa_Z$) and three CP violating ($\tilde{\kappa}_Z$, $\tilde{\lambda}_Z$ and g_4^Z) shifting the WWZ vertex.

This is a typical problem of this type of approach, where the model independence is traded off by a large number of free parameters the sources of which are not known. However, simplification is possible, when considering rare B and K decays. we can neglect the contribution of $\Delta\kappa_Z$, λ_Z, as well as the three WWZ CP violating anomalous TGC, since their effects are suppressed by powers of the

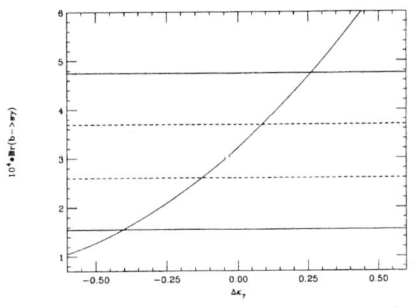
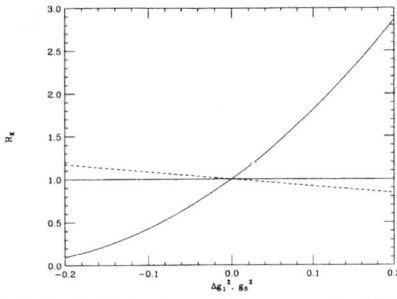

FIGURE 1. a) The $Br(b \to s\gamma)$ vs. $\Delta\kappa_\gamma$. The solid (dashed) horizontal lines are the $3(1)\sigma$ CLEO measurement. b) The $Br(K^+ \to \pi^+\nu\bar\nu)$ normalized to the SM expectation, vs. Δg_1^Z (solid) and g_5^Z (dashed). From Ref. [21].

small external momenta over m_Z. This selective sensitivity is an advantage, rather than a handicap, when we view these measurements as complement of other ones made at higher energies and sensitive to all the Z TGC.

Thus up to this point, we have 6 coefficients left. However, we can ignore λ_γ and $\tilde\lambda_\gamma$ if we assume that the dynamics producing these non-SM effects resides at a scale parametrically larger than the weak scale, say $\Lambda \simeq \mathcal{O}(1)$ TeV. Although this is is not general, I believe this is a reasonable scenario, since if this was not the case we should take into account the states that are present with weak scale masses (e.g. superpartners, weakly coupled scalars, etc.) and not integrate them out as we do in an effective coupling approach. When we accept this, we see that in an effective Lagrangian approach, these coefficients can only be generated by next to leading order operators, which can be ignored since they are suppressed by (M_W^2/Λ^2) with respect to the leading order ones [‡].

Thus, at the energies at hand in rare decays, the only relevant coefficients are the three CP conserving parameters ($\Delta\kappa_\gamma$, Δg_1^Z, g_5^Z); and a CP violating one, $\tilde\kappa_\gamma$. Their FCNC effects are most interesting in B and K decays. For instance, in Fig. 1a we see the sensitivity of the current measurements of the $b \to s\gamma$ rate to the presence of $\Delta\kappa_\gamma$, whereas in Fig. 1b the branching ratio for $K^+ \to \pi^+\nu\bar\nu$, normalized to the SM, is plotted against both Δg_1^Z and g_5^Z. The effect of $\Delta\kappa_\gamma$ in $b \to s\gamma$ is obtained without any assumption other than the suppression of λ_γ. Similarly, from Fig. 1b we see that the only significant contribution of anomalous TGC to $K^+ \to \pi^+\nu\bar\nu$ is given by Δg_1^Z, which could give effects as large as factors of $(2-3)$ in the branching ratio. The CP violating parameter $\tilde\kappa_\gamma$ gives smaller effects that its CP conserving counterpart since it does not interfere with the SM. In $b \to s\ell^+\ell^-$ decays again these two coefficients ($\Delta\kappa_\gamma$, Δg_1^Z) give the dominant contributions. In Fig. 2a we see the effects of the WWZ couplings on the SM

[‡] This corresponds to the so called non-linear realization of the EWSB sector. However, it is also possible to imagine a scenario where there is a light scalar similar to the SM Higgs, with all other new states above the scale Λ. In these linear realization scenarios, the power counting requires the consideration of λ_N and $\tilde\lambda_N$ on the same footing with the other anomalous TGC.

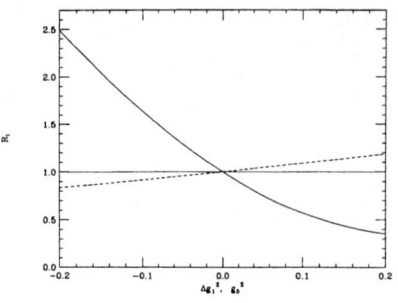 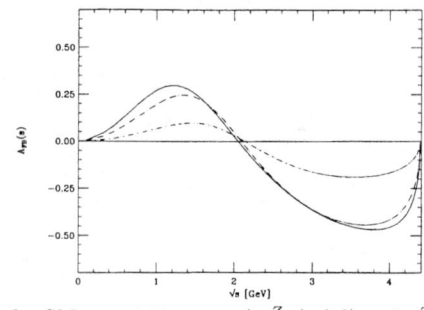

FIGURE 2. a) The $Br(b \to s\ell^+\ell^-)$, normalized to the SM expectation, vs. Δg_1^Z (solid) and g_5^Z (dashed). b) The forward-backward lepton asymmetry in $B \to K^*\ell^+\ell^-$ vs. the dilepton mass $\sqrt{s}$, for $\Delta g_1^Z = 0$, 0.1 and 0.2 (solid, dashed, dot-dashed respectively). From Ref. [21].

normalized branching ratio, taking $\Delta \kappa_\gamma = 0$. Somewhat less dramatic effects are given by this $WW\gamma$ coupling by itself. However, an interesting feature of these decay modes, is that the additional information given by the lepton asymmetry can be used to disentangle the two contributions. As it can be seen in Fig. 2b, the effect of Δg_1^Z on the forward-backward lepton asymmetry $A_{FB}(s)$ in $B \to K^*\ell^+\ell^-$ is such that it does not change the position of the zero [21], but affects the shape of the asymmetry as well as the rate. On the other hand, the zero of $A_{FB}(s)$ is shifted significantly by accessible values of $\Delta \kappa_\gamma$. Thus, a measurement of this asymmetry, as well as the rate, provides enough information to constrain the two relevant anomalous TGC without assuming that one of them vanishes. Then, the bounds obtained from FCNC decays are not only competitive with those from high energy colliders, but also complementary to them due to their rather selective sensitivity. For comparison, we show in Table III the projected sensitivities of LEPII [22] at $\sqrt{s} = 190$ GeV and $500 pb^{-1}$ integrated luminosity, the upgraded Tevatron [23] with $1 fb^{-1}$, and a guess of the 3σ sensitivity to be reached in the next round of B and K experiments for FCNC decays.

	Table III. Comparison of bounds on Anomalous TGC.		
	LEPII 190 GeV	Tevatron RunII 1 fb^{-1}	FCNC Decays
$\Delta\kappa_\gamma$	(-0.25,0.40)	(-0.38,0.38)	(-0.20,0.20)
Δg_1^Z	(-0.08,0.08)	(-0.18,0.48)	(-0.10,0.10)
$\tilde{\kappa}_\gamma$	-	(-0.33,0.33)	(-0.50,0.50)

REFERENCES

1. L. Littenberg and G. Valencia, *Ann. Rev. Nucl. Part. Sci.* **43**, 729 (1993) and references therein.

2. J. F. Donoghue and F. Gabbiani, *Phys. Rev.* **D51**, 2187 (1995).
3. G. Buchalla, A. J. Buras and M. E. Lautenbacher, *Rev. Mod. Phys.* **68**, 1125 (1996); A. J. Buras, hep-ph/9806471, to appear in *"Probing the Standard Model of Particle Interactions"*, F.David and R. Gupta, eds., 1998, Elsevier Science B.V.
4. S. Adler et al., the BNL 787 collaboration, *Phys. Rev. Lett.* **79**, 2204 (1997).
5. G. Burdman, E. Golowich, J. L. Hewett and S. Pakvasa, *Phys. Rev.* **D52**, 6383 (1995).
6. G. Burdman, E. Golowich, J. L. Hewett and S. Pakvasa, in preparation.
7. For a recent discussion of the $b \to s\gamma$ prediction see C. Greub and T. Hurth, hep-ph/9809468, proceedings of the *"International Euroconference on Quantum Chromodynamics"*, Montpellier, France, 2-8 Jul 1998.
8. E. Golowich and S. Pakvasa, *Phys. Rev.* **D51**, 1215 (1995); N.G. Deshpande, X.-G. He and J. Trampetic, *Phys. Lett.* **B367**, 362 (1996); J. M. Soares, *Phys. Rev.* **D53**, 241 (1996).
9. Petr Gaidarev, for the CLEO collaboration, these proceedings; R. Barate et al., the ALEPH collaboration, *Phys. Lett.* **B429**, 169 (1998).
10. G. Buchalla and G. Isisdori, *Nucl. Phys.* **B525**, 333 (1998), and references therein.
11. For reviews see, for example, A. Ali, hep-ph/9709507, in the *"Proceedings of 7th International Symposium on Heavy Flavor Physics"*, Santa Barbara, CA, 7-11 Jul 1997; also see Ref. [3].
12. N. Isgur and M. B. Wise, *Phys. Rev.* **D42**, 2388 (1990).
13. G. Burdman, *Phys. Rev.* **D57**, 4254 (1998).
14. A. Kronfeld, these proceedings.
15. S. Ryan et al., hep-lat/9810041, presented at *"The 16th International Symposium on Lattice Field Theory"*, (LATTICE 98), Boulder, CO, 13-18 Jul 1998.
16. For a recent review on FCNC effects in Supersymmetry, see M. Misiak, S. Pokorski and J. Rosiek, hep-ph/9703442, in *"Heavy Flavors II"*, eds. A.J. Buras and M. Lindner, Advanced Series on Directions in High-Energy Physics, World Scientific Publishing Co., Singapore.
17. A. J. Buras, A. Romanino and L. Silvestrini, *Nucl. Phys.* **B520**, 3 (1998).
18. A. Masiero and L. Silvestrini, hep-ph/9711401, Lectures given at the *"International School of Physics, 'Enrico Fermi': Heavy Flavor Physics - A Probe of Nature's Grand Design*, Varenna, Italy, 8-18 Jul 1997; J. L. Hewett and J. D. Wells, *Phys. Rev.* **D55**, 5549 (1997). See also Ref. [16].
19. G. Colangelo and G. Isidori, *J. High Energy Phys.* **9809**, 9 (1998).
20. K. Hagiwara, K. Hikasa, R. D. Peccei and D. Zeppenfeld, *Nucl. Phys.* **B282**, 253 (1987).
21. G. Burdman, hep-ph/9806360, to appear in *Phys. Rev. D*.
22. G. Gounaris et al., report of the *Triple Gauge Boson Couplings* working group at the *"Physics at LEPII"* Workshop, hep-ph/9601233. For a recent update see G. Abbiendi et al., the OPAL collaboration, hep-ex/9811028.
23. *Future Electroweak Physics at the Tevatron*, D. Amidei and R. Brock editors, FERMILAB-PUB-96/082.

Rare decays of the K_L and π^0: New results from E799-II and E832[1]

E. D. Zimmerman[2,3]
for the E799/E832 (KTeV) collaboration

*Enrico Fermi Institute, The University of Chicago,
Chicago, Illinois 60637*

Abstract. Recent rare decay results from Fermilab experiments E799-II and E832 (the KTeV experiments) are shown. Results of searches for the direct CP-violating decay $K_L \to \pi^0 \nu \bar{\nu}$ are given, as well as measurements of the decays $K_L \to \pi^0 \gamma \gamma$ and $K_L \to \pi^0 e^+ e^- \gamma$, which provide information on CP-conserving contributions to $K_L \to \pi^0 e^+ e^-$. A new measurement of the rare decay $\pi^0 \to e^+ e^-$ is also presented; this result is the first significant measurement of the excess rate above the unitarity limit for this mode.

INTRODUCTION

We present rare decay results from Fermilab experiments E799-II and E832 (the KTeV experiments). A measurement of the rare decay $\pi^0 \to e^+ e^-$ is presented. Direct CP-violation in rare kaon decays is discussed, including measurement of the CP-conserving decay $K_L \to \pi^0 \gamma \gamma$ and searches for $K_L \to \pi^0 e^+ e^-$ and $K_L \to \pi^0 \nu \bar{\nu}$.

E799-II and E832, two neutral kaon experiments, took data in the 1996-97 Tevatron fixed target run. The two experiments used the same beamline and spectrometer in different configurations. E799-II was a high-rate experiment designed to search for rare K_L decays, primarily those with two leptons and one or more photons in the final state. The primary goal of the experiment was to search for $K_L \to \pi^0 e^+ e^-$ with about 50 times the sensitivity of its predecessor experiment, E799-I. The primary goal of E832 was to measure $\text{Re}(\varepsilon'/\varepsilon)$ to a precision of $\sim 10^{-4}$. The current status of this measurement is described in a separate paper from this conference [1]. Certain rare decay measurements were performed in E832 as well.

---

[1] Presented at the Workshop on Heavy Quarks at Fixed Target (HQ98), Batavia, Ill., October 10-12, 1998.
[2] Electronic address: edz@fnal.gov
[3] Address after December 15, 1998: Columbia University, New York, NY 10027

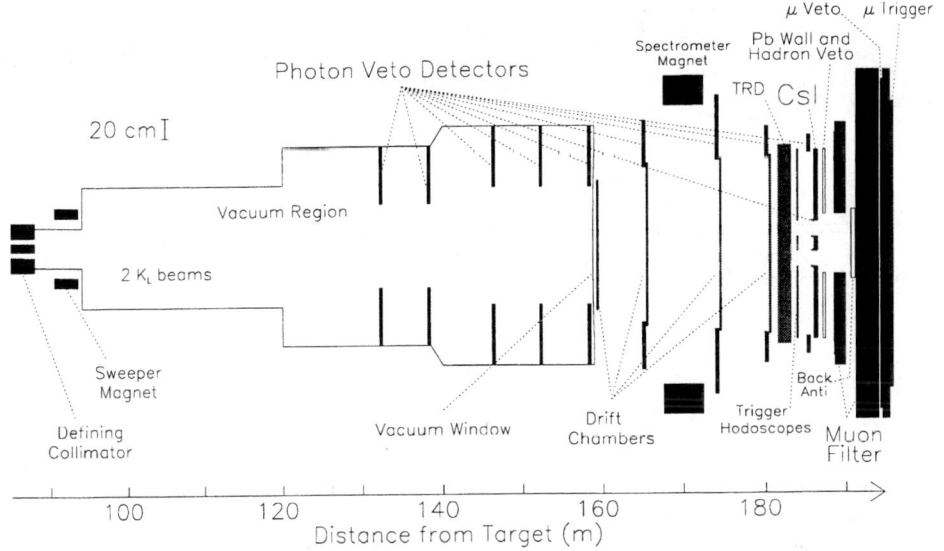

FIGURE 1. Plan view of the decay region and the spectrometer configured for E799-II. The horizontal scale is compressed.

THE E799-II/E832 SPECTROMETER

The experiments used two nearly-parallel neutral kaon beams produced by 800 GeV protons which struck a BeO target. The neutral beams were defined by a series of collimators; photons were converted by a lead absorber, and charged particles were removed by a series of sweeping magnet. The beams traveled through a 65 meter vacuum decay region, where approximately 5% of the K_L decayed. The beams in the decay region were composed mostly of neutrons and K_L, with small numbers of K_S, Λ^0, $\bar{\Lambda}^0$, Ξ^0, $\bar{\Xi}^0$. These short-lived particles tended to decay upstream. The K_L momentum ranged from ~ 20 to ~ 200 GeV/c.

The decay region and the detector are shown in Fig. 1, in the E799-II configuration. The charged-particle spectrometer consisted of four drift chambers, each with two orthogonal views, and a spectrometer magnet between the second and third chamber which bent charged particles in the horizontal direction. A pure cesium iodide calorimeter (CsI) provided electron identification and photon energy measurement. Photon veto detectors lined the vacuum region and the perimeter of each drift chamber; these detectors defined the fiducial region. The primary difference between E799-II and E832 was a regenerator in the decay region which reintroduced a K_S component to one beam in E832 only. In addition, an extra photon veto detector called the "mask anti" (MA) was placed immediately upstream of the regenerator in E832 to reject upstream decays. The E832 configuration is shown in Ref. [1]. In E799-II, a set of transition radiation detectors (TRD) provided

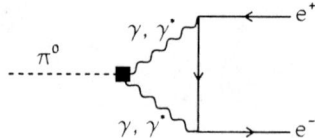

FIGURE 2. Feynman diagram of the Standard Model electromagnetic contribution to $\pi^0 \to e^+e^-$.

additional electron identification.

THE DECAY $\pi^0 \to e^+e^-$

E799-II has performed rare π^0 decay studies using the copious $K_L \to 3\pi^0$ decay. Here we present a new result on the decay $\pi^0 \to e^+e^-$. This decay has received theoretical and experimental attention for nearly four decades, since Drell first calculated its branching ratio in 1959 [2]. The decay is helicity suppressed, and is expected to proceed predominantly through a two-photon intermediate state (Fig. 2). The decay rate can be expressed in terms of the $\pi^0\gamma^*\gamma^*$ form factor, and contains a contribution from off-shell internal photons as well as from on-shell photons. The contribution from on-shell photons is exactly calculable in QED, and yields the *unitarity bound* [3], a lower limit on the decay rate:

$$\frac{\text{BR}(\pi^0 \to e^+e^-)}{\text{BR}(\pi^0 \to \gamma\gamma)} \geq 4.75 \times 10^{-8}$$

ignoring final-state radiative corrections. The off-shell photon contribution should manifest itself as an excess above unitarity, providing a constraint on the $\pi^0\gamma^*\gamma^*$ form factor. Corrections from electron bremsstrahlung can produce a $e^+e^-\gamma$ final state with $m_{e^+e^-} < m_{\pi^0}$, indistinguishable from the Dalitz decay (tree-level $\pi^0 \to e^+e^-\gamma$). We therefore impose a cutoff $(m_{e^+e^-}/m_{\pi^0})^2 > 0.95$, selecting the region where the Dalitz decay rate is very small. The branching ratio for $\pi^0 \to e^+e^-$ with this cutoff is 13.5% lower than it would be in the absence of all radiative corrections [4].

The E799-II analysis identified events with two electron candidates (charged tracks with $E/p \approx 1$) and four photon clusters (CsI clusters with no tracks pointing to them), and no more than 0.4 GeV in any photon veto counter. The four photons were paired as two $\pi^0 \to \gamma\gamma$ decays: of the three possible pairing combinations, we selected the one with the best χ^2 for the hypothesis that the two π^0 decays occured at the same distance from the calorimeter, using the π^0 mass as a constraint. This χ^2 was required to be less than 4.5 (with one degree of freedom). Events were required to reconstruct with total transverse momentum $|p_T| < 30$ MeV/c, and total mass $|m_{\pi^0\pi^0 e^+e^-} - m_{K^0}| < 10$ MeV/c^2. Four-electron events (multiple

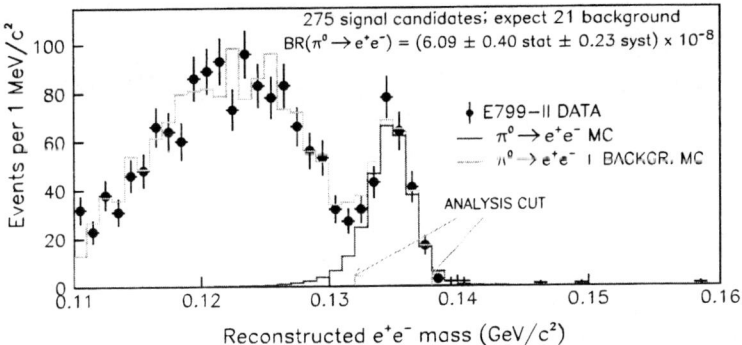

FIGURE 3. $m_{e^+e^-}$ spectrum for $\pi^0 \to e^+e^-$ candidate events. Background MC is normalized to data in the region $0.110 < m_{e^+e^-} < 0.125$ GeV/c^2. There are 275 candidates in the signal region.

Dalitz decays, or photon conversions in the vacuum window), where two electrons were removed by the spectrometer magnet, were a major background; these were removed by cutting events which had extra in-time activity in the second drift chamber. Finally, $\pi^0 \to e^+e^-$ candidates were required to have electron pair mass $|m_{e^+e^-} - m_{\pi^0}| < 6$ MeV/c^2.

After all cuts, there were 275 $\pi^0 \to e^+e^-$ candidates in the data.; the Monte Carlo (MC) estimated a residual background of 21.4±6.2 events. About 85% of the background was from high-$m_{e^+e^-}$ Dalitz decays, with four-electron modes accounting for the rest. The $\pi^0 \to e^+e^-$ sample was normalized to the $\pi^0 \to e^+e^-\gamma$ sample, canceling out any uncertainty in the $K_L \to 3\pi^0$ branching ratio. The branching ratio for $\pi^0 \to e^+e^-$ was found to be:

$$\mathrm{BR}(\pi^0 \to e^+e^-, (m_{e^+e^-}/m_{\pi^0})^2 > 0.95) = (6.09 \pm 0.40_{(\mathrm{stat})} \pm 0.24_{(\mathrm{syst})}) \times 10^{-8}$$

Note that this is a preliminary result.

The largest component of the systematic error was the 2.7% uncertainty in the branching ratio of the normalization mode, followed by a 2.4% systematic error resulting from the uncertainty in the background. This measurement represents a factor of four improvement over the current PDG average. Fig. 4 summarizes the existing measurements.

The number stated above is the physical branching ratio for the decay. In order to compare this result to the unitarity bound and to models which ignore radiative corrections, we invert the radiative corrections and extrapolate this measurement to the "lowest-order" rate (what the branching ratio would be if there were no

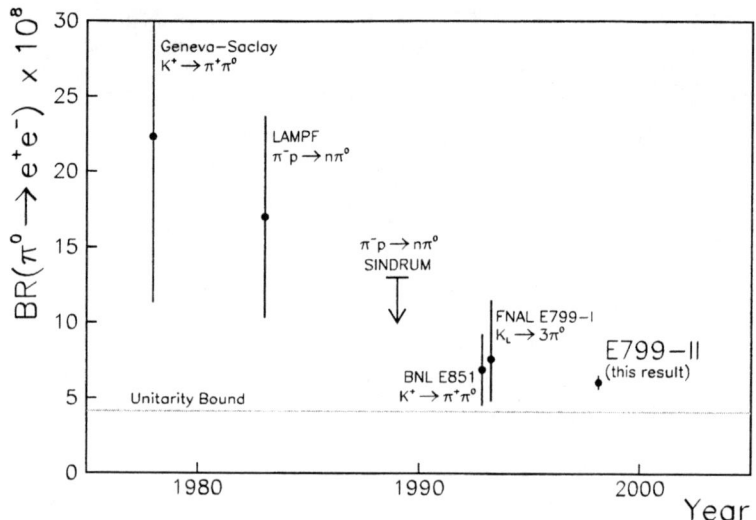

FIGURE 4. Measurements of the branching ratio of $\pi^0 \to e^+e^-$, including this preliminary result.

final-state radiation). This yields

$$\frac{\Gamma_{e^+e^-}^{\text{lowest order}}}{\Gamma_{\text{all}}} = (7.04 \pm 0.46_{\text{(stat)}} \pm 0.28_{\text{(syst)}}) \times 10^{-8}$$

which is over four standard deviations above the unitarity bound. This result is in reasonably good agreement with recent Chiral Perturbation Theory (χPT) [5] [6] and Vector Meson Dominance (VMD) [7] models.

CP-VIOLATION SEARCHES IN FCNC K_L DECAYS

The flavor-changing neutral current decays $K_L \to \pi^0 \ell \bar{\ell}$ have been considered promising avenues to search for direct CP-violation. Although these decays are very rare, direct CP-violating weak penguin and W box diagrams are expected to make significant contributions to them. In addition, it has recently been suggested [8] that supersymmetric extensions to the Standard Model could produce large enhancements to these decay rates.

E799-II is searching for all three of these decays ($K_L \to \pi^0 \mu^+ \mu^-$, $K_L \to \pi^0 e^+ e^-$, $K_L \to \pi^0 \nu \bar{\nu}$). The status of the latter two is described here, as well as a preliminary measurement from E832 of $K_L \to \pi^0 \gamma \gamma$, a CP-conserving decay which is used to calculate long-distance contributions to $K_L \to \pi^0 e^+ e^-$.

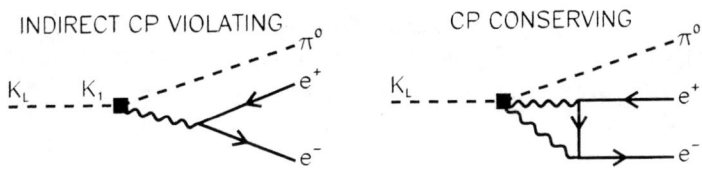

FIGURE 5. Long-distance contributions to $K_L \to \pi^0 e^+ e^-$.

The decay $K_L \to \pi^0 e^+ e^-$

Direct CP-violating diagrams are expected to contribute to the decay $K_L \to \pi^0 e^+ e^-$ at the level of $\sim 4 \times 10^{-12}$ [9]. Here we present the status of the search for direct CP-violation in this mode.

Experimental situation on $K_L \to \pi^0 e^+ e^-$

The current experimental limit on $K_L \to \pi^0 e^+ e^-$ is [10] BR($K_L \to \pi^0 e^+ e^-$) < 4.3×10^{-9}; this result is from E799-I. E799-II expects to achieve a single-event sensitivity of $\mathcal{O}(10^{-10})$ from the data collected in 1997. The analysis is currently in progress. The major background to this mode [11] is expected to be the radiative Dalitz decay $K_L \to e^+ e^- \gamma\gamma$ ($K_L \to e^+ e^- \gamma$ with a hard bremsstrahlung photon). This mode has a much higher branching ratio than $K_L \to \pi^0 e^+ e^-$, and can be indistinguishable from the signal if the two photons have invariant mass $m_{\gamma\gamma} \approx m_{\pi^0}$.

Other contributions to $K_L \to \pi^0 e^+ e^-$

Unfortunately, direct CP-violation is not the only Standard Model contribution to $K_L \to \pi^0 e^+ e^-$. Long-distance processes (Fig. 5) can contribute to the decay at comparable rates.

An indirect CP-violating component can arise from an intermediate $\pi^0 \gamma^*$ state, which proceeds through the CP-even (K_1) component of the K_L. Theoretical expectations for this process are currently imprecise and model-dependent [12]. This contribution to $K_L \to \pi^0 e^+ e^-$ could, however, be limited or measured by searching for the decay $K_S \to \pi^0 e^+ e^-$ with a sensitivity of $\mathcal{O}(10^{-9})$. The current limit on this decay is BR($K_S \to \pi^0 e^+ e^-$) < 1.1×10^{-6} [13]. It is hoped that E832, KLOE, and NA48 can make substantial improvements to this limit.

The $K_L \to \pi^0 \gamma\gamma$ intermediate state produces a CP-conserving component to $K_L \to \pi^0 e^+ e^-$. The two photons can be in either a relative S-wave or a D-wave. When the photons are in an S-wave, the $\gamma\gamma \to e^+ e^-$ process is helicity suppressed (as in $\pi^0 \to e^+e^-$), making this contribution to $K_L \to \pi^0 e^+ e^-$ very small. However, when the photons are in a D-wave, the process is helicity-allowed and may contribute significantly ($\gtrsim 10^{-12}$) to $K_L \to \pi^0 e^+ e^-$.

$$K_L \to \pi^0 \gamma\gamma$$

The decay $K_L \to \pi^0\gamma\gamma$ has been calculated in χPT to $\mathcal{O}(p^6)$ [14]. The D-wave component of the $\gamma\gamma$ spectrum in this decay is dominated by vector-meson exchange, which is parameterized by the quantity a_V. In E832, we have measured the branching ratio of the $K_L \to \pi^0\gamma\gamma$ decay, and analyzed its decay spectrum to estimate a_V and the CP-conserving contribution to $\pi^0 \to e^+e^-$.

The analysis required events with exactly four CsI clusters and no tracks. The four photons were constrained to form the kaon mass, giving the decay distance from the calorimeter. Assuming this decay position, the $\gamma\gamma$ invariant mass of each pair of photons was calculated. The pair which came closest to the π^0 mass was selected. If the other pair also formed a π^0 mass, the event was considered $K_L \to 2\pi^0$ and rejected from the $K_L \to \pi^0\gamma\gamma$ sample.

The major background came from $K_L \to 3\pi^0$ decays with either missing decay products or "fused" photons (photons which struck the CsI very close together, and merged to form a single cluster). These were rejected by requiring very little energy deposited in the photon vetoes, and by requiring that the transverse shape of each photon cluster was consistent with a single electromagnetic shower. Another background was multiple π^0 and η production in detector material. Events were rejected if they were consistent with $\pi^0\pi^0$ or $\pi^0\eta$ production at the vacuum window or farther downstream.

The $m_{\gamma\gamma}$ distribution of the remaining 907 events is shown in Fig. 6. The estimated background was 111 ± 14 events. Normalizing to $K_L \to 2\pi^0$, we found the branching ratio:

$$\text{BR}(K_L \to \pi^0\gamma\gamma) = (1.76 \pm 0.06_{(\text{stat})} \pm 0.09_{(\text{syst})}) \times 10^{-6}.$$

This preliminary result is in agreement with the previous measurement of $(1.7 \pm 0.3) \times 10^{-6}$ from the NA31 experiment [15].

Our result provides the first firm evidence for the low-$m_{\gamma\gamma}$ tail ($m_{\gamma\gamma} < 0.240$ GeV/c^2) which is dominated by vector meson exchange. A fit to the distribution of the kinematic parameters of the decay gives $a_V = -0.77 \pm 0.06$. Given the model of Ref. [14], this translates into a CP-conserving component to $K_L \to \pi^0 e^+ e^-$ of $(1-2) \times 10^{-12}$, somewhat smaller than the expected direct CP-violating component.

The decay $K_L \to \pi^0 \nu\bar{\nu}$

The decay $K_L \to \pi^0\nu\bar{\nu}$ is expected to proceed via direct CP-violating weak penguin and W box diagrams, analagous to $K_L \to \pi^0 e^+ e^-$ [16]. However, CP-

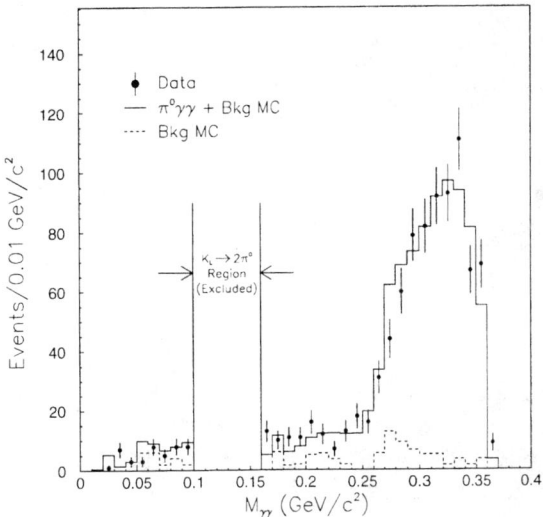

FIGURE 6. $m_{\gamma\gamma}$ distribution for candidate $K_L \to \pi^0\gamma\gamma$ events in E832 data and MC.

conserving and indirect CP-violating long-distance physics cannot contribute significantly to this decay because the neutrinos cannot couple directly to photons. This makes the decay theoretically straightforward. In the Wolfenstein parameterization of the CKM matrix, the branching ratio of $K_L \to \pi^0\nu\bar{\nu}$ is directly proportional to the phase η^2. The decay is expected to occur at the level of $(1-5) \times 10^{-11}$. The uncertainty is in the measured parameters; the theoretical uncertainty is only about 1% [17], making this decay one of the best potential measurements of η.

Of course, although $K_L \to \pi^0\nu\bar{\nu}$ is theoretically very clean, it presents a formidable experimental challenge. The signature of the decay is an isolated π^0 with high p_T (up to 231 MeV/c); there are, of course, many potential backgrounds.

$K_L \to \pi^0\nu\bar{\nu}$ searches using $\pi^0 \to e^+e^-\gamma$

The most sensitive searches to date for $K_L \to \pi^0\nu\bar{\nu}$ have used the $\pi^0 \to e^+e^-\gamma$ decay. This method is expensive, as the $\pi^0 \to e^+e^-\gamma$ branching ratio is only 1.2%, and the experimental acceptance is low due to the small opening angle between the tracks in a Dalitz decay. However, being able to reconstruct a charged vertex, and hence both the π^0 mass *and* the decay position, allows much improved background rejection.

In E799-II, we identified events with an isolated $\pi^0 \to e^+e^-\gamma$ decay inside the vacuum decay region. The major backgrounds to this search included the K_{e3}

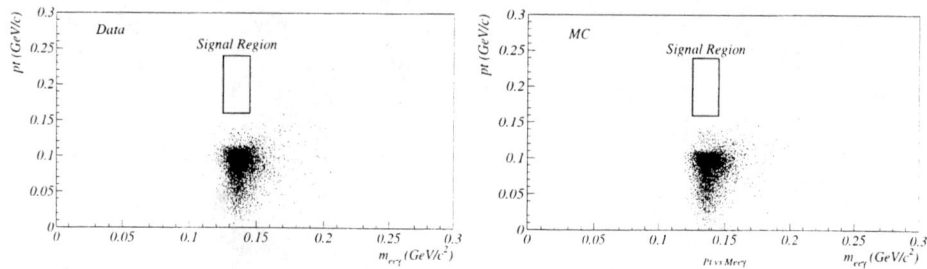

FIGURE 7. $e^+e^-\gamma$ mass vs. $|p_T|$ for $K_L \to \pi^0\nu\bar{\nu}$, $\pi^0 \to e^+e^-\gamma$ data (left) and background MC (right).

($K_L \to \pi^\pm e^\mp \nu$) decay with an extra photon and the pion misidentified as an electron. This background was reduced by making tight particle identification requirements using the CsI and TRD, as well as kinematic cuts which rejected high-$m_{e^+e^-}$ events. Other backgrounds came from $K_L \to 3\pi^0$ and $K_L \to 2\pi^0$, with one π^0 Dalitz decay and multiple lost particles. These were removed by tight cuts requiring no energy in the photon veto detectors. Hadronic interactions in the detector could produce multiple π^0's with high p_T; these were rejected by requiring the reconstructed decay vertex to be at least 8 meters upstream of the vacuum window. Finally, hyperon decays ($\Lambda^0 \to n\pi^0$ and $\Xi^0 \to \Lambda^0\pi^0$) with secondary or tertiary $\pi^0 \to e^+e^-\gamma$ decays presented backgrounds, as the detector did not have efficient neutron detection. These were removed by requiring $|p_T| > 160$ MeV/c, above the kinematic cutoff for hyperon decays. This cut cost nearly half of the signal acceptance.

After all cuts, our single event sensitivity (SES) for $K_L \to \pi^0\nu\bar{\nu}$, $\pi^0 \to e^+e^-\gamma$ was 3.1×10^{-9}. Dividing by the Dalitz decay branching ratio, the SES for all $K_L \to \pi^0\nu\bar{\nu}$ decays was 2.6×10^{-7}. The expected background was 0.12 events. The data and background MC are shown in Fig. 7. No events were observed in the data; this yields the preliminary E799-II limit:

$$\text{BR}(K_L \to \pi^0\nu\bar{\nu}) < 5.9 \times 10^{-7} \quad (90\% \text{ C. L.})$$

which is a factor of 100 improvement over the current PDG value.

$K_L \to \pi^0\nu\bar{\nu}$ search using $\pi^0 \to \gamma\gamma$

For the same number of kaon decays in our experiment, our single-event sensitivity for $K_L \to \pi^0\nu\bar{\nu}$ would have been over three orders of magnitude better using the $\pi^0 \to \gamma\gamma$ decay mode than using $\pi^0 \to e^+e^-\gamma$. However, the backgrounds (and the trigger rate) would have been prohibitive. To search for $K_L \to \pi^0\nu\bar{\nu}$ using the $\pi^0 \to \gamma\gamma$ mode, a one-day special run was taken with one beam closed off, and the remaining beam reduced in solid angle by a factor of 6.

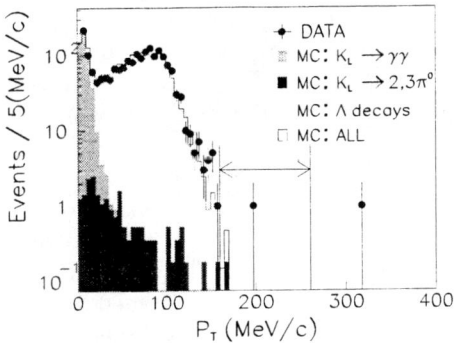

FIGURE 8. $|p_T|$ distribution of $K_L \to \pi^0 \nu \bar{\nu}$, $\pi^0 \to \gamma\gamma$ sample.

The analysis required events with two isolated CsI clusters (photons) and no in-time activity in any other detector element. The two photons were constrained to form a π^0 mass, and the decay position of the π^0 had to fall within the vacuum decay region. The π^0 had to have $160 < p_T < 260$ MeV/c to reject background from hyperons and $K_L \to \gamma\gamma$ decays. Very tight cuts were applied to remove events with energy in the photon vetoes. Figure 8 shows the distribution of p_T for events which pass all other cuts.

The single event sensitivity for $K_L \to \pi^0 \nu \bar{\nu}$ was 4.0×10^{-7}. One event was observed, which was consistent with the expectation from neutron interactions producing multiple π^0's or η's in the detector material. However, in calculating the limit, we made the conservative assumption that the event is signal, obtaining

$$\mathrm{BR}(K_L \to \pi^0 \nu \bar{\nu}) < 1.6 \times 10^{-6} \quad (90\% \text{ C. L.}).$$

This result is described in detail in Ref. [18].

CONCLUSIONS

The KTeV experiments have a broad program of rare kaon and pion decay studies. Many interesting decay modes, which have not been described here, are being studied or searched for. A partial list includes $K_L \to \pi^0 \mu^+ \mu^-$, $K_L \to \pi^0 e^+ e^- \gamma$, $K_L \to \pi^0 e^{\pm} \mu^{\mp}$, and the Dalitz and double Dalitz decays of the K_L and π^0.

Although the observation of Standard Model direct CP-violation in rare K_L decays is not yet within reach, the situation for current and future experiments is far from grim. The current limits on $K_L \to \pi^0 \nu \bar{\nu}$ are four to five orders of magnitude above Standard Model expectations, and the $\pi^0 \to \gamma\gamma$ mode is background-limited in our current experiment. However, our detector was optimized to measure $\mathrm{Re}(\varepsilon'/\varepsilon)$, not $K_L \to \pi^0 \nu \bar{\nu}$. The most serious backgrounds affecting our searches can be removed easily in any new experiments: keeping the neutral beam in vacuum up to the face of the calorimeter will remove nearly all beam interaction backgrounds, and lowering the primary beam energy (for example, using 120 GeV Main Injector beam) will remove hyperon contamination from the beams. Obtaining the necessary kaon flux should not be a problem; experiments at BNL have already produced $\mathcal{O}(10^{13})$ kaon decays. The major challenge for future experiments will be the design of hermetic photon veto systems to reject $K_L \to 3\pi^0$ and $K_L \to 2\pi^0$ backgrounds.

A Tevatron fixed-target run is scheduled for 1999, in which we expect to accumulate 2 to 4 times more statistics on all the modes described here. This new data set will allow us to search for exotic enhancements to these decays, as well as provide a valuable next step toward the search for these decays at Standard Model sensitivities.

ACKNOWLEDGMENT

We thank the technical staffs of the participating institutions, as well as the FNAL accelerator staff, for their contributions to the experiments. This work was supported in part by the U. S. Department of Energy, the National Science Foundation, and the Ministry of Education and Science of Japan.

REFERENCES

1. M. Arenton, *Proceedings of the Workshop on Heavy Quarks at Fixed Target (HQ98)*, Batavia, Ill. (1998)
2. S. Drell, *Nuovo Cim.* **XI**, 693 (1959)
3. L. Bergström et al., *Phys. Lett.* **B126**, 117 (1983)
4. L. Bergström, *Z. Phys.* **C20**, 135 (1983)
5. M. J. Savage, M. Luke, and M. B. Wise, *Phys. Lett.* **B291**, 481 (1992)
6. D. Gomez Dumm and A. Pich, *Phys. Rev. Lett.* **80**, 4633 (1998)
7. Ll. Ametller, A. Bramon, and E. Massó, *Phys. Rev.* **D48**, 3388 (1993)
8. G. Colangelo and G Isidori, *J. High Energy Phys.* **9809**, 009 (1998)
9. J. F. Donoghue and F. Gabbiani, Phys. Rev. **D51**, 2187 (1995)
10. D. A. Harris et al., *Phys. Rev Lett.* **71**, 3918 (1993)
11. H. B. Greenlee, *Phys. Rev.* **D42**, 3724 (1990)
12. G. D'Ambrosio et al., *J. High Energy Phys.* **9808**, 004 (1998)
13. Particle Data Group, *Eur. Phys. J* **C3**, 1 (1998)
14. G. D'Ambrosio and J. Portolés, *Nucl. Phys.* **B492**, 417 (1997)
15. G. Barr et al., *Phys. Lett.* **B284**, 440 (1992)
16. L. Littenberg, *Phys. Rev.* **D39**, 3322 (1989)
17. A. Buras, *Phys. Lett.* **B333**, 476 (1994)
18. J. Adams, et al., e-print hep-ex/9806007 (1998)

CONFERENCE SUMMARY

Perspectives on Heavy Quark 98

Chris Quigg

Fermi National Accelerator Laboratory[1]
P.O. Box 500, Batavia, Illinois 60510 USA
E-mail: quigg@fnal.gov

Abstract. I summarize and comment upon some highlights of *HQ98*, the Workshop on Heavy Quarks (strange, charm, and beauty) at Fixed Target.

HISTORICAL PROLOGUE

Half a century has passed since George Rochester and Clifford Butler announced their discovery of "vee particles," penetrating products of cosmic-ray showers that proved to be K mesons, the first strange particles [1]. Through the years, it is striking how thoroughly the study of heavy flavors has defined our progress toward an elegant and comprehensive picture of the fundamental constituents and their interactions. Understanding heavy flavors has been essential to understanding the ordinary stuff of everyday matter.

To kick off *HQ98*, we had the pleasure of hearing reminiscences from Lincoln Wolfenstein [2], Jon Rosner [3], and Tony Sanda [4] on the beginnings of our understanding of strangeness, charm, and beauty. They offered interesting lessons in where we have been, and where we hope to go.

Not all history lies so far in the past. *HQ98* weekend witnessed an important event for the future of Fermilab and of heavy-quark physics. At 17:21 on Saturday, October 10, circulating beam was established for the first time in the Main Injector. This new element in Fermilab's cascade of accelerators, a proton synchrotron precisely π km around, will play a double role as injector to the Tevatron and as the high-intensity source for a new 120-GeV fixed-target program at Fermilab. On behalf of the participants in *HQ98*, it is my pleasure to salute our colleagues for this fine achievement, and to wish them continued success in commissioning our newest accelerator.

---

[1] Fermilab is operated by Universities Research Association Inc. under Contract No. DE-AC02-76CH03000 with the United States Department of Energy.

$B - \bar{B}$ MIXING AND RELATED TOPICS

Mixing Phenomenology and Experiment

I gave a similar conference summary at *Heavy Flavors'87* at Stanford [5], so it was natural to look through my transparencies from that meeting while preparing this talk. Although I agree with the many speakers here at *HQ98* who have said that we are just at the beginning of the study of this or the serious study of that, what struck me most was the very dramatic progress we have made in nearly every aspect of heavy-quark physics.

To take a prominent example, in 1987 we were just digesting the first evidence for particle-antiparticle mixing in the neutral B mesons. For some time, there had been provocative indications from the UA1 experiment in the form of an excess of same-sign dimuon events over what can be accounted for in the absence of mixing [6]. Because of theoretical prejudice that $B_s - \bar{B}_s$ mixing might be large, and because of published upper bounds on the rate of $B_d - \bar{B}_d$ mixing, this result was taken as the scent of $B_s - \bar{B}_s$ mixing. Since this interpretation relied on simulations, and UA1 had not reconstructed any B mesons, the case for mixing was not proved.

The needed proof was supplied by the ARGUS Collaboration working at the DORIS storage ring [7]. The demonstration that $B_d - \bar{B}_d$ mixing takes place came in the form of a single (nearly) reconstructed $B^0 B^0$ event produced in the chain

$$e^+ e^- \rightarrow \Upsilon(4S) \rightarrow B^0 \bar{B}^0$$
$$\hookrightarrow B^0 \ .$$

The two neutral B^0s, which must be nonstrange because the B_s cannot be pair-produced at the $\Upsilon(4S)$, were identified in the decay chains

$$\begin{aligned} B^0 &\rightarrow D^{*-} \mu^+ \nu \\ &\hookrightarrow \pi^- \bar{D}^0 \\ &\qquad \hookrightarrow K^+ \pi^- \end{aligned} \quad (1)$$

and

$$\begin{aligned} B^0 &\rightarrow D^{*-} \mu^+ \nu \\ &\hookrightarrow \pi^0 \bar{D}^- \\ &\qquad \hookrightarrow K^+ \pi^- \pi^- \ , \\ &\hookrightarrow \gamma\gamma \end{aligned} \quad (2)$$

both fully reconstructed, except for the undetected neutrinos. Inspired by this event, the ARGUS experimenters carried out two statistical analyses using dilepton events or incompletely reconstructed $B^0 \rightarrow D^* \ell \nu$ events to determine the degree of

$B_d - \bar{B}_d$ mixing. They estimated $x_d \equiv \Delta M_B/\Gamma_B \simeq 0.7$ (where Γ_B is the average lifetime of the heavy and light B^0 states, to be compared with $\Delta M_K/\Gamma_K \simeq 0.5$.

At *HQ98*, we have seen two lovely examples of *time-dependent* $B^0 - \bar{B}^0$ oscillations in the talks by Kevin Pitts [8], representing CDF, and Achille Stocchi [9], representing the four LEP collaborations. These plots, which are based on thousands of clean events, will take their place in textbooks alongside the classic plots of the time evolution of $K^0 - \bar{K}^0$ oscillations. They represent phenomenal progress since the discovery of $B^0 - \bar{B}^0$ mixing in 1987. And this is not all. Through the combined efforts of ALEPH, DELPHI, OPAL, and L3 at LEP, CDF at the Tevatron, and SLD at the Stanford Linear Collider, we now can quote a very precise world average for the mass difference [8]

$$\Delta M_B = 0.475 \pm 0.010_{\text{stat}} \pm 0.014_{\text{sys}} \text{ ps}^{-1}. \tag{3}$$

so that $x_d \approx 0.74$.

This much is solid achievement, but a great deal more is in the works. We expect the frequency of $B_s - \bar{B}_s$ oscillations to be much more rapid than that of $B^0 - \bar{B}^0$. The LEP experiments plus CDF and SLD can now set a lower bound of $\Delta M_{B_s} > 12.4$ ps^{-1} [10], which implies $x_s \equiv \Delta M_{B_s}/\Gamma_{B_s} \gtrsim 18.5$. An observation of $B_s - \bar{B}_s$ mixing would fix the ratio of the quark-mixing matrix elements V_{td} and V_{ts}. While the LEP experiments continue to press their analyses, the greatest reach in the near future will come from CDF and D, using the 2 fb^{-1} of data each experiment will accumulate in Run II of the Tevatron Collider. For the baseline detector, CDF anticipates a reach in the range $x_s \approx 30 - 40$; additional upgrades could extend the reach to $x_s \approx 55 - 65$ [8].

In the standard electroweak theory, the dominant contributions to $B - \bar{B}$ mixing come from box diagrams involving loops of W bosons and quarks, most importantly top quarks. These lead to expressions for the mixing parameters,

$$x_d = \frac{\Delta M_B}{\Gamma_B} \propto f_{B_d}^2 |V_{td}^* V_{tb}|^2 B_{B_d} \tau_{B_d} m_t^2 \tag{4}$$

and

$$x_s = \frac{\Delta M_{B_s}}{\Gamma_{B_s}} \propto f_{B_s}^2 |V_{ts}^* V_{tb}|^2 B_{B_s} \tau_{B_s} m_t^2, \tag{5}$$

that contain many parameters. I think the worst moment in my career as a summary speaker came during that talk at *Heavy Flavors - '87*. When I flashed these formulas on the screen, I suddenly became aware that I could pronounce the name of every parameter, but I didn't know the value of a single one! Our ignorance in 1987 ranged from the uncertain relationship between quark matrix elements and hadronic matrix elements subsumed in the infamous B parameters to the mass of the top quark and the B-meson lifetimes.

I'm very happy that a decade of progress means I do not have to relive that unsettling moment. Harry Cheung's review of charm and beauty lifetimes [11] presents us with wonderfully precise values for $\tau_{B_d} = (1.556 \pm 0.027)$ ps and $\tau_{B_s} = (1.489 \pm 0.058)$ ps. The top mass, which was entirely unknown in 1987, is now known to very impressive precision. My informal average of the latest results from CDF [12] and D [13] yields $m_t = (173.8 \pm 4.8)$ GeV/c^2. Andreas Kronfeld reported on the development of lattice QCD calculations of the pseudoscalar decay constants and related parameters [14]. The study of heavy (b and c) quarks on the lattice, which coincidentally began in 1987, is now a mature subject. I think it is fair to say that almost definitive calculations of f_B and f_{B_s} (as well as f_D and f_{D_s}) are in hand, and that convergence on the B_B parameters is on the horizon. It would be incautious of me to record "best values," but I think it is useful to quote *representative values* drawn from Kronfeld's compilation: $f_B \approx 165 \pm 15$ MeV, $f_{B_s} \approx 188 \pm 15$ MeV, $f_D \approx 195 \pm 15$ MeV, and $f_{D_s} \approx 220 \pm 15$ MeV, not including the estimated effects of quark loops, which are thought to increase the values about 10%. Our best experimental test of these calculations comes from the purely leptonic decays $D_s \to \mu\nu$ and $\tau\nu$, which yield a world average value, $f_{D_s} = 254 \pm 31$ MeV [9]. This is encouragingly close to the calculated value adjusted for quark loops. The "bag factors" B_B are not in such settled condition; the scatter among calculated values still exceeds the uncertainties attributed to the calculations. More work is needed, but I do hope that convergence on reliable values is near. I am encouraged to note that the suitably defined bag factor for the neutral kaon system does seem to have converged to a value near 0.62 (see Ref. [14] for the precise definition), with little sensitivity to the omission of light-quark loops.

The remaining quantities, the quark-mixing matrix elements, are known to reasonable precision if we assume the three-generation picture and invoke unitarity of the CKM matrix to fix their values. (The Particle Data Group advises $|V_{td}| = 0.004$ to 0.013, $|V_{ts}| = 0.035$ to 0.042, and $|V_{tb}| = 0.9991$ to 0.9994 [15].) But if we demand a measurement, it is from the study of $B - \bar{B}$ oscillations that we get our best information about $|V_{td}|$ and $|V_{ts}|$. Single-top production in $\bar{p}p$ collisions will give us our first measurement of $|V_{tb}|$.

CP Violation in the B System

One of our very important near-term goals is the observation and detailed study of CP violation in B meson decays. If the observed CP violation in the neutral kaon system indeed arises from the phase in the Cabibbo–Kobayashi–Maskawa quark mixing matrix, then the manifestations of CP violation in B decays will be rich and informative. Distinguishing B^0 from $\bar{B}^0$ by means of a same-side–tagging technique, the CDF Collaboration has performed a first measurement of the asymmetry

$$\mathcal{A} = \frac{N(\bar{B}^0 \to \psi K_S) - N(B^0 \to \psi K_S)}{N(\bar{B}^0 \to \psi K_S) + N(B^0 \to \psi K_S)}, \tag{6}$$

using 200 events in which both muons from the decay $\psi \to \mu^+\mu^-$ are reconstructed in the silicon vertex detector. In the standard model, this time-varying CP-violating asymmetry is given by

$$A(t) = \sin(\Delta M_B t) \cdot \sin 2\beta, \qquad (7)$$

where β is the angle in the complex plane between V_{td} and V_{cb}^*. The CDF measurement [8,16],

$$\sin 2\beta = 1.8 \pm 1.1_{\text{stat}} \pm 0.3_{\text{sys}}, \qquad (8)$$

is statistics limited. The dominant contribution to the systematic error comes from the uncertainty in the dilution factor. While it is more in the nature of a dress rehearsal than an informative measurement, this CDF exercise is an important step on the path toward the discovery of CP violation in the B system [17].

The search for CP-violating effects in the B system will soon begin in earnest. The fixed-target experiment HERA–B has begun to take data, as we heard from B. Schwingenheuer [18]. Commissioning is well under way at the SLAC B factory, and the BABAR experiment described by G. Bonneaud [19] is approaching completion, with first data expected in 1999. Also in 1999, we expect the first run of BELLE at the KEK B factory [20]. The following year, the upgraded CDF and D detectors will profit from the greatly enhanced luminosity that the Main Injector will bring to the Tevatron Collider [8]. Farther in the future are LHCb [21] and the BTeV proposal at Fermilab [22]. A general survey of the promise of forthcoming experiments was presented at HQ98 by Marina Artuso [23].

While the new experiments prepare themselves for serious data-taking, the highly successful Cornell Electron Storage Ring, which we may regard as a stationary center-of-momentum B factory, continues to produce a rich harvest of physics results in the upgraded CLEO detector. Among the CLEO results presented at HQ98, those reported by Peter Gaidarev on rare decays [24] are of special interest to the search for CP violation. The CLEO measurements of the branching fractions for $B^0 \to K\pi, \pi\pi$, and KK will inform the search for, and interpretation of, a CP asymmetry in the decays $(B^0, \bar{B}^0) \to \pi^+\pi^-$. The current values are $B(B^0 \to K^\pm \pi^\mp) = (1.4 \pm 0.3_{\text{stat}} \pm 0.2_{\text{sys}}) \times 10^{-5}$ and $B(B^0 \to \pi^+\pi^-) < 0.8 \times 10^{-5}$ at 90% CL. The small $\pi^+\pi^-$ branching fraction complicates the program to extract the angle α between V_{ub}^* and V_{td}.

Before leaving the CLEO data on rare decays, let us note that the new measurement of the inclusive $b \to s\gamma$ branching fraction,

$$B(b \to s\gamma) = (3.15 \pm 0.35_{\text{stat}} \pm 0.32_{\text{sys}} \pm 0.26_{\text{model}}) \times 10^{-4}, \qquad (9)$$

is in good agreement with standard-model expectations, and limits the phase space of models for new physics.

Baryogenesis

Together with the existence of fundamental processes that violate baryon number and a departure from thermal equilibrium during the epoch in which baryon-number–violating processes were important, microscopic CP violation is a necessary condition for generatng a nonvanishing baryon number from an initially symmetrical universe. As Peter Arnold reviewed in his talk at $HQ98$ [25], the CP violation we attribute to a phase in the quark mixing matrix does not appear capable of generating a baryon-to-photon ratio nearly as large as the value

$$n_B/n_\gamma \approx 10^{-9\pm 1} \tag{10}$$

inferred from astronomical observations. Nevertheless, we hope that lessons learned from the study of CP-violating phenomena in the domain of heavy quarks will inform our eventual understanding of the baryon number of the universe. One of the most active areas of recent theoretical work has been to elaborate the possibility that baryogenesis occurs on the scale of electroweak symmetry breaking. In this scenario, it is not possible to generate a large enough baryon-to-photon ratio in the minimal electroweak theory with a single Higgs doublet. The feat can be accomplished in a supersymmetric theory, but only if the lightest Higgs boson is very light, $M_h \lesssim 105 \text{ GeV}/c^2$, and the stop squark weighs no more than the top quark [26]. Under these conditions, both h and $\tilde{t}$ should be accessible soon at LEP200 and the Tevatron Collider, and we can expect departures from the standard-model CP phenomenology in the B mesons.

HEAVY-FLAVOR SPECTROSCOPY

Excited Mesons

For many years, the principal focus of charm and beauty physics has been on the weak decays of states stable against strong or electromagnetic decay. From these, we have learned important lessons about the structure of the weak charged current and about the interplay between the strong and weak interactions. Over the past five years or so, the study of excited states—especially excited meson resonances—has taken on a new interest, as high-statistics experiments with excellent mass resolution have attained maturity. We now know a good deal about the meson states $c\bar{q}$, $c\bar{s}$, $b\bar{q}$, and $b\bar{s}$ beyond the ground-state (0^{-+} and 1^{--}) doublet.

The gross structure of the spectra of heavy-light mesons is rather well understood, from a combination of potential-inspired intuition and heavy-quark effective theory. The fine structure of the spectra is not an unambiguous prediction of HQET; thus, experimental results may provide some surprises and some new insights. To give an example, the separation of the centroids of the $j_q = 3/2$ (1^{++} and 2^{++}) and $j_q = 1/2$ (0^{++} and 1^{++}) doublets is not a robust prediction of the heavy-quark theory [27]. We believe that HQET and the chiral quark model do give us the tools we need to describe the strong-interaction transitions among mesonic states with precision, as described by Estia Eichten in his talk at *HQ98* [28]. The $j_q = 3/2$ states are expected to be narrow; the charmed states are thoroughly known, and the beauty states are under investigation in a number of experiments. Franz Muheim [29] showed what the LEP experiments have been able to achieve in analyses that rely to varying degrees on the predictions [30] of heavy-quark effective theory. The experimental observations are in broad agreement with HQET, and indicate that a significant fraction of B mesons (25 to 40%) are produced through the p-wave B^{**} states. This conclusion holds great interest for the same-side tagging of B flavor for studies of CP violation.

The $j_q = 1/2$ levels have not been established yet. They are expected to be broad, but theorists are only beginning to endow that label with a numerical meaning. At *HQ98*, Jorge Rodriguez [31] presented CLEO's evidence for the $D_1^*(j_q = 1/2)$ state near 2461 MeV/c^2, with a total width of about 200 – 400 MeV. This seems considerably broader than the theoretical expectation [28,32] that $\Gamma(0^+ \to 0^-\pi) \approx \Gamma(1^+ \to 1^-\pi) \approx 85$ MeV. In the B system, an L3 analysis reported by Muheim [29] suggests that the B_1^* has a mass of $5675 \pm 12 \pm 4$ MeV/c^2 and a width of $78 \pm 28 \pm 15$ MeV, in the range suggested by chiral-quark-model calculations. We can expect both theoretical and experimental progress by the time of *HQ2000*.

The DELPHI Collaboration [33] recently reported an excess of events in the $D^{*+}\pi^+\pi^-$ mass spectrum at a mass of $2637 \pm 2 \pm 6$ MeV, which is consistent with expectations for the first radial excitation of the D^* meson. The width of the excess is quite small, consistent with their experimental resolution. Both OPAL (see Muheim, Ref. [29]) and CLEO (see Rodriguez, Ref. [31]) have presented upper limits that are inconsistent with the DELPHI observation.

Charmed Baryons

In SU(4)$_{\text{flavor}}$ symmetry, the ground-state baryons include 20 $1/2^+$ states, of which 12 contain one or more charmed quarks, and 20 $3/2^+$ states, of which 10 contain one or more charmed quarks. All nine of the singly-charmed $1/2^+$ states and four of the six singly-charmed $3/2^+$ states have been observed. Only the Σ_c^{*+} and Ω_c^* remain undetected. Sajjad Alam [34] reviewed CLEO's extensive contributions to charmed-baryon spectroscopy, while Eric Vaandering [35] summarized the achievements of Fermilab Experiment E687 and the promise of its successor, FOCUS. The multiply charmed baryons (and, more generally, heavy-heavy-light baryons) remain tempting experimental targets, in part for the analogy in HQET to heavy-light mesons [28].

Charmonium Spectroscopy

Todd Pedlar reported on Fermilab experiment E835 [36], the study of charmonium spectroscopy in resonant $\bar{p}p$ annihilations. E835 is conducted in what we think of as the Fermi National *Decelerator* Laboratory, when the Antiproton Accumulator decelerates antiprotons from 8.9 GeV/c to the momenta required for resonant formation of $c\bar{c}$ states, in the range 4 – 7 GeV/c. E835 reports the first evidence for formation of the 1^3P_0 state χ_0 in $\bar{p}p$ annihilations. The resonance parameters, $M(\chi_0) = (3415^{+2.1}_{-1.7})$ MeV/c^2 and $\Gamma(\chi_0) = (13.9^{+5.3}_{-3.9})$ MeV, are in good agreement with Particle Data Group averages [15]. The $\bar{p}p$ branching fraction, while consistent with the preëxisting upper bound, is tantalizingly large:

$$B(\chi_0 \to \bar{p}p) = (4.24^{+0.96}_{-0.70} \pm 1.16) \times 10^{-4}. \qquad (11)$$

The BES Collaboration has just published the first determination of this branching fraction in e^+e^- collisions at the $\psi(2S)$ [37]; their value is $(1.59\pm0.43\pm0.53)\times 10^{-4}$.

Despite assiduous efforts, the E835 Collaboration has not been able to find any sign of the expected pseudoscalar radial excitation η'_c (2^1S_0). The old Crystal Ball claim of a state at 3594 MeV/c^2, which was too distant from the ψ' for theoretical comfort, is rather decisively ruled out, but it is somewhat maddening that the real η'_c has not shown itself. What are we missing?

Production of Heavy Flavors

A wealth of information about the production of heavy quarks was presented at *HQ98*. Fred Olness [38] presented an excellent summary of the outstanding theoretical and experimental issues. A key point is that heavy-quark production processes are both challenging and interesting for our understanding of perturbative QCD because they typically involve two large scales: the heavy-quark mass, or threshold, and the typical large-momentum scale that encourages the application of perturbative techniques. Once we have reliable predictions for the kinematic distributions of the heavy quarks, we need to understand how those are reflected in the kinematic distributions of the hadrons that contain those quarks. The current state of understanding in the Lund string-fragmentation approach was reported to *HQ98* by Emanuel Norrbin [39]. Fermilab Experiment E791's new measurements of the production asymmetries $A \equiv (\sigma_X - \sigma_{\bar{X}})/(\sigma_X + \sigma_{\bar{X}})$ for $D^{\pm}$, $D_s^{\pm}$, and Λ_c baryons and antibaryons, presented by Kevin Stenson [40], offer empirical insight into the fragmentation process. The HERMES experiment at HERA has completed three years of successful data taking that includes a measurement of the cross section for J/ψ photoproduction and a clear open charm signal [41], still under analysis. We also look forward to the operation of the COMPASS experiment at CERN [42].

Collider data on charmonium production have forced us to look more broadly for the right physical picture of the process than the color-singlet mechanism in which the observed charmonium state has the quantum numbers of the produced $c\bar{c}$ pair. Since the discovery by the CDF Collaboration [43] that the direct production of both J/ψ and ψ' occur at some fifty times the color-singlet–model rate, our attention has been drawn to a color-octet mechanism in which the produced $c\bar{c}$ pair evolves into a color-singlet hadron by emitting a soft gluon. For this reason, it is sometimes called the color-evaporation mechanism. At *HQ98*, Andrzej Zieminski [44] presented recent results from the D Collaboration that test the color-octet picture of J/ψ production into the regime of large rapidity and small transverse momentum. Within uncertainties, the color-octet model describes the pseudorapidity dependence of J/ψ production at all angles. An interesting production-mechanism diagnostic is the polarization of the produced charmonium state. In a thorough report on charm and beauty production in CERN Experiment WA92, Dario Barberis [45] showed that in 350-GeV/c π^-A collisions, the J/ψ polarization is small, in agreement with the color-octet picture. Ting-Hua Chang [46] similarly reported that in 800-GeV/c pCu collisions, the NUSEA Collaboration (Fermilab Experiment E866) observes no polarization in the J/ψ decay angular distribution integrated over all production angles (represented by x_F).

Models for charmonium production need also to confront the photoproduction data shown by Beate Naroska [47] in her summary of heavy-flavor work at H1 and ZEUS. The HERA measurements, taken in a kinematic regime where diffraction dominates, are described very well by a next-to-leading-order color singlet model, and do not demand a significant color-octet contribution. The link between nonperturbative parameters set by CDF data and the consequences for diffractive production at HERA involves some subtleties that need further work to resolve.

In heavy-ion collisions, charmonium suppression—beyond the normal nuclear absorption long observed in hadron–nucleus and light-ion collisions—has been predicted as a diagnostic for the creation of a quark-gluon plasma, *i.e.*, a deconfined state of hadronic matter. A sudden drop in the J/ψ yield as a function of the product of target and projectile mass numbers has been observed in Pb–Pb collisions by the NA50 Collaboration at CERN, as we heard in the heavy-ion summary by Carlos Lourenço [48]. The effect is large: the measured point lies some 4.5σ below the extrapolation from smaller values of the mass product. It is clearly a tantalizing hint.

The CCFR Collaboration at Fermilab has used the production of one heavy quark, charm, to investigate the population of another heavy quark, strange, in the nucleon. Todd Adams [49] presented the results of their next-to-leading-order fits to the charm production cross section in $(\nu_\mu, \bar{\nu}_\mu)N$ deeply inelastic scattering. They confirm the expectation (and the conclusion of earlier experiments) that the nucleon sea is not SU(3)-symmetric, and find that the inferred value of the charm quark mass is analysis dependent. The successor experiment, NuTeV, has an improved data sample that will permit more incisive analyses.

STRANGE-PARTICLE DECAYS

Search for Direct *CP* Violation

CP violation can arise from a small impurity in the mass eigenstates of the K^0–$\bar{K}^0$ complex, or from a "direct" *CP*-violating contribution to the decay amplitudes, or from interference between the two. If CPT is a good symmetry, the mass eigenstates can be written as

$$|K_S\rangle = p|K^0\rangle + q|\bar{K}^0\rangle, \qquad |K_L\rangle = p|K^0\rangle - q|\bar{K}^0\rangle. \qquad (12)$$

If CP invariance held, we would have $q = p = 1/\sqrt{2}$, so that K_S would be CP-even and K_L CP-odd. In a convenient phase convention [50], we can express CP violation in K^0–$\bar{K}^0$ mixing through the parameter ε ($|\varepsilon| \approx 2.28 \times 10^{-3}$, with a phase near $45°$),

$$\frac{p}{q} = \frac{(1+\varepsilon)}{(1-\varepsilon)}. \qquad (13)$$

CP violation in the decay amplitudes gives rise to an inequality—a phase difference—in the amplitudes $A(K^0 \to \pi\pi(I))$ and $A(\bar{K}^0 \to \pi\pi(I))$, where I is the isospin of the $\pi\pi$ system. It is conventional to express the CP-violating observables in terms of the parameters ε and ε', the latter defined through

$$\eta_{+-} \equiv \frac{A(K_L \to \pi^+\pi^-)}{A(K_S \to \pi^+\pi^-)} = \varepsilon + \varepsilon',$$

$$\eta_{00} \equiv \frac{A(K_L \to \pi^0\pi^0)}{A(K_S \to \pi^0\pi^0)} = \varepsilon - 2\varepsilon'. \qquad (14)$$

The observable $|\eta_{+-}|^2/|\eta_{00}|^2 \approx 1 + 6\,\mathrm{Re}(\varepsilon'/\varepsilon)$ is very close to unity. In the electroweak theory, a tiny deviation from one arises from the phase in the quark mixing matrix. In the electroweak theory, it seems most likely that ε'/ε should be on the order of 10^{-3}, or perhaps smaller [51].

Published measurements of ε'/ε have already reached a remarkable level of precision, with E731 at Fermilab reporting $(0.74 \pm 0.52 \pm 0.29) \times 10^{-3}$ [52] and the NA31 experiment at CERN quoting $(2.3 \pm 0.65) \times 10^{-3}$ [53]. If we take both of these beautiful results at face value, then both the existence and perforce the magnitude of a direct CP-violating amplitude remain in doubt. Three experiments now in progress aim at a precision that will settle the issue. NA48 at CERN and KTeV at Fermilab have already logged very significant data sets. Augusto Ceccucci reported [54] that NA48 anticipates a result based on their 1997 data in time for the winter conferences. The statistical uncertainty on $\mathrm{Re}(\varepsilon'/\varepsilon)$ should be around $(4-5) \times 10^{-4}$, and the systematic error should be still smaller. According to Mike Arenton [55], KTeV is nearing a final result based on 20% of the data they recorded in 1996 and 1997. They expect a statistical uncertainty of about 3×10^{-4}, and are currently studying systematic effects. And we learned from G. Bencivenni that when the ϕ factory DAΦNE operates at full luminosity, the KLOE experiment should be able to probe $\mathrm{Re}(\varepsilon'/\varepsilon)$ to 10^{-4} [56].

CP Violation in Hyperon Decay?

To understand the origin of CP violation, it is a matter of urgent interest to find CP-violating phenomena outside the neutral-kaon system. A natural place to look is in the decays of other strange particles, notably the hyperons of the baryon octet. HYPERCP at Fermilab, described at $HQ98$ by Cat James [57], is the first experiment dedicated to the search for CP violation in hyperon decay. HYPERCP uses a high-rate spectrometer to compare the decay chains

$$\Xi^- \to \Lambda\pi^-$$
$$\hookrightarrow p\pi^-$$

and

$$\bar{\Xi}^+ \to \bar{\Lambda}\pi^+$$
$$\hookrightarrow \bar{p}\pi^+.$$

The decay angular distribution of the proton in the Λ rest frame,

$$\frac{dN}{d\cos\theta} = \tfrac{1}{2}(1 + \alpha_\Lambda \alpha_\Xi \cos\theta), \tag{15}$$

where θ is the angle between the proton momentum and the Λ polarization vector, is characterized by the asymmetry parameters of the sequential hyperon decays. HYPERCP aims to measure the CP-violating asymmetry,

$$\mathcal{A} = \frac{\alpha_\Lambda \alpha_\Xi - \alpha_{\bar{\Lambda}} \alpha_{\bar{\Xi}}}{\alpha_\Lambda \alpha_\Xi + \alpha_{\bar{\Lambda}} \alpha_{\bar{\Xi}}} \approx A_\Lambda + A_\Xi , \tag{16}$$

with a sensitivity of one part in 10^4, adding data from the 1999 run to the data in hand from 1997. Published predictions for the joint asymmetry range from about 10^{-5} to a few times 10^{-4}, whereas the superweak model predicts no asymmetry.

Direct Observation of T-Violation

Perhaps the most satisfying new result presented at $HQ98$ was the KTeV observation [55] of a time-reversal–violating asymmetry in the rare decay $K_L \to \pi^+\pi^- e^+ e^-$. The best evidence that the $\pi^+\pi^- e^+ e^-$ mode qualifies as rare is that it had not been reported until this year [58]. An analysis of about 60% of KTeV's 1997 data set now allows a precise determination of the branching fraction as $(3.32\pm 0.14\pm 0.28)\times 10^{-7}$. The interest in this decay mode derives from the fact that the underlying process, $K_L \to \pi^+\pi^-\gamma$, proceeds through both CP-conserving and CP-violating mechanisms. A Bremsstrahlung component is associated with the CP-violating $K_L \to \pi^+\pi^-$ decay, while a direct-emission component arises from a CP-conserving M1 transition. The interference between amplitudes with different CP properties can lead to a CP-violating effect in the photon polarization. The

$K_L \to \pi^+\pi^-e^+e^-$ channel, which represents the internal conversion of the photon to an electron-positron pair, analyzes the virtual photon polarization through the orientation of the e^+e^- plane relative to the $\pi^+\pi^-$ plane [59].

To be explicit, let $\hat{\mathbf{n}}_\pi = (\mathbf{p}_{\pi^+} \times \mathbf{p}_{\pi^-})/|\mathbf{p}_{\pi^+} \times \mathbf{p}_{\pi^-}|$ be the normal to the pion plane and $\hat{\mathbf{n}}_e = (\mathbf{p}_{e^+} \times \mathbf{p}_{e^-})/|\mathbf{p}_{e^+} \times \mathbf{p}_{e^-}|$ be the normal to the electron plane, and define the azimuthal angle φ through $\cos\varphi = \hat{\mathbf{n}}_\pi \cdot \hat{\mathbf{n}}_e$. The decay angular distribution is of the form $d\Gamma/d\varphi = \Gamma_1 \cos^2\varphi + \Gamma_2 \sin^2\varphi + \Gamma_3 \sin\varphi\cos\varphi$. We can express $\sin\varphi = (\hat{\mathbf{n}}_\pi \times \hat{\mathbf{n}}_e) \cdot \hat{\mathbf{z}}$, where $\hat{\mathbf{z}} = (\mathbf{p}_{\pi^+} + \mathbf{p}_{\pi^-})/|\mathbf{p}_{\pi^+} + \mathbf{p}_{\pi^-}|$. Under the action of the charge conjugation operator C, we have $\mathbf{p}_{\pi^\pm} \to \mathbf{p}_{\pi^\mp}$ and $\mathbf{p}_{e^\pm} \to \mathbf{p}_{e^\mp}$, so that $\hat{\mathbf{n}}_\pi \to -\hat{\mathbf{n}}_\pi$, $\hat{\mathbf{n}}_e \to -\hat{\mathbf{n}}_e$, and $\hat{\mathbf{z}} \to \hat{\mathbf{z}}$. Either the parity operator P or the time-reversal operator T takes $\mathbf{p}_{\pi^\pm} \to -\mathbf{p}_{\pi^\pm}$ and $\mathbf{p}_{e^\pm} \to -\mathbf{p}_{e^\pm}$, so that $\hat{\mathbf{n}}_\pi \to \hat{\mathbf{n}}_\pi$, $\hat{\mathbf{n}}_e \to \hat{\mathbf{n}}_e$, and $\hat{\mathbf{z}} \to -\hat{\mathbf{z}}$. Accordingly, P and T take $\sin\varphi \to -\sin\varphi$ and $\cos\varphi \to \cos\varphi$, while C leaves both $\sin\varphi$ and $\cos\varphi$ unchanged. The presence of a $\sin\varphi\cos\varphi$ term in the decay angular distribution, i.e., of a nonzero value of Γ_3, is direct evidence for time-reversal noninvariance and, since C leaves the decay angular distribution unchanged, for CP violation. Indirect CP violation—the same physics that produces a nonzero value of ε—induces a T-violating asymmetry in the decay-angular distribution whose size is determined by $\text{Im}(\varepsilon)$. The effect is large, because the CP-violating contribution to the $K_L \to \pi^+\pi^-e^+e^-$ decay amplitude occurs at a lower order in chiral perturbation theory than the CP-conserving contribution. Sehgal and Wanninger [60] have computed a forward-backward asymmetry $A = (14.3 \pm 1.3)\%$.

KTeV's full 1997 data set leads to a sample of 1811 ± 42 events, enough to study the decay angular distributions in detail. The preliminary asymmetry presented at *HQ98* is

$$A = (13.5 \pm 2.5_{\text{stat}} \pm 3.0_{\text{sys}})\%, \qquad (17)$$

which represents direct evidence for a violation of time-reversal symmetry. This is the largest particle-antiparticle asymmetry so far observed. The measured value is in good agreement with theoretical expectations.

During the week of *HQ98*, the CPLEAR Collaboration at CERN reported on the first observation of time-reversal symmetry violation through a comparison of the probabilities for the transformations $\bar{K}^0 \leftrightarrow K^0$ as a function of the neutral-kaon proper time [61]. In their experiment, the strangeness of the neutral kaon at the moment of its creation, $t = 0$, was tagged by observing the kaon charge in the formation reaction $\bar{p}p \to K^\pm \pi^\mp (K^0, \bar{K}^0)$ at rest, while the strangeness of the neutral kaon at the time of its semileptonic decay, $t = \tau$, was tagged by the charge of the final-state lepton. The time-average decay-rate asymmetry, measured over the interval $1 \times \tau_s < \tau < 20 \times \tau_s$, is

$$\left\langle \frac{\Gamma(\bar{K}^0|_0 \to e^+\pi^-\nu|_\tau) - \Gamma(K^0|_0 \to e^-\pi^+\bar{\nu}|_\tau)}{\Gamma(\bar{K}^0|_0 \to e^+\pi^-\nu|_\tau) + \Gamma(K^0|_0 \to e^-\pi^+\bar{\nu}|_\tau)} \right\rangle = (6.6 \pm 1.3_{\text{stat}} \pm 1.0_{\text{sys}}) \times 10^{-3}.$$

(18)

This asymmetry is a direct manifestation of T-violation. If CPT is a good symmetry in semileptonic decays and the $\Delta S = \Delta Q$ rule is exact, then the observed asymmetry (18) is identical to

$$\frac{\mathcal{P}(\bar{K}^0 \to K^0) - \mathcal{P}(K^0 \to \bar{K}^0)}{\mathcal{P}(\bar{K}^0 \to K^0) + \mathcal{P}(K^0 \to \bar{K}^0)}, \tag{19}$$

where $\mathcal{P}$ is a probability for strangeness oscillation. The observed result is in good agreement with the theoretical expectation, $4\,\mathrm{Re}(\varepsilon) = (6.63 \pm 0.06) \times 10^{-3}$.

These two new results confirm our expectation that time-reversal invariance is violated in neutral-kaon decays, as must be the case if CPT holds and CP is not respected. In quantitative terms, the newly observed T-violations occur at just the level required to compensate for the CP violation known since 1964 to occur in the decay $K_L \to \pi^+\pi^-$.

Plenty of Nothing

One of the most beautiful results of the year past was the observation by Brookhaven Experiment 787 [62] of a single, very clean, example of the decay

$$K^+ \to \pi^+ \nu \bar{\nu}, \tag{20}$$

corresponding to a branching fraction of $(4.2^{+9.7}_{-3.5}) \times 10^{-10}$ that is consistent with the standard-model expectation, $0.6 \times 10^{-10} \leq B(K^+ \to \pi^+\nu\bar{\nu}) \leq 1.5 \times 10^{-10}$. What is most impressive to me is not the one beautiful candidate event, but the extremely low level of background: the event occurs on an empty field. In the report presented to *HQ98* by Steve Kettell [63], we learned that preliminary indications from the analysis of 1995–1997 data are that the background rejection is three times better, with an increased acceptance. Over the next two years, the E787 sensitivity should provide a thorough survey of the standard-model regime. Brookhaven proposal 949 [64] would increase the sensitivity to 10^{-11}, and the CKM proposal at Fermilab [65] aims at a sensitivity of 10^{-12}. As long as the experimental sensitivity fell far short of standard-model expectations, the principal interest of searching for $K^+ \to \pi^+\nu\bar{\nu}$ was to probe non–standard-model physics. With detection achieved near the band of standard-model predictions, the branching ratio takes on additional importance as a determination of $|V_{td}|$.

A little less nothing has been achieved by KTeV in its search for the companion process, $K_L \to \pi^0 \nu \bar{\nu}$. With a small expected background of 0.12 event in the signal region for the Dalitz-decay final state $(e^+e^-\gamma)\nu\bar{\nu}$, they observe no events, and can quote an upper limit $B(K_L \to \pi^0\nu\bar{\nu}) < 4.9 \times 10^{-7}$ at 90% confidence level [66]. Dedicated experiments to measure the $K_L \to \pi^0\nu\bar{\nu}$ branching fraction are being planned at Fermilab and Brookhaven [67,68]. These would provide an unambiguous determination of $\mathrm{Im}(V_{td})$.

Rarest of Them All

Two formerly rare decays are now being studied with impressive statistical power. KTeV has detected 275 candidates for the decay $\pi^0 \to e^+e^-$ over an expected background of 21 events, for a preliminary branching fraction $B(\pi^0 \to e^+e^-) = (6.09 \pm 0.40_{\text{stat}} \pm 0.23_{\text{sys}}) \times 10^{-8}$ [66]. This is in good agreement with theoretical expectations [69]. Brookhaven Experiment E871 now has accumulated over 6200 candidates for the decay $K_L \to \mu^+\mu^-$ [70]. That, too, is in good agreement with theoretical expectations [71]. They also hold the record sensitivity for the forbidden decay $K_L \to \mu^\pm e^\mp$, and can quote an upper limit on the branching fraction, $B(K_L \to \mu^\pm e^\mp) < 4.8 \times 10^{-12}$.

E871 holds the further distinction of measuring the smallest branching fraction of them all, with their observation of four candidates for the decay $K_L \to e^+e^-$ [70]. These four events lead to a branching fraction $B(K_L \to e^+e^-) = (8.7^{+5.7}_{-4.1}) \times 10^{-12}$, in close agreement with modern calculations based on chiral perturbation theory [71]. This is a very impressive achievement indeed.

Since 1995, Brookhaven experiment E865 has collected data on many rare and formerly rare decays [72]. Among their targets is the lepton-flavor-violating decay $K^+ \to \pi^+\mu^+e^-$, for which they have already set a 90% CL upper limit of 2.1×10^{-10}. The projected single-event sensitivity is 10^{-11}.

Spinoffs

Although the focus of KTeV is the precision study of neutral kaon decays, the KTeV spectrometer is also well-matched to a number of other important physics goals. Doug Jensen [73] presented a preliminary measurement of the branching fraction $B(\Xi^0 \to \Sigma^+ e^- \bar{\nu}_e)$ based on 153 ± 13 events that fit the pattern

$$\Xi^0 \to \Sigma^+ e^- \bar{\nu}_e$$
$$\hookrightarrow p\pi^0,$$

upon a background of 6 ± 2 events. The preliminary branching fraction is $(2.5 \pm 0.2 \pm 0.3) \times 10^{-4}$, in excellent accord with the theoretical expectation of 2.61×10^{-4}.

In a similar spirit, the SELEX experiment, which is mainly concerned with the study of charmed particles produced in a Σ^- beam, has obtained interesting new results on hyperon properties [74]. They have determined the Σ^- charge radius to be $\langle r^2 \rangle = (0.60 \pm 0.08_{\text{stat}} \pm 0.08_{\text{sys}})$ fm^2, measured the $\Sigma^- p$ total cross section to be about 36 mb at 600 GeV/c, and set a new upper bound on the U-spin–forbidden transition rate, $\Gamma(\Sigma(1385)^- \to \Sigma^-\gamma) < 12$ keV at 95% CL.

WEAK INTERACTIONS OF CHARM AND BEAUTY

Lifetimes

The lifetimes of hadrons containing c and b quarks have important engineering value and give us insight into the interplay between the strong and weak interactions. With the development of the heavy-quark expansion, theorists now have well-defined expectations for the hierarchy of b-hadron lifetimes that high-precision data can confront. At *HQ98*, Harry Cheung [11] presented a survey of recent progress in lifetime measurements. Speaking for the E791 Collaboration at Fermilab, Nader Copty [75] presented a new precise measurement of the D_s lifetime, $\tau_{D_s} = 0.518 \pm 0.014 \pm 0.007$ ps. This is considerably larger than the Particle Data Group average, $\langle \tau_{D_s} \rangle_{\rm PDG} = 0.467 \pm 0.017$ ps [15]. Combined with the PDG average lifetime for the D^0, $\langle \tau_{D^0} \rangle_{\rm PDG} = 0.415 \pm 0.004$ ps, the E791 value gives a ratio $\tau_{D_s}/\tau_{D^0} = 1.25 \pm 0.04$, a six-standard-deviation difference from unity. [Harry Cheung's world average, including the new data from E791, is $\tau_{D_s}/\tau_{D^0} = 1.193 \pm 0.027$.] This represents a substantial change from the PDG98 value of 1.125 ± 0.042. We expect further improvements in our knowledge of charm lifetimes from CLEO analyses using the new silicon vertex detector and the forthcoming FOCUS data described by Jonathan Link [76] and Eric Vaandering [35], which will be statistically dominant over the next few years.

The CDF Collaboration has recently used semileptonic decays to determine the lifetime of the B_c meson as $\tau_{B_c} = 0.46^{+0.18}_{-0.16} \pm 0.03$ ps [77], very close to the value expected in the spectator picture [78].

Semileptonic decays

The semileptonic decays of charm and beauty are of interest for the light they can shed on quark mixing matrix elements and on the dynamics embodied in hadronic form factors. We want to both test and exploit the predictions of heavy-quark effective theory for the behavior of form factors and for the connection between D and B decays.

The study of semileptonic B decays constrains the parameters $|V_{cb}|$ and $|V_{ub}|$ that will be crucial for interpreting CP-violating effects in B decays. Karl Ecklund [79] reported recent CLEO results on the exclusive reconstruction of $B \to D\ell\nu$ and $B \to \rho\ell\nu$, and presented a new moment analysis of inclusive semileptonic B decays that may reduce the uncertainties in extracting V_{cb}.

Fermilab Experiment E791 has made important strides in the study of form factor ratios in the decays $D^+ \to \bar{K}^*\ell^+\nu_\ell$ and $D_s^+ \to \phi\ell^+\nu_\ell$. Daniel Mihalcea [80] showed the evolution of these measurements, which now offer a worthy challenge for theory based on lattice QCD. Fermilab Experiment E687 has made competitive measurements of semileptonic form factors in the past. Will Johns [81] reviewed these contributions and demonstrated that the successor experiment, FOCUS, will yield thirty to forty times the number of events used in the E687 semileptonic analyses. The gigantic event sample raises the prospect of high-precision studies of Cabibbo-suppressed semileptonic decays of charm.

More Promises and Prospects

The SELEX experiment at Fermilab is a new spectrometer that took data in 1996–1997 with 600-GeV/c Σ^- and π^- and 540-GeV/c proton beams. They are just beginning to produce preliminary results on their large sample of charm decays. Alex Kushnirenko [82] reported the first observation of the Cabibbo-suppressed decay $\Xi_c^+ \to pK^-\pi^+$.

Mitsuhiro Nakamura [83] presented new limits on $\nu_\mu \to \nu_\tau$ oscillations from the emulsion experiment CHORUS at CERN. They have been able to move the exclusion plot (in the Δm^2 – $\sin 2\theta$ plane) near $\sin 2\theta = 10^{-3}$, a significant increase in sensitivity over previous experiments. Of particular interest to the heavy-quark community are the remarkable advances that have been achieved in automated emulsion scanning. Another order of magnitude in scanning power should be in hand by *HQ2000*.

Rare decays

Gustavo Burdman [84] presented an elegant summary of the potential of rare K, D, and B decays to probe the structure of the electroweak theory at one-loop level. By looking for effects that derive from higher-order processes in the standard model, we may hope to probe momentum scales at or above the scale of electroweak symmetry breaking. The theoretical art lies in identifying processes in which the dominant contributions are from short-distance (high momentum scale) processes.

The experimental status of searches for rare decays and CP violation in the charm system was summarized by Simon Kwan [85]. No CP violation has been observed, with a sensitivity of a few percent. The current limits on rare and forbidden D decays are at the level of one part in 10^5. The current experimental limits on flavor changing neutral current processes are still orders of magnitude above the standard-model expectation, so there is a large window for the discovery of new physics. In the immediate future, the greatest sensitivity to CP violation in the charm sector will come from the B factory experiments, BABAR, BELLE, and CLEO III.

SUMMARY REMARKS

It is a glorious time for heavy-quark physics. The results presented at *HQ98* reflect dramatic progress over the past decade and offer immense promise for the years ahead. For each of the heavy quarks—strange, charm, and beauty—that have occupied our attention at this workshop, experiments in progress and under construction will decisively improve the quality and amount of information available to us. And let us not forget the torrent of new information about the top quark that the next run of the Tevatron Collider will bring [86]. Theoretical advances make it ever clearer that we will be able to interpret the new experimental findings to get at the essence of the interactions of heavy quarks. I look forward, with eager anticipation, to *HQ2000* and beyond.

ACKNOWLEDGEMENTS

It is a pleasure to thank Joel Butler and the local organizing committee for the stimulating and pleasant atmosphere of *HQ98*. I am grateful to my scientific secretaries, Erik Gottschalk, Rob Kutschke, and Erik Ramberg, for providing me with insightful advice and timely copies of transparencies. I thank the workshop staff for performing many small miracles, and Harry Cheung for his attention to the *HQ98 Proceedings*. Andreas Kronfeld, Zoltan Ligeti, and Bruce Winstein made helpful comments on the manuscript.

REFERENCES

1. G. D. Rochester and C. C. Butler, *Nature (London)* **160**, 855 (1947). See also the earlier indication of an intermediate-mass charged particle in L. Leprince-Ringuet and M. L'héritier, *J. Phys. Radium* (sér. 8) **7**, 66, 69 (1946).
2. L. Wolfenstein, "CP Violation: The Kaon Problem," *These Proceedings*.
3. Jonathan L. Rosner, "The Arrival of Charm" (hep-ph/9811359), *These Proceedings*.
4. A. I. Sanda, "High Promises of Beauty," *These Proceedings*.
5. C. Quigg, "Heavy Flavors - '87: Conference Summary," in *Proceedings of the International Symposium on the Production and Decay of Heavy Flavors*, edited by E. Bloom and A. Fridman, *Annals of the New York Academy of Sciences* **535**, 617 (1988).
6. C. Albajar, et al. (UA1 Collaboration), *Phys. Lett.* **186B**, 247 (1987); ibid. **197B**, 565E (1987).
7. H. Albrecht, et al. (Argus Collaboration), *Phys. Lett.* **B192**, 245 (1987).
8. K. T. Pitts, "CP Violation, Mixing, and Rare Decays at the Tevatron, Now and in Run II," FERMILAB–CONF–98/380–E and *These Proceedings*.
9. A. Stocchi, "B^0 – $\bar{B}^0$ Oscillation and Measurements of $|V_{ub}|/|V_{cb}|$ at LEP," *These Proceedings*.
10. The LEP B Oscillation Working Group, "Combined Results on B^0 Oscillations: update for the summer 1998 conferences," LEPBOSC–98/2.
11. H. W. K. Cheung, "Review of Charm and Beauty Lifetimes," *These Proceedings*.
12. F. Abe, et al. (CDF Collaboration), *Phys. Rev. Lett.* **80**, 2767, 2773, 2779 (1998), quote an average value $m_t = (176.0 \pm 6.5)$ GeV/c^2.
13. B. Abbott, et al. (D Collaboration), *Phys. Rev.* **D58**, 052001 (1998), give an average value $m_t = (172.1 \pm 5.2_{\text{stat}} \pm 4.9_{\text{sys}})$ GeV/c^2.
14. A. Kronfeld, "Lattice QCD Calculations of Leptonic and Semileptonic Decays," FERMILAB–CONF–98/393–T (hep-ph/9812288) and *These Proceedings*.

15. C. Caso, et al. (Particle Data Group), *Eur. Phys. J.* **C3**, 1 (1998).
16. F. Abe, et al. (CDF Collaboration), "Measurement of the CP Violation Parameter $\sin(2\beta)$ in $B_d^0/\bar{B}_d^0 \to J/\psi K_S^0$ Decays," FERMILAB-PUB-98-189-E (hep-ex/9806025).
17. K. Ackerstaff, et al. (OPAL Collaboration), *Euro. Phys. J.* **C5**, 379 (1998) (hep-ex/9801022) used a sample of 24 $B^0 \to \psi K_S$ candidates to determine $\sin 2\beta = 3.2^{+1.8}_{-2.0} \pm 0.5$.
18. B. Schwingenheuer, "Status of the HERA-B Experiment, *These Proceedings*. For current information, see http://www-hera-b.desy.de/.
19. G. Bonneaud, "The PEP–II B Factory and the BABAR Detector," *These Proceedings*. For current information, see the B Factory home page at http://www.slac.stanford.edu/BF/doc/www/bfHome.html.
20. Information about BELLE and the KEK B-Factory can be found at http://bsunsrv1.kek.jp/.
21. For information about the LHCb Experiment, consult http://lhcb.cern.ch/.
22. For detailed information about the BTeV proposal, consult http://www-btev.fnal.gov/btev.html.
23. M. Artuso, "B Physics in the Next Millennium," *These Proceedings*.
24. P. B. Gaidarev, "Rare B Decays at CLEO," *These Proceedings*.
25. P. Arnold, "One Reason Why CP Violation is Way Radically Cool," *These Proceedings*.
26. M. Carena, M. Quirós and C. E. M. Wagner, *Nucl. Phys.* **B524**, 3 (1998).
27. We use j_q (q for light quark) to denote the vector sum of the light-quark spin and the orbital angular momentum between the quark and antiquark.
28. E. Eichten, "Heavy Quark Spectroscopy," not included in *These Proceedings*.
29. F. Muheim, "B and D Spectroscopy at LEP," *These Proceedings*.
30. E. J. Eichten, C. T. Hill, and C. Quigg, *Phys. Rev. Lett.* **71**, 4116 (1993); see also "Orbitally Excited Heavy-Light Mesons Revisited," in *The Future of High-Sensitivity Charm Experiments,* Proceedings of the CHARM2000 Workshop, Fermilab, June 1994, edited by D. M. Kaplan and S. Kwan, FERMILAB-Conf-94/190, p. 355, available at http://fnphyx-www.fnal.gov/conferences/ftp_home/charm2000/eichten_2.ps.Z.
31. J. Rodriguez, "Hadronic Decays of Beauty and Charm from CLEO," *These Proceedings*.
32. J. L. Goity and W. Roberts, "A relativistic chiral quark model for pseudoscalar emission from heavy mesons," JLAB-THY-98-36 (hep-ph/9809312).
33. C. Bourdarios, "Study of D^{**} and $D^{*\prime}$ Production in b and c Jets, with the DELPHI Detector," LAL-98-72, (hep-ex/9811014).
34. S. Alam, "Charmed Baryon Spectroscopy from CLEO at CESR," *These Proceedings*.
35. E. W. Vaandering, "Dalitz Analysis, Rare Decays, and Charmed Baryons at FOCUS and E687," *These Proceedings*.
36. T. K. Pedlar, "Charmonium Spectroscopy from Fermilab E835," *These Proceedings*.
37. J. Z. Bai, et al. (BES Collaboration), *Phys. Rev. Lett.* **81**, 3091 (1998).
38. F. Olness, "Heavy Quark Production" (hep-ph/9812270) *These Proceedings*.
39. E. H. Norrbin, "Heavy Quark Fragmentation," *These Proceedings*.
40. K. Stenson, "E791 Results on Hadroproduction of Charm from 500-GeV $\pi^- N$ Interactions," *These Proceedings*.

41. E.-C. Aschenauer, "Charm Detection at HERMES," *These Proceedings.*
42. A. Bravar, "The COMPASS Experiment at CERN," *These Proceedings.*
43. F. Abe, *et al.* (CDF Collaboration), *Phys. Rev. Lett.* **69**, 3704 (1992).
44. A. Zieminski, "B Production and Onium Production at the Tevatron," *These Proceedings.*
45. D. Barberis, C. Gemme, and L. Malferrari, "Charm and Beauty Production in Experiment WA92," *These Proceedings.*
46. T. Chang, "New Results from NUSEA on Onium Production," not included in *These Proceedings.*
47. B. Naroska, "Heavy Flavour Physics at HERA," *These Proceedings.*
48. C. Lourenço, "Heavy Flavour Production in Heavy Ion Collisions," *These Proceedings.*
49. T. Adams, "Heavy Quark Production in Neutrino Deep-Inelastic Scattering," *These Proceedings.*
50. For a thorough explanation of phase conventions, see B. Winstein and L. Wolfenstein, *Rev. Mod. Phys.* **65**, 1113 (1993).
51. For a recent review, see G. Buchalla, A. J. Buras, and M. E. Lautenbacher, *Rev. Mod. Phys.* **68**, 1125 (1996).
52. L. K. Gibbons, *et al.*, *Phys. Rev. Lett.* **70**, 1203 (1993).
53. G. D. Barr, *et al.*, *Phys. Lett.* **B317**, 233 (1993).
54. A. Ceccucci, "Towards Results on Direct CP Violation in Kaon Decays from the CERN–SPS Experiment NA48," *These Proceedings.*
55. M. Arenton, "Results on CP Violation from KTeV," *These Proceedings.*
56. G. Bencivenni, "The Status of the KLOE Experiment," *These Proceedings.* For updated information, see http://www.lnf.infn.it/kloe/kloedef.html.
57. C. James, "CP Violation in Strange Baryon Decays: A Report from Fermilab Experiment 871," *These Proceedings.*
58. J. Adams, *et al.* (KTeV Collaboration), *Phys. Rev. Lett.* **80**, 4123 (1998), measured the branching ratio using a sample of 36.6 ± 6.8 events. During *HQ98*, the KEK group reported their observation of 13.5 ± 4.0 events: Y.Takeuchi, *et al.*, "Observation of the decay mode $K_L \to \pi^+\pi^- e^+ e^-$," Kyoto preprint KUNS-1537 (hep-ex/9810018), submitted to *Phys. Lett. B.*
59. Not long after the discovery of CP violation in K_L decays, A. D. Dolgov and L. A. Ponomarev, *Yad. Fiz.* **4**, 367 (1966) [English translation: *Sov. J. Nucl. Phys.* **4**, 262 (1967)] remarked that $K_L \to \pi^+\pi^- e^+ e^-$ decays could allow a further study of the phenomenon.
60. L. M. Sehgal and M. Wanninger, *Phys. Rev.* **D46**, 1035 (1992); an important correction appears in *ibid.* **46**, 5209E (1992). P. Heilinger and L. M. Sehgal, *ibid.* **48**, 4146 (1993) have examined the implications of direct CP violation. See also J. K. Elwood, M. B. Wise, and M. J. Savage, *Phys. Rev.* **D52**, 5095 (1995) for an analysis in chiral perturbation theory; J. K. Elwood, M. B. Wise, M. J. Savage, and J. W. Walden, *ibid.* **53**, 4078 (1996).
61. A. Angelopoulos, *et al.* (CPLEAR Collaboration), "First direct observation of time-reversal non-invariance in the neutral-kaon system," CERN–EP/98–153 (October 7, 1998), submitted to *Phys. Lett. B.*
62. S. Adler, *et al.* (E787 Collaboration), *Phys. Rev. Lett.* **79**, 2204 (1997).

63. S. Kettell, "Evidence for $K^+ \to \pi^+ \nu \bar{\nu}$," *These Proceedings*.
64. M Aoki, *et al.* "P949: An experiment to measure the branching ratio $B(K^+ \to \pi^+ \nu \bar{\nu})$," available at http://www.phy.bnl.gov/e949/.
65. E. Coleman, *et al.* (CKM Collaboration), "P905: Charged Kaons at the Main Injector, A Proposal for a Precision Measurement of the Decay $K^+ \to \pi^+ \nu \bar{\nu}$ and Other Rare K^+ Processes at Fermilab Using the Main Injector," available at http://www.fnal.gov/projects/ckm/Welcome.html.
66. Eric Zimmerman, "Rare Decays of the K_L and π^0: New Results from E799–II and E832," *These Proceedings*.
67. E. Cheu, *et al.* (The KAMI Collaboration), "An Expression of Interest to Detect and Measure the Direct CP Violating Decay $K_L \to \pi^0 \nu \bar{\nu}$ and Other Rare Decays at Fermilab Using the Main Injector" (hep-ex/9709026), available at http://fnalpubs.fnal.gov/archive/proposals/KAMI.html.
68. I-Hung Chiang, *et al.*, "P926: Measurement of $K_L \to \pi^0 \nu \bar{\nu}$," available at http://sitka.triumf.ca/e926/.
69. C. Quigg and J. D. Jackson, "Decays of Neutral Pseudoscalar Mesons into Lepton Pairs," Lawrence Radiation Laboratory Report UCRL-18487 (1968, unpublished). For the latest word, see M. Savage, M. Luke, and M. B. Wise, *Phys. Lett.* **B291**, 481 (1992).
70. D. Ambrose, "Rare Kaon Decays: Brookhaven E871," *These Proceedings*. See also D. Ambrose, *et al.*, *Phys. Rev. Lett.* **81**, 4309 (1998).
71. G. Valencia, *Nucl. Phys.* **B517**, 339 (1998); D. Gomez Dumm and A. Pich, *Phys. Rev. Lett.* **80**, 4633 (1998).
72. A. Sher, *et al.*, "Recent Results from BNL E865," *These Proceedings*.
73. D. Jensen, "An Overview of Hyperon Physics in KTeV," *These Proceedings*.
74. I. Eschrich, "Hyperon Physics Results from SELEX," *These Proceedings*.
75. N. Copty, "Hadronic Charm Decays and Lifetimes from E791," *These Proceedings*.
76. J. M. Link, "The FOCUS Spectrometer and Hadronic Decays in FOCUS," *These Proceedings*.
77. F. Abe, *et al.* (CDF Collaboration), *Phys. Rev. Lett.* **81**, 2432 (1998); *Phys. Rev.* **D58**, 112004 (1998).
78. M. Beneke and G. Buchalla, *Phys. Rev.* **D53**, 4991 (1996); I. Bigi, *Phys. Lett.* **B371**, 105 (1996).
79. K. Ecklund, "Semileptonic B Decays from CLEO," *These Proceedings*.
80. D. Mihalcea, "Charm Semileptonic Decays from E791," *These Proceedings*.
81. W. E. Johns, "Charm Semileptonic Decays from FOCUS/E687: A Survey of Previous E687 Results, and Expectations from FOCUS Data," *These Proceedings*.
82. A. Kushnirenko, "Charm Physics Results from SELEX," *These Proceedings*.
83. M. Nakamura, "Study of Neutrino Production of Charm Particles in CHORUS," not included in *These Proceedings*.
84. G. Burdman, "Theoretical Issues in Rare K, D, and B Decays" (hep-ph/9811457), *These Proceedings*.
85. S. Kwan, "Review of Rare Decays and CP Violation in Charm," *These Proceedings*.
86. For an informal tour of prospects for top physics at the Tevatron Collider, consult http://lutece.fnal.gov/thinkshop/.

COLLABORATIONS

BNL E871 Collaboration

D. Ambrose[1], C. Arroyo[2], M. Bachman[3], P. de Cecco[3], D. Connor[3], M. Eckhause[4], K. M. Ecklund[2], S. Graessle[1], A. D. Hancock[4], K. Hartman[2], M. Hebert[2], C. H. Hoff[4], G. W. Hoffmann[1], G. M. Irwin[2], J. R. Kane[4], N. Kanematsu[3], Y. Kuang[4], K. Lang[1], R. Lee[3], R. D. Martin[4], J. McDonough[1], A. Milder[1], W. R. Molzon[3], M. Pommot-Maia[2], P. J. Riley[1], J. L. Ritchie[1], P. D. Rubin[5], V. I. Vassilakopoulos[1], C. B. Ware[1], R. E. Welsh[4], S. G. Wojcicki[2], E. Wolin[4], S. Worm[1]

[1] *University of Texas, Austin, Texas, 78712,*
[2] *Stanford University, Stanford, California, 94305,*
[3] *University of California, Irvine, California, 92697,*
[4] *College of William and Mary, Williamsburg, Virginia, 23187,*
[5] *University of Richmond, Richmond, Virginia, 23173*

Fermilab E791 Collaboration

E. M. Aitala,[9] S. Amato,[1] J. C. Anjos,[1] J. A. Appel,[5] D. Ashery,[14]
S. Banerjee,[5] I. Bediaga,[1] G. Blaylock,[8] S. B. Bracker,[15]
P. R. Burchat,[13] R. A. Burnstein,[6] T. Carter,[5] H. S. Carvalho,[1]
N. K. Copty,[12] L. M. Cremaldi,[9] C. Darling,[18] K. Denisenko,[5]
A. Fernandez,[11] G.F. Fox,[12] P. Gagnon,[2] C. Gobel,[1] K. Gounder,[9]
A. M. Halling,[5] G. Herrera,[4] G. Hurvits,[14] C. James,[5] P. A. Kasper,[6]
S. Kwan,[5] D. C. Langs,[12] J. Leslie,[2] B. Lundberg,[5] S. MayTal-Beck,[14]
B. Meadows,[3] J. R. T. de Mello Neto,[1] D. Mihalcea,[7] R. H. Milburn,[16]
J. M. de Miranda,[1] A. Napier,[16] A. Nguyen,[7] A. B. d'Oliveira,[3,11]
K. O'Shaughnessy,[2] K. C. Peng,[6] L. P. Perera,[3] M. V. Purohit,[12]
B. Quinn,[9] S. Radeztsky,[17] A. Rafatian,[9] N. W. Reay,[7] J. J. Reidy,[9]
A. C. dos Reis,[1] H. A. Rubin,[6] D. A. Sanders,[9] A. K. S. Santha,[3]
A. F. S. Santoro,[1] A. J. Schwartz,[3] M. Sheaff,[17] R. A. Sidwell,[7]
A. J. Slaughter,[18] M. D. Sokoloff,[3] J. Solano,[1] N. R. Stanton,[7]
R. J. Stefanski,[5] K. Stenson,[17] D. J. Summers,[9] S. Takach,[18]
K. Thorne,[5] A. K. Tripathi,[7] S. Watanabe,[17] R. Weiss-Babai,[14]
J. Wiener,[10] N. Witchey,[7] E. Wolin,[18] D. Yi,[9] S. M. Yang,[7]
S. Yoshida,[7] R. Zaliznyak,[13] and C. Zhang[7]

[1] *Centro Brasileiro de Pesquisas Físicas, Rio de Janeiro, Brazil*
[2] *University of California, Santa Cruz, California 95064*
[3] *University of Cincinnati, Cincinnati, Ohio 45221*
[4] *CINVESTAV, Mexico*
[5] *Fermilab, Batavia, Illinois 60510*
[6] *Illinois Institute of Technology, Chicago, Illinois 60616*
[7] *Kansas State University, Manhattan, Kansas 66506*
[8] *University of Massachusetts, Amherst, Massachusetts 01003*
[9] *University of Mississippi, University, Mississippi 38677*
[10] *Princeton University, Princeton, New Jersey 08544*
[11] *Universidad Autonoma de Puebla, Mexico*
[12] *University of South Carolina, Columbia, South Carolina 29208*
[13] *Stanford University, Stanford, California 94305*
[14] *Tel Aviv University, Tel Aviv, Israel*
[15] *Box 1290, Enderby, BC, V0E 1V0, Canada*
[16] *Tufts University, Medford, Massachusetts 02155*
[17] *University of Wisconsin, Madison, Wisconsin 53706*
[18] *Yale University, New Haven, Connecticut 06511*

FOCUS (FNAL-E831) COLLABORATION

J. M. Link, M. Reyes, P. M. Yager
University of California-Davis

J. C. Anjos, I. Bediaga, C. Gobel, J. Magnin, I. M. Pepe, A. C. Reis,
A. Sánchez-Hernández, F. R. A. Simão
CBPF, Rio de Janeiro, Brazil

S. Carrillo, E. Casimiro, H. Mendez, M. Sheaff, C. Uribe, F. Vázquez
CINVESTAV, México City, Mexico

L. Cinquini[1], J. P. Cumalat, B. O'Reilly, E. W. Vaandering
University of Colorado-Boulder

J. N. Butler, H. W. K. Cheung, I. Gaines, P. H. Garbincius,
L. A. Garren, S. A. Gourlay[2], P. H. Kasper, A. E. Kreymer,
R. Kutschke
Fermi National Accelerator Laboratory

S. Bianco, F. L. Fabbri, S. Sarwar, A. Zallo
Laboratori Nazionali di Frascati dell'INFN, Italy

C. Cawlfield, R. Gardner[3], E. Gottschalk[4], K. S. Park, A. Rahimi,
J. Wiss
University of Illinois-Champaign

Y. S. Chung, J. S. Kang, B. R. Ko, J. W. Kwak, K. B. Lee,
S. S. Myung, H. Park
Korea University, Seoul, Korea

G. Alimonti, M. Boschini, B. Caccianiga, A. Calandrino,
P. D'Angelo, M. DiCorato, P. Dini, M. Giammarchi, P. Inzani,
F. Leveraro, S. Malvezzi, D. Menasce, M. Mezzadri, L. Milazzo,
L. Moroni, D. Pedrini, F. Prelz, A. Sala, S. Sala

INFN and University of Milano, Italy

T. F. Davenport III

University of North Carolina-Asheville

V. Arena, G. Boca, G. Bonomi[5], G. Gianini, G. Liguori, M. Merlo,
D. Pantea[6], S. P. Ratti, C. Riccardi, L. Viola, P. Vitulo

Dipartimento di Fisica Nucleare e Teorica and INFN, Pavia, Italy

A. M. Lopez, L. Mendez, A. Mirles, E. Montiel, D. Olaya[7],
J. E. Ramirez[7], C. Rivera, Y. Zhang[8]

University of Puerto Rico-Mayaguez

N. Copty, W. E. Johns[9], M. V. Purohit, J. R. Wilson

University of South Carolina-Columbia

K. Cho, T. Handler

University of Tennessee-Knoxville

D. Engh, M. Hosack, M. Nehring[10], M. Sales, P. D. Sheldon,
M. Webster

Vanderbilt University

K. Stenson[11]

University of Wisconsin-Madison

Y. Kwon

Yonsei University, Korea

---

[1] Present Address: NCAR, Boulder, CO
[2] Present Address: Lawrence Berkeley Lab
[3] Present Address: Indiana University, Bloomington
[4] Present Address: Fermilab
[5] Present Address: University of Brescia, Italy
[6] Present Address: Nat Inst. of Phys and Nucl. Eng., Bucharest, Romania
[7] Present Address: Univ. of Colorado, Boulder
[8] Present Address: Lucent Technology
[9] Present Address: Univ. of Puerto Rico, Mayaguez
[10] Present Address: Adams State College, Alamosa, CO
[11] Present Address: Vanderbilt University

KTeV Collaboration

A. Affolder[4], A. Alavi-Harati[12], I.F. Albuquerque[10], T. Alexopoulos[12],
M. Arenton[11], K. Arisaka[2], S. Averitte[10], A.R. Barker[5], L. Bellantoni[7],
A. Bellavance[9], J. Belz[10], R. Ben-David[7], D.R. Bergman[10],
E. Blucher[4], G.J. Bock[7], C. Bown[4], S. Bright[4], E. Cheu[1],
S. Childress[7], R. Coleman[7], M.D. Corcoran[9], G. Corti[11], B. Cox[11],
M.B. Crisler[7], A.R. Erwin[12], R. Ford[7], A. Golossanov[11], G. Graham[4],
J. Graham[4], K. Hagan[11], E. Halkiadakis[10], K. Hanagaki[8], S. Hidaka[8],
Y.B. Hsiung[7], V. Jejer[11], J. Jennings[2], D.A. Jensen[7], R. Kessler[4],
H.G.E. Kobrak[3], J. LaDue[5], A. Lath[10], A. Ledovskoy[11],
P.L. McBride[7], A.P. McManus[11], P. Mikelsons[5], E. Monnier[4,*],
T. Nakaya[7], U. Nauenberg[5], K.S. Nelson[11], H. Nguyen[7], V. O'Dell[7],
M. Pang[7], R. Pordes[7], V. Prasad[4], C. Qiao[4], B. Quinn[4],
E.J. Ramberg[7], R.E. Ray[7], A. Roodman[4], M. Sadamoto[8],
S. Schnetzer[10], K. Senyo[8], P. Shanahan[7], P.S. Shawhan[4], W. Slater[2],
N. Solomey[4], S.V. Somalwar[10], R.L. Stone[10], I. Suzuki[8],
E.C. Swallow[4,6], R.A. Swanson[3], S.A. Taegar[1], R.J. Tesarek[10],
G.B. Thomson[10], P.A. Toale[5], A. Tripathi[2], R. Tschirhart[7],
Y.W. Wah[4], J. Wang[1], H.B. White[7], J. Whitmore[7], B. Winstein[4],
R. Winston[4], J.-Y. Wu[5], T. Yamanaka[8], E.D. Zimmerman[4]

[1] *University of Arizona, Tucson, Arizona 85721*
[2] *University of California at Los Angeles, Los Angeles, California 90095*
[3] *University of California at San Diego, La Jolla, California 92093*
[4] *The Enrico Fermi Institute, The University of Chicago, Chicago, Illinois 60637*
[5] *University of Colorado, Boulder, Colorado 80309*
[6] *Elmhurst College, Elmhurst, Illinois 60126*
[7] *Fermi National Accelerator Laboratory, Batavia, Illinois 60510*
[8] *Osaka University, Toyonaka, Osaka 560 Japan*
[9] *Rice University, Houston, Texas 77005*
[10] *Rutgers University, Piscataway, New Jersey 08855*
[11] *The Department of Physics and Institute of Nuclear and Particle Physics, University of Virginia, Charlottesville, Virginia 22901*
[12] *University of Wisconsin, Madison, Wisconsin 53706*
* *On leave from C.P.P. Marseille/C.N.R.S., France*

The SELEX (E781) Collaboration

G. P. Thomas
Ball State University, Muncie, IN 47306, U.S.A.

E. Gülmez
Bogazici University, Bebek 80815 Istanbul, Turkey

R. Edelstein, E. Gottschalk[12], S. Y. Jun, A. Kushnirenko, D. Mao[13],
P. Mathew[14], M. Mattson, M. Procario, J. Russ, J. You
Carnegie-Mellon University, Pittsburgh, PA 15213, U.S.A.

A. M. F. Endler
Centro Brasiliero de Pesquisas Físicas, Rio de Janeiro, Brazil

P. S. Cooper, J. Engelfried[15], J. Kilmer, S. Kwan, J. Lach, G. Oleynik,
E. Ramberg, D. Skow, L. Stutte
Fermilab, Batavia, IL 60510, U.S.A.

Y. M. Goncharenko, O. A. Grachov[16], V. P. Kubarovsky,
A. I. Kulyavtsev[17], V. F. Kurshetsov, A. A. Kozhevnikov,
L. G. Landsberg, V. V. Molchanov, V. A. Mukhin, S. B. Nurushev,
A. N. Vasiliev, D. V. Vavilov, V. A. Victorov
Institute for High Energy Physics, Protvino, Russia

Li Yunshan, Li Zhigang, Mao Chensheng, Zhao Wenheng, He
Kangling, Zheng Shuchen, Mao Zhenlin
Institute of High Energy Physics, Beijing, PR China

M. Y. Balatz, G. V. Davidenko, A. G. Dolgolenko, G. B. Dzyubenko,
A. V. Evdokimov, A. D. Kamenskii, M. A. Kubantsev, I. Larin,
V. Matveev, A. P. Nilov, V. A. Prutskoi, V. K. Semyatchkin,
A. I. Sitnikov, V. S. Verebryusov, V. E. Vishnyakov
Institute of Theoretical and Experimental Physics, Moscow, Russia

U. Dersch, I. Eschrich, K. Königsmann[18], I. Konorov[19], H. Krüger,

B. Povh, J. Simon, K. Vorwalter[20]
Max-Planck-Institut für Kernphysik, 69117 Heidelberg, Germany

I. S. Filimonov, E. M. Leikin, A. V. Nemitkin, V. I. Rud
Moscow State University, Moscow, Russia

V. A. Andreev, A. G. Atamantchouk, N. F. Bondar, V. L. Golovtsov,
V. T. Kim, L. M. Kochenda, A. G. Krivshich, N. P. Kuropatkin,
V. P. Maleev, P. V. Neoustroev, S. Patrichev, B. V. Razmyslovich,
V. Stepanov, M. Svoiski, N. K. Terentyev[21], L. N. Uvarov,
A. A. Vorobyov
Petersburg Nuclear Physics Institute, St. Petersburg, Russia

I. Giller, M. A. Moinester, A. Ocherashvili, V. Steiner
Tel Aviv University, 69978 Ramat Aviv, Israel

A. Morelos
Universidad Autonoma de San Luis Potosí, San Luis Potosí, Mexico

M. Luksys
Universidade Federal da Paraíba, Paraíba, Brazil

S. L. McKenna, V. J. Smith
University of Bristol, Bristol BS8 1TL, United Kingdom

C. Kenney, S. Parker
University of Hawaii, Honolulu, HI 96822, U.S.A.

N. Akchurin, M. Aykac, M. Kaya, D. Magarrel, E. McCliment,
K. Nelson, C. Newsom, Y. Onel, E. Ozel, S. Ozkorucuklu, P. Pogodin
University of Iowa, Iowa City, IA 52242, U.S.A.

L. J. Dauwe
University of Michigan-Flint, Flint, MI 48502, U.S.A.

T. Ferbel, C. Ginther, C. Hammer, P. Slattery, M. Zielinski
University of Rochester, Rochester, NY 14627, U.S.A.

M. Gaspero, M. Iori

University of Rome "La Sapienza" and INFN, Rome, Italy

L. Emediato, C. Escobar[22], F. Garcia, P. Gouffon, T. Lungov[23],
M. Srivastava, R. Zukanovich Funchal

University of São Paulo, São Paulo, Brazil

A. Bravar, D. Dreossi, A. Lamberto, A. Penzo, G. F. Rapazzo,
P. Schiavon

University of Trieste and INFN, Trieste, Italy

---

[12] *Present address: Fermilab, Batavia, IL 60510, U.S.A.*
[13] *Present address: AT&T, Chicago, IL*
[14] *Present address: Motorola Inc., Chicago, IL*
[15] *Now at Universidad Autonoma de San Luis Potosi, Mexico*
[16] *Present address: Dept. of Physics, Wayne State University, Detroit, MI 48201*
[17] *Present address: Carnegie-Mellon University, Pittsburgh, PA 15213, U.S.A.*
[18] *Present address: Universität Freiburg, 79104 Freiburg, Germany*
[19] *Present address: Physik-Department, Technische Universität München, 85748 Garching, Germany*
[20] *Present address: Deutsche Bank AG, 65760 Eschborn, Germany*
[21] *Present address: Carnegie-Mellon University, Pittsburgh, PA 15213, U.S.A.*
[22] *Current Address: Instituto de Fisica da Universidade Estadual de Campinas, UNICAMP, SP, Brazil.*
[23] *Current Address: Instituto de Fisica Teorica da Universidade Estadual Paulista*

LIST OF PARTICIPANTS

Achiman, Yoav	University of Wuppertal
Adams, Todd	Kansas State University
Alam, Sajjad	State University of New York
Alavi, Ashkan	University of Wisconsin, Madison
Alton, Andrew	Kansas State University
Ambrose, David	University of Texas, Austin
Appel, Jeffrey	Fermi National Accelerator Laboratory
Arenton, Michael	University of Virginia
Arnold, Peter	University of Virginia
Artuso, Marina	Syracuse University
Aschenauer, Elke-Caroline	DESY, Zeuthen
Barberis, Dario	INFN/University of Genova
Bellini, Gianpaolo	Milano University/INFN
Ben-David, Ram	Fermi National Accelerator Laboratory
Bencivenni, Giovanni	INFN, Frascati
Berger, Edmond	Argonne National Laboratory
Bianco, Stefano	INFN, Frascati
Bigi, Ikaros	University of Notre Dame
Bishai, Mary	Fermi National Accelerator Laboratory
Bock, Greg	Fermi National Accelerator Laboratory
Bonneaud, Gerard	Ecole Polytechnique
Bravar, Alessandro	University of Mainz
Brown, Chuck	Fermi National Accelerator Laboratory
Burdman, Gustavo	University of Wisconsin, Madison
Butler, Joel N.	Fermi National Accelerator Laboratory
Ceccucci, Augusto	CERN
Cester, Rosanna	University of Torino
Chang, Ting-Hua	New Mexico State University
Cheung, Harry W. K.	Fermi National Accelerator Laboratory
Christian, David	Fermi National Accelerator Laboratory
Coleman, Rick	Fermi National Accelerator Laboratory
Cooper, Peter	Fermi National Accelerator Laboratory
Copty, Nader	University of South Carolina
Cox, Brad	University of Virginia
Crisler, Michael	Fermi National Accelerator Laboratory
Dauwe, Loretta	University of Michigan, Flint
Devmal, Shiral	University of Cincinnati
Ecklund, Karl	Cornell University
Eichten, Estia	Fermi National Accelerator Laboratory
Engelfried, Jurgen	Universidad Autonoma de San Luis Potosi
Eschrich, Ivo	Max Planck Institute fuer Kernphysik
Fabbri, Franco L.	INFN, Frascati

Fan, Xiaoling	Northwestern University
Gaidarev, Peter	Cornell University
Gaines, Irwin	Fermi National Accelerator Laboratory
Garbincius, Peter H.	Fermi National Accelerator Laboratory
Garren, Lynn	Fermi National Accelerator Laboratory
Goldman, Jesse	Kansas State University
Goncharov, Maxim	Kansas State University
Gottschalk, Erik	Fermi National Accelerator Laboratory
Grancagnolo, Francesco	Universita' di Lecce
Handler, Thomas	University of Tennessee
Hill, Tony	Cornell University
Hosack, Michael	Vanderbilt University
Hsiung, Yee Bob	Fermi National Accelerator Laboratory
Jackson, Gerald	Fermi National Accelerator Laboratory
James, Catherine	Fermi National Accelerator Laboratory
Jensen, Douglas A.	Fermi National Accelerator Laboratory
Johns, Will	University of Puerto Rico
Kaplan, Daniel	Illinois Institute of Technology
Kasper, Penelope	Fermi National Accelerator Laboratory
Kasper, Peter	Fermi National Accelerator Laboratory
Kettell, Steve	Brookhaven National Laboratory
Klinksiek, Stephen A.	University of New Mexico
Kretzer, Stefan	Dortmund University
Kronfeld, Andreas	Fermi National Accelerator Laboratory
Kushnirenko, Alexander	Carnegie Mellon University
Kutschke, Robert	Fermi National Accelerator Laboratory
Kwan, Simon	Fermi National Accelerator Laboratory
Lebrun, Paul	Fermi National Accelerator Laboratory
Link, Jonathan	University of California
Lipkin, Harry	Argonne National Laboratory
Lourenco, Carlos	CERN
McBride, Patricia	Fermi National Accelerator Laboratory
Mihalcea, Daniel	Kansas State University
Muheim, Franz	University of Geneva
Nakamura, Mitsuhiro	Nagoya University
Naroska, Beate	University Hamburg
Nguyen, Hogan	Fermi National Accelerator Laboratory
Norrbin, Emanuel	Lund University
Olness, Fredrick	Southern Methodist University
Papavassiliou, Vassili	New Mexico State University
Pedlar, Todd	Northwestern University
Peoples, John	Fermi National Accelerator Laboratory
Petrov, Alexey	Johns Hopkins University
Pitts, Kevin	Fermi National Accelerator Laboratory

Pordes, Stephen	Fermi National Accelerator Laboratory
Quigg, Chris	Fermi National Accelerator Laboratory
Rahimi, Amir	University of Illinois, Urbana-Champaign
Ramberg, Erik	Fermi National Accelerator Laboratory
Ray, Ron	Fermi National Accelerator Laboratory
Reis, Alberto	Centro Brasileiro de Pesquisas Fisicas
Reyes, Marco A.	University of California
Rodriguez, Jorge L.	University of Hawaii
Rosner, Jonathan	University of Chicago
Russ, James	Carnegie Mellon University
Sala, Silvano	INFN, Milano
Sanda, Anthony	Nagoya University
Schwingenheuer, B.	Max-Planck Institut fuer Kernphysik
Seth, Kamal	Northwestern University
Shanahan, Peter	Fermi National Accelerator Laboratory
Shawhan, Peter	University of Chicago
Sher, Alexander	University of Pittsburgh
Simone, James	Fermi National Accelerator Laboratory
Solomey, Nickolas	University of Chicago
Spiegel, Leonard	Fermi National Accelerator Laboratory
Stefanski, Ray	Fermi National Accelerator Laboratory
Stenson, Kevin	University of Wisconsin
Stocchi, Achille	Los Alamos National Laboratory
Stone, Sheldon	Syracuse University
Swallow, Earl	Elmhurst College/University of Chicago
Taegar, Sydney	University of Arizona
Tkabladze, Avtandil	DESY, Zeuthen
Tschirhart, Robert	Fermi National Accelerator Laboratory
Uribe, Cecilia	IFUAP, Mexico
Vaandering, Eric	University of Colorado
Velasco, Mayda M.	CERN
Webster, Medford	Vanderbilt University
White, Herman B.	Fermi National Accelerator Laboratory
Whitmore, Julie	Fermi National Accelerator Laboratory
Wilson, Jeffrey	University of South Carolina
Wolfenstein, Lincoln	Carnegie Mellon University
Yager, Philip	University of California
Zieminski, Andrzej	Indiana University
Zimmerman, Eric	University of Chicago

WORKSHOP PROGRAM

Special Joint Experimental/Theoretical Seminar
Friday October 9th 1998 3:30pm - 5:00pm
STRANGENESS, CHARM, BEAUTY

Lincoln Wolfenstein - Carnegie Mellon University
Jonathan Rosner - University of Chicago
Anthony Sanda - Nagoya University

Session I - CP Violation and Mixing I
Saturday 10th October 8:30am - 10:15am
Session Chair: Stefano Bianco

WELCOME
John Peoples, Director, Fermi National Accelerator Laboratory

THE HERAB EXPERIMENT AT DESY
Bernhard Schwingenheuer, MPI Heidelberg

CP VIOLATION, MIXING AND RARE DECAYS OF B'S AT THE TEVATRON, NOW AND IN TEVATRON RUN II
Kevin Pitts, Fermi National Accelerator Laboratory

RECENT RESULTS ON B MIXING AND RARE B DECAYS FROM LEP
Achille Stocchi, Laboratoire de l'Accélérateur Linéaire, Orsay

Session II - CP Violation and Mixing II and Heavy Ions
Saturday 10th October 10:35am - 12:35pm
Session Chair: Chuck Brown

THE BABAR EXPERIMENT AT SLAC
Gerard Bonneaud, Ecole Polytechnique - LPNHE, Palaiseau

THE DEVELOPMENT OF OUR KNOWLEDGE OF CP VIOLATION AND RARE B DECAYS OVER THE NEXT DECADE
Marina Artuso, Syracuse University

MECHANISMS FOR GENERATING THE BARYON ASYMMETRY OF THE UNIVERSE
Peter Arnold, University of Virginia, Charlottesville

HEAVY QUARK SIGNATURES IN HEAVY ION PHYSICS – PRESENT STATE OF OUR UNDERSTANDING AND FUTURE PROSPECTS
   Carlos Lourenço, CERN

Session III - Production Dynamics and Structure Functions I
Saturday 10th October 1:35pm - 3:30pm
Session Chair: Peter H. Garbincius

CHARM PRODUCTION FROM E791
   Kevin Stenson, University of Wisconsin, Madison

CHARM PHYSICS RESULTS FROM SELEX
   Alexander Kushnirenko, Carnegie Mellon University

CHARM AND BEAUTY PRODUCTION FROM WA92
   Dario Barberis, University and INFN, Genova

RESULTS FROM HERMES
   Elke-Caroline Aschenauer, DESY-Zeuthen, Zeuthen

Session IV - Production Dynamics and Structure Functions II
Saturday 10th October 3:50pm - 5:45pm
Session Chair: Franco Luigi Fabbri

HEAVY QUARK PRODUCTION IN NEUTRINO DEEP-INELASTIC SCATTERING
   Todd Adams, Kansas State University

RESULTS FROM ZEUS AND H1
   Beate Naroska, Univ. of Hamburg

NEW RESULTS FROM NUSEA ON ONIUM PRODUCTION
   Ting Chang, New Mexico State University

NEUTRINO PRODUCTION OF CHARM AND STRANGE PARTICLES
   Mitsuhiro Nakamura, Nagoya University

Session V - Production Dynamics and Structure Functions III and Hadronic Decays I
Sunday 11th October 8:30am - 9:55am
Session Chair: Dan Kaplan

B PRODUCTION AND ONIUM PRODUCTION AT THE TEVATRON
Andrzej Zieminski, Indiana University

HEAVY QUARK FRAGMENTATION
Emanuel Norrbin, Lund University

HADRONIC CHARM DECAYS AND LIFETIMES FROM E791
Nader Copty, University of South Carolina

Session VI - Hadronic Decays II and Production Dynamics IV
Sunday 11th October 10:15am - 12:25pm
Session Chair: Marina Artuso

CHARM DECAYS FROM FOCUS/E687-I
Jonathan Link, University of California, Davis

CHARM DECAYS FROM FOCUS/E687-II
Eric Vaandering, University of Colorado

CHARM AND B DECAYS CLEO
Jorge Rodriguez, University of Hawaii

REVIEW OF CHARM AND B LIFETIMES
Harry W. K. Cheung, Fermi National Accelerator Laboratory

HEAVY QUARK PRODUCTION ISSUES
Fred Olness, Southern Methodist University

Session VII - Semileptonic and Leptonic Decays I
Sunday 11th October 1:25pm - 3:20pm
Session Chair: Erik Ramberg

HYPERON PHYSICS RESULTS FROM SELEX
Ivo Eschrich, Max Planck Institute for Nuclear Physics, Heidelberg

HYPERON SEMILEPTONIC AND RADIATIVE DECAYS FROM KTEV
Doug Jensen, Fermi National Accelerator Laboratory

CHARM SEMILEPTONIC DECAYS FROM E791
Daniel Mihalcea, Kansas State University

CHARM SEMILEPTONIC DECAYS FROM FOCUS/E687
Will E. Johns, University of Puerto Rico

Session VIII - Semileptonic and Leptonic Decays II and Spectroscopy I
Sunday 11th October 3:40pm - 5:30pm
Session Chair: Robert Kutschke

CHARM AND B SEMILEPTONIC DECAYS FROM CLEO
Karl Ecklund, Cornell University

LATTICE GAUGE CALCULATIONS OF SEMILEPTONIC AND LEPTONIC DECAYS
Andreas Kronfeld, Fermi National Accelerator Laboratory

CHARMONIMUM SPECTROSCOPY FROM E835
Todd Pedlar, Northwestern University

CHARM SPECTROSCOPY FROM CLEO
Saj Alam, SUNY, Albany

Session IX - Spectroscopy II
Monday 12th October 8:30am - 9:50am
Session Chair: Gianpaolo Bellini

PROSPECTS FOR CHARM AND STRUCTURE FUNCTION MEASUREMENTS FROM COMPASS
Alessandro Bravar, University of Mainz

B SPECTROSCOPY FROM LEP
Franz Muheim, University of Geneva

HEAVY FLAVOR SPECTROSCOPY
Estia Eichten, Fermi National Accelerator Laboratory

Session X - Rare and Forbidden Decays I
Monday 12th October 10:00am - 12:00pm
Session Chair: Brad Cox

NEW RESULTS FROM BNL E871 ON RARE K-LONG DECAYS TO TWO LEPTONS
Dave Ambrose, University of Texas, Austin

RECENT RESULTS ON $K^+ \to \pi^+ \nu \bar{\nu}$ FROM BNL E787
Steve Kettell, Brookhaven National Laboratory

RECENT RESULTS FROM BNL E865
Alexander Sher, University of Pittsburgh

REVIEW OF RARE DECAYS AND CP VIOLATION IN CHARM
  Simon Kwan, Fermi National Accelerator Laboratory

RARE B DECAYS AND CP VIOLATION FROM CLEO
  Petr Gaidarev, Cornell University

Session XI - Rare Decays II and CP Violation and Mixing II
Monday 12th October 1:00pm - 2:55pm
Session Chair: Jeffrey Appel

THEORETICAL ISSUES IN RARE DECAYS OF STRANGE, CHARMED AND BOTTOM PARTICLES
  Gustavo Burdman, University of Wisconsin, Madison

CP VIOLATION IN STRANGE BARYON DECAYS FROM HYPERCP
  Catherine James, Fermi National Accelerator Laboratory

CP VIOLATION AT PHI FACTORIES
  Giovanni Bencivenni, Frascati/INFN

RARE K DECAYS FROM KTEV
  Eric Zimmerman, University of Chicago

Session XII - CP Violation and Mixing III
Monday 12th October 3:15pm - 4:15pm
Session Chair: Harry W.K. Cheung

RESULTS ON CP VIOLATION IN KAON DECAYS FROM NA48
  Augusto Ceccucci, CERN

RESULTS ON CP VIOLATION IN KAON DECAYS FROM KTEV
  Mike Arenton, University of Virginia

Conference Summary
Monday 12th October 4:15pm - 5:00pm
Session Chair: Joel N. Butler

CONFERENCE SUMMARY
  Chris Quigg, Fermi National Accelerator Laboratory

Photo by Reidar Hahn

Photo by Harry Cheung

Photo by Jenny Mullins

Photo by Harry Cheung

Photo by Harry Cheung

Photo by Harry Cheung

Photo by Medford Webster

Photo by Harry Cheung

Photo by Harry Cheung

AUTHOR INDEX

A

Adams, T., 198
Alam, M. S., 379
Alton, A., 198
Ambrose, D. A., 411
Appel, R., 431
Arenton, M., 135
Arnold, P., 97
Arroyo, C. G., 198
Artuso, M., 87
Aschenauer, E.-C., 189
Atoyan, G. S., 431
Avvakumov, S., 198

B

Barberis, D., 179
Bassaleck, B., 431
Bazarko, A. O., 198
Bencivenni, G., 116
Bergman, D. R., 431
Bernstein, R. H., 198
Bodek, A., 198
Bolton, T., 198
Bonneaud, G., 77
Brau, J., 198
Bravar, A., 390
Brown, D. N., 431
Buchholz, D., 198
Budd, H., 198
Bugel, L., 198
Burdman, G., 461
Burnstein, R. A., 107

C

Ceccucci, A., 125
Chakravorty, A., 107
Chan, A., 107
Chen, Y. C., 107
Cheung, H. W. K., 291
Cheung, N., 431
Choong, W. S., 107
Clark, K., 107

Conrad, J. M., 198
Copty, N., 251

D

de Barbaro, L., 198
de Barbaro, P., 198
Dhawan, S., 431
Do, H., 431
Drucker, R. B., 198
Dukes, E. C., 107
Durandet, C., 107

E

Ecklund, K. M., 344
Egger, J., 431
Eilerts, S., 431
Eschrich, I., 303

F

Felder, C., 431
Felix, J., 107
Fischer, H., 431
Formaggio, J. A., 198
Frey, R., 198
Fuzesy, R., 107

G

Gach, M., 431
Gaidarev, P. B., 451
Gemme, C., 179
Gidal, G., 107
Goldman, J., 198
Goncharov, M., 198
Gustafson, H. R., 107

H

Harris, D. A., 198
Herold, W. D., 431
Ho, C., 107
Holmstrom, T., 107
Huang, M., 107

I

Isakov, V. V., 431

J

James, C., 107
Jenkins, M., 107
Jensen, D. A., 314
Johns, W. E., 334
Johnson, R. A., 198

K

Kaplan, D. M., 107
Kaspar, H., 431
Kettell, S., 421
Kim, J. H., 198
King, B. J., 198
Kinnel, T., 198
Koutsoliotas, S., 198
Kraus, D., 431
Kronfeld, A. S., 355
Kushnirenko, A. Y., 168
Kwan, S. W. L., 441

L

Lamm, M. J., 198
Lazarus, D. M., 431
Lederman, L. M., 107
Leipuner, L., 431
Leros, N., 107
Link, J. M., 261
Longo, M. J., 107
Lopez, F., 107
Lourenço, C., 147
Lowe, J., 431

Lozano, J., 431
Luebke, W., 107
Luk, K. B., 107

M

Ma, H., 431
Majid, W., 431
Malferrari, L., 179
Marsh, W., 198
Mason, D., 198
McFarland, K. S., 198
McNulty, C., 198
Menzel, W., 431
Mihalcea, D., 324
Mishra, S. R., 198
Moreno, G., 107
Muheim, F., 399

N

Naples, D., 198
Naroska, B., 207
Nelson, K., 107
Nienaber, P., 198
Norrbin, E. H., 228

O

Olness, F. I., 238

P

Papavassiliou, V., 107
Pedlar, T. K., 369
Perroud, J. P., 107
Pislak, S., 431
Pitts, K. T., 55
Poblaguev, A. A., 431
Proskurjakov, A. L., 431

Q

Quigg, C., 485

R

Rajaram, D., 107
Rehak, P., 431
Robmann, P., 431
Rodriguez, J. L., 280
Romosan, A., 198
Rosner, J. L., 9
Rubin, H. A., 107

S

Sakumoto, W. K., 198
Sanda, A. I., 28
Schellman, H. M., 198
Schwingenheuer, B., 45
Sculli, F. J., 198
Seligman, W. G., 198
Shaevitz, M. H., 198
Sher, A., 431
Smith, W. H., 198
Sosa, M., 107
Spentzouris, P., 198
Stenson, K., 159
Stern, E. G., 198
Stocchi, A., 66
Stotzer, R., 431

T

Tamminga, B. M., 198
Teng, P. K., 107
Thompson, J. A., 431

Truöl, P., 431
Turko, B., 107

V

Vaandering, E. W., 270
Vaitaitis, A., 198
Vakili, M., 198
Volk, J., 107

W

Weyer, H., 431
White, C. G., 107
White, S. L., 107
Wolfenstein, L., 3
Wu, V., 198

Y

Yang, U. K., 198
Yu, J., 198

Z

Zeller, G. P., 198
Zeller, M. E., 431
Zieminski, A., 218
Zimmerman, E. D., 472
Zyla, P., 107